Tomato Leaf Curl New Delhi Virus (ToLCNDV)

Ravinder Kumar • Rahul Kumar Tiwari
Milan Kumar Lal • Anirban Roy
Editors

Tomato Leaf Curl New Delhi Virus (ToLCNDV)

Insights into Virome Dynamics and Management

Springer

Editors
Ravinder Kumar
ICAR Indian Agricultural Research Institute
New Delhi, Delhi, India

Milan Kumar Lal
ICAR-ICAR-Central Rice Research Institute
Cuttack, Odisha, India

Rahul Kumar Tiwari
ICAR-Indian Institute of Sugarcane Research
Lucknow, Uttar Pradesh, India

Anirban Roy
ICAR-Indian Agricultural Research Institute
New Delhi, Delhi, India

ISBN 978-981-95-2851-6 ISBN 978-981-95-2852-3 (eBook)
https://doi.org/10.1007/978-981-95-2852-3

This Springer imprint is published by the registered company Springer Nature Singapore Pte Ltd.
The registered company address is: 152 Beach Road, #21-01/04 Gateway East, Singapore 189721, Singapore

Preface

Tomato leaf curl New Delhi virus (ToLCNDV) has emerged as one of the most economically destructive plant viruses of the twenty-first century, posing a serious threat to global horticultural and agricultural productivity. First reported from the Indian subcontinent, ToLCNDV has since transcended geographical boundaries, infecting a wide range of hosts, particularly solanaceous and cucurbitaceous crops. The virus's ability to cause severe yield losses, coupled with its adaptability to diverse climatic conditions and expansion into new agro-ecosystems, has elevated it to a major concern for growers, researchers, and policymakers alike. Its complex interactions with whitefly vectors, seed transmission potential, genetic variability, and wide host range make ToLCNDV a challenging pathogen to manage.

Over the last two decades, remarkable advances have been made in understanding ToLCNDV biology, genome organization, epidemiology, and host–pathogen interactions. Cutting-edge molecular techniques, including omics approaches, genome editing, and RNA interference, have opened new avenues for virus detection, characterization, and sustainable management. At the same time, emerging challenges such as climate change-driven shifts in vector populations, evolving viral strains, and transboundary spread necessitate a consolidated resource that integrates current knowledge with future perspectives.

This edited volume, *Tomato Leaf Curl New Delhi Virus (ToLCNDV): Insights into Virome Dynamics and Management*, brings together contributions from leading scientists across diverse disciplines, offering a comprehensive overview of the virus from multiple angles. The book encompasses the latest developments in the occurrence, genome biology, molecular mechanisms of infection, vector ecology, seed transmission, host adaptation, and evolutionary dynamics of ToLCNDV. It highlights recent innovations in diagnostics, nanotechnology-based delivery systems, RNA-based vaccines, and CRISPR–Cas genome editing, alongside sustainable disease management strategies. Importantly, the chapters also address broader themes such as the impact of climate change, the potential for beneficial virus–host interactions, and policy-relevant considerations for disease control.

By integrating fundamental research with applied solutions, this book aims to serve as a definitive reference for plant virologists, molecular biologists, plant breeders, epidemiologists, extension professionals, and students. It is also intended to guide agricultural policymakers and industry stakeholders in formulating informed strategies to mitigate ToLCNDV-related crop losses and safeguard global food security.

We express our sincere gratitude to all contributing authors for their meticulous work and commitment to advancing the scientific understanding of ToLCNDV. We also thank our respective institutions for their support and the publishing team for their cooperation in bringing this book to fruition. We hope that this volume will not only serve as a scholarly resource but also inspire new research directions, collaborations, and innovative solutions to address the ongoing and future challenges posed by this formidable plant pathogen.

New Delhi, Delhi, India — Ravinder Kumar
Lucknow, Uttar Pradesh, India — Rahul Kumar Tiwari
Cuttack, Odisha, India — Milan Kumar Lal
New Delhi, Delhi, India — Anirban Roy

Contents

Editors and Contributors

About the Editors

Ravinder Kumar, PhD, is a Senior Scientist in Plant Pathology at the ICAR-Indian Agricultural Research Institute (IARI), New Delhi, with over 15 years of experience in research, teaching, and extension. He earned his master's and doctoral degrees in Plant Pathology from Sardar Vallabhbhai Patel University of Agriculture & Technology, Meerut, Uttar Pradesh, and earlier worked at the Central Potato Research Institute, Shimla, Himachal Pradesh. His research focuses on understanding host–pathogen interactions, plant virus epidemiology, and developing management strategies using both biotechnological and conventional approaches. He has made significant contributions in molecular characterization, genome sequencing, and diagnostics of plant pathogens, particularly Tomato Leaf Curl New Delhi Virus (ToLCNDV). He has also developed transgenic lines resistant to ToLCNDV, dsRNA-based formulations, and contributed to the management of plant genetic resources. He has designed several advanced diagnostic tools for potato pathogens, including uniplex/multiplex RT-PCR, real-time RT-PCR, LAMP, and RT-RPA protocols. With more than 150 publications, he has also edited books with renowned publishers such as Elsevier, Springer, and CRC Press. He serves on the editorial boards of many international journals and has received several awards for his scientific contributions.

Rahul Kumar Tiwari, a doctor in Plant Pathology, is currently working on melatonin-mediated mitigation of soil and tuber-borne diseases and physiological disorders in crop plants. He has also worked on signaling mechanism of pathogens, pathogenomics, and the management of hemibiotrophic fungi. He has explored the genetic diversity of pathogenic microbes in field and horticultural crops. Additionally, his area of interest is on multiple aspects of pathophysiological alterations in food crops under the influence of diverse biotic and abiotic stresses. He has published more than 100 publications in national and internationally reputed journal.

Milan Kumar Lal, Doctor in Plant Physiology, working in the area of abiotic, abiotic stress, and nutritional aspect of potato and other starchy crops. He is an expert worker in the area of abiotic stress such as heat, drought salinity, and heavy metal. Moreover, he is also working on the aspect of the effect of biotic stress such as fungus, virus, and bacteria on plant physiological, biochemical, and molecular responses. Apart from this, he also has expertise in the nutritional and quality aspects of starchy crops, including resistant starch, glycemic index, phytonutrients, functional fermented foods and beverages, bioactive compounds, and various processing techniques to enhance these components in food products of starchy crops. His findings have generated more than 160 publications in international peer-reviewed journals.

Anirban Roy, a doctor in Plant Pathology, is currently working on the diagnosis, characterization, and management of DNA and RNA viruses, with a particular emphasis on begomoviruses including ToLCNDV of horticultural crops. Resistant germplasm resources, diagnostic kits for begomoviruses, and virus-based gene expression vectors have been developed by him. He worked on genome editing during his postdoctoral research at Washington State University to combat virus diseases in chilli. He has been a faculty member at the Post Graduate School of IARI since 2009 and guided 10 post graduate students. He has published over 80 research articles in national/international journals of repute. He is the recipient of awards like INSA Medal for Young Scientist and the Lal Bahadur Shastri Young Scientist Award of ICAR, and is a Fellow of the different scientific professional societies

Contributors

P. Anbazhagan ICAR National Bureau of Plant Genetic Resources, Regional Station, Hyderabad, Telangana, India

Mohammad Ansar Department of Plant Pathology, Bihar Agricultural University, Sabour, Bhagalpur, Bihar, India

Gulab Chand Arya School of Biotechnology, Kalinga Institute of Industrial Technology, Deemed to be University, Bhubaneswar, Odisha, India

Kumari Shashi Bala Department of Molecular Biology and Genetic Engineering, Bihar Agricultural University, Sabour, Bhagalpur, Bihar, India

Akanksha Behera Department of Life Sciences, Central University of Karnataka, Kalaburagi, Karnataka, India

Bibhu Prasad Behera Institute of Life Sciences, Bhubaneswar, Odisha, India

Tanmaya Kumar Bhoi Forest Protection Division, ICFRE-Arid Forest Research Institute (ICFRE-AFRI), Jodhpur, Rajasthan, India

Swati Bhuria Plant Molecular Virology Laboratory, CSIR – National Botanical Research Institute, Lucknow, Uttar Pradesh, India

Academy of Scientific and Innovative Research (AcSIR), Ghaziabad, Uttar Pradesh, India

N. Boopathi Department of Plant Pathology, Tamil Nadu Agricultural University, Coimbatore, Tamil Nadu, India

V. Celia Chalam ICAR-National Bureau of Plant Genetic Resources, New Delhi, India

G. V. Chaitra ICAR National Bureau of Plant Genetic Resources, Regional Station, Hyderabad, Telangana, India

Dharmappa Chavan Division of Plant Pathology, ICAR-Indian Agricultural Research Institute, New Delhi, India

Ankita Das Institute of Life Sciences, Bhubaneswar, Odisha, India

Tuyelee Das University Centre of Research and Development, Chandigarh University, Mohali, Punjab, India

Komal Devi Department of Botany, Kurukshetra University, Kurukshetra, Haryana, India

M. Gayathri Department of Plant Pathology, Centre for Plant Protection Studies, Tamil Nadu Agricultural University, Coimbatore, Tamil Nadu, India

Jyoti Jha Division of Plant Pathology, ICAR-Indian Agricultural Research Institute, New Delhi, India

Krishnaveer Singh Jhala Nanoparticle Synthesis and Bioinformatics Laboratory, Department of Biotechnology, Vigyan Bhawan, Block – B, New Campus, Mohanlal Sukhadia University, Udaipur, Rajasthan, India

G. Karthikeyan Department of Plant Pathology, Tamil Nadu Agricultural University, Coimbatore, Tamil Nadu, India

Haripriya Kavidayal Genetics and Tree Improvement Division, ICFRE-Forest Research Institute, Dehradun, Uttarakhand, India

Chandan Kishore Department of Plant Breeding and Genetics, Bihar Agricultural University, Sabour, Bhagalpur, Bihar, India

A. Kokila Department of Plant Pathology, Centre for Plant Protection Studies, Tamil Nadu Agricultural University, Coimbatore, Tamil Nadu, India

S. Kousalya Department of Plant Pathology, Centre for Plant Protection Studies, Tamil Nadu Agricultural University, Coimbatore, Tamil Nadu, India

Ravi Ranjan Kumar Department of Molecular Biology and Genetic Engineering, Bhola Paswan Shastri Agricultural College, Purnea, Bihar Agricultural University, Sabour, Bhagalpur, Bihar, India

Ravinder Kumar Advanced Centre for Plant Virology, Division of Plant Pathology, ICAR-Indian Agricultural Research Institute, New Delhi, India

Susheel Kumar Plant Molecular Virology Laboratory, CSIR – National Botanical Research Institute, Lucknow, Uttar Pradesh, India

Academy of Scientific and Innovative Research (AcSIR), Ghaziabad, Uttar Pradesh, India

Milan Kumar Lal Division of Crop Physiology & Biochemistry, ICAR-Central Rice Research Institute, Cuttack, Odisha, India

Vinut Anant Madival Division of Plant Pathology, ICAR-Indian Agricultural Research Institute, New Delhi, India

Deepak Kumar Mahanta Forest Entomology Discipline, Forest Protection Division, Indian Council of Forestry Research and Education (ICFRE)-Forest Research Institute (ICFRE-FRI), Dehradun, Uttarakhand, India

Yogita Maheshwari Plant Molecular Virology Laboratory, CSIR – National Botanical Research Institute, Lucknow, Uttar Pradesh, India

Academy of Scientific and Innovative Research (AcSIR), Ghaziabad, Uttar Pradesh, India

Anik Majumdar Advanced Centre for Plant Virology, Division of Plant Pathology, ICAR-Indian Agricultural Research Institute, New Delhi, India

Sayanti Guha Majumdar ICAR-Indian Institute of Sugarcane Research, Lucknow, Uttar Pradesh, India

V. G. Malathi Retired Scientist, Division of Plant Pathology, ICAR-Indian Agricultural Research Institute, New Delhi, Delhi, India

Ayushi Malik Nanoparticle Synthesis and Bioinformatics Laboratory, Department of Biotechnology, Vigyan Bhawan, Block – B, New Campus, Mohanlal Sukhadia University, Udaipur, Rajasthan, India

Ashutosh Kumar Mall ICAR-Indian Institute of Sugarcane Research, Lucknow, Uttar Pradesh, India

Aditya Mallick Department of Life Sciences, Central University of Karnataka, Kalaburagi, Karnataka, India

Krishnamayee Mallick Institute of Life Sciences, Bhubaneswar, Odisha, India

Chinnakaruppan Marimuthu Plant Molecular Virology Laboratory, CSIR – National Botanical Research Institute, Lucknow, Uttar Pradesh, India

Avinash Marwal Nanoparticle Synthesis and Bioinformatics Laboratory, Department of Biotechnology, Vigyan Bhawan, Block – B, New Campus, Mohanlal Sukhadia University, Udaipur, Rajasthan, India

Prantik Mazumder Advanced Centre for Plant Virology, Division of Plant Pathology, ICAR-Indian Agricultural Research Institute, New Delhi, India

Niranjana Prem Minipreman Plant Molecular Virology Laboratory, CSIR – National Botanical Research Institute, Lucknow, Uttar Pradesh, India

Academy of Scientific and Innovative Research (AcSIR), Ghaziabad, Uttar Pradesh, India

Mobassareen Department of Molecular Biology and Genetic Engineering, Bihar Agricultural University, Sabour, Bhagalpur, Bihar, India

R. K. Mohana Pradeep Department of Plant Pathology, Tamil Nadu Agricultural University, Coimbatore, Tamil Nadu, India

Yeluru Mohanbabu Advanced Centre for Plant Virology, Division of Plant Pathology, ICAR-Indian Agricultural Research Institute, New Delhi, India

Auroshikha Mohanty Department of Molecular Biology and Biotechnology, Acharya Narendra Deva University of Agriculture and Technology, Ayodhya, Uttar Pradesh, India

Firoz Mondal Advanced Centre for Plant Virology, Division of Plant Pathology, ICAR-Indian Agricultural Research Institute, New Delhi, India

K. Nagendran Division of Crop Protection, ICAR-National Research Centre for Banana, Tiruchirappalli, Tamil Nadu, India

Rajit Kumar Pal Institute of Life Sciences, Bhubaneswar, Odisha, India

Vijay Vardhan Pandey Jawaharlal Nehru Custom House Laboratory, Raigarh, Maharashtra, India

B. Parameswari ICAR National Bureau of Plant Genetic Resources, Regional Station, Hyderabad, Telangana, India

Varshini Pillai Department of Genetic Engineering, SRM Institute of Science and Technology, Kattankulathur, Tamil Nadu, India

Ashish Prasad Department of Botany, Kurukshetra University, Kurukshetra, Haryana, India

Ansh Raj ICAR-National Research Centre on Litchi, Muzaffarpur, Bihar, India

Tushar Ranjan Department of Molecular Biology and Genetic Engineering, Bihar Agricultural University, Sabour, Bhagalpur, Bihar, India

Sheikh Aafreen Rehman ICAR-Central Institute of Temperate Horticulture, Rangreth, Srinagar, Jammu and Kashmir, India

P. Renukadevi Department of Plant Pathology, Centre for Plant Protection Studies, Tamil Nadu Agricultural University, Coimbatore, Tamil Nadu, India

Anirban Roy Advanced Centre for Plant Virology, Division of Plant Pathology, ICAR-Indian Agricultural Research Institute, New Delhi, India

Bishal Roy Advanced Centre for Plant Virology, Division of Plant Pathology, ICAR-Indian Agricultural Research Institute, New Delhi, India

Bimal Kumar Sahoo Department of Agricultural Entomology, Tamil Nadu Agricultural University, Coimbatore, Tamil Nadu, India

Ramswaroop Saini Department of Biotechnology, School of Life and Allied Health Sciences, Joy University, Tirunelveli, Tamil Nadu, India

Ipsita Samal ICAR-National Research Centre on Litchi, Muzaffarpur, Bihar, India

B. Sangeetha Department of Plant Pathology, Centre for Plant Protection Studies, Tamil Nadu Agricultural University, Coimbatore, Tamil Nadu, India

N. Saranya Department of Plant Molecular Biology and Bioinformatics, Centre for Plant Molecular Biology and Biotechnology, Tamil Nadu Agricultural University, Coimbatore, Tamil Nadu, India

Mehulee Sarkar Advanced Centre for Plant Virology, Division of Plant Pathology, ICAR-Indian Agricultural Research Institute, New Delhi, India

Vijayanandraj Selvaraj Plant Molecular Virology Laboratory, CSIR – National Botanical Research Institute, Lucknow, Uttar Pradesh, India

Academy of Scientific and Innovative Research (AcSIR), Ghaziabad, Uttar Pradesh, India

Mohd Abas Shah ICAR-Central Institute of Temperate Horticulture, Rangreth, Srinagar, Jammu and Kashmir, India

R. Sharanya Department of Plant Pathology, Centre for Plant Protection Studies, Tamil Nadu Agricultural University, Coimbatore, Tamil Nadu, India

Namisha Sharma Institute of Life Sciences, Bhubaneswar, Odisha, India

Rohit Sharma Department of Botany, Kurukshetra University, Kurukshetra, Haryana, India

Sanskriti Shiwani Department of Molecular Biology and Genetic Engineering, Bihar Agricultural University, Sabour, Bhagalpur, Bihar, India

Abhilasha Singh Plant Molecular Virology Laboratory, CSIR – National Botanical Research Institute, Lucknow, Uttar Pradesh, India

Academy of Scientific and Innovative Research (AcSIR), Ghaziabad, Uttar Pradesh, India

Ayushi Singh Department of Genetics & Plant Breeding, G. B. Pant University of Agriculture and Technology, Pantnagar, Udham Singh Nagar, Uttarakhand, India

Dinesh Singh ICAR-Indian Institute of Sugarcane Research, Lucknow, Uttar Pradesh, India

Gyanendra Singh Seed Production Unit (SPU), ICAR-Indian Agricultural Research Institute, New Delhi, India

K. Sivagnanapazham Department of Plant Pathology, Tamil Nadu Agricultural University, Coimbatore, Tamil Nadu, India

A. Swapna Geetanjali Department of Genetic Engineering, SRM Institute of Science and Technology, Kattankulathur, Tamil Nadu, India

Kirtika Swaraj ICAR-Indian Institute of Sugarcane Research, Lucknow, Uttar Pradesh, India

Rahul Kumar Tiwari Division of Crop Protection, ICAR-Indian Institute of Sugarcane Research, Lucknow, Uttar Pradesh, India

Hafsa Ajaz Tramboo ICAR-Central Institute of Temperate Horticulture, Rangreth, Srinagar, Jammu and Kashmir, India

Manoj Kumar Tripathi ICAR-Indian Institute of Sugarcane Research, Lucknow, Uttar Pradesh, India

Nimmala Sree Valli Advanced Centre for Plant Virology, Division of Plant Pathology, ICAR-Indian Agricultural Research Institute, New Delhi, India

Namrata Verma Advanced Centre for Plant Virology, Division of Plant Pathology, ICAR-Indian Agricultural Research Institute, New Delhi, India

Nikeshun Vivekananthan Plant Molecular Virology Laboratory, CSIR – National Botanical Research Institute, Lucknow, Uttar Pradesh, India

List of Abbreviations

AAP	Acquisition access period
ABA	Abscisic acid
ABE	Adenine base editor
ABI5	ABA-insensitive 5
AC1	Replication initiator protein gene of Geminiviruses
AC1, AC2, AC3, AC4	Viral genes of DNA-A (Replication initiator protein, Transcriptional activator, Replication enhancer, Symptom determinant)
AC1/C1 (Rep, RepA)	Replication-associated protein
AC1/Rep	Replication-associated protein
AC2/TrAP	Transcriptional activator protein (begomoviruses)
AC3/C3	Replication enhancer protein
AC3/REn	Replication enhancer protein
AC4	Begomovirus-encoded protein interfering with silencing
AC4	Open reading frame AC4 protein
AC4/C4	Pathogenicity-related protein
ACC	1-Aminocyclopropane-1-carboxylate
ACCD	ACC deaminase
ACMV	African cassava mosaic virus
ADF	Actin depolymerizing factor
ADK	Adenosine kinase
AGO	Argonaute protein
AGO1	Argonaute 1
AGO1, AGO4, AGO7	Argonaute protein isoforms
AI	Artificial intelligence
amiRNA	Artificial microRNA
Amp	Ampicillin
AMPs	Antimicrobial peptides
ANN	Artificial neural network
AP2	APETALA2 (transcription factor family)
ARF	Auxin response factor
ATG8f/SlATG8f	Autophagy-related protein 8f (tomato homolog)
AuNPs	Gold nanoparticles
AV1, AV2	Viral genes of DNA-A (Coat protein, Pre-coat protein)

AV1/V1	Coat protein gene
AV1/V1, AV2/V2	Coat protein genes on DNA-A
AV2	Open reading frame AV2 protein
AV2/V2	Pre-coat protein/movement-associated gene
BC1 (MP)	Movement protein
BC1, BV1	Genes encoded on DNA-B
BGMV	Bean golden mosaic virus
BgYMV	Bitter gourd yellow mosaic virus
BMC	Biological and Medical Centre (context-specific)
BMV	Brome mosaic virus
BSMV	Barley stripe mosaic virus
BT	Bacillus thuringiensis
BV1 (NSP)	Nuclear shuttle protein
BYDV	Barley yellow dwarf virus
CAB	Chlorophyll a/b-binding protein
CaM/rgs-CaM	Calmodulin-like protein
Cas	CRISPR-associated protein
Cas9	CRISPR-associated protein 9
CAT	Catalase
CatBoost	Categorical boosting
CBE	Cytosine base editor
CKs	Cytokinins
CLCuAV	Cotton leaf curl Alabad virus
CLCuD	Cotton leaf curl disease
CLCuKoV	Cotton leaf curl Kokhran virus
CLCuMB	Cotton leaf curl Multan betasatellite
CLCuMBBur	Recombinant Burewala strain of CLCuMB
CLCuMuV	Cotton leaf curl Multan virus
CLCuV	Cotton leaf curl virus
CMV	Cucumber mosaic virus
CNN	Convolutional neural network
Coat protein (CP)	Structural protein of the virus
CP	Capsid protein/coat protein
CPs/CP	Coat proteins/coat Protein
CRESS	Circular rep-encoding single-stranded DNA (domain of replicase gene)
CRISPR	Clustered regularly interspaced short palindromic repeats
CRISPR/Cas	Clustered regularly interspaced short palindromic repeats/CRISPR-associated protein
crRNA	CRISPR RNA
CThTV	Cocksfoot tolerant heat virus (likely, confirm context)
Cy-1 QTL	Resistance locus in cucumber
DAC-ELISA	Direct antigen coating enzyme-linked immunosorbent assay
DAMPs	Damage-associated molecular patterns

DAS-ELISA	Double antibody sandwich enzyme-linked immunosorbent assay
DBT	Department of Biotechnology (India)
DC	Dendritic cell/diffusion coefficient (depends on context)
DCL	Dicer-like protein
DCL(s)	Dicer-like protein(s)
DDH	Direct DNA hybridization
DET1	De-etiolated 1 (tomato gene)
DIBA	Dot immuno-binding assay
DL	Deep learning
DLT	Distributed ledger technology
DNA	Deoxyribonucleic acid
DNA A	Genomic component A of bipartite begomoviruses
DNA B	Genomic component B of bipartite begomoviruses
dNAAT	Digital nucleic acid amplification testing
DnaSP	DNA Sequence Polymorphism software
DoYMV	Dolichos yellow mosaic virus
dpi	Days post-infiltration
DRB	Double-stranded RNA-binding protein
DREB2	Dehydration-responsive element-binding protein 2
DRM2	Domains rearranged methyltransferase-2
DSB	Double-strand break
dsDNA	Double-stranded DNA
dsRNA	Double-stranded RNA
DSS	Decision support system
DT	Decision tree
EC	European Commission (context-specific)
EFSA	European Food Safety Authority
ELISA	Enzyme-linked immunosorbent assay
EOs	Essential oils
EPA	Eicosapentaenoic acid (or Environmental Protection Agency, depends on text)
EPPO	European and Mediterranean Plant Protection Organization
ERF1B	Ethylene response factor 1B
ET	Ethylene
ETI	Effector-triggered immunity
FAO	Food and Agriculture Organization
FT	Flowering locus T
GBM	Gradient boosting machine
GCA	General combining ability
GFLV	Grapevine fanleaf virus
GFP	Green fluorescent protein
GIS	Geographic information system
GMOs	Genetically modified organisms

GPS	Global positioning system
gRNA	guide RNA
GRSPaV	Grapevine *rupestris* stem pitting-associated virus
GST	Glutathione S-transferase
GUS	β-glucuronidase
GVs	Geminiviruses
GWAS	Genome-wide association studies
HASTY	Nuclear export factor for miRNA
HC-Pro	Helper component proteinase
hc-siRNA	Heterochromatic small interfering RNA
HDR	Homology-directed repair
HEJ	Non-homologous end joining
HEN1	Hua enhancer 1 methyltransferase
HESO1	HEN1 suppressor 1
HIGS	Host-induced gene silencing
HKTMV	Hibiscus chlorotic ringspot virus (context: heat-killed)
HM-RNAi	Host-mediated RNA interference
hpRNAi	Hairpin RNA interference
HR	Homologous recombination
HR	Hypersensitive response
HS	Hydrosols
HSF	Heat shock factor
HSFA1a/HSFA2	Heat shock transcription factors
HTS	High-throughput sequencing
HVX	Hosta virus X
HY5	Elongated hypocotyl 5 (light-regulated transcription factor)
HYL1	Hyponastic leaves 1
IAA	Indole-3-acetic acid (auxin)
IARI	Indian Agricultural Research Institute
IC-ELISA	Immuno capture enzyme-linked immunosorbent assay
IC-LAMP	Immuno capture loop-mediated isothermal amplification
ICMR	Indian Council of Medical Research
ICRISAT	International Crops Research Institute for the Semi-Arid Tropics
ICT	Information and communication technology
IFP	Inoculation feeding period
IFS	International Foundation for Science
IL	Interleukin/introgression line (depends on usage)
IoT	Internet of things
IPM	Integrated pest management
IPTG	Isopropyl β-D-1-thiogalactopyranoside
IR	Intergenic region
IRD	Iteron-related domain
IRES	Internal ribosome entry site

ISR	Induced systemic resistance
JA	Jasmonic acid
kb	Kilobase (1000 base pairs)
KNN	K-nearest neighbor
KNOX	KNOTTED1-like homeobox
LAMP	Loop-mediated isothermal amplification
LIR	Long intergenic region
LMV	Lettuce mosaic virus
LRR	Leucine-rich repeat
LSPR	Localized surface plasmon resonance
LSTM	Long short-term memory
LTP	Lipid transfer protein
mAbs	Monoclonal antibodies
MAMPs	Microbe-associated molecular patterns
MAPK/MPK	Mitogen-activated protein kinase
MAPKs	Mitogen-activated protein kinases
MDMV	Maize dwarf mosaic virus
MEGA	Molecular Evolutionary Genetics Analysis (software)
MeJA	Methyl jasmonate
miRNA	MicroRNA
ML	Machine learning
ML	Maximum likelihood (phylogenetic method)
MP	Movement protein
MP (BC1)	Movement protein gene of DNA-B
mRNA	Messenger RNA
MYMIV	Mungbean yellow mosaic India virus
MYMV	Mungbean yellow mosaic virus
NAC	NAM, ATAF1/2, and CUC2 (plant TF family)
NASH	Nucleic acid spot hybridization
nat-siRNA	Natural antisense small interfering RNA
NB	Naïve Bayes
NDVI	Normalized Difference Vegetation Index
NES	Nuclear export signal
NGS	Next-generation sequencing
NIK	NSP-interacting kinase
NLP	Natural language processing
NLR	Nucleotide-binding leucine-rich repeat (immune receptors)
NLS	Nuclear localization signal
NPR1	Nonexpressor of pathogenesis-related Genes 1
NSP	Nuclear shuttle protein
NSP (BV1)	Nuclear shuttle protein gene of DNA-B
NSs	Non-structural protein (silencing suppressor)
NTOs	Non-target organisms
OD	Optical density

OD600	Optical density at 600 nm
ORF	Open reading frame
ORF(s)	Open reading frame(s)
PAL	Phenylalanine ammonia-lyase
PaLCuV	Papaya leaf curl virus
PAM	Protospacer adjacent motif
PAMP	Pathogen-associated molecular pattern
PCA	Principal component analysis
PCR	Polymerase chain reaction
PCV	Peanut clump virus
PEBV	Pea early browning virus
PepYLCIV	Pepper yellow leaf curl Indonesia virus
PGPR	Plant growth-promoting rhizobacteria
pha-siRNA	Phased siRNA
Pi	Nucleotide diversity
PI3K	Phosphoinositide 3-kinase
piRNA	PIWI-interacting RNA
PKC	Protein kinase C
PME/PMEI	Pectin methylesterase/PME inhibitor
Pol IV/Pol V	Plant-specific RNA polymerases IV and V
PPV	Plum pox virus
PR proteins	Pathogenesis-related proteins
PR1	Pathogenesis-related protein 1
pri-miRNA	Primary microRNA
PRRs	Pattern recognition receptors
PSbMV	Pea seed-borne mosaic virus
PSTVd	Potato spindle tuber viroid
PTGS	Post-transcriptional gene silencing
PTI	PAMP-triggered immunity
PVY	Potato virus Y
PYL	Pyrabactin resistance-like (ABA receptor)
qPCR	Quantitative polymerase chain reaction
QTL	Quantitative trait locus
R gene	Resistance gene
rasiRNA	Repeat-associated small interfering RNA
RCA	Rolling circle amplification
RCA-PCR	Rolling circle amplification polymerase chain reaction
RCR	Rolling circle replication
RdDM	RNA-directed DNA methylation
RDR	RNA-dependent RNA polymerase
RDR(s)	RNA-dependent RNA polymerase(s)
RDR1	RNA-dependent RNA polymerase 1
RDR1, RDR2, RDR6	Specific RNA-dependent RNA polymerases involved in vsiRNA production
RDR6	RNA-dependent RNA polymerase 6

RDRP	RNA-dependent RNA polymerase
RDV	Rice dwarf virus
REn	Replication enhancer protein
Rep	Replication-associated protein
RFLP	Restriction fragment length polymorphism
RIN4	RPM1-interacting protein 4
RISC	RNA-induced silencing complex
RNAi	RNA interference
RNN	Recurrent neural network
ROS	Reactive oxygen species
RPA	Recombinase polymerase amplification
RRSV	Rice ragged stunt virus
RS	Remote sensing
RSS	RNA silencing suppressor
RT-PCR	Reverse transcription polymerase chain reaction
SA	Salicylic acid
SAG	Senescence-associated gene
SAM	S-adenosyl methionine
SAR	Systemic acquired resistance
SCE1	SUMO-conjugating enzyme 1
SCLCrV	Squash leaf curl virus
SDGs	Sustainable development goals
SDH	Sandwich DNA hybridization
SDN	Site-directed nuclease
SDT	Sequence Demarcation Tool (software)
SE	Serrate
sgRNA	single guide RNA
SGS3	Suppressor of gene silencing 3
siRNA	Small interfering RNA
siRNA clusters	Loci producing siRNAs
siRNA(s)	Small interfering RNA(s)
SLR1	Slender rice 1 (DELLA protein)
SnRK2	SNF1-related kinase 2
SpCas9	*Streptococcus pyogenes* Cas9
SPL6	Squamosa promoter-binding protein-like 6
SPLCV	Sweet potato leaf curl virus
SPSMV-1	Sweet potato symptomless virus 1
SPU	Seed production unit
SRBSDV	Southern rice black-streaked dwarf virus
sRNA	Small RNA
SSB	Single-stranded DNA binding protein
ssDNA	Single-stranded DNA
SSR	Simple sequence repeat
ssRNA	Single-stranded RNA
STM	Shoot meristemless

STMV	Satellite tobacco mosaic virus
STNV	Satellite tobacco necrosis virus
SUMO	Small ubiquitin-like modifier
SVBV	Strawberry vein banding virus
SVM	Support vector machine
syn-tasiRNA	Synthetic trans-acting small interfering RNA
TALEN	Transcription activator-like effector nuclease
TAR1	Tryptophan aminotransferase 1-like protein
tasiRNA	Trans-acting small interfering RNA
TBSV	Tomato bushy stunt virus
TCV	Turnip crinkle virus
TEs	Transposable elements
Tet	Tetracycline
TGMV	Tomato golden mosaic virus
TGS	Transcriptional gene silencing
Ti-plasmid	Tumor-inducing plasmid
TLPs	Thaumatin-like proteins
ToLCD	Tomato leaf curl disease
ToLCNDV	Tomato leaf curl New Delhi virus
ToLCNDV-ES	Tomato leaf curl New Delhi virus—European strain (Mediterranean basin)
ToLCNDV-In	Tomato leaf curl New Delhi virus—Indian strain
ToLCV	Tomato leaf curl virus
ToMV	Tobacco mosaic virus
TORNADO1 (SlTRN1)	Tomato TORNADO1 gene
ToRSV	Tomato ringspot virus
ToTV	Tomato torrado virus
tracrRNA	Trans-activating CRISPR RNA
TrAP	Transcriptional activator protein
TRV	Tobacco rattle virus
TSWV	Tomato spotted wilt Virus
TYLCD	Tomato yellow leaf curl disease
TYLCSV	Tomato yellow leaf curl Sardinia Virus
TYLCTHV	Tomato yellow leaf curl Thailand virus
TYLCV	Tomato yellow leaf curl virus
UAV	Unmanned aerial vehicle
UN	United Nations
URT1	UTP: RNA uridylyltransferase 1
vasiRNA	Virus-activated small interfering RNA
VIGS	Virus-induced gene silencing
VOCs	Volatile organic compounds
v-ori	Viral origin of replication
vsiRNA	Virus-derived small interfering RNA
VSR	Viral suppressor of RNA silencing
VSR(s)	Viral suppressor(s) of RNA silencing

WCR	Western corn rootworm
WDV	Wheat dwarf virus
WSMV	Wheat streak mosaic virus
XGBoost	Extreme gradient boosting
Xrn4	Exoribonuclease 4
ZFN	Zinc finger nuclease
βC1	Betasatellite pathogenicity determinant protein

Contemporary Status of Occurrence, Spread, and Yield Loss Due to Tomato Leaf Curl New Delhi Virus in Global Crop Cultivation

1

Ravinder Kumar, Rahul Kumar Tiwari, Milan Kumar Lal, Sayanti Guha Majumdar, Kirtika Swaraj, Ashutosh Kumar Mall, Manoj Kumar Tripathi, Dinesh Singh, and Gyanendra Singh

Abstract

Viruses are a significant challenge in global agriculture. The emergence and re-emergence of viruses in horticultural and field crops continuously threaten crop cultivation in tropical, subtropical, and temperate regions worldwide. Tomato leaf curl New Delhi virus (ToLCNDV), identified in 1995, has emerged as a significant challenge in crop cultivation. This virus has been reported in 23 countries, with Asia and Europe accounting for the majority of reported occurrences. The virus has been extensively intercepted from horticultural crops such as cucurbits, tomato, and potato. The recent outbreak of ToLCNDV in China on the cucurbit host caused a loss of $15 million. In potatoes, apical leaf curl disease caused by ToLCNDV-potato is a significant threat to processing and table potatoes. The wider spread and adaptability of ToLCNDV is due to its nature of transmission through the whitefly vector, seed, mechanically, and pollen. Extensive work has been done on its pathogenicity and host-pathogen interactions. Amino acid substitution-based study confirmed the role of proteins in mechanical transmission ability. This chapter highlights the occurrence of ToLCNDV in diverse hosts, the yield loss, spread in continents and the mechanism of transmission.

R. Kumar
Division of Plant Pathology, ICAR-Indian Agricultural Research Institute, New Delhi, India

R. K. Tiwari (✉) · S. G. Majumdar · K. Swaraj · A. K. Mall · M. K. Tripathi · D. Singh
ICAR-Indian Institute of Sugarcane Research, Lucknow, Uttar Pradesh, India

M. K. Lal
Division of Crop Physiology and Biochemistry, ICAR-Central Rice Research Institute, Cuttack, Odisha, India

G. Singh
Seed Production Unit (SPU), ICAR-Indian Agricultural Research Institute, New Delhi, India

R. Kumar et al. (eds.), *Tomato Leaf Curl New Delhi Virus (ToLCNDV)*,
https://doi.org/10.1007/978-981-95-2852-3_1

Keywords

Whitefly · Apical · Enation · Movement proteins · Zucchini · Begomovirus

1 Introduction

In tropical and subtropical regions worldwide, crop cultivation is severely affected by Begomoviruses (family Geminiviridae) (Kumar et al., 2022; Thi et al., 2025). These viruses are circular, single-stranded DNA viruses that exhibit vast genomic and antigenic variations through recombination, mutation, attainment of extra DNA components, pseudo-recombination, and other mechanisms (Pratap et al., 2011; Thi et al., 2025). Begomoviruses are responsible for up to 100% yield losses in important food, feed and fiber crops globally. A diverse range of cultivated crop families, including *Cucurbitaceae*, *Solanaceae*, *Malvaceae*, *Fabaceae* and *Euphorbiaceae* are affected by these white-fly transmitted viruses (Kumar et al., 2021, 2023). During the last two decades, Tomato leaf curl New Delhi virus (ToLCNDV) has emerged as one of the most significant viruses affecting diverse crops in the Indian and Mediterranean regions (Thi et al., 2025). The virus was first reported in 1995, and since then, several reports of its occurrence have been documented in commercial cultivated crops such as tomato, potato, chilli, okra, pumpkin, cucumber, mungbean, black gram, French bean, pigeonpea, oilseed and cotton, jute, mesta (Moriones et al., 2017). In tomato, severe yield losses and crop failure have been previously reported, which may go up to 100%. Similarly, the sponge gourd crop had been severely affected, and a major yield loss has been documented from India (Patil et al., 2017; Tomar et al., 2018; Kumar et al., 2021). The virus has affected cucurbit crops in Mediterranean countries, such as Spain and Italy, resulting in a 20% crop loss in zucchini and melons, and an 80% loss in pumpkins due to a severe attack on the standing crop (Crespo et al. 2015, 2020; Panno et al., 2016). A recent report from China (Zhejiang Province, and Shanghai) indicates a monetary loss of $15 million in a total 650-hectare planted cucurbit crop infected with ToLCNDV in the autumn season (Cai et al., 2023). This recent episode of ToLCNDV emergence in China, along with the detection of such a virulent strain, highlights the need for an exhaustive investigation of host-pathogen interaction, epidemiological factors involved in the severity of disease, and the deployment of suitable integrated disease management strategies alongside technological interventions (Kumar & Singh, 2023; Lal et al., 2021). The influence of host species is less evident compared to regional variations in terms of genetic relationship among the isolates of ToLCNDV. Molecular studies categorize ToLCNDV isolates into three distinct groups: ToLCNDV-ES (populations in the Mediterranean basin), ToLCNDV-In (Indian strain), and Southeast Asian isolates (Thi et al., 2025).

ToLCNDV comes under the genus begomovirus, having a bipartite nature and DNA A (2.7 kb) and DNA B (2.6 kb) as two genomic components (Phaneendra et al., 2012). The DNA-A segment contains six open reading frames (ORFs): AC1, AC2, AC3, AC4, AV1, and AV2. These ORFs govern the virus encapsulation, virulence and replication. Besides that, two proteins encoded by BC1 and BV1 ORFs facilitate virus movement (Hanley-Bowdoin et al., 2013; Panno et al., 2016; Srivastava et al., 2016). It causes discrete symptoms in several horticultural and weed host plants. The prominent symptoms of leaf curling, mosaic, yellowing, stunted growth, fruit malformation and poor fruit setting have been reported in tomatoes (Moriones et al., 2017; Patil et al., 2017; Jones, 2021). In cucurbit crops such as zucchini, melons, and pumpkins, typical symptoms include conspicuous yellow mosaic patterns, vein thickening, blisters on leaves, and malformed or deformed fruits (Sharma et al., 2015; Orfanidou et al., 2019). Other cucurbit plants, viz., cucumber and watermelon, show upward leaf curling, yellowing, mosaic and mottling. The major devastation has been reported in the potato crop, where this virus causes potato apical leaf curl disease. The initial symptoms of chlorosis and spotting appear on leaves, but more damaging effects occur in the form of leaf crinkling, blotching and stunting of the plants (Patil et al., 2017; Siskos et al., 2022). The virus's ability to infect a wide range of hosts, including weeds, increases the risk of sustained outbreaks. Weeds serve not only as alternate hosts but also support populations of insect vectors, enhancing the virus's transmission potential (Sivalingam and Varma, 2007). These weed hosts can persist across growing seasons, making them effective reservoirs for disease recurrence. Various disease control strategies, including chemical, biological, and cultural approaches, have been combined to maximize effectiveness in limiting virus spread. Since no single strategy is capable of managing viral pathogens, including ToLCNDV, developing resistant varieties has been recognized as one of the most viable and effective methods for ToLCNDV management (Jeevalatha et al., 2021; Majid et al., 2023). Therefore, several researchers are actively working to identify hosts that exhibit resistance or tolerance to this virus (Bhardwaj et al., 2019). This chapter highlights the widespread concern about the emergence of this virus in cultivated crops and associated losses in production and productivity. The discussion is supported by global spread, including new strains, associated crops, weed hosts, and the ultimate losses in cultivated crops.

2 Global Distribution and Economic Loss Due to ToLCNDV

ToLCNDV has emerged as a crucial concern for the crop growers due to its significant damage to major vegetable crops. First detected in 1995 on tomatoes in India (Padidam et al., 1995) and later on other solanaceous crops (brinjal, chilli pepper, and potato), cucurbits (gourds, melon), okra, and papaya (Naga et al., 2021a, b; Kumar et al., 2021). Since its first report, *Tomato* ToLCNDV has been detected in a wide range of cultivated crops across different countries. The movement of seeds and planting materials primarily led to the spread of this virus across Asia, Europe, Africa, and the Middle East (Fig. 1.1). Currently, the various isolates of this virus are grouped into three genetic clusters: ToLCNDV-ES (Western Mediterranean and Europe), ToLCNDV-In (Indian subcontinent and East Asia), and Southeast Asia ToLCNDV (Thi et al., 2025). The highest number of incidence reports is from Asian

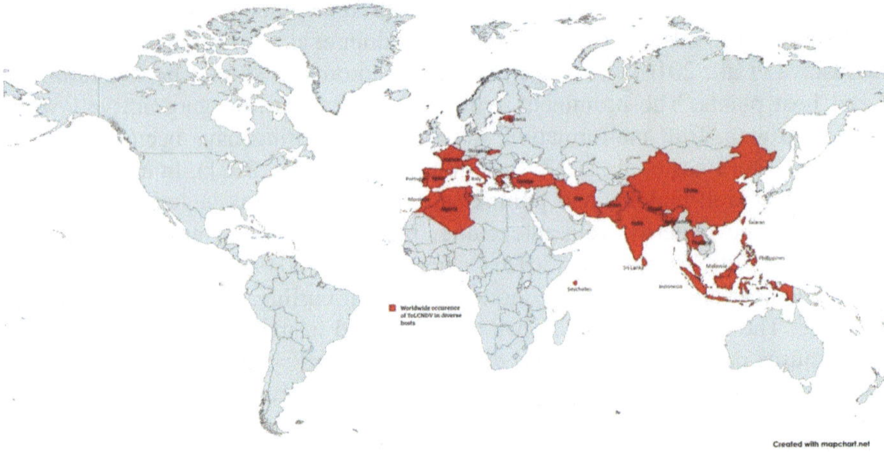

Fig. 1.1 The worldwide distribution of ToLCNDV

countries. Before 2000, the virus had been reported from India, Pakistan, Bangladesh, Nepal, and Sri Lanka, which are part of the South Asian region (Zaidi et al., 2017; Cai et al., 2023). In China and the Taiwan region of East Asia, this virus has been documented. In Southeast Asia, the virus has been reported in Thailand, Malaysia, Indonesia, and the Philippines (as reviewed by Thi et al., 2025). The report of ToLCNDV also comes from Iran, which can be observed as a transition zone of ToLCNDV-In and ToLCNDV-ES clusters. In Europe, Spain, France, Italy, Greece, Portugal, Slovenia, and Estonia have reports of ToLCNDV. The crops in Morocco, Algeria, and Tunisia are affected by this virus in North Africa. Seychelles documented the report on ToLCNDV. Several reports have shown the serious threat caused by ToLCNDV in cucurbits and solanaceous crop cultivation worldwide. In chayote (*Sechium edule*), a yield loss of 60% was previously reported in India, with maximum disease incidence up to 100% (Nagendran et al., 2017). The fruits were deformed, which affected the marketable value of the virus-affected crop. In the hilly region of Tamil Nadu, 50.3–100% disease incidence and 69.9% yield loss were reported (Sangeetha et al., 2018).

In the Zucchini crop cultivated in Almeria, up to 22% incidence of ToLCNDV was reported under greenhouse conditions (Crespo et al. 2020). In Spain, the cultivated melon crops showed an estimated yield loss of 20% under open field conditions (Sáez et al., 2016, 2017). In *Cucurbita moschata* plants, the incidence of this disease ranged from 80% to 100% in the Campania region of Italy (Parrella et al., 2022; Troiano and Parrella, 2023). In the Indo-Gangetic plains, a 40–100% incidence has been reported in the potato crop where this virus causes apical leaf curl disease (Lakra, 2010; Kumar et al., 2021; Jeevalatha et al., 2023). In the West Bengal state of India, up to 40% of the incidence is reported from potato (Sharma et al., 2015). In some of the prominent potato varieties, such as Kufri Pukhraj and Kufri Khyati, up to 60% yield loss has been reported (Chandel et al., 2020; Naga et al., 2021b; Jeevalatha et al., 2023). In the tomato crop in India, ToLCNDV causes yield losses in the range of 17.6–99.7% in different seasons and depending upon the incidence in severity (Reddy et al., 2011). Singh and Sastry (1973) documented 92.3% losses in 1-month-old tomato seedling infected by ToLCNDV (Kumar et al., 2021).

3 Host Range and Weed Reservoirs

The prominent crop families affected by this virus include Cucurbitaceae, Solanaceae, Malvaceae, Fabaceae, Caricaceae and Apiaceae (Kumar et al., 2021; Thi et al., 2025). The family Cucurbitaceae is significantly affected, with the majority of hosts exhibiting a range of symptoms and notable yield loss. In Cucumber (*Cucumis sativus*), leaf curling, leaf mosaic, and blistering symptoms are observed (Patil et al., 2017). In bottle gourd (*Lagenaria siceraria*) and Bitter gourd (*Momordica charantia*), leaf yellowing and blotching symptoms have been reported (Rajeshwari, 2014; Tahir & Haider, 2006). In chayote (*Sechium edule*), conspicuous leaf mosaic and fruit malformations are observed under severe infection (Nagendran et al., 2017). In Pumpkin, the primary symptoms appear as leaf curling, mosaic, and stunting as a result of ToLCNDV infection (*Cucurbita maxima*). Likewise, ivy gourd (*Coccinia grandis*), musk gourd (*Cucurbita moschata)*, and other cucurbit hosts exhibit symptoms of leaf yellowing, curling, and stunting when ToLCNDV infection occurs in a severe form (Ito et al., 2008; Juárez et al., 2014). Protuberant leaf curling and yellowing have been reported in Spine gourd (*Momordica dioica*), Tinda (*Praecitrullus fistulosus*), water melon (*Citrullus lanatus*) (Zaidi et al., 2017; Charoenvilaisiri et al., 2020), whereas in Zucchini, thickening of veins appears under ToLCNDV infection (Panno et al., 2016). Family *Solanaceae* crops, such as tomato, potato, pepper, and brinjal, are intensively affected by this begomovirus. Apical leaf curl disease in potato caused by ToLCNDV-potato is a widely reported challenge in potato cultivation in Asian countries, specifically India. In potato apical leaf curl disease, leaves exhibit yellowing or spotting, mild mosaic, reduced leaf area, leaf crinkling, and curling from the top portions, among other symptoms. Meanwhile, the plant stem becomes weak, and there is substantial stunting in the plants. A number of tubers and starch quality parameters are severely hampered due to ToLCNDV-potato infection (Kumar et al., 2021; Naga et al., 2021b). The common symptoms of yellow mosaic and chlorosis are reported from brinjal (*Solanum melongena*). In tomatoes, slight curling, chlorotic patches on leaves, mottling, and mosaic are reported. In the family Apiaceae, carrot (*Daucus carota*), a 1-week infection leads to the spread of the virus to the entire leaf lamina. This leads to the curling of leaves from the margin, reduced lamina size and ultimate drying of the virus-affected plants (Sivalingam et al., 2011). The typical symptoms, including leaf curling and severe mosaic, are observed in papaya, which produces fewer fruits due to infection (Raj et al., 2008). In the Soybean crop, ToLCNDV is reported to affect leaves and cause severe leaf curling, vein thickening and leaf yellowing. Another pulse crop in the *Fabaceae* family is the Mungbean (*Vigna radiata*), which

Table 1.1 Non-crop plants affected by ToLCNDV

Country	Host	Scientific name	Year	References/Accession
India	Ti plant	*Cordyline fruticose*	2019	Lager et al. (2022)
	False daisy	*Eclipta prostrata*	2008	EU688961
	Santa maria	*Parthenium hysterophorus*	2013	KF002409
	Indian acalypha	*Acalypha indica*	2014	Mall et al. (2014)
	Chanca piedra	*Phyllanthus niruri*	2018	Pant et al. (2018)
	Sunberry	*Physalis minima*	2019	Juárez et al. (2019)
	Thornapple	*Datura stramonium*	2009	JN676054
Pakistan	Goosefoot	*Chenopodium album*	2012	KC914896
	Common sowthistle	*Sonchus oleraceus*	2006	Haider et al. (2006)
	Ageratum spp.	*Ageratum* spp.	2013	KC960492
	Toothed dock	*Rumex dentatus*	2013	HG316125
	Nightshade	*Solanum nigrum*	2021	Ansar et al. (2021)
Spain	Field bindweed	*Convolvulus arvensis*	2019	Zammouri et al. (2017)
	Squirting cucumber	*Ecballium elaterium*	2019	Juárez et al. (2019)
	Sunberry	*Physalis minima*	2019	Juárez et al. (2019)

is affected by ToLCNDV. The infection causes vein banding and leaf curling symptoms in this crop, which leads to reduced yield. In the Malvaceae family, crops cotton (*Gossypium hirsutum*), Kenaf (*Hibiscus cannabis*) and okra (*Abelmoschus esculentus*) are affected by this virus and a range of symptoms, viz., vein thickening, vein net, enation, and leaf curl are observed due to ToLCNDV virus infection (Raj et al., 2007; Venkataravanappa et al., 2018). Apart from these main crops, the incidence of this virus in high-value ornamentals, such as Chrysanthemum and Dahlia pinnata, is also concerning, as the market value and trade of these crops are affected. Some other ornamentals of family *Crossandra infundibuliformis* (Acanthaceae), *Catharanthus roseus and Calotropis procera* (Apocynaceae), *Jatropha species* (Euphorbiaceae), *Papaver somniferum* (Papaveraceae), *Sauropus androgynus* (Phyllanthaceae), and *Cestrum nocturnum* (Solanaceae) (Table 1.1).

4 Mechanisms of Spread

The primary reason for this virus's widespread occurrence is its multiple transmission modes in the cropping system (Chang et al., 2010; Vo et al., 2025). As per the reports, this virus can be transmitted mechanically, through the virus vector, seeds and pollen. The capability of spreading through multiple means denotes its greater adaptability in diverse environments and ability to infect several hosts. The initial virus records in Asia, in parallel, documented the association of whitefly *Bemisia tabaci* (Hemiptera, family Aleyrodidae) as a significant vector of ToLCNDV. Whiteflies, acting as vectors and sucking pests, are abundant in the tropics, subtropics, and Mediterranean basin, and are considered a significant threat to agricultural and field crops (Sharma et al., 2021). This vector is reported to infest more than 600 plant species in the world. Garg et al. (2001) hypothesized that

B. tabaci might act as a primary vector for ToLCNDV transmission in potatoes (Garg et al., 2001). The circulative and non-propagative manner of the whitefly-mediated ToLCNDV transmission in potato was documented by Chandel et al. (2010). In the absence of cultivated crop hosts in fields, weeds and shrubs favor the survival of *B. tabaci*.

Furthermore, greenhouse whitefly *Trialeurodes vaporariorum* has been previously reported as a ToLCNDV vector in (Sangeetha et al., 2018). However, in Spain, another study reported the lack of infection in susceptible Zucchini hosts of ToLCNDV in the presence of *T. vaporariorum* (Farina et al., 2023). The outbreak and rapid emergence of the virus are also influenced by differences in the whitefly species within the *B. tabaci* species complex. Seed transmission of ToLCNDV was a fascinating report, while earlier, there was no report of Begomovirus transmission through seed (Kil et al., 2020; Thi et al., 2025). Seed transmission of ToLCNDV was first recorded in India on the cucurbit crop chayote (*Sechium edule*), and about 25% of seedlings were found infected with this virus (Sangeetha et al., 2018). The young seedlings germinated from fallen fruits tested positive for ToLCNDV in 2017 in Italy. In the virus-infected zucchini squash seed plants, the infection rate was reported to be 60%. This confirms the vertical transmission of ToLCNDV in the cucurbit host (Kil et al., 2020). Interestingly, it was observed that bleach treatment could remove the ToLCNDV infection from seeds, and it was concluded that the virus is seed-borne but does not transmit through seeds (Fortes et al., 2023). Likewise, in cucumber, ToLCNDV was intercepted with an infection rate of 79% in seedlings and seed coats (Chang et al., 2023). Additional research on virus-vector interactions in Pakistan reported ToLCNDV as seed-transmissible in tomato plants. The seed and pollen transmissibility of ToLCNDV, along with other begomoviruses like ToLCTV and TYLCTHV, was reported in cucumber and tomatoes (Chang et al., 2023). The infection rates were 79% for tomatoes and 36% for cucumbers. The pollen transmission rate was 70% in cucumbers and 77% in tomatoes. On oriental melon, the ToLCNDV-OM isolate was mechanically transmissible, while another isolate, ToLCNDV-CB, on cucumber was not mechanically transmissible. An attempt was made to identify the molecular mechanisms of mechanical transmission. The infectious clone of the virus was prepared, and artificial inoculation was done on *Nicotiana benthamiana*, oriental melon, and cucumber plants to understand the infection mechanisms. Mechanical transmission was performed through direct rub inoculation with sap extracted from *N. benthamiana*. The presence of the virus was validated through PCR, Southern blotting, and in situ hybridization (Jeevalatha et al., 2014). The movement protein of viral DNA-B in ToLCNDV-OM was observed to be involved in virus transmission.

Additionally, it was confirmed that the nuclear shuttle protein of DNA-B was not responsible for mechanical transmission. Single amino acid substitution of glutamate at amino acid position 19 of MP in ToLCNDV-OM was correlated with mechanical transmissibility. The mechanical transformation of ToLCNDV-OM was hampered when glycine was substituted in place of glutamate. Interestingly, ToLCNDV-CB was mechanically transmitted after substitution of glycine with glutamate at the 19th amino acid position (Lee et al., 2020).

5 Conclusion

The emergence of ToLCNDV in diverse crop families and non-crop hosts is a matter of great concern. The recent outbreaks in cucurbit crops and potatoes in Asian countries have raised alarms to prioritize the research and strategies the management of ToLCNDV. Large-scale screening to identify the resistant sources is required. Screening can be done by synthesizing infectious clones, maintaining vector populations, and using mechanical means in the reported host plants. Exhaustive studies are necessary for the development of rapid on-site diagnostic tools and to plan integrated disease management for this virus. In the era of genome editing, the identification of susceptibility genes and subsequent editing through CRISPR-based technology might emerge as a modern solution to manage the outbreak of this pathogen.

References

Ansar, M., Agnihotri, A. K., Ranjan, T., Karn, M., Srinivasaraghavan, A., Kumar, R. R., et al. (2021). Nightshade (*Solanum nigrum)*, an intermediate host between tomato and cucurbits of tomato leaf curl New Delhi virus. *Phytopathologia Mediterranea, 60*, 409–419. https://doi.org/10.36253/PHYTO-12745

Bhardwaj, V., Sood, S., Kumar, A., Vanishree, G., Sharma, S., Sundaresha, S., Raigond, B., Kumar, R., Bairwa, A., Lal, M., & Chakrabarti, S. K. (2019). Efficiency and reliability of marker-assisted selection for resistance to major biotic stresses in potato. *Potato Journal, 46*, 56–66.

Cai, L., Mei, Y., Ye, R., Deng, Y., Zhang, X., Hu, Z., et al. (2023). Tomato leaf curl New Delhi virus: an emerging plant begomovirus threatening cucurbit production. *aBIOTECH, 4*, 257–266. https://doi.org/10.1007/S42994-023-00118-4

Chandel, R. S., Vashisth, S., Soni, S., Kumar, R., & Kumar, V. (2020). The potato tuber moth, *Phthorimaea operculella* (Zeller), in India: Biology, ecology, and control. *Potato Research, 63*, 15–39. https://doi.org/10.1007/S11540-019-09426-Z

Chang, H. H., Ku, H. M., Tsai, W. S., Chien, R. C., & Jan, F. J. (2010). Identification and characterization of a mechanical transmissible begomovirus causing leaf curl on oriental melon. *European Journal of Plant Pathology, 127*, 219–228. https://doi.org/10.1007/s10658-010-9586-0

Chandel, R. S., Banyal, D. K., Singh, B. P., Kamlesh, M., Lakra, B. S. (2010). Integrated management of whitefly, Bemisia tabaci (Gennadius) and potato apical leaf curl virus in India. Potato Res 53:129–139.

Chang, H. H., Gustian, D., Chang, C. J., & Jan, F. J. (2023). Seed and pollen transmission of tomato leaf curl New Delhi virus, tomato leaf curl Taiwan virus, and tomato yellow leaf curl Thailand virus in cucumbers and tomatoes. *Plant Disease, 107*, 2002–2008. https://doi.org/10.1094/PDIS-09-22-2164-RE

Charoenvilaisiri, S., Seepiban, C., Phironrit, N., Phuangrat, B., Yoohat, K., Deeto, R., et al. (2020). Occurrence and distribution of begomoviruses infecting tomatoes, peppers and cucurbits in Thailand. *Crop Protection, 127*. https://doi.org/10.1016/J.CROPRO.2019.104948

Crespo, I., San-Miguel, B., Fernández, A., De Urbina, J. O., González-Gallego, J., & Tuñón, M. J. (2015). *Melatonin limits the expression of profibrogenic genes and ameliorates the progression of hepatic fibrosis in mice*. Elsevier. Available at: https://www.sciencedirect.com/science/article/pii/S1931524414003466 [Accessed March 29, 2025]

Crespo, O., Robles, C., Ruiz, L., & Janssen, D. (2020). Antagonism of cucumber green mottle mosaic virus against tomato leaf curl New Delhi virus in zucchini and cucumber. *The Annals of Applied Biology, 176*, 147–157. https://doi.org/10.1111/aab.12535

Farina, A., Rapisarda, C., Fiallo-Olivé, E. and Navas-Castillo, J. (2023). Tomato leaf curl New Delhi virus spain strain is not transmitted by Trialeurodes vaporariorum and is inefficiently transmitted by Bemisia tabaci Mediterranean between zucchini and the wild cucurbit Ecballium elaterium. Insects, 14(4), p.384.

Fortes, I. M., Pérez-Padilla, V., Romero-Rodríguez, B., Fernández-Muñoz, R., Moyano, C., Castillo, A. G., et al. (2023). Begomovirus tomato leaf curl New Delhi virus is Seedborne but not seed transmitted in melon. *Plant Disease, 107*, 473–479. https://doi.org/10.1094/PDIS-09-21-1930-RE

Garg, I. D., Khurana, S. M. P., Shiv, K., & Lakra, B. S. (2001). Association of a geminivirus with potato apical leaf curl in India and its immuno-electron microscopic detection. *Journal of Indian Potato Association, 2001*(28), 227–232.

Haider, M. S., Tahir, M., Latif, S., & Briddon, R. W. (2006). First report of tomato leaf curl New Delhi virus infecting Eclipta prostrata in Pakistan. *Plant Pathology, 55*, 285. https://doi.org/10.1111/j.1365-3059.2005.01278.x

Hanley-Bowdoin, L., Bejarano, E. R., Robertson, D., & Mansoor, S. (2013). Geminiviruses: Masters at redirecting and reprogramming plant processes. *Nature Reviews. Microbiology, 11*, 777–788. https://doi.org/10.1038/nrmicro3117

Ito, T., Sharma, P., Kittipakorn, K., & Ikegami, M. (2008). Complete nucleotide sequence of a new isolate of tomato leaf curl New Delhi virus infecting cucumber, bottle gourd and muskmelon in Thailand. *Archives of Virology, 153*, 611–613. https://doi.org/10.1007/s00705-007-0029-y

Jeevalatha, A., Singh, B. P., Kaundal, P., Kumar, R., & Raigond, B. (2014). RCA-PCR: A robust technique for the detection of tomato leaf curl New Delhi virus-[potato] at ultra-low virus titre. *Potato Journal, 41*, 76–80.

Jeevalatha, A., Vanishree, G., Siddappa, S., Kumar, R., Kaundal, P., Kumar, A., & Chakrabarti, S. K. (2021). Molecular characterization and infectivity analysis of tomato leaf curl New Delhi virus isolates infecting potato. *3 Biotech, 11*, 203. https://doi.org/10.1007/s13205-021-02752-5

Jeevalatha, A., Siddappa, S., Kumar, R., Tiwari, R. K., Lal, M. K., Sharma, S., Chakrabarty, S. K., & Singh, B. P. (2023). RNA-seq analysis reveals an early defense response to tomato leaf curl New Delhi virus in potato cultivar Kufri Bahar. *Functional & Integrative Genomics, 23*, 215. https://doi.org/10.1007/s10142-023-01138-5

Jones, R. A. C. (2021). Global plant virus disease pandemics and epidemics. *Plants, 10*, 233. https://doi.org/10.3390/PLANTS10020233

Juárez, M., Tovar, R., Fiallo-Olivé, E., Aranda, M. A., Gosálvez, B., Castillo, P., et al. (2014). First detection of tomato leaf curl New Delhi virus infecting zucchini in Spain. *Plant Disease, 98*, 857. https://doi.org/10.1094/PDIS-10-13-1050-PDN

Juárez, M., Rabadán, M. P., Martínez, L. D., Tayahi, M., Grande-Pérez, A., & Gómez, P. (2019). Natural hosts and genetic diversity of the emerging tomato leaf curl New Delhi virus in Spain. *Frontiers in Microbiology, 10*, 140.

Kil, E. J., Bich Vo, T. T., Fadhila, C., Thi Ho, P., Lal, A., Troiano, E., et al. (2020). Seed transmission of tomato leaf curl New Delhi virus from zucchini squash in Italy. *Plants, 9*, 563. https://doi.org/10.3390/PLANTS9050563

Kumar, R., & Singh, B. P. (2023). Diagnostics for production of disease-free planting material of potato. *International Journal of Innovative Horticulture, 12*, 31–49. https://doi.org/10.5958/2582-2527.2023.00003.9

Kumar, R., Tiwari, R. K., Jeevalatha, A., Siddappa, S., Shah, M. A., Sharma, S., et al. (2021). Potato apical leaf curl disease: Current status and perspectives on a disease caused by tomato leaf curl New Delhi virus. *Journal of Plant Diseases and Protection, 128*, 897–911. https://doi.org/10.1007/S41348-021-00463-W

Kumar, R., Kaundal, P., Tiwari, R. K., Siddappa, S., Kumari, H., Lal, M. K., Naga, K. C., Sharma, S., Sagar, V., & Kumar, M. (2022). Establishment of a one-step reverse transcription recombinase polymerase amplification assay for the detection of potato virus S. *Journal of Virological Methods, 307*, 114568.

Kumar, R., Lal, M. K., Tiwari, R. K., Chourasia, K. N., Kumar, A., Kumar, R., Sharma, S., & Singh, B. (2023). Investigating the interplay between tomato leaf curl New Delhi virus infection, starch metabolism and antioxidant defence system in potato (*Solanum tuberosum* L.). *Antioxidants, 12*, 1447. https://doi.org/10.3390/antiox12071447

Lager, P., Sharma, J., & Kumar, Y. (2022). First report of a begomovirus infecting Cordyline fruticosa. *New Disease Reports, 45*. https://doi.org/10.1002/NDR2.12068

Lakra, B. S. (2010). Degeneration of potato cultivars due to potato apical leaf curl virus disease under Hisar ecological conditions. *Potato Journal, 37*, 164–166.

Lal, M. K., Tiwari, R. K., Kumar, R., Naga, K. C., Kumar, A., Singh, B., et al. (2021). Effect of potato apical leaf curl disease on glycemic index and resistant starch of potato (*Solanum tuberosum* L.) tubers. *Food Chemistry, 359*. https://doi.org/10.1016/j.foodchem.2021.129939

Lee, C. H., Zheng, Y. X., Chan, C. H., Ku, H. M., Chang, C. J., & Jan, F. J. (2020). A single amino acid substitution in the movement protein enables the mechanical transmission of a geminivirus. *Molecular Plant Pathology, 21*, 571–588. https://doi.org/10.1111/MPP.12917

Majid, A., Shakeel, M. T., Aslam, M. N., Ashraf, W., Raheel, M., Maqsood, A., et al. (2023). Unraveling the devastating impact of tomato leaf curl New Delhi virus (ToLCNDV): Understanding its characteristics, host range, and management strategies. *Advances in Virology, 1*, 10–19. https://doi.org/10.58801/2k23rv2utd

Mall, S., Gupta, S., & Upadhyaya, P. P. (2014). Identification of tomato leaf curl virus infecting Acalypha indica: An ethnomedicinal weed in North-Eastern Uttar Pradesh. *Microbial Diversity and Biotechnology in Food Security, 177–181*. https://doi.org/10.1007/978-81-322-1801-2_14

Moriones, E., Praveen, S., & Chakraborty, S. (2017). Tomato leaf curl New Delhi virus: An emerging virus complex threatening vegetable and fiber crops. *Viruses, 9*. https://doi.org/10.3390/v9100264

Naga, K. C., Subhash, S., Bhatnagar, A., Tiwari, R. K., Kumar, R., Verma, G., Sharma, S., & Chakrabarti, S. K. (2021a). Influence of host plants on virus acquisition and endosymbionts of whitefly, Bemisia tabaci (Gennadius). *Indian Journal of Entomology, 83*, 371–375. https://doi.org/10.5958/0974-8172.2020.00206.0

Naga, K. C., Siddappa, S., Kumar, R., Tiwari, R. K., Subhash, S., Verma, G., et al. (2021b). A new record of Asia II 5 genetic group of *Bemisia tabaci* (Gennadius) in the major potato growing areas of India and its relationship with tomato leaf curl New Delhi virus infecting potato. *3 Biotech, 11*, 421. https://doi.org/10.1007/s13205-021-02966-7

Nagendran, K., Mohankumar, S., Mohammed Faisal, P., Bagewadi, B., & Karthikeyan, G. (2017). Molecular evidence for the occurrence of tomato leaf curl New Delhi virus on chayote (*Sechium edule*) in southern India. *VirusDisease, 28*, 425–429. https://doi.org/10.1007/S13337-017-0403-7

Orfanidou, C. G., Malandraki, I., Beris, D., Kektsidou, O., Vassilakos, N., Varveri, C., et al. (2019). First report of tomato leaf curl New Delhi virus in zucchini crops in Greece. *Journal of Plant Pathology, 101*, 799. https://doi.org/10.1007/s42161-019-00265-y

Padidam, R., Beachy, R. N., & Fauquet, C. M. (1995). Tomato leaf curl geminivirus from India has a bipartite genome and coat protein is not essential for infectivity. *The Journal of General Virology, 76*, 25–35. https://doi.org/10.1099/0022-1317-76-1-25

Panno, S., Iacono, G., Davino, M., Marchione, S., Zappardo, V., Bella, P., et al. (2016). First report of tomato leaf curl New Delhi virus affecting zucchini squash in an important horticultural area of southern Italy. *New Disease Reports, 33*, 6. https://doi.org/10.5197/j.2044-0588.2016.033.006

Pant, R. P., Bhatnagar, A., & Lal, M. (2018). Role of alternate host plants in the transmission of apical leaf curl disease of potato caused by tomato leaf curl New Delhi virus-potato (ToLCNDV-pot.) in northern India. *Indian Journal of Agricultural Sciences, 88*, 1258–1262. https://doi.org/10.56093/IJAS.V88I8.82565

Parrella, G., Elbeaino, T., & Guy, P. L. (2022). Editorial: Emerging and reemerging plant viruses in a context of global change. *Frontiers in Plant Science, 13*. https://doi.org/10.3389/FPLS.2022.1108211/FULL

Patil, C. V., Ramdas, S. V., Premchand, U., & Shankarappa, K. S. (2017). Survey, symptomatology, transmission, host range and characterization of begomovirus associated with yellow mosaic disease of ridge gourd in southern India. *VirusDisease, 28*, 146–155. https://doi.org/10.1007/s13337-017-0376-6

Phaneendra, C., Rao, K. R. S. S., Jain, R. K., & Mandal, B. (2012). Tomato leaf curl New Delhi virus is associated with pumpkin leaf curl: A new disease in northern India. *Indian Journal of Virology, 23*, 42–45. https://doi.org/10.1007/s13337-011-0054-z

Pratap, D., Kashikar, A. R., & Mukherjee, S. K. (2011). Molecular characterization and infectivity of a tomato leaf curl New Delhi virus variant associated with newly emerging yellow mosaic disease of eggplant in India. *Virology Journal, 8*, 305. https://doi.org/10.1186/1743-422X-8-305

Raj, S. K., Khan, M. S., Snehi, S. K., & Roy, R. K. (2007). Yellow vein netting of Bimili jute (*Hibiscus cannabinus* L.) in India caused by a strain of tomato leaf curl New Delhi virus containing DNA β. *Australasian Plant Disease Notes, 2*, 45. https://doi.org/10.1071/DN07021

Raj, S. K., Snehi, S. K., Khan, M. S., Singh, R., & Khan, A. A. (2008). Molecular evidence for association of tomato leaf curl New Delhi virus with leaf curl disease of papaya (*Carica papaya* L.) in India. *Australasian Plant Disease Notes, 3*, 152–155. https://doi.org/10.1007/BF03211275

Reddy, A. B., Patti, M. S., Reddy, K. M., & Venkataravanappa, V. (2011). Detection and diagnosis of tomato leaf curl virus infecting tomato in Northern Karnataka. *African Journal of Agricultural Research, 6*(5), 1051–1057.

Sáez, C., Martínez, C., Ferriol, M., Manzano, S., Velasco, L., Jamilena, M., et al. (2016). Resistance to tomato leaf curl New Delhi virus in *Cucurbita* spp. *The Annals of Applied Biology, 169*, 91–105. https://doi.org/10.1111/AAB.12283

Sáez, C., Esteras, C., Martínez, C., Ferriol, M., Dhillon, N. P. S., López, C., et al. (2017). Resistance to tomato leaf curl New Delhi virus in melon is controlled by a major QTL located in chromosome 11. *Plant Cell Reports, 36*, 1571–1584. https://doi.org/10.1007/S00299-017-2175-3

Sangeetha, B., Malathi, V. G., Alice, D., Suganthy, M., & Renukadevi, P. (2018). A distinct seed-transmissible strain of tomato leaf curl New Delhi virus infecting Chayote in India. *Virus Research, 258*, 81–91. https://doi.org/10.1016/J.VIRUSRES.2018.10.009

Sharma, S., Kang, S. S., & Sharma, A. (2015). First report of mixed infection of zucchini yellow mosaic virus and tomato leaf curl New Delhi virus in bittergourd in India. *Journal of Plant Pathology, 97*, 397. https://doi.org/10.4454/JPP.V97I2.045

Sharma, N., Sahu, P. P., Prasad, A., Muthamilarasan, M., Waseem, M., Khan, Y., et al. (2021). The Sw5a gene confers resistance to ToLCNDV and triggers an HR response after direct AC4 effector recognition. *Proceedings of the National Academy of Sciences of the United States of America, 118*. https://doi.org/10.1073/PNAS.2101833118

Siskos, L., Cui, L., Wang, C., Visser, R. G. F., Bai, Y., & Schouten, H. J. (2022). A new challenge in melon resistance breeding: The ToLCNDV case. *Euphytica, 218*. https://doi.org/10.1007/S10681-022-03081-1

Sivalingam, P. N., & Varma, A. (2007). Non-tomato natural hosts of tomato infecting begomoviruses in North-Western India. *Indian Journal of Virology, 18*, 20–27.

Sivalingam, P. N., Sumiya, K. V., & Malathi, V. G. (2011). Carrot as a new host for a begomovirus: Yellow mosaic disease of carrot reported in India. *New Disease Reports, 23*, 34–34. https://doi.org/10.5197/J.2044-0588.2011.023.034

Srivastava, A., Kumar, S., Jaidi, M., Raj, S. K., & Shukla, S. K. (2016). First report of tomato leaf curl New Delhi virus on opium poppy (*Papaver somniferum*) in India. *Plant Disease, 100*, 232. https://doi.org/10.1094/PDIS-08-15-0883-PDN

Singh, S. J., Sastry, K. S. M., & Sastry, K. S. (1973). Effect of oil spray on the control of tomato leaf-curl virus in field. *Indian J. Agric. Sci, 43*, 669–672.

Thi, T. P., Vo, B., Tabassum, M., Nattanong, B., Qureshi, M. A., Im, H.-J., et al. (2025). The insidious threat: Assessing the dangers and spread of tomato leaf curl New Delhi virus. *Plant Pathology Journal, 41*, 1–16. https://doi.org/10.5423/PPJ.RW.11.2024.0177

Tahir, M., & M. S. Haider. (2006). Naturally occurring strains of bipartite begomoviruses affect some members of Cucurbitaceae family inside and outside the cotton zone in Pakistan 527–533.

Tomar, G., Chakrabarti, S. K., Sharma, N. N., Jeevalatha, A., Sundaresha, S., Vyas, K., et al. (2018). RNAi-based transgene conferred extreme resistance to the geminivirus causing apical leaf curl disease in potato. *Plant Biotechnology Reports, 12*, 195–205. https://doi.org/10.1007/s11816-018-0485-8

Troiano, E., & Parrella, G. (2023). First report of tomato leaf curl New Delhi virus in *Lagenaria siceraria* var. longissima in Italy. *Phytopathologia Mediterranea, 62*, 25–28. https://doi.org/10.36253/PHYTO-14147

Venkataravanappa, V., Reddy, L. R. C. N., Saha, S., Subbanna, S. K., & Manem, K. R. (2018). Detection and characterization of tomato leaf curl New Delhi virus association with mosaic disease of ivy gourd (*Coccinia grandis* (L.) Voigt) in North India. *Archives of Biological Sciences, 70*, 339–347. https://doi.org/10.2298/ABS170616051V

Vo, T. T. B., Tabassum, M., Nattanong, B., Qureshi, M. A., Im, H. J., Parrella, G., et al. (2025). The insidious threat: Assessing the dangers and spread of *tomato leaf curl New Delhi virus*. *Plant Pathology Journal, 41*, 1–16. https://doi.org/10.5423/PPJ.RW.11.2024.0177

Zaidi, S. S. E. A., Martin, D. P., Amin, I., Farooq, M., & Mansoor, S. (2017). Tomato leaf curl New Delhi virus: A widespread bipartite begomovirus in the territory of monopartite begomoviruses. *Molecular Plant Pathology, 18*, 901–911. https://doi.org/10.1111/mpp.12481

Zammouri, S., Zaagueri, T., Eddouzi, J., Belkhadhi, M. S., Hajlaoui, M. R., & Mnari-Hattab, M. (2017). First report of tomato leaf curl New Delhi virus on tomato crop in Tunisia. *Journal of Plant Pathology, 99*, 813. https://doi.org/10.4454/jpp.v99i3.3975

Recent Insights on Genome Structure and Biology of Tomato Leaf Curl New Delhi Virus

2

B. Parameswari, G. V. Chaitra, P. Anbazhagan, and V. Celia Chalam

Abstract

Tomato leaf curl New Delhi virus (ToLCNDV) is a devastating virus that severely affects the economically important crops. Understanding the genetic mechanisms of ToLCNDV is crucial for developing efficient management strategies for combating its global impact on agriculture. This chapter focuses on the recent trends in deciphering the genetic structure and transmission nature of the virus. The role of DNA alpha and betasatellites in the biology of the virus is well documented, along with studies on taxonomic position and host range. Key insights are provided into recombination; pseudo-recombination contributes to the evolutionary diversity of the virus. The chapter describes the present achievements in the context of improving the insights into this newly emerging *Begomovirus*. Despite conventional management practices, the need for innovative transgenic solutions to control the spread of disease provides a sustainable approach. Recent achievements in genetic resistance and RNA interference (RNAi)-based strategies for ToLCNDV management are explored. Understanding of ToLCNDV biology and genetic diversity guides future research for sustainable disease management.

Keywords

Begomovirus · Genetic structure · Host virus interactions · RNA interference · Whitefly transmission

B. Parameswari (✉) · G. V. Chaitra · P. Anbazhagan
ICAR National Bureau of Plant Genetic Resources, Regional Station, Hyderabad, Telangana, India

V. Celia Chalam
ICAR-National Bureau of Plant Genetic Resources, New Delhi, India

R. Kumar et al. (eds.), *Tomato Leaf Curl New Delhi Virus (ToLCNDV)*,
https://doi.org/10.1007/978-981-95-2852-3_2

1 Introduction

Geminiviruses are a significant class of plant viruses, mainly affecting horticultural crops across tropical and subtropical regions. Among these, Tomato leaf curl New Delhi virus (ToLCNDV) causes a major challenge to the production of the tomato crop in South and Southeast Asia (Chakraborty, 2009; Kumar et al., 2021). ToLCNDV is a bipartite, single-stranded DNA virus transmitted by whitefly, *Bemisia tabaci* (order *Hemiptera*, family *Aleyrodidae*), in a circulative, persistent mechanism (Zaidi et al., 2017). ToLCNDV dispersal might also be facilitated through contaminated seeds, as seed transmission of TYLCV has been recently reported to occur in tomato plants (Kil et al., 2016; Sangeetha et al., 2018). Long-distance spread is mediated by winds and through human activities. Several studies reported that monopartite begomoviruses play a key role in causing disease complexes and affecting the crop yield in the Old World. ToLCNDV is the bipartite begomovirus that causes diseases in many crop plants belonging to the families of *Cucurbitaceae*, *Euphorbiaceae*, *Fabaceae*, *Malvaceae*, and importantly, *Solanaceae* (Moriones et al., 2017).

The virus was reported in solanaceous crops for the first time, and later, significant damage was observed in cucurbit crops as well. ToLCNDV symptoms vary for the host plant species and growing conditions. Infected tomato plants exhibit typical symptoms of upward/downward curling and crinkling, vein-clearing, mottling, leaf puckering, and blistering. At the later stages of the infection, plants appear stunted with shortened internodes and show impaired fruit setting, leading to the complete loss of the crop (Chakraborty, 2009). In cucurbits, leaf distortion, vein swelling, and mosaic symptoms are caused by ToLCNDV, which negatively affects food productivity and quality (Yazdani-Khameneh et al., 2016). The emergence of epidemics of ToLCNDV causes huge yield losses in tomatoes and other economically important crops. ToLCNDV is spreading rapidly and is listed as a quarantine disease by the European and Mediterranean Plant Protection Organization (EPPO). A deeper understanding of genetic structure and its interactions with hosts and vectors is essential for effective disease management. Such insights will be beneficial in the development of resistant crop varieties, particularly in the context of a changing climate. Given the significant impact of these viruses on economically and socially essential crops, extensive research has been carried out to understand the biology of these viruses, resulting in significant advancements in the field.

2 Taxonomy and Classification

Begomovirus is a prominent group of newly emerging plant viruses (Fiallo-Olive & Navas-Castillo, 2023). Based on the genetic diversity and geographical origin, the genus is divided into two phylogenetic groups: New World (NW) which encompasses viruses such as desmodium leaf distortion virus (DesLDV), and corchorus yellow vein Cuba virus (CoYVCUV) and the Old World (OW) includes tomato yellow leaf curl Thailand virus (TYLCTHV), cotton leaf curl virus (CLCuV) and tomato yellow leaf curl virus (TYLCV) (Farina et al., 2023). Among the Old World viruses, ToLCNDV is a highly destructive virus belonging to the genus *Begomovirus* family *Geminiviridae.* The virus was first identified in India in 1995, affecting solanaceous crops (Padidam et al., 1995). It also spread to the Mediterranean basin, where it was first detected in Spain in 2012 (Juarez et al., 2019). The Mediterranean ToLCNDV isolates belong to a novel Spain strain—ToLCNDV-ES, which infects cucurbitaceous crops but is poorly adapted to the tomato crop. Phylogenetic analysis revealed that the Mediterranean virus population is distinct from the Asian populations, with no clear clustering pattern.

3 Genome Structure

ToLCNDV is characterized by quasi-icosahedral particles measuring approximately 18×30 nm^2 in size and protects the circular, single-stranded DNA genomes, i.e., DNA-A and DNA-B, ranging from 2.5 to 2.7 kb. DNA-A component is capable of independent replication, and DNA-B component is essential for systemic infection in plants (Zaidi et al., 2017; Kumar et al., 2021). Significant research over the past decade has focused on unravelling the genome structure and protein functions of viruses to gain deeper insights into the complex mechanisms underlying their interactions with host plants in causing diseases (Fig. 2.1).

The genomic components DNA A and B each encode distinct sets of genes with specific functions. DNA A codes for five different proteins including (i) coat protein (AV1/V1 and AV2/V2), which plays a vital role in the movement of virions, whereas the complementary-sense strand codes for (ii) replication-associated protein (RepA, AC1/C1), (iii) a transcriptional activator protein (AC2/C2), (iv) a replication enhancer protein (AC3/C3), and (v) C4 protein (AC4/C4). DNA B codes for two proteins that are (i) the nuclear shuttle protein (BV1). (ii) the movement protein (BC1). DNA-A can replicate itself and code for viral replication and encapsidation. DNA-B synthesizes two proteins, both intercellular and intracellular, associated with virus movement. DNA-B needs DNA-A for replication, systemic infection,

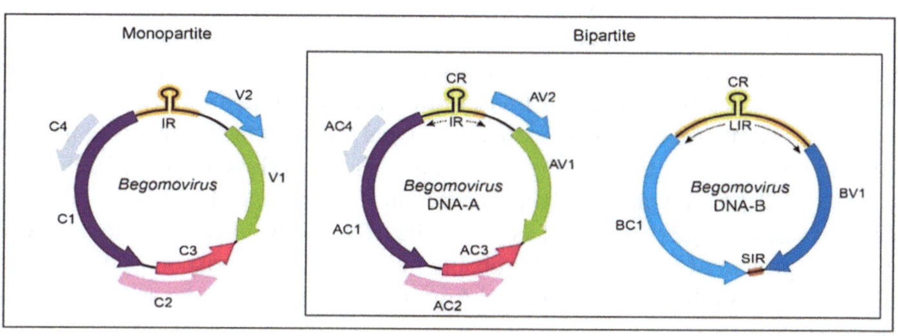

Fig. 2.1 Genome organization of tomato leaf curl New Delhi virus, having two DNA components, DNA-A and DNA-B, containing all the ORFs for proteins (Kadi et al., 2024)

movement, manifestation, and nuclear localization (Sanderfoot & Lazarowitz, 1996; Kumar et al., 2022). ToLCNDV is a bipartite begomovirus that uniquely contains an AV2 gene in its DNA, distinguishing it from other bipartite begomoviruses identified in the New World. ToLCNDV is also often associated with betasatellites (Akhter et al., 2014).

Betasatellites are half the size of the virus DNA-A and DNA-B components of their helper begamoviruses, measuring 1.4 kb in length. A key feature of betasatellites is the *βC1* gene, which encodes the βC1 protein—a pathogenicity determinant protein that induces the disease's characteristic symptoms in the host plants. It is situated between the adenine-rich region and the conserved hairpin structure, both of which are distinctive features of betasatellites. This strategic positioning allows the βC1 protein to interact effectively with both the helper begomovirus and the host plant's cellular machinery (Hanley-Bowdoin et al., 2013). The *βC1* protein has been shown to exhibit novel ATPase activity, which is thought to play a role in modulating the replication and movement of the helper virus. Mutations in the *βC1* gene that disrupt this ATPase activity have been linked to alterations in symptom development and viral DNA accumulation (Gnanasekaran et al., 2021).

Betasatellites and DNA B depend on the Rep region for their replication. Rep is encoded by DNA-A, containing three main domains: a DNA-binding domain, an oligomerization domain, and a helicase domain (Fondong, 2013). The interaction between the oligomerization domain and the DNA-binding domain facilitates the accumulation of the virus (Chatterji et al., 2001). Two main factors determine the specificity of the ToLCNDV Rep region, which can replicate DNA A and B, as well as betasatellite molecules (Kumar et al., 2023). An iteron-related domain (IRD) in the N-terminus of the Rep region and the sequences of iterons within the Rep-binding region 5′ of v-ori. Upon recognition of appropriate iteron sequences by the Rep IRD, Rep binds near v-ori and initiates rolling circle replication (RCR).

Geminiviruses predominantly depend on the DNA polymerases of the host and other accessory factors to replicate within the host nucleus. After replication, they are transferred throughout the plant by the systemic movement of the virus. Virus-encoded proteins mediate the inter- and intracellular trafficking of viral DNA. The proteins encoded by the virus influence several cellular pathways of their host, including RNA silencing, cell differentiation, cell cycle control, and plasmodesmata function (Hanley-Bowdoin et al., 2013). The ToLCNDV NSP is the pathogenicity determinant and triggers a hypersensitive response (HR) in infected plants (Hussain et al., 2005). The N-terminus of transcription activator protein (TrAP) contains a zinc finger domain and nuclear localization signal (NLS), which is involved in overcoming host defenses through the inhibition of the HR (Hussain et al., 2005). The CP and AV2 proteins are involved in virus movement, and the absence of these proteins causes reduced levels of ssDNA. The CP of ToLCNDV also interacts with a midgut protein of *B. tabaci* and probably facilitates virus transport both from the digestive tract to the hemolymph, and from the hemolymph to the salivary glands (Rana et al., 2016).

4 Pathogenicity Determinant

The intra- and intercellular movement is carried out by Movement Protein (MP) and Nuclear Suttle Protein (NSP), encoded on the component designated DNA B in the case of bipartite begomoviruses. The MP plays a prominent role in pathogenicity and both proteins determine the viral host range. NSP is an avirulence determinant that interacts with the product of a resistance gene (R gene product) encoded by a host defense system, triggering a host defense response involving a Hypersensitive Response (HR) reaction. Suppose a lower level of NSP is expressed. In that case, it leads to a lack of HR, and the virus has evolved mechanisms mediated by other virus-encoded proteins to overcome the HR elicited by NSP. Studies reported that the deletion of 100 amino acids from the C terminus did not compromise the HR response, suggesting that this region has no role in HR and the deletion of 60 or 100 amino acids from the N terminus of NSP abolished the HR response, suggesting that these sequences are required for the HR response (Hussain et al., 2005).

5 Pseudo Recombination and Component Capture

The re-assortment of genetic components among the multipartite viruses is known as pseudo-recombination. The capture of one or more genomic components or satellites of a multipartite virus by another virus is component capture. Re-assortment and component capture are the primary mechanisms whereby begomoviruses adapt to infect new host plants. ToLCNDV DNA A and B components have been known to interact with several viruses and betasatellites during the time of mixed infections in host species. Pseudo-recombination of ToLCNDV with tomato leaf curl Palampur virus (ToLCPalV) has been observed both in the field and in experimental conditions (Malik et al., 2017). It has also been shown that the CP of ToLCNDV mediates the transmission of ToLCPalV DNA-B by whiteflies and trans-encapsidation, which is the transfer of one virus genome within the capsid encoded by a different virus genome (Kanakala et al., 2013).

6 Transmission Biology

ToLCNDV is transmitted by *Bemisia tabaci* in a non-propagative, persistent, circulative manner. The *Bemisia tabaci* complex is a group of cryptic species that are morphologically similar but genetically distinct, and their global distribution and behavior vary significantly. These viruses do not replicate in vector tissues but traverse the insect gut, hemolymph, and salivary tissue membranes to reach the salivary glands for transmission. The virus is limited to the plant phloem. Adult insects acquire the virus by sucking on phloem sap and circulate within their body upon ingestion. It can further transmit the virus to another plant during feeding. The virus does not replicate in the insect, and transmission efficiency decreases with the age of the vector. The vector remains viruliferous throughout its life span. The virus can persist in the vector even in the temporary absence of host plants. The whitefly species such as Asia 1, Asia II 1/5/7, and MEAM1 (Middle East-Asia Minor 1) (earlier known as B biotype) are the vectors of ToLCNDV in different parts of South Asia. Mediterranean species (MED, also commonly known as biotype Q is the most prevalent and devastating biotype across horticultural crops in Mediterranean countries. The Q biotype is resistant to insecticides, which is displacing the B biotype. *Trialeurodes vaporariorum* (Westwood), the greenhouse whitefly, was able to transmit a ToLCNDV isolate found infecting chayote (*Sechium edule* Jacq.), a cucurbit, in India with a 4-h acquisition access period (AAP) and a 24-h inoculation feeding period (IFP) (Sangeetha et al., 2018). Whereas, the transmission of ToLCNDV by *Bemisia tabaci* requires a minimum acquisition period of about 30 min and an inoculation access period of about 10 min.

7 RNA Silencing Suppressors

In addition to aiding in virus multiplication and movement, the viral proteins have developed to suppress plant defense mechanisms, preventing plants from responding to viral infection. In host species, RNA interference (RNAi) is the primary defense mechanism against DNA viruses. Numerous viral proteins have been reported as critical factors in counteracting host defenses, particularly by targeting and suppressing the RNAi-mediated silencing machinery. Initially, AC2 acts as the RNAi silencing suppressor (RSS); it interacts with Adenosine kinase (ADK) enzyme, which is necessary to produce the S-adenosyl methionine (SAM) and inhibit the methylation cycle. It attenuates post-transcriptional gene silencing (PTGS) by inhibiting the activity of RNA-dependent RNA polymerase 6 (RDR6), Argonaute1 (AGO1), and calmodulin-like protein (rugs-CaM), which are required for siRNA biogenesis, thus making the host plants susceptible to viral infection. AV2 is hypothesized to suppress the PTGS-mediated defense mechanism in the host plant and demonstrated that AV2 interacts with SGS3, which prevents the conversion of ssRNA to dsRNA, thus inhibiting the production of siRNAs.

AC4 also acts as an RNA silencing suppressor (RSS) by binding the single-stranded siRNA and miRNA, thus inhibiting the cleavage of target mRNA and inducing the viral spread. AC5 represses both the PTGS and TGS of the host plant. It interacts with ssRNA to inhibit the PTGS machinery. It was hypothesized that AC5 suppresses chromatin methylation by reducing the expression of Domains Rearranged Methyltransferase-2 (DRM2; CpHpH cytosine methyltransferase) in *N. benthamiana.* A recent study suggests the role of Rep in TGS. Rep associates with small ubiquitin-like modifiers (SUMO) conjugating enzyme 1 (SCE1) and modifies the sumoylation pattern of host factors to create conditions suitable for viral infection. Thus, by targeting these virus proteins through a transgenic approach, resistance can be developed against ToLCNDV. Numerous plant miRNAs are predicted to play significant roles during viral infections; however, the exact functions of most remain unknown. While various studies on different plant species have highlighted the importance of miRNAs in response to viral infections, a thorough examination into the specific roles of miRNAs associated with viral infections in tomato crops has yet to be conducted.

8 Epidemiology

Tomato yellow leaf curl virus causes severe epidemics in tomato crops across different geographical regions in both the Old and New World countries (Pico et al., 1996).It affects both Old and New World countries, leading to significant yield losses. The virus spreads rapidly, making it a serious threat to tomato production. Effective management is crucial to control its impact on global tomato farming. The disease is found to be very severe in hot and humid areas where crops are grown during the summer and rainy seasons. Moderate-to-high temperatures and humidity predispose the population to the buildup of the whitefly vectors, leading to efficient transmission of the virus. After the acquistion feeding period by whitefly, healthy tomato plants exhibited similar symptoms within 14–16 days under an insect-proof glasshouse. The majority of begomoviruses cause 10–30% yield loss in different environmental conditions in various crops. TYLCD was first identified in Israel in 1959 and sequenced in 1988 as Tomato yellow leaf curl virus Israel (TYLCV-Is) (Navot et al., 1991; Czosnek & Laterrot, 1997). It causes severe damage, with 100% yield loss reported in areas with high whitefly populations (Cohen & Antignus, 1994; Polston & Lapidot, 2007). The virus unknowingly spread to the Dominican Republic in the late 1980s through greenhouse transplants. This marked the first entry of Old World Begomovirus into the New World, earning it the name "virus of the New World." Presently, there are 11 species reported of TYLCV virus, viz., TYLCV, TYLC Axarquia virus (TYLCAxV), TYLC China virus (TYLCChV), TYLC Guangdong virus (TYLCGuV), TYLC Indonesia virus (TYLCIDV), TYLC Kanchanaburi virus (TYLCKaV), TYLC Malaga virus (TYLCMalV), TYLC Mali virus (TYLCMLV), TYLC Sardinia virus (TYLCSV), TYLC Thailand virus (TYLCThV), and TYLC Vietnam virus (TYLCVNV). ToLCNDV was found in all cultivated cucurbit crops except watermelon and tomato. Weeds such as *Ecballium elaterium* and *Datura stramonium* were identified as alternative hosts, potentially serving as new sources of virus inoculum.

9 Host Range Studies

ToLCNDV infects majorly tomato and chili, potato, besides having a broad host range that includes economically essential crops with different families such as *Cucurbitaceae* (squash, cucumber, pumpkin), *Malvaceae* (cotton, okra) and *Euphorbiaceae* (castor, poinsettia), highlighting the adaptability and economic significance of the virus (Jeevalatha et al., 2021). The key differences between ToLCNDV infecting tomato and cucumber isolates are mentioned in Table 2.1. Recent research has highlighted pathogenic differences between ToLCNDV isolates from the Mediterranean region (ToLCNDV-ES) and those from Southeast Asia. For

Table 2.1 Comparative difference between ToLCNDV infecting tomato and cucurbits as host plants

Aspects	ToLCNDV in tomato	ToLCNDV in cucurbits	References
Symptoms	Leaf curling, yellowing, stunting, and reduced fruit yield	Yellow mosaic, blistering, leaf curling, vein thickening, and malformed fruits	Fortes et al. (2016)
Family of host	Solanaceae (tomato, pepper, brinjal)	*Cucurbitaceae* (squash, pumpkin, zucchini, cucumber)	Reddy et al. (2011)
Economic impact	Smaller, unmarketable fruits cause a greater impact on production	Malformed fruits lead to reduced market quality	Singh et al. (2019)
Replication efficiency	High replication efficiency due to strong compatibility with host replication machinery	Often associated with mixed infections	Zaidi et al. (2017)
Resistance mechanism	Resistant varieties with Ty genes provide partial tolerance	Resistance poorly documented; no widely available resistant varieties	Prasanna et al. (2015)
DNA satellite association	Associated mainly with DNA-A and DNA-B components; beta satellites are less critical for pathogenicity	Often associated with beta satellites, which increase symptom severity and adaptability	Jyothsna et al. (2013)
Viral suppressors	Suppression of RNA silencing primarily *via* AC2 (C2) and AC4 proteins	Similar suppression mechanism but with variations in host factor targeting	Chakraborty et al. (2008)
Epidemiology	Outbreaks common in tomato monoculture regions under high whitefly populations	Widespread in mixed cropping systems; high incidence in cucurbit-producing areas	Singh et al. (2019)
Genetic studies	Well-documented studies on host-virus interaction and resistance gene deployment	Limited studies on host interaction; and transgenic approaches for resistance are in early stages	Jyothsna et al. (2013)

instance, cucumbers infected with ToLCNDV India exhibited severe leaf curling and mosaic yellowing (Jeevalatha et al., 2023). In contrast, those infected with ToLCNDV-ES showed milder or no symptoms, despite the presence of viral DNA. Similarly, in zucchini plants, ToLCNDV-India caused more pronounced curling and mosaic symptoms compared to ToLCNDV-ES, suggesting that ToLCNDV-ES has lower pathogenicity in specific cucurbit hosts (Vo et al., 2025). Understanding these pathogenic differences is crucial for developing effective disease management strategies. The varying symptom severity and host responses necessitate tailored approaches, including breeding for resistance and implementing region-specific control measures. For example, identifying and utilizing quantitative trait loci

(QTLs) associated with ToLCNDV resistance, such as the Cy-1 QTL in cucumbers, can aid in developing resistant cultivars (Koeda et al., 2024).

10 Host Range and Disease Symptoms in Cucurbits

ToLCNDV infects a diverse range of host plants, particularly among cucurbits. These hosts include melon, cucumber, watermelon, zucchini, pumpkin, sponge gourd, wax gourd, bottle gourd (*Lagenaria siceraria*), ridge gourd (*Luffa acutangula*), bitter gourd (*Momordica charantia*), ivy gourd (*Coccinia grandis*), and ash gourd (Bragard et al., 2020; Sarkar et al., 2024). Infected plants typically exhibit symptoms such as pronounced yellowing and mosaic patterns on young leaves, along with leaf curling, swollen veins, shortened internodes, rough fruit surfaces, and longitudinal cracks (Bragard et al., 2020; Juarez et al., 2019; Lopez et al., 2015).

11 Host Range of ToLCNDV in Cotton

The cotton leaf curl disease (CLCuD) was found to be linked to at least four monopartite begomoviruses: Cotton leaf curl Multan virus (CLCuMuV), Cotton leaf curl Alabad virus (CLCuAV), Cotton leaf curl Kokhran virus (CLCuKoV), and Papaya leaf curl virus (PaLCuV) (Zhou et al., 1998; Mansoor et al., 2006; Chauhan et al., 2023). Among these, only CLCuMuV, CLCuKoV, and PaLCuV have been experimentally confirmed to induce CLCuD in cotton when co-infected with a specific betasatellite, Cotton leaf curl Multan betasatellite (CLCuMB) (Briddon et al., 2001; Mansoor et al., 2006). The CLCuD outbreaks were recorded across Punjab, Pakistan, and were primarily attributed to a single monopartite virus, the "Burewala" strain of CLCuKoV (CLCuKoV-Bu), formerly known as Cotton leaf curl Burewala virus. This strain is a recombinant virus incorporating segments from CLCuMuV. Notably, CLCuKoV-Bu, associated with the resurgence of the disease, lacks one of the typical viral genes (Briddon et al., 2014) and is accompanied by a recombinant variant of CLCuMB (CLCuMBBur), which contains sequences from another betasatellite (Haider et al., 2006). More recently, an isolate of Tomato leaf curl virus (ToLCV) possessing a DNA-B component was detected in cotton plants showing CLCuD symptoms in Punjab, marking the first discovery of a bipartite begomovirus in cotton within the Indian subcontinent (Zaidi et al., 2017). The present study further documents the presence of bipartite ToLCNDV in cotton displaying CLCuD symptoms. The virus was found to be widespread across major cotton-growing areas of Punjab and Sindh in Pakistan.

12 Management of ToLCNDV

Bemisia tabaci can acquire and transmit viruses from several genera, such as *Begomovirus*, *Crinivirus*, *Ipomovirus*, *Caulimovirus*, and *Torradovirus* (Navas-Castillo et al., 2011). Flupyradifurone, a recently approved insecticide, is recognized for its favorable safety profile in terms of toxicity and environmental impact. It has been developed for diverse applications worldwide, including drenching, foliar spraying, and seed treatments, primarily targeting sap-feeding pests in agricultural and horticultural crops. Flupyradifurone also showed the highest effectiveness in reducing ToLCNDV transmission by *B. tabaci*. Glasshouse experiments conducted 24 h after insecticide application revealed complete prevention of virus transmission in all plants treated with flupyradifurone (Naga et al., 2021; Raigond et al., 2022). In field trials, flupyradifurone effectively suppressed whitefly-mediated ToLCNDV transmission for up to 14 days after application, which indicates that flupyradifurone can effectively manage the vector population. Genome editing using CRISPR/Cas9 has demonstrated significant potential in creating crops that are resistant to diseases (Ali et al., 2015). Five crucial targets within the viral genome, necessary for replication and pathogenicity, were chosen. Cas9-expressing constructs, along with the ToLCNDV infectious clone, were agroinfiltrated into tomato plants. Three of the constructs successfully disrupted the ToLCNDV genome. In contrast, plants infiltrated with three Cas9 constructs showed mild leaf yellowing, which resolved after around 21 days post-infiltration (dpi). It highlights the effectiveness of the CRISPR/Cas9 system in achieving durable resistance against ToLCNDV, with a 30% transformation efficiency in tomato plants (Ali et al., 2025). Additionally, it opens potential pathways for applying this technology to confer resistance to both single and multiple infectious viruses across various crop species (Borrelli et al., 2018).

13 Conclusion and Future Prospects

Recent studies have highlighted the intricate genome structure and biological mechanisms of ToLCNDV, which significantly contribute to the widespread yield losses in numerous staple crops. This virus is characterized by its broad host range, high transmissibility through its vector *Bemisia tabaci*, and rapid evolution through recombination and mutations in its single-stranded DNA genome, leading to the appearance of new, highly virulent strains. Early management strategies, such as vector control using insecticides, have become increasingly ineffective due to the virus's evolutionary plasticity and the development of resistance in *B. tabaci*. Comprehensive research is imperative to understand the molecular basis of ToLCNDV's evolution and its interactions with host plants, co-infecting viruses, and satellite DNA molecules. High-throughput sequencing, comparative genomics, and advanced bioinformatics have shed light on the structural features of the ToLCNDV genome, including its overlapping open reading frames, regulatory sequences, and unique replication mechanisms. Deciphering how ToLCNDV manipulates host factors to facilitate replication, suppress host immunity, and enhance its pathogenicity remains a critical area of investigation.

Furthermore, the ability of ToLCNDV to establish synergistic interactions with other viruses and satellites complicates its management and necessitates a holistic approach to disease control. Future research should focus on the identification of host resistance genes through genome-wide association studies (GWAS) and the development of resistant crop varieties using CRISPR-Cas9 and other genome-editing tools. RNA interference (RNAi)-based strategies, targeting key viral genes or vector pathways, offer promising avenues for sustainable viral control. Addressing these challenges will require interdisciplinary collaboration and the integration of cutting-edge technologies to mitigate the global impact of ToLCNDV and safeguard agricultural productivity.

References

Akhter, A., Akhtar, S., Saeed, M., & Mansoor, S. (2014). Chili leaf curl beta satellite enhances symptoms induced by tomato leaf curl new Delhi virus a bipartite *Begomovirus*. *International Journal of Agriculture and Biology, 16*(6), 1225–1228.

Ali, Z., Abulfaraj, A., Idris, A., Ali, S., Tashkandi, M., & Mahfouz, M. M. (2015). CRISPR/Cas9-mediated viral interference in plants. *Genome Biology, 16*, 1–11.

Ali, M. A., Khan, M. A. U., Naz, H., Malik, M. A. M., Rashid, U., Ashraf, N. M., Afroz, A., Shafiq, M., & Rao, A. Q. (2025). Cas9 mediated in-planta defence strategy against tomato leaf curl new Delhi virus in tomato. *Journal of Phytopathology, 173*, e70026.

Borrelli, V. M., Brambilla, V., Rogowsky, P., Marocco, A., & Lanubile, A. (2018). The enhancement of plant disease resistance using CRISPR/Cas9 technology. *Frontiers in Plant Science, 9*, 1245.

Bragard, C., et al. (2020). Pest categorisation of tomato leaf curl New Delhi virus. *European Food Safety Authority Journal, 18*, e06179.

Briddon, R. W., Mansoor, S., Bedford, I. D., Pinner, M. S., Saunders, K., Stanley, J., et al. (2001). Identification of DNA components required for induction of cotton leaf curl disease. *Virology, 285*, 234–243.

Briddon, R. W., Akbar, F., Iqbal, Z., Amrao, L., Amin, I., Saeed, M., & Mansoor, S. (2014). Effects of genetic changes to the begomovirus/betasatellite complex causing cotton leaf curl disease in South Asia post-resistance breaking. *Virus Research, 186*, 114–119.

Chakraborty, S. (2009). *Tomato leaf curl viruses from India. Desk encyclopedia of plant and fungal virology* (pp. 339–347). Elsevier.

Chakraborty, S., Vanitharani, R., Chattopadhyay, B., & Fauquet, C. M. (2008). Super virulent pseudo recombination and asymmetric synergism between genomic components of two distinct species of begomovirus associated with severe tomato leaf curl disease in India. *Journal of General Virology, 89*(3), 818–828.

Chatterji, A., Beachy, R. N., & Fauquet, C. M. (2001). Expression of the oligomerization domain of the replication-associated protein (rep) of tomato leaf curl New Delhi virus interferes with DNA accumulation of heterologous geminiviruses. *Journal of Biological Chemistry, 276*(27), 25631–25638.

Chauhan, P. K., Mehta, N., Chauhan, R. S., Kumar, A., Singh, H., Lal, M. K., Tiwari, R. K., & Kumar, R. (2023). Utilization of primary and secondary biochemical compounds in cotton as diagnostic markers for measuring resistance to cotton leaf curl virus (CLCuV). *Frontiers in Plant Science, 14*, 1185337. https://doi.org/10.3389/fpls.2023.1185337

Cohen, S., & Antignus, Y. (1994). Tomato yellow leaf curl virus, a whitefly-borne geminivirus of tomatoes. *Advances in Disease Vector Research, 10*(259), 288.

Czosnek, H., & Laterrot, H. (1997). A worldwide survey of tomato yellow leaf curl viruses. *Archives of Virology, 142*(1391), 1406.

Farina, A., Rapisarda, C., Fiallo-Olive, E., & Navas-Castillo, J. (2023). Tomato leaf curl New Delhi virus Spain strain is not transmitted by *Trialeurodes vaporariorum* and is inefficiently transmitted by *Bemisia tabaci* mediterranean between zucchini and the wild cucurbit *Ecballium elaterium*. *Insects, 14*(4), 384.

Fiallo-Olive, E., & Navas-Castillo, J. (2023). Begomoviruses: What is the secret (s) of their success? *Trends in Plant Science, 28*(6), 715–727.

Fondong, V. N. (2013). Geminivirus protein structure and function. *Molecular Plant Pathology, 14*(6), 635–649.

Fortes, I. M., Sánchez-Campos, S., Fiallo-Olivé, E., Díaz-Pedon, J. A., Navas-Castillo, J., & Moriones, E. (2016). A novel strain of tomato leaf curl New Delhi virus has spread to the Mediterranean basin. *Viruses, 8*(11), 307.

Gnanasekaran, P., Gupta, N., Ponnusamy, K., & Chakraborty, S. (2021). Geminivirus betasatellite-encoded βC1 protein exhibits novel ATP hydrolysis activity that influences its DNA-binding activity and viral pathogenesis. *Journal of Virology, 95*(17), 10–1128.

Haider, M. S., Tahir, M., Latif, S., & Briddon, R. W. (2006). First report of tomato leaf curl New Delhi virus infecting *Eclipta prostrata* in Pakistan. *Plant Pathology, 53*, 285.

Hanley-Bowdoin, L., Bejarano, E. R., Robertson, D., & Mansoor, S. (2013). Geminiviruses: Masters at redirecting and reprogramming plant processes. *Nature Reviews Microbiology, 11*(11), 777–788.

Hussain, M., Mansoor, S., Iram, S., Fatima, A. N., & Zafar, Y. (2005). The nuclear shuttle protein of the tomato leaf curl New Delhi virus is a pathogenicity determinant. *Journal of Virology, 79*(7), 4434–4439.

Jeevalatha, A., Vanishree, G., Siddappa, S., Kumar, R., Kaundal, P., Kumar, A., & Chakrabarti, S. K. (2021). Molecular characterization and infectivity analysis of tomato leaf curl New Delhi virus isolates infecting potato. *3 Biotech, 11*, 203. https://doi.org/10.1007/s13205-021-02752-5

Jeevalatha, A., Siddappa, S., Kumar, R., Tiwari, R. K., Lal, M. K., Sharma, S., Chakrabarty, S. K., & Singh, B. P. (2023). RNA-seq analysis reveals an early defense response to tomato leaf curl New Delhi virus in potato cultivar Kufri Bahar. *Functional & Integrative Genomics, 23*, 215. https://doi.org/10.1007/s10142-023-01138-5

Juarez, M., Rabadan, M. P., Martínez, L. D., Tayahi, M., Grande-Pérez, A., & Gómez, P. (2019). Natural hosts and genetic diversity of the emerging tomato leaf curl New Delhi virus in Spain. *Frontiers in Microbiology, 10*, 140.

Jyothsna, P., Haq, Q. M. I., Singh, P., Sumiya, K. V., Praveen, S., Rawat, R., & Malathi, V. G. (2013). Infection of tomato leaf curl New Delhi virus (ToLCNDV), a bipartite begomovirus with beta satellites, results in an enhanced level of helper virus components and antagonistic interaction between DNA B and beta satellites. *Applied Microbiology and Biotechnology, 97*, 5457–5471.

Kadi, A. S., Gunnaiah, R., Karthik, C. S., & Ramanna, M. (2024). Molecular Insights into the Adaptability and Pathogenicity of Tomato Leaf Curl New Delhi Virus. https://doi.org/10.2139/ssrn.4971303

Kanakala, S., Jyothsna, P., Shukla, R., Tiwari, N., Veer, B. S., Swarnalatha, P., & Malathi, V. G. (2013). Asymmetric synergism and hetero-encapsidation between two bipartite begomoviruses, tomato leaf curl New Delhi virus and tomato leaf curl Palampur virus. *Virus Research, 174*(1–2), 126–136.

Kil, E. J., Kim, S., Lee, Y. J., Byun, H. S., Park, J., Seo, H., et al. (2016). Tomato yellow leaf curl virus (TYLCV-IL): A seed-transmissible geminivirus in tomatoes. *Scientific Reports, 6*(1), 19013.

Koeda, S., Yamamoto, C., Yamamoto, H., et al. (2024). *Cy-1*, a major QTL for tomato leaf curl New Delhi virus resistance, harbours a gene encoding a DFDGD-class RNA-dependent RNA polymerase in cucumber (*Cucumis sativus*). *BMC Plant Biology, 24*, 879.

Kumar, R., Tiwari, R. K., Jeevalatha, A., Sundaresha, S., Shah, M. A., Sharma, S., Sagar, V., Kumar, M., & Chakrabarti, S. K. (2021). Potato apical leaf curl disease: Current status and perspectives on a disease caused by tomato leaf curl New Delhi virus. *Journal of Plant Diseases and Protection, 128*, 897–911. https://doi.org/10.1007/s41348-021-00463-w

Kumar, R., Tiwari, R. K., Sundaresha, S., Kaundal, P., & Raigond, B. (2022). Potato viruses and their management. In S. K. Chakrabarti, S. Sharma, & M. A. Shah (Eds.), *Sustainable management of potato pests and diseases*. Springer. https://doi.org/10.1007/978-981-16-7695-6_12

Kumar, R., Lal, M. K., Tiwari, R. K., Chourasia, K. N., Kumar, A., Kumar, R., Sharma, S., & Singh, B. (2023). Investigating the interplay between tomato leaf curl New Delhi virus infection, starch metabolism and antioxidant defence system in potato (*Solanum tuberosum* L.). *Antioxidants, 12*, 1447. https://doi.org/10.3390/antiox12071447

Lopez, C., Ferriol, M., & Pico, M. B. (2015). Mechanical transmission of tomato leaf curl New Delhi virus to cucurbit germplasm: Selection of tolerance sources in *Cucumis melo*. *Euphytica, 204*, 679–691.

Mansoor, S., Briddon, R. W., Bull, S. E., Bedford, I. D., Bashir, A., Hussain, M., et al. (2006). Cotton leaf curl disease is associated with multiple monopartite begomoviruses supported by single DNA β. *Archives of Virology, 148*, 1969–1986.

Moriones, E., Praveen, S., & Chakraborty, S. (2017). Tomato leaf curl New Delhi virus: An emerging virus complex threatening vegetable and fibre crops. *Viruses, 9*(10), 264.

Naga, K. C., Sundaresha, S., Kumar, R., Tiwari, R. K., Subhash, S., Verma, G., Buckseth, T., Bairwa, A., Sharma, S., Katare, S., & Srivastava, R. M. (2021). A new record of Asia II 5 genetic group of *Bemisia tabaci* (Gennadius) in the major potato growing areas of India and its relationship with tomato leaf curl New Delhi virus infecting potato. *3 Biotech, 11*, 421. https://doi.org/10.1007/s13205-021-02966-7

Navas-Castillo, J., Fiallo-Olivé, E., & Sánchez-Campos, S. (2011). Emerging virus diseases transmitted by whiteflies. *Annual Review of Phytopathology, 49*(1), 219–248.

Navot, N., Pichersky, E., Zeidan, M., Zamir, D., & Czosnek, H. (1991). Tomato yellow leaf curl virus: A whitefly-transmitted geminivirus with a single genomic component. *Virology, 185*(151), 161.

Padidam, M., Beachy, R. N., & Fauquet, C. M. (1995). Tomato leaf curl geminivirus from India has a bipartite genome, and the oat protein is not essential for infectivity. *Journal of General Virology, 76*(1), 25–35.

Pico, B. M., Di'ez, J., & Nuez, F. (1996). Viral diseases causing the greatest economic losses to the tomato crop II. The tomato yellow leaf curl virus—A review. *Scientia Horticulturae, 67*, 151–196.

Polston, J. E., & Lapidot, M. (2007). Management of tomato yellow leaf curl virus: US and Israel perspectives. In H. Czosnek (Ed.), *Tomato yellow leaf curl virus disease* (pp. 251–262). Springer.

Prasanna, H. C., Kashyap, S. P., Krishna, R., Sinha, D. P., Reddy, S., & Malathi, V. G. (2015). Marker-assisted selection of ty-2 and ty-3 carrying tomato lines and their implications in breeding tomato leaf curl disease resistant hybrids. *Euphytica, 204*, 407–418.

Raigond, B., Verma, G., Kumar, R., & Tiwari, R. K. (2022). Serological and molecular diagnosis of potato viruses: An overview. In S. K. Chakrabarti, S. Sharma, & M. A. Shah (Eds.), *Sustainable management of potato pests and diseases* (pp. 337–352). Springer. https://doi.org/10.1007/978-981-16-7695-6_13

Rana, V. S., Popli, S., Saurav, G. K., Raina, H. S., Chaubey, R., Ramamurthy, V. V., & Rajagopal, R. (2016). A *Bemisia tabaci* midgut protein interacts with begomoviruses and plays a role in virus transmission. *Cellular Microbiology, 18*(5), 663–678.

Reddy, A. B., Patti, M. S., Reddy, K. M., & Venkataravanappa, V. (2011). Detection and diagnosis of tomato leaf curl virus infecting tomato in northern Karnataka. *African Journal of Agricultural Research, 6*(5), 1051–1057.

Sanderfoot, A. A., & Lazarowitz, S. G. (1996). Getting it together in plant virus movement: Cooperative interactions between bipartite geminivirus movement proteins. *Trends in Cell Biology, 6*, 353–358.

Sangeetha, B., Malathi, V. G., Alice, D., Suganthy, M., & Renukadevi, P. A. (2018). Distinct seed-transmissible strain of tomato leaf curl New Delhi virus infecting chayote in India. *Virus Research, 258*, 81–89.

Sarkar, M., Gupta, D., Singh, O. W., Paul, S., Kumar, R., Mandal, B., & Roy, A. (2024). Identification of pathogenicity determinants in ToLCNDV and their RNAi-based knock-down for disease management in *Nicotiana benthamiana* and tomato plants. *Frontiers in Microbiology, 15*, 1481523. https://doi.org/10.3389/fmicb.2024.1481523

Singh, R. K., Rai, N., Singh, A. K., Kumar, P., & Singh, B. (2019). A critical review on tomato leaf curl virus resistance in tomato. *International Journal of Vegetable Science, 25*(4), 373–393.

Vo, T. T. B., Tabassum, M., Nattanong, B., Qureshi, M. A., Im, H. J., Parrella, G., & Lee, S. (2025). The insidious threat: Assessing the dangers and spread of tomato leaf curl New Delhi virus. *The Plant Pathology Journal, 41*(1), 1–16.

Yazdani-Khameneh, S., Aboutorabi, S., Shoori, M., Aghazadeh, A., Jahanshahi, P., Golnaraghi, A., & Maleki, M. (2016). Natural occurrence of tomato leaf curl New Delhi virus in Iranian cucurbit crops. *Plant Pathology Journal, 32*, 201–208.

Zaidi, S. S. S., Martin, D. P., Amin, I., Farooq, M., & Mansoor, S. (2017). Tomato leaf curl New Delhi virus: A widespread bipartite begomovirus in the territory of monopartite begomoviruses. *Molecular Plant Pathology, 18*, 901–911.

Zhou, X., Liu, Y., Robinson, D. J., & Harrison, B. D. (1998). Four DNA-A variants among Pakistani isolates of cotton leaf curl virus and their affinities to DNA-A of geminivirus isolates from okra. *The Journal of General Virology, 79*, 915–923.

Omics-Based Approaches in Deciphering Molecular Mechanisms of Tomato Leaf Curl New Delhi Virus Infection

3

Ravinder Kumar, Rahul Kumar Tiwari, Milan Kumar Lal, Ayushi Singh, Jyoti Jha, Vinut Anant Madival, Anirban Roy, and Gyanendra Singh

Abstract

Tomato leaf curl New Delhi virus (ToLCNDV) is a highly adaptable bipartite begomovirus causing severe yield losses in solanaceous and cucurbitaceous crops across Asia, the Mediterranean, and beyond. Its wide host range, efficient whitefly transmission, and recombination-driven genetic diversity complicate management. This chapter synthesizes advances in omics-based approaches—genomics, transcriptomics, proteomics, metabolomics, lipidomics, epigenomics, and small RNA profiling that have elucidated ToLCNDV's molecular mechanisms of infection, host responses, and evolutionary dynamics. Genomic studies reveal recombination hotspots, phylogeographic patterns, and population structure, while transcriptomic and proteomic analyses identify defense-related pathways, viral suppressors of RNA silencing, and host–virus protein interactions. Metabolomic and lipidomic profiling uncover shifts in primary and secondary metabolism, hormone signaling, and membrane composition, and epigenomic research highlights viral manipulation of DNA methylation and histone modifi-

R. Kumar · J. Jha · V. A. Madival · A. Roy
Division of Plant Pathology, ICAR-Indian Agricultural Research Institute, New Delhi, India

R. K. Tiwari (✉)
Division of Crop Protection, ICAR-Indian Institute of Sugarcane Research, Lucknow, Uttar Pradesh, India

M. K. Lal
Division of Crop Physiology and Biochemistry, ICAR-Central Rice Research Institute, Cuttack, Odisha, India

A. Singh
Department of Genetics and Plant Breeding, G. B. Pant University of Agriculture and Technology, Pantnagar, Udham Singh Nagar, Uttarakhand, India

G. Singh
Seed Production Unit (SPU), ICAR-Indian Agricultural Research Institute, New Delhi, India

R. Kumar et al. (eds.), *Tomato Leaf Curl New Delhi Virus (ToLCNDV)*,
https://doi.org/10.1007/978-981-95-2852-3_3

cations. Small RNA studies demonstrate the role of RNA silencing in antiviral defense and viral counter-defense strategies. Integration of multi-omics datasets through systems biology enables the identification of key regulatory hubs and predictive biomarkers, informing genome editing, RNAi-based biopesticides, and molecular breeding for durable resistance. The chapter emphasizes translational applications, challenges in field deployment, and the potential of artificial intelligence-assisted omics integration to develop predictive and preventive ToLCNDV management strategies.

Keywords

ToLCNDV · Genomics · Transcriptomics · Proteomics · Recombination · Host–virus genome editing · CRISPR/Cas

1 Introduction

Tomato leaf curl New Delhi virus (ToLCNDV) is a bipartite *begomovirus* belonging to the family *Geminiviridae*. First identified in India during the 1990s, the virus initially remained confined to the Indian subcontinent, where it caused widespread epidemics in countries such as Pakistan, Bangladesh, Thailand, Indonesia, and Taiwan. However, in 2012, ToLCNDV made an unexpected westward expansion, first detected in Iran and subsequently triggering frequent outbreaks across the western Mediterranean Basin. Severe infections in cucurbit crops marked this expansion, with notable epidemics reported in Tunisia and Italy (Kumar et al., 2021b, 2023). ToLCNDV represents a major global threat to horticultural production due to its extensive host range, efficient transmission by the whitefly (*Bemisia tabaci*), and high adaptability. The virus severely affects solanaceous crops, such as tomato (*Solanum lycopersicum*), and in cucurbitaceous crops, including zucchini, cucumber, and bitter gourd, leading to substantial economic losses across Asia, the Middle East, and parts of Europe (EFSA, 2020; Kumar et al., 2022). In tomato, yield losses during severe outbreaks can reach up to 90% (Sattar, M.N., 2012), while in the Mediterranean region, cucurbit production has suffered significant disruptions due to recent incursions into zucchini and melon cultivation (Juarez et al., 2014). The ToLCNDV genome consists of circular single-stranded DNA components, DNA-A and DNA-B, which act synergistically in replication, systemic movement, and pathogenesis. The virus is equipped with a mechanism to suppress RNA silencing and manipulate host defense networks, enabling efficient colonization and systemic spread. Associated DNA satellites, especially alphasatellites and betasatellites, further influence pathogenicity, vector transmission, and host adaptation. Frequent recombination and pseudo-recombination events have accelerated virus diversity and facilitated its adaptation to new hosts, complicating conventional resistance breeding efforts (Bhardwaj et al., 2019). During the last decade, research on ToLCNDV has evolved from basic studies in molecular biology and epidemiology to advanced explorations of virus–host interactions and integrated management strategies. Investigations have focused on molecular determinants of virulence and host range, virome dynamics, genome editing, nanotechnology application, phytoprotectants, botanicals, and beneficial endophytes. Yet, the rapid spread and adaptability of ToLCNDV underscore the urgent need for innovative, molecularly aided control strategies.

Omics technologies, including genomics, transcriptomics, proteomics, metabolomics, lipidomics, and epigenomics, have transformed our understanding of plant–virus interactions. These high-throughput platforms allow comprehensive profiling of host responses, the identification of virus-responsive genes and proteins, and the mapping of metabolic and regulatory pathways altered during infection (Samarskaya et al., 2025). For example, transcriptomic studies in *Nicotiana benthamiana* and tomato have revealed differentially expressed genes associated with RNA silencing, hormone signaling, and stress responses during ToLCNDV infection. Proteomic and metabolomic analyses have identified host factors that either facilitate viral replication or contribute to antiviral defence (Gaur et al., 2021). Integrating these datasets through systems biology approaches enables the construction of predictive models of infection dynamics, providing valuable insights for targeted intervention strategies. The combination of multi-omics data with genome editing tools such as CRISPR–Cas systems is opening new avenues in translational plant virology. These strategies facilitate the identification of critical molecular targets, the engineering of durable virus resistance, and the design of sustainable crop protection frameworks. This chapter consolidates current knowledge on ToLCNDV biology, host–virus interactions, and epidemiology while highlighting the pivotal role of omics-based approaches in uncovering infection mechanisms. By linking fundamental virology with cutting-edge biotechnology, it aims to establish a foundation for developing innovative, science-driven management solutions for one of the most damaging plant viruses affecting global horticulture.

2 Genomics Approaches

Whole Genome Sequencing of ToLCNDV Isolates: Whole genome sequencing (WGS) has emerged as an essential tool for elucidating the molecular characteristics of ToLCNDV. This bipartite begomoviruses possess two genomic components, DNA-A and DNA-B, each approximately 2.5–2.7 kb in size, which have been extensively sequenced from isolates collected across Asia, the Middle East, and the Mediterranean region. Sequencing efforts have revealed considerable genetic diversity within ToLCNDV, including recombinant strains with altered pathogenicity and expanded host ranges (Zaidi et al., 2017; Mnari-Hattab et al., 2015). Advancements in high-throughput sequencing technologies, such as Illumina and Oxford Nanopore platforms, have enabled the rapid and cost-effective assembly of complete viral genomes, providing critical data for epidemiological surveillance and the development of reliable diagnostic tools (Jeevalatha et al., 2017; Chen et al., 2013). The recent next-generation sequencing (NGS) studies have obtained complete genome sequences from ToLCNDV isolates infecting symptomatic plants in countries including China, Spain, and Indonesia. These analyses have reported high sequence identity amongst isolates, identified recombination events, and successfully assembled full-length DNA-A and DNA-B components (Li et al., 2023; Mei et al., 2023; Wilisiani et al., 2019a, b; Juárez et al., 2019). Comparative studies generally indicate low nucleotide variation between isolates from different hosts and regions; however, specific mutations and recombination patterns have been detected in some instances (Arif, 2024; Wilisiani et al., 2019a, b; Juárez et al., 2019; Moriones et al., 2017). WGS data also reveal the rapid geographical expansion of ToLCNDV across Asia, the Mediterranean, and other regions, with infections reported in tomatoes, cucurbits, and even medicinal plants. Investigations in Spain and Indonesia have highlighted the virus's potential to infect both cultivated and wild hosts, with genetic evidence indicating a large part of a homogeneous Mediterranean population distinct from Asian isolates (Wilisiani et al., 2019a, b; Juárez et al., 2019; Moriones et al., 2017). Furthermore, the availability of complete genome sequences has facilitated the development of infectious clones, supporting pathogenicity studies, reverse genetics, and resistance screening in economically important crops such as melon and tomato.

Variability, Recombination, and Phylogenetic Analyses: Comparative genomic analyses of ToLCNDV isolates reveal notable intra- and inter-species variability, primarily driven by mutation and recombination. DNA-A is a component particularly prone to recombination within the replication-associated protein (Rep) and coat protein (CP) coding regions essential for viral replication and vector-mediated transmission (Mansoor et al., 2003). Phylogenetic studies have classified ToLCNDV isolates into multiple clades, often reflecting their geographic origin and host specialization. Indian isolates show the most extraordinary genetic diversity and form the ancestral lineage, supporting the hypothesis that the Indian subcontinent is the virus's center of origin (Gaur et al., 2021). Recombination detection tools such as RDP4, SimPlot, and SplitsTree have been instrumental in identifying genetic exchange events, many of which correlate with modifications in host range or vector efficiency. ToLCNDV is a rapidly emerging begomovirus posing a severe threat to a wide range of vegetable and fiber crops internationally. Its genetic adaptability, high diversity, and capability for recombination allow it to infect new hosts and expand into new geographic areas, complicating management strategies. Globally, ToLCNDV populations display distinct patterns of diversity. Asian populations are genetically heterogeneous, whereas Mediterranean populations such as those in Spain and Italy are comparatively uniform, likely due to recent introductions and founder effects (Moriones et al., 2017; Juárez et al., 2019; Fortes et al., 2016; Ruiz et al., 2017). Wild plants often harbor greater within-host diversity, serving as potential reservoirs for novel variants (Juárez et al., 2019). Recombination plays a central role in ToLCNDV evolution, producing new strains with altered host specificity and pathogenicity. Several recombinant isolates have been documented, including inter-strain and inter-species recombinants with other begomoviruses such as Squash leaf curl China virus (SLCCNV) (Moriones et al., 2017; Rauniyar et al., 2023; Wilisiani et al., 2019a, b; Renukadevi et al., 2024; Mastrochirico et al.,

2023; Fortes et al., 2016). These events can influence symptom severity, epidemiological dynamics, and host adaptation. Phylogenetic reconstructions consistently divide ToLCNDV into two major clades: Asian and Mediterranean. European isolates generally form a well-defined subgroup strongly adapted to cucurbits rather than tomatoes, whereas Asian isolates typically exhibit a broader host range (Rauniyar et al., 2023; Juárez et al., 2019; Troiano & Parrella, 2023; Mastrochirico et al., 2023; Fortes et al., 2016; Cai et al., 2023). Whole-genome-based phylogenies remain invaluable for tracing virus introductions, detecting recombination, and elucidating evolutionary relationships.

Insights into Viral Evolution and Spread: The genomic diversity of ToLCNDV facilitates rapid adaptation to new hosts and environments, thereby enhancing its epidemic potential. Recombination has been a major driver in the emergence of new variants, such as ToLCNDV-ES in Europe, underscoring the virus's capacity to evolve quickly under selection pressure (Juárez et al., 2019). Whole-genome sequencing (WGS) has identified cases of interspecies recombination with related begomoviruses, broadening ToLCNDV's genetic repertoire for survival and dissemination. Global genomic analyses have also revealed patterns of transboundary movement, often linked to the international trade of infected plant materials and the involvement of *Bemisia tabaci* cryptic species in transmitting distinct viral lineages (Brown et al., 2015). These findings provide critical guidance for quarantine protocols and resistance breeding programs. ToLCNDV is a bipartite begomovirus comprising two single-stranded DNA components (DNA-A and DNA-B), conferring high genetic flexibility. The genome displays significant variability, with notable lineages such as ToLCNDV-ES (Spain) and ToLCNDV-BG (bitter gourd) arising through recombination and host adaptation (Moriones et al., 2017; Juárez et al., 2019; Zaidi et al., 2017; Mnari-Hattab et al., 2015; Renukadevi et al., 2024). Both recombination and pseudo-recombination contribute to diversification and adaptive evolution, enabling infection of a broad range of plant species across diverse environments (Moriones et al., 2017; Renukadevi et al., 2024). Phylogenetic analyses consistently show a marked genetic separation between Asian and Mediterranean populations, with wild plant hosts acting as reservoirs that enhance within-host diversity and emergence of new variants (Juárez et al., 2019; Rauniyar et al., 2023; Renukadevi et al., 2024). ToLCNDV infects more than 40 plant species spanning multiple families, including tomatoes, cucurbits, and several weeds, facilitating persistence and spread in diverse agroecosystems (Moriones et al., 2017; Juárez et al., 2019; Zaidi et al., 2017; Mnari-Hattab et al., 2015, Li et al., 2023). Its ability to suppress host gene silencing and interact with DNA satellites further enhances its adaptability, pathogenicity, and epidemiological success (Moriones et al., 2017; Majid et al., 2023).

Genome Editing Targeting ToLCNDV Genes: Recent advances in genome editing have opened new avenues for developing resistance to ToLCNDV. The CRISPR-Cas9 system, derived initially from bacterial adaptive immunity, has been successfully employed to target and cleave viral DNA within host plant cells. For example, CRISPR-mediated disruption of the Rep and CP genes in *Nicotiana benthamiana* led to reduced viral accumulation and milder symptoms (Ali et al., 2015). Multiplex CRISPR approaches, which simultaneously target multiple viral genes, have demonstrated stronger resistance and reduced potential for escape mutants. These strategies mark a significant shift in plant virus management and hold strong potential for application in commercial tomato and cucurbit cultivars. Genome editing technologies, particularly CRISPR-Cas systems, enable precise modification, mutagenesis, and regulation of target genes in plants and animals. While CRISPRCas has been widely applied for crop improvement and enhancing disease resistance, its direct use against plant viruses such as ToLCNDV remains a developing field. Available studies describe general methodologies, tools, and challenges relevant to targeting plant viral genomes, though ToLCNDV-specific work is limited. CRISPRCas enables targeted mutagenesis, nucleotide editing, and transcriptional modulation, making it a powerful tool for disrupting viral genes or enhancing plant defense. Successful application depends on the design of highly specific sgRNAs, efficient delivery of CRISPR components, and accurate screening for edits. Advances in computational tools have improved sgRNA design and off-target prediction, which are essential for precise viral targeting (Li et al., 2023; Guo et al., 2023). Off-target activity remains a significant concern, as it can lead to unintended mutations. To address this, recent innovations include high-fidelity Cas9 variants, base and prime editing, and stepwise delivery strategies. Additional measures such as truncated gRNAs, paired nickases, and ribonucleoprotein delivery can further minimize off-target effects, improving the safety and efficiency of CRISPR applications (Guo et al., 2023; Naeem et al., 2020; Chen et al., 2017). Although most reviewed studies focus on CRISPRCas for general crop improvement and functional genomics, their approaches to sgRNA design and off-target control are directly applicable to the development of CRISPR-based resistance against ToLCNDV in plants (Li et al., 2023; Naeem et al., 2020).

3 Transcriptomics

Transcriptomic profiling of RNA sequencing (RNA-seq) has emerged as an effective technique to decipher global gene expression changes in the course of ToLCNDV infection (Jeevalatha et al., 2021, 2023). Comparative RNA-seq studies among ToLCNDV-inflamed and ridicule-inoculated host plants, together with tomato and *Nicotiana benthamiana*, have discovered considerable alterations in transcript abundance associated with pressure, signaling, and protection (Gorovits et al., 2016; Jeevalatha et al., 2017; Chen et al., 2013). Excessive-throughput sequencing permits the identification of virus-responsive genes with excessive sensitivity and precision, including both coding and non-coding RNAs. These datasets shape the idea for exploring the dynamic molecular landscape in the course of viral infection. Differential gene expression (DGE) evaluation of RNA-seq data from inflamed tissues commonly highlights masses of upregulated and downregulated genes. These genes are involved in key metabolic pathways together with hormone signaling (salicylic acid, jasmonic acid, and ethylene), oxidative stress response, secondary metabolism, and protein degradation through the ubiquitin-proteasome system (Gupta et al., 2019; Sahu et al., 2021; Ramesh et al., 2017). Gene Ontology (GO) enrichment and KEGG pathway mapping, in addition, suggest modulation of plant-pathogen interplay pathways, photosynthesis-related genes, and transcription factor networks, substantially WRKY, MYB, and NAC families (Sharma and Prasad, 2017). Such analyses provide insights into the host's attempts to spark off protection responses and the viral techniques to counteract them. ToLCNDV infection suppresses a massive spectrum of innate immune responses, including sample-triggered Immunity (PTI) and Effector-triggered Immunity (ETI). Transcriptome studies monitor downregulation of numerous protection-related genes, including PR proteins, pathogenesis-related kinases, and reactive oxygen species (ROS) scavengers. Concurrently, additives of RNA interference (RNAi), together with Dicer-like (DCL) and Argonaute (in the past) genes, frequently display decreased expression stages, suggesting viral suppression of gene silencing pathways (Samarskaya et al., 2025). The virus is known to modulate transcription elements like ERFs and bZIPs that regulate hormone crosstalk, leading to altered signaling outputs and extended susceptibility (Jangra et al., 2024; Gupta et al., 2021). These molecular shifts are associated with viral replication, movement, and symptom improvement. Transcriptomics additionally aids in the identity and functional evaluation of viral suppressors of RNA silencing (VSRs), which are essential for successful infection. In ToLCNDV, proteins like AC2 (lure) and AC4 are recognized VSRs that interfere with the host RNAi machinery (Bisaro, D.M., 2006; Aregger et al., 2012; Jeevalatha et al., 2021, 2023). RNA-seq can identify their expression in the early stages of infection and their impact on host gene silencing pathways. Functional assays, coupled with transcriptomic readouts, have shown that these VSRs inhibit the buildup of small RNAs and suppress silencing signal transduction. Such discoveries not only enhance our understanding of virus-host interplay but also provide objectives for RNAi-mediated resistance techniques.

4 Proteomics

Proteomic analyses provide essential insights into the dynamic modifications in host and viral protein expression throughout ToLCNDV infection. The use of superior mass spectrometry (MS)-based platforms, including LC-MS/MS, enables researchers to identify changes in protein abundance, providing a functional perspective that complements transcriptomic statistics (Prakash et al., 2020). Comparative proteomic studies among ToLCNDV-inflamed and uninfected tomato or *Nicotiana benthamiana* flowers have identified proteins involved in photosynthesis, redox balance, protection signaling, and metabolism, which might be either upregulated or suppressed throughout infection (Gaur et al., 2021). These proteomic shifts reflect the host's reaction to viral invasion and the pathogen's efforts to subvert host mobile procedures for replication and motion. ToLCNDV-encoded proteins, including Rep (AC1), lure (AC2), and C4, engage with several host proteins to govern mobile pathways. Proteomic gear is instrumental in figuring out these host interactors. For instance, co-immunoprecipitation (Co-IP) coupled with MS has been used to capture host proteins associated with ToLCNDV C4, revealing its role in modulating host cellular division and protection pathways (Sarkar et al., 2024). Further, viral suppressors of RNA silencing (VSRs) target Argonaute proteins and other RNAi components to suppress antiviral defenses, as demonstrated through proteomic interplay studies (Vanitharani et al., 2004). These findings provide goals for engineering resistance through the disruption of crucial virus-host protein interactions. Post-translational changes (PTMs), including phosphorylation, ubiquitination, and SUMOylation, drastically impact host-pathogen interactions through regulating protein function, balance, and localization. ToLCNDV infection is associated with altered phosphorylation states of key signaling proteins involved in hormone law and innate immunity (Gupta et al., 2019; Sahu et al., 2021; Ramesh et al., 2017). Ubiquitin-mediated degradation of host proteins, frequently hijacked through viral factors, is another mechanism through which the virus promotes its replication and motion. In some instances, viral proteins themselves are modified post-translationally to optimize their functionality or avoid host defenses. Superior proteomics, which includes phosphoproteomics and ubiquitinome profiling, is being increasingly utilized to dissect these complicated regulatory mechanisms. Mapping protein-protein interactions (PPIs) is crucial for expertise in ToLCNDV pathogenesis. Yeast two-hybrid (Y2H) monitors have identified several host proteins that engage with viral effectors, which include kinases, transcription factors, and RNA-binding proteins (Wang et al., 2022). Co-IP and tandem affinity purification observed through MS are widely used to validate these interactions in planta. Latest studies additionally leverage proximity labeling techniques, including BioID and APEX, to capture temporary or susceptible interactions that may be missed through conventional techniques. These PPI maps offer a structural-stage understanding of the viral infection process and highlight key nodes for intervention in disease control.

5 Metabolomics

ToLCNDV contamination results in tremendous metabolic reprogramming in host flowers, affecting both primary and secondary metabolism. Primary metabolites, including sugars, amino acids, and natural acids, are vital for power metabolism and are frequently hijacked by the virus to aid replication and motion. Infected flowers commonly exhibit altered ranges of sucrose, glucose, and tricarboxylic acid (TCA) cycle intermediates, reflecting a redirection of host power waft (Sade et al., 2015). Secondary metabolites, which include phenolics, flavonoids, and alkaloids, which play key roles in plant protection, also undergo dynamic modifications. as an example, in tomato flowers infected with begomoviruses like ToLCNDV, decreased ranges of phenylpropanoids and elevated lignin biosynthesis had been pronounced, indicating a host effort to enhance mobile walls and restrict viral motion (Rossouw, L.T., 2016). Plant hormones are vital regulators of protection signaling, and their homeostasis is frequently disrupted throughout ToLCNDV contamination. Salicylic acid (SA), jasmonic acid (JA), abscisic acid (ABA), and ethylene (ET) are especially essential in antiviral protection. SA is commonly associated with resistance in opposition to biotrophic pathogens and systemic acquired resistance (SAR), whilst JA and ET are involved in responses to necrotrophs and insect vectors, including whiteflies (Sharma et al., 2025). Throughout ToLCNDV contamination, a common strategy employed by the virus is to suppress SA-mediated responses whilst simultaneously manipulating JA and ABA pathways to its advantage. Quantitative metabolomics using LC-MS or GC-MS has confirmed tremendous reductions in SA ranges and elevated ABA accumulation, suggesting a viral effort to dampen host immunity and enhance susceptibility (Zhao et al., 2021). Metabolomics holds promise for the identification of biomarkers that may be used for early analysis of ToLCNDV contamination or to predict resistance in host cultivars. Unique changes in amino acid profiles (e.g., elevated proline, tryptophan), sugar metabolism (e.g., accumulation of raffinose or trehalose), and particular secondary metabolites (e.g., unique flavonoids or phytoalexins) have been associated with tolerant or resistant phenotypes (Ali et al., 2015; Sade et al., 2015). Such metabolic signatures, tested throughout distinctive cultivars and conditions, can serve as non-invasive, fast tools for screening resistant germplasm in breeding applications or for discipline diagnostics using handheld metabolite sensors. The genuine strength of metabolomics is discovered when it is included with transcriptomic and proteomic datasets. Multi-omics methods can link transcriptional regulation and protein expression with metabolic fluxes; for this reason, presenting a systems-level knowledge of host-virus interactions, as an example, decreased expression of genes encoding enzymes inside the phenylpropanoid pathway, confirmed at the protein level, correlates with reduced accumulation of phenolic compounds in infected flowers (Gaur et al., 2021). Such integrative studies help pinpoint key regulatory nodes and pathways that may be focused through genetic engineering or breeding for superior resistance. Network analysis and pathway enrichment tools like MetaboAnalyst, MapMan, and Cytoscape are broadly used to visualize and interpret these complicated datasets.

6 Epigenomics

Epigenetic law is a pivotal layer of host protection in plants and is dynamically altered throughout viral infections, together with those caused by ToLCNDV. DNA methylation, especially at cytosine residues in CG, CHG, and CHH contexts, serves as a transcriptional silencing mechanism toward invading viral genomes (Mahomed, N.B., 2022). Upon ToLCNDV infection, host DNA methylation machinery is regularly activated, resulting in methylation of viral DNA to suppress replication and transcription. However, begomoviruses, together with ToLCNDV, encode proteins along with lure (transcriptional activator protein) and Rep (replication-associated protein), which can counteract this protection by interfering with methyltransferase activity or demethylating viral genomes (Raja et al., 2008). Histone changes, along with methylation (e.g., H3K9me2, H3K27me3) and acetylation (e.g., H3K9ac), additionally play essential roles in chromatin remodeling and gene expression. Throughout ToLCNDV infection, worldwide shifts in histone marks were discovered in host plants, especially on genes associated with protection, hormone signaling, and strain response. Extended acetylation of histones may cause the activation of susceptibility genes, whilst repressive methylation marks can be deposited on key protection genes, aiding in viral pathogenesis (Rosas-Díaz et al., 2018). Epigenetic changes can tightly alter the expression of host protection genes in response to ToLCNDV infection. Numerous pathogenesis-associated (PR) genes, R genes (resistance genes), and transcription elements involved in salicylic acid (SA)-mediated immunity exhibit differential methylation and histone modification patterns upon infection. For instance, demethylation of promoters of positive WRKY transcription elements or NPR1-like genes is associated with improved expression throughout early stages of infection, activating protection responses (Lang et al., 2017). Conversely, viral interference may restore methylation and repress these genes at later degrees to suppress systemic infection. The RNA-directed DNA methylation (RdDM) pathway, which is predicated on small interfering RNAs (siRNAs), is imperative to organizing methylation marks on each viral and host genome. Disruption of RdDM components like AGO4, DRM2, or RDR2 by ToLCNDV-encoded suppressors can impair protection-associated methylation and facilitate viral replication and movement (Zhou et al., 2018). ToLCNDV actively manipulates the host's epigenetic machinery to create a more favorable environment for its replication. The lure protein is a known suppressor of transcriptional gene silencing (TGS), capable of binding to and inhibiting key host methyltransferases along with MET1 or DRM2, leading to hypomethylation of viral and host genes (Raja et al., 2008). Furthermore, Rep can have interaction with histone deacetylases (HDACs), disrupting chromatin compaction and enhancing viral gene expression. A few current studies additionally recommend that ToLCNDV may recruit or hijack host chromatin remodeling complexes, thereby changing the histone panorama and chromatin accessibility in its favor (Ramesh et al., 2017). Such sophisticated interactions mirror an ongoing evolutionary arms race between virus and host, highlighting epigenetic law as a crucial frontier for developing long-lasting resistance strategies. Epigenome editing tools, along with dCas9-fusion proteins concentrated on methylation or demethylation pathways, represent promising future guidelines for manipulating host susceptibility and resistance to ToLCNDV.

7 Lipidomics and Cell Wall Profiling

Lipids are essential cellular components that serve not only as structural factors of membranes but also as signaling molecules and platforms for viral replication complexes (VRCs). Within the context of ToLCNDV contamination, lipid metabolism undergoes considerable remodeling to facilitate viral replication and intercellular motion. Geminiviruses like ToLCNDV exploit host endomembrane systems—inclusive of the endoplasmic reticulum (ER), chloroplasts, and peroxisomes—to establish VRCs enriched in specific lipid species (Jiang et al., 2023). These lipid microenvironments, especially the ones enriched in phosphatidylinositol phosphates and sterols, are essential for anchoring viral proteins, which include Rep and motion protein (MP), permitting efficient viral genome replication and intracellular trafficking (Rosas-Díaz et al., 2018). Alterations in fatty acid biosynthesis and desaturation pathways have also been noted in host flowers infected with geminiviruses. Upregulation of lipid biosynthetic genes, inclusive of the ones involved in glycerolipid and sphingolipid metabolism, contributes to the rearrangement of membrane domain names to favor virus-host protein interactions (Gullner et al., 2018). Recent lipidomic research in inflamed *Nicotiana benthamiana* and tomato flowers reveals better levels of linolenic acid and its oxylipin derivatives, which may also play dual roles in viral facilitation and signaling protection. The plant cuticle and plasma membrane are the primary strains of protection against biotic stresses. Viral contamination often ends in disruption of those obstacles, both passively through cellular damage and actively via manipulation by using viral effectors. ToLCNDV contamination has been proven to affect the biosynthesis and composition of cuticular waxes and membrane lipids. A decline in wax biosynthetic gene expression, which includes CER1 and LACS1, results in reduced cuticle integrity, making host tissues more susceptible to viral ingress and vector (whitefly) feeding (Wang et al., 2020; Naga et al., 2021). Further, the composition of membrane lipids, which include phosphatidylcholine (pc), phosphatidylethanolamine (PE), and galactolipids (MGDG and DGDG), is altered throughout contamination, which affects membrane fluidity, permeability, and signaling. These changes not only facilitate the formation of VRCs but additionally modulate vesicular trafficking pathways for virus transport (Pan et al., 2025). Mobile wall change is a key aspect of the plant's antiviral protection approach. Viral infections, including those caused by ToLCNDV, set off the expression of genes involved in cell wall reinforcement, which include the ones coding for lignin biosynthesis enzymes (friend, CCoAOMT), cellulose synthases, and callose synthases. Accelerated deposition of lignin and callose can create a physical barrier that restricts virus motion and whitefly feeding (Koziel et al., 2021). Conversely, viruses often hijack the host's mobile wall-loosening enzymes, including expansins and pectin methylesterases (PMEs), to facilitate their motion through plasmodesmata. Transcriptomic and mobile wall proteomics research has proven upregulation of expansin-like proteins and PMEs throughout ToLCNDV contamination, which may also weaken wall integrity and facilitate intercellular viral spread (Shakir et al., 2023; Devendran et al., 2022). Targeting these enzymes through RNAi or genome editing may also provide novel resistance strategies. Integrating lipidomics with mobile wall profiling and transcriptomics can offer a holistic view of how host cellular architecture and lipid biochemistry are rewired to favor or face up to ToLCNDV contamination (Sarkar et al., 2024).

8 Small RNA and miRNA Profiling

One of the most crucial innate immune responses in vegetation toward viruses is RNA silencing, a system that generates virus-derived small interfering RNAs (vsiRNAs) from the viral genome. During ToLCNDV contamination, the host Dicer-like (DCL) enzymes understand viral double-stranded RNA (dsRNA) intermediates and cleave them into vsiRNAs, which might be then loaded onto Argonaute (in the past) proteins to form RNA-caused silencing complexes (RISCs). These RISCs manually manipulate the viral RNAs, thereby preventing contamination (Kryovrysanaki et al., 2021). Excessive-throughput small RNA sequencing has enabled the profiling of vsiRNAs generated from ToLCNDV-infected tomato and *Nicotiana benthamiana* vegetation, revealing hotspots inside the AV1 (coat protein) and AC1 (Rep) regions of the viral genome (Gaur et al., 2021). The abundance, size distribution (predominantly 21- and 22-nt), and strand polarity of vsiRNAs reflect the pastime of host DCLs and their specialization during contamination. Such insights have contributed to the improvement of artificial siRNA-based RNAi constructs focused on conserved viral sequences for better resistance in transgenic vegetation (Tiwari et al., 2022). MicroRNAs (miRNAs) are endogenous small RNAs that play key roles in post-transcriptional gene regulation, including plant protection against viruses. During ToLCNDV contamination, the expression patterns of host miRNAs undergo full-size modulation. Several miRNAs are upregulated to govern viral replication and movement, not directly by focusing on host genes concerned in hormone signaling, cell cycle regulation, and RNA silencing pathways (Naqvi et al., 2010). For example, miR6024, miR482, and miR2118 households, which adjust NBS-LRR resistance genes, are differentially expressed in virus-infected tomato tissues (Shivaprasad et al., 2012). In addition, miR398, a known regulator of oxidative stress responses, is frequently suppressed, leading to ROS accumulation and a hypersensitive reaction-like phenotype that can contribute to viral restriction. The profiling of miRNA–target interactions using degradome sequencing has similarly clarified the mechanistic hyperlinks among host miRNAs and their downstream targets during ToLCNDV contamination. ToLCNDV, like different geminiviruses, has advanced suppressors of RNA silencing (VSRs) to neutralize the host's small RNA-mediated protection. The AC2 (trap) and AC4 proteins of ToLCNDV have been recognized as robust VSRs that intervene with multiple steps of the RNA silencing pathway. AC2 interacts with and inhibits the expression of key silencing genes together with RDR6 and SGS3, even as AC4, without delay binds to siRNAs and miRNAs, thereby preventing their incorporation into the past complexes (Vanitharani et al., 2004; Sarkar et al., 2024). The latest research has additionally discovered that ToLCNDV AC4 mimics host miRNAs and acts as a molecular decoy to hijack the silencing equipment (Gupta et al., 2019; Sahu et al., 2021; Ramesh et al., 2017). These counter-protection strategies allow the virus to subvert host immunity and establish systemic contamination. Integrating small RNA sequencing with purposeful genomics can provide a holistic understanding of host-virus interactions and reveal novel targets for RNAi-based resistance breeding or genome editing processes to combat ToLCNDV.

9 Multi-omics Integration and Systems Biology

Expertise in the complex mechanisms of ToLCNDV contamination requires a holistic approach that integrates multiple layers of biological statistics. Multi-omics integration combines genomics, transcriptomics, proteomics, and metabolomics to create a complete picture of host-virus interactions. As an instance, integrating transcriptome and metabolome profiles of *Nicotiana benthamiana* and tomato through ToLCNDV contamination has revealed synchronized modifications in gene expression and metabolic shifts related to protection and strain response pathways (Jangra et al., 2024; Khalil et al., 2025). Viral genome statistics (genomics) offer insights into variability and mutation hotspots, even as transcriptomics enables the identification of responsive host genes and pathways. Proteomics refines this view by mapping the dynamic protein machinery and post-translational modifications, while metabolomics highlights the physiological effects of contamination. Systems biology leverages this multidimensional statistic to construct interplay networks that map the complicated crosstalk among host and viral components. For example, gene regulatory networks (GRNs) and protein-protein interaction (PPI) networks can be evolved using transcriptomic and proteomic datasets. Those networks are regularly visualized using systems including Cytoscape or STRING, permitting identification of significant nodes that may represent capability objectives for genetic engineering or resistance breeding (Chen et al., 2021). In ToLCNDV-infected plant life, co-expression community evaluation has exposed modules enriched in protection signaling, hormone signaling (salicylic acid, jasmonic acid), redox procedures, and RNA silencing components. The virus appears to subvert those hubs to set up contamination and spread. For example, the viral AC2 and AC4 proteins disrupt multiple regulatory cascades concerned with antiviral protection and hormonal homeostasis (Sarkar et al., 2024). Through integration and community modeling, several regulatory hubs had been identified that orchestrate the host's response to ToLCNDV. Transcription factors, including WRKY, NAC, and MYB families, are differentially expressed throughout contamination and are significant in activating downstream protection genes. In addition, MAPK cascades, calcium signaling components, and hormonal regulators, including NPR1 (for SA) and COI1 (for JA), serve as essential signaling hubs affected throughout contamination (Naqvi et al., 2010). Those findings underscore the virus's focus on high-impact nodes for green suppression of host immunity. Furthermore, integrated omics statistics have shown that modifications in gene expression regularly align with shifts in metabolite abundance and protein accumulation, confirming practical correlations. For example, downregulation of genes concerned with photosynthesis corresponds with reduced chlorophyll content and visible signs and symptoms, including leaf curling and yellowing. Multi-omics integration has additionally enabled the construction of predictive models to forecast sickness development and host susceptibility. Machine learning algorithms applied to massive omics datasets can classify resistant vs. prone genotypes, expect contamination severity, and endorse biomarkers for early detection or resistance breeding (Jangra et al., 2024). Dynamic modeling of host-pathogen interplay networks also aids in simulating viral perturbation results and identifying key intervention factors. In the future, such integrative systems biology tactics might be necessary for designing durable resistance techniques, such as genome enhancing, dsRNA-based therapeutics, informed breeding, and the use of omics-derived markers.

10 Functional Validation and Translational Applications

The combination of omics technologies has not only multiplied our information on ToLCNDV pathogenesis but also opened avenues for translational programs in plant protection, diagnostics, and breeding. Validating and utilizing omics-derived applications through advanced molecular technology has emerged as a practical approach to developing sustainable solutions for ToLCNDV management. High-throughput omics tactics, including transcriptomics and proteomics, frequently yield an extended list of potential genes and proteins involved in antiviral responses or susceptibility. Practical validation is vital to verify the organic relevance of those objectives. Virus-induced gene silencing (VIGS) has been effectively employed in *Nicotiana benthamiana* and tomato to knock down host susceptibility (S) genes identified through transcriptomic facts. For instance, silencing of genes like SnRK1, HSP70, or eIF4E has tested altered responses to begomoviruses, together with ToLCNDV (Sahu et al., 2012). RNA interference (RNAi) generation has also been implemented to target each host and viral gene. Host genes implicated in viral movement and replication can be knocked down using hairpin constructs, whilst viral genes including AC1 (Rep), AC2 (lure), and AV2 (movement protein) can be specifically targeted to dam contamination development. In parallel, overexpression research using transgenic tactics helps determine the role of defense-related genes like PR1, WRKY33, or NPR1 in conferring resistance. CRISPR-Cas9 genome editing is increasingly being used to knock out host S genes or immediately target viral genomes. For instance, Cas9-mediated knockdown of the AC1 gene of ToLCNDV has been proven to interfere with viral replication (Tiwari et al., 2022). Omics-guided identification of resistance-associated genes and metabolites permits the development of molecular markers for marker-assisted selection (MAS). Differential expression of genes like Ty-1, RDR1, or AGO1 and the presence of defense-associated metabolites (e.g., flavonoids, phytoalexins) throughout ToLCNDV contamination can serve as biomarkers. These biomarkers can be tested across diverse germplasm to perceive evident resistant traces. Genomic selection strategies incorporating these markers can enhance the breeding of ToLCNDV-resistant cultivars. RNAi-based total biopesticides, especially double-stranded RNA (dsRNA) formulations concentrated on vital viral genes, represent an environmentally safe and particular approach to viral disease control. Omics facts are a valuable resource for identifying highly conserved viral gene sequences suitable for dsRNA design. As an instance, dsRNA concentrated on Rep or CP genes of ToLCNDV has proven promising effects in lowering viral load and symptom severity under glasshouse conditions (Kumar et al., 2022). These formulations are being explored for their applicability in various disciplines, including the use of foliar sprays, trunk injections, or nanoparticle delivery structures. Transcriptomic and proteomic profiling also contribute to the development of early diagnostic gear. Precise expression signatures or metabolic fingerprints throughout early ranges of ToLCNDV contamination can serve as biomarkers for speedy detection. Strategies like RT-qPCR, ELISA, or CRISPR-based diagnostics (e.g., SHERLOCK, DETECTR) can be custom-designed to detect omics-diagnosed viral transcripts or protein objectives. Additionally, portable biosensors that use aptamers or nanomaterials conjugated with omics-derived probes are being developed for on-web page detection in the discipline (Gupta et al.,2019; Sahu et al., 2021; Ramesh et al., 2017).

11 Challenges and Future Perspectives

The utility of omics technologies has revolutionized our understanding of plant-virus interactions, especially in the context of ToLCNDV. However, several challenges remain in absolutely exploiting those processes for realistic packages. Bridging the distance between lab-based discoveries and field-based answers calls for overcoming technical, organic, and computational hurdles. One of the most pressing challenges in omics studies is the combination of multidimensional datasets generated from genomics, transcriptomics, proteomics, metabolomics, epigenomics, and other platforms. Every dataset varies in resolution, format, and dynamics, making fact harmonization complicated. Moreover, organic interpretation requires robust statistical and computational pipelines, which are often lacking in plant virology-specific packages. The limited availability of person-friendly equipment for multi-omics integration poses a barrier for researchers without superior bioinformatics education. Furthermore, inconsistencies in annotation databases, reference genomes, and pathway assets for non-version flowers complicate the correct mapping and interpretation of omics signals (Jangra et al., 2024; Gupta et al., 2021). Addressing those bottlenecks demands funding in open-supply platforms, standardized metadata protocols, and go-omics visualization equipment. While tomato and *Nicotiana benthamiana* benefit from well-annotated genomes and transcriptomes, many natural hosts of ToLCNDV, especially wild relatives and cucurbit species, have incomplete or poorly annotated genomic assets. These boundaries limit the scope of comparative genomics and hinder the identification of resistance alleles or regulatory factors. Efforts, including pan-genome assemblies, de novo transcriptome sequencing, and community genome annotation tasks, are vital for expanding genomic insurance across host species. Without those assets, the extrapolation of findings to diverse agroecological contexts stays restricted. Every other key task is the translation of laboratory findings into field-prepared technologies. For instance, RNAi or CRISPR-based resistance techniques regularly show promise under controlled situations but face regulatory, ecological, and delivery challenges in open environments. Moreover, expression profiles or biomarkers recognized under greenhouse infections won't hold under field strain situations regarding more than one pathogen, abiotic stressors, and variable vector populations. To bridge this gap, omics-pushed targets ought to be established in diverse agro-climatic zones through the use of marker-assisted breeding, genome editing, or synthetic biology processes. Field trials, biosafety evaluations, and farmer-centric participatory checking out are essential to ensure the real international applicability of those technologies (Tiwari et al., 2022). Synthetic Intelligence (AI) and machine learning (ML) offer promising avenues to solve the complexity of omics facts. Those processes can discover hidden patterns, expect gene–phenotype associations, and prioritize candidate genes or biomarkers from large datasets. For example, ML models were used to classify resistant and prone genotypes based entirely on transcriptomic signatures and to predict viral infection degrees from metabolomic profiles. Destiny advancements may encompass real-time ML-based diagnostic equipment, predictive modeling of virus outbreaks, the use of omics and epidemiological facts, and automatic identification of resistance genes for breeding applications (Samarskay et al., 2025). Integrating AI with omics not only enhances fact interpretation but also quickens decision-making in plant sickness management.

12 Conclusion

- Omics technology has drastically improved our understanding of ToLCNDV pathogenesis, host responses, and molecular interactions. Through the application of genomics, transcriptomics, proteomics, metabolomics, lipidomics, epigenomics, and small RNA profiling, researchers have uncovered a comprehensive view of the virus-host interface. These studies have enabled the identification of critical viral genes, host susceptibility and resistance elements, differentially regulated pathways, and molecular signatures associated with infection and protection.
- The integration of entire genome sequencing has helped monitor the genetic diversity, recombination hotspots, and evolutionary dynamics of ToLCNDV, assisting in phylogeographic tracking of its unfold. Transcriptomic analyses have provided insights into host gene regulation and immune responses, while proteomic studies have uncovered key interactions among viral proteins and host machinery. Metabolomics and lipidomics have highlighted sizable changes in cellular homeostasis and hormone signaling in the course of infection, contributing to symptom expression and viral replication. Epigenomic changes, including DNA methylation and histone modifications, suggest that ToLCNDV may also manage host chromatin states to ensure its survival. Moreover, the identification of small RNAs and microRNAs emphasizes the function of RNA silencing as a key antiviral protection. It illustrates the complex counter-protection techniques employed with the aid of the virus.
- The importance of integrated omics strategies in plant virology cannot be overstated. Singular omics layers often provide fragmented insights, but their integration creates a structure-stage perspective, allowing researchers to assemble dynamic interaction networks and pick out regulatory hubs that govern sickness progression. This holistic view is instrumental in unraveling complex biological phenomena along with cross-communication among signaling pathways, coordi-

nated transcriptional and post-translational responses, and epigenetic reprogramming in the course of viral infection. Moreover, integrating omics with equipment like CRISPR/Cas genome editing, virus-caused gene silencing (VIGS), and RNAi-based functional validation accelerates the discovery-to-application pipeline in virus control.

- Shifting forward, the strategic use of multi-omics facts in tandem with artificial intelligence (AI) and system mastering (ML) fashions will revolutionize plant virus diagnostics and resistance breeding. Predictive modeling of sickness susceptibility, identification of resistance biomarkers, and precision-centered biotechnological interventions, along with dsRNA sprays or CRISPR-edited cultivars, are foreseeable results of this integrative approach. Translating laboratory findings to discipline packages would require harmonized efforts across omics studies, crop development, and agroecosystem modeling.
- For sustainable ToLCNDV control, future studies need to emphasize: (i) increasing omics assets for underexplored vegetation and wild relatives, (ii) validating candidate genes and proteins through functional studies, (iii) growing excessive-throughput discipline-deployable diagnostic and RNAi-based management equipment, and (iv) fostering collaborations among molecular biologists, breeders, virologists, and facts scientists. Integrative omics, supported by robust bioinformatics and translational pipelines, holds the promise to transform plant virology from a reactive to a predictive and preventive technology.

References

Arif, M. (2024). Cross-species substitution matrix comparison of tomato leaf curl New Delhi virus (ToLCNDV) with medicinal plant isolates. *Journal of Plant Diseases and Protection.* https://doi.org/10.1007/s41348-024-00980-4

Aregger, M., Borah, B. K., Seguin, J., Rajeswaran, R., Gubaeva, E. G., Zvereva, A. S., Windels, D., Vazquez, F., Blevins, T., Farinelli, L., & Pooggin, M. M. (2012). Primary and secondary siRNAs in geminivirus-induced gene silencing.

Ali, Z., Abulfaraj, A., Idris, A., Ali, S., Tashkandi, M., & Mahfouz, M. M. (2015). CRISPR/Cas9-mediated viral interference in plants. *Genome Biology, 16*(1), 238.

Bhardwaj, V., Sood, S., Kumar, A., Vanishree, G., Sharma, S., Sundaresha, S., Raigond, B., Kumar, R., Bairwa, A., Lal, M., & Chakrabarti, S. K. (2019). Efficiency and reliability of marker-assisted selection for resistance to major biotic stresses in potato. *Potato Journal, 46*, 56–66.

Brown, J. K., Zerbini, F. M., Navas-Castillo, J., Moriones, E., Ramos-Sobrinho, R., Silva, J. C., et al. (2015). Revision of begomovirus taxonomy based on pairwise sequence comparisons. *Archives of Virology, 160*(6), 1593–1619. https://doi.org/10.1007/s00705-015-2398-y

Bisaro, D. M. (2006). Silencing suppression by geminivirus proteins. *Virology, 344*(1), 158–168.

Cai, L., Mei, Y., Ye, R., Deng, Y., Zhang, X., Hu, Z., Zhou, X., Zhang, M., & Yang, J. (2023). Tomato leaf curl New Delhi virus: An emerging plant begomovirus threatening cucurbit production. *aBIOTECH, 4*, 257–266. https://doi.org/10.1007/s42994-023-00118-4

Chen, F., Ding, X., Feng, Y., Seebeck, T., Jiang, Y., & Davis, G. (2017). Targeted activation of diverse CRISPR–Cas systems for mammalian genome editing via proximal CRISPR targeting. *Nature Communications, 8*, 14958. https://doi.org/10.1038/ncomms14958

Chen, T., Lv, Y., Zhao, T., Li, N., Yang, Y., Yu, W., He, X., Liu, T., & Zhang, B. (2013). Comparative transcriptome profiling of a resistant vs. susceptible tomato (Solanum lycopersicum) cultivar in response to infection by tomato yellow leaf curl virus. *PloS one, 8*(11), e80816.

Chen, Q., Li, Y., Tan, K., Qiao, Y., Pan, S., Jiang, T., & Chen, Y. P .P. (2021). Network-based methods for gene function prediction. *Briefings in Functional Genomics, 20*(4), 249–257.

Devendran, R., Namgial, T., Reddy, K. K., Kumar, M., Zarreen, F., & Chakraborty, S. (2022). Insights into the multifunctional roles of geminivirus-encoded proteins in pathogenesis. *Archives of Virology, 167*(2), 307–326.

EFSA Panel on Plant Health (PLH). (2020). Pest categorisation of tomato leaf curl New Delhi virus. *EFSA Journal, 18*(7), e06114. https://doi.org/10.2903/j.efsa.2020.6114

Fortes, I., Sánchez-Campos, S., Fiallo-Olivé, E., Díaz-Pendón, J., Navas-Castillo, J., & Moriones, E. (2016). A novel strain of tomato leaf curl New Delhi virus has spread to the Mediterranean Basin. *Viruses, 8*. https://doi.org/10.3390/v8110307

Guo, C. X., Gao, F., & Guo, Y. (2023). Off-target effects in CRISPR/Cas9 gene editing. *Frontiers in Bioengineering and Biotechnology, 11*. https://doi.org/10.3389/fbioe.2023.1143157

Gupta, N., Reddy, K., Bhattacharyya, D., & Chakraborty, S. (2021). Plant responses to geminivirus infection: guardians of the plant immunity. *Virology Journal, 18*(1), 143.

Gaur, R. K., Khurana, S. P., Sharma, P., & Hohn, T. eds. (2021). Plant virus-host interaction: molecular approaches and viral evolution. Academic Press.

Gupta, D., & Mukherjee, S. K. (2019). Antiviral RNAi mediated Plant defense versus its suppression by viruses. *Journal of Plant Science and Phytopathology, 3*(1), 001–008.

Gullner, G., Komives, T., Király, L., & Schröder, P. (2018). Glutathione S-transferase enzymes in plant-pathogen interactions. *Frontiers in Plant Science, 9*, 1836.

Gorovits, R., Liu, Y., & Czosnek, H. (2016). The involvement of HSP70 and HSP90 in tomato yellow leaf curl virus infection in tomato plants and insect vectors. In Heat Shock Proteins and Plants (pp. 189–207). Cham: Springer International Publishing.

Jeevalatha, A., Vanishree, G., Siddappa, S., Kumar, R., Kaundal, P., Kumar, A., & Chakrabarti, S. K. (2021). Molecular characterization and infectivity analysis of tomato leaf curl New Delhi virus isolates infecting potato. *3 Biotech, 11*, 203. https://doi.org/10.1007/s13205-021-02752-5

Jeevalatha, A., Siddappa, S., Kumar, R., Tiwari, R. K., Lal, M. K., Sharma, S., Chakrabarty, S. K., & Singh, B. P. (2023). RNA-seq analysis reveals an early defense response to tomato leaf curl New Delhi virus in potato cultivar Kufri Bahar. *Functional & Integrative Genomics, 23*, 215. https://doi.org/10.1007/s10142-023-01138-5

Juarez, M., Tovar, R., Fiallo-Olivé, E., Aranda, M. A., Gosálvez, B., & Navas-Castillo, J. (2014). First detection of tomato leaf curl New Delhi virus infecting zucchini in Spain. *Plant Disease, 98*(6), 857.

Juárez, M., Rabadán, M., Martínez, L., Tayahi, M., Grande-Pérez, A., & Gómez, P. (2019). Natural hosts and genetic diversity of the emerging tomato leaf curl New Delhi virus in Spain. *Frontiers in Microbiology, 10*, 140. https://doi.org/10.3389/fmicb.2019.00140

Jangra, S., Chinnaiah, S., Patil, S. R., Shukla, B., Devendran, R., & Kumar, M. (2024). Deciphering the role of virus receptors in plant–virus–vector interactions. *Receptors, 3*(2), 255–279.

Jeevalatha, A., Siddappa, S., Kumar, A., Kaundal, P., Guleria, A., Sharma, S., Nagesh, M., & Singh, B. P. (2017). An insight into differentially regulated genes in resistant and susceptible genotypes of potato in response to tomato leaf curl New Delhi virus-[potato] infection. *Virus Research, 232*, 22–33.

Jiang, T. & Zhou, T. (2023). Unraveling the mechanisms of virus-induced symptom development in plants. *Plants, 12*(15), 2830.

Kumar, R., Tiwari, R. K., Jeevalatha, A., Sundaresha, S., Shah, M. A., Sharma, S., Sagar, V., Kumar, M., & Chakrabarti, S. K. (2021b). Potato apical leaf curl disease: Current status and perspectives on a disease caused by tomato leaf curl New Delhi virus. *Journal of Plant Diseases and Protection, 128*, 897–911. https://doi.org/10.1007/s41348-021-00463-w

Kumar, R., Lal, M. K., Tiwari, R. K., Chourasia, K. N., Kumar, A., Kumar, R., Sharma, S., & Singh, B. (2023). Investigating the interplay between tomato leaf curl New Delhi virus infection, starch metabolism and antioxidant defence system in potato (*Solanum tuberosum* L.). *Antioxidants, 12*, 1447. https://doi.org/10.3390/antiox12071447

Kryovrysanaki, N., James, A., Tselika, M., Bardani, E. and Kalantidis, K. 2021. RNA silencing pathways in plant development and defense. *The International Journal of Developmental Biology, 66*(1-2-3), 163–175.

Khalil, A., Fatima, K., Sadaqat, M., & Tahir ul Qamar, M. (2025). Disease resistance at the molecular level: Omics-based approaches for tomato defense. In *Omics Approaches for Tomato Yield and Quality Trait Improvement* (pp. 323–345). Springer Nature Singapore.

Kozieł, E., Otulak-Kozieł, K., & Bujarski, J. J. (2021). Plant cell wall as a key player during resistant and susceptible plant-virus interactions. *Frontiers in Microbiology, 12*, 656809.

Kumar, R., Tiwari, R. K., Sundaresha, S., Kaundal, P., & Raigond, B. (2022). Potato viruses and their management. In *Sustainable management of potato pests and diseases* (pp. 309–335). Springer Singapore.

Li, C., Chu, W., Gill, R., Sang, S., Shi, Y., Hu, X., Yang, Y., Zaman, Q., & Zhang, B. (2023). Computational tools and resources for CRISPR/Cas genome editing. *Genomics, Proteomics & Bioinformatics, 21*, 108–126. https://doi.org/10.1016/j.gpb.2022.02.006

Lang, Z., Wang, Y., Tang, K., Tang, D., Datsenka, T., Cheng, J., Zhang, Y., Handa, A.K., & Zhu, J. K. (2017). Critical roles of DNA demethylation in the activation of ripening-induced genes and inhibition of ripening-repressed genes in tomato fruit. *Proceedings of the National Academy of Sciences, 114*(22), E4511–E4519.

Majid, A., Shakeel, M., Aslam, M., Ashraf, W., Raheel, M., Maqsood, A., Khan, A., Moosa, A., & Shafiq, M. (2023). Unraveling the devastating impact of tomato leaf curl New Delhi virus (ToLCNDV): Understanding its characteristics, host range, and management strategies. *Advanced Virology*. https://doi.org/10.58801/2k23rv2utd

Mansoor, S., Briddon, R. W., Zafar, Y., & Stanley, J. (2003). Geminivirus disease complexes: An emerging threat. *Trends in Plant Science, 8*(3), 128–134.

Mastrochirico, M., Spanò, R., De Miccolis Angelini, R., & Mascia, T. (2023). Molecular characterization of a recombinant isolate of tomato leaf curl New Delhi virus associated with severe outbreaks in zucchini squash in southern Italy. *Plants, 12*. https://doi.org/10.3390/plants12132399

Mei, Y., Cai, L., Wang, Y., Li, F., Yang, X., Yang, J., & Zhou, X. (2023). Molecular characterization and pathogenicity of an infectious clone of tomato leaf curl New Delhi virus isolate infecting *Cucumis melo*. *Stress Biology, 3*. https://doi.org/10.1007/s44154-023-00128-8

Moriones, E., Praveen, S., & Chakraborty, S. (2017). Tomato leaf curl New Delhi virus: An emerging virus complex threatening vegetable and fiber crops. *Viruses, 9 (10):264* https://doi.org/10.3390/v9100264

Mahomed, N. B. (2022). Investigation of histone methylation of tomato curly stunt virus (ToCSV) in resistant and susceptible tomato lines. University of Johannesburg (South Africa).

Mnari-Hattab, M., Zammouri, S., Belkadhi, M. S., Doña, D. B., Ben Nahia, E.. & Hajlaoui, M. (2015). First report of Tomato leaf curl New Delhi virus infecting cucurbits in Tunisia. *New Dis. Rep, 31*(21), 2044–0588.

Naeem, M., Majeed, S., Hoque, M., & Ahmad, I. (2020). Latest developed strategies to minimize the off-target effects in CRISPR-Cas-mediated genome editing. *Cells, 9*, 1608. https://doi.org/10.3390/cells9071608

Naga, K. C., Sundaresha, S., Kumar, R., Tiwari, R. K., Subhash, S., Verma, G., Buckseth, T., Bairwa, A., Sharma, S., Katare, S., & Srivastava, R. M. (2021). A new record of Asia II 5 genetic group of Bemisia tabaci (Gennadius) in the major potato growing areas of India and its relationship with tomato leaf curl New Delhi virus infecting potato. *3 Biotech, 11*, 421. https://doi.org/10.1007/s13205-021-02966-7

Naqvi, A. R., Haq, Q. M. I., & Mukherjee, S. K. (2010). MicroRNA profiling of tomato leaf curl New Delhi virus (ToLCNDV)-infected tomato leaves indicates that deregulation of mir159/319 and mir172 might be linked with leaf curl disease. *Virology Journal, 7*(1), 281.

Prakash, V., Singh, A., Singh, A. K., Dalmay, T. & Chakraborty, S. (2020). Tobacco RNA-dependent RNA polymerase 1 affects the expression of defence-related genes in Nicotiana benthamiana upon Tomato leaf curl Gujarat virus infection. *Planta, 252*(1), 11.

Pan, Z., Yu, Y., Ping, Y., Liu, M., Ji, H., Han, Y., Li, Z., & Sun, J. (2025). High temperature influences on virus–host interaction in plants: A case study of sweet potato virus disease and kost lipid metabolism. *Physiologia Plantarum, 177*(4), e70442.

Raja, P., Sanville, B. C., Buchmann, R. C., & Bisaro, D. M. (2008). Viral genome methylation as an epigenetic defense against geminiviruses. *Journal of Virology, 82*(18), 8997–9007.

Rauniyar, N., Singhal, P., Diksha, D., Srivastava, D., & Baranwal, V. K. (2023). Characterization of a recombinant tomato leaf curl New Delhi virus (ToLCNDV) in a perennial medicinal climber host (*Ipomoea cairica* (L.) sweet). *3 Biotech, 13*, 3. https://doi.org/10.1007/s13205-022-03418-6

Renukadevi, P., Devi, R., Jothika, C., Karthikeyan, G., Malathi, V., Balakrishnan, N., Rajagopal, B., Nakkeeran, S., & Abd-Allah, E. (2024). Genomic distinctiveness and recombination in tomato leaf curl New Delhi virus (ToLCNDV-BG) isolates infecting bitter gourd. *3 Biotech, 14*(8), 184. https://doi.org/10.1007/s13205-024-04009-3

Ruiz, L., Simón, A., Velasco, L., & Janssen, D. (2017). Biological characterization of tomato leaf curl New Delhi virus from Spain. *Plant Pathology, 66*, 376–382. https://doi.org/10.1111/ppa.12587

Rosas-Diaz, T., Zhang, D., Fan, P., Wang, L., Ding, X., Jiang, Y., Jimenez-Gongora, T., Medina-Puche, L., Zhao, X., Feng, Z., & Zhang, G. (2018). A virus-targeted plant receptor-like kinase promotes cell-to-cell spread of RNAi. *Proceedings of the National Academy of Sciences, 115*(6), 1388–1393.

Rossouw, L. T. (2016). Metabolite profiling of defence-related secondary metabolites in Solanum lycopersicum, in response to whitefly mediated begomovirus infection. University of Johannesburg (South Africa).

Ramesh, S. V., Sahu, P. P., Prasad, M., Praveen, S., & Pappu, H. R. (2017). Geminiviruses and plant hosts: a closer examination of the molecular arms race. *Viruses, 9*(9), 256.

Sade, D., Shriki, O., Cuadros-Inostroza, A., Tohge, T., Semel, Y., Haviv, Y., Willmitzer, L., Fernie, A. R., Czosnek, H., & Brotman, Y. (2015). Comparative metabolomics and transcriptomics of plant response to Tomato yellow leaf curl virus infection in resistant and susceptible tomato cultivars. *Metabolomics, 11*(1), 81–97.

Sarkar, M., Gupta, D., Singh, O. W., Paul, S., Kumar, R., Mandal, B., & Roy, A. (2024). Identification of pathogenicity determinants in ToLCNDV and their RNAi-based knockdown for disease management in *Nicotiana benthamiana* and tomato plants. *Frontiers in Microbiology, 15*, 1481523. https://doi.org/10.3389/fmicb.2024.1481523

Shivaprasad, P. V., Chen, H. M., Patel, K., Bond, D. M., Santos, B. A., & Baulcombe, D. C. (2012). A microRNA superfamily regulates nucleotide binding site–leucine-rich repeats and other mRNAs. *Plant Cell, 24*(3), 859–874.

Samarskaya, V., Spechenkova, N., Kalinina, N. O., Love, A. J., & Taliansky, M. (2025). The emerging role of omics-based approaches in plant virology. *Viruses, 17*(7), 986.

Sattar, M. N. (2012). Diversity and Interactions of Begomoviruses and their associated DNA-Satellites. Uppsala, Sweden: Department of Plant Biology and Forest Genetics, Swedish University of Agricultural Sciences.

Sahu, A. K., Mishra, N. S., & Gaur, R. K. (2021). Suppressor to survival: RNAi as a molecular weapon in arms race between virus and host. In *RNA-based technologies for functional genomics in plants* (pp. 131–154). Cham.

Sharma, S., & Prasad, M. (2025). Diverse roles of phytohormonal signaling in modulating plant–virus interactions. *Journal of Experimental Botany, 76*(7), 1921–1940.

Sharma, N., & Prasad, M. (2017). An insight into plant–Tomato leaf curl New Delhi virus interaction. *The Nucleus, 60*(3), 335–348.

Sahu, P. P., Puranik, S., Khan, M., & Prasad, M. (2012). Recent advances in tomato functional genomics: utilization of VIGS. *Protoplasma, 249*(4), 1017–1027.

Shakir, S., Mubin, M., Nahid, N., Serfraz, S., Qureshi, M. A., Lee, T. K., Liaqat, I., Lee, S., & Nawaz-ul-Rehman, M. S. (2023). REPercussions: how geminiviruses recruit host factors for replication. *Frontiers in Microbiology, 14*, 1224221.

Troiano, E., & Parrella, G. (2023). First report of tomato leaf curl New Delhi virus in Lagenaria siceraria var. longissima in Italy. *Phytopathologia Mediterranea*. https://doi.org/10.36253/phyto-14147

Tiwari, J. K., A, J., Tuteja, N., & Khurana, S. P. (2022). Genome editing (CRISPR-Cas)-mediated virus resistance in potato (Solanum tuberosum L.). *Molecular Biology Reports 49*(12), 12109–12119.

Vanitharani, R., Chellappan, P., & Fauquet, C. M. (2004). Geminiviruses and RNA silencing. *Trends in Plant Science, 9*(4), 144–151.

Wilisiani, F., Mashiko, T., Wang, W., Suzuki, T., Hartono, S., Neriya, Y., Nishigawa, H., & Natsuaki, T. (2019a). New recombinant of tomato leaf curl New Delhi virus infecting melon in Indonesia. *Journal of General Plant Pathology, 85*, 306–310. https://doi.org/10.1007/s10327-019-00849-7

Wilisiani, F., Neriya, Y., Tagami, M., Kaneko, M., Hartono, S., Nishigawa, H., & Natsuaki, T. (2019b). Complete genome sequence of tomato leaf curl New Delhi virus from Luffa in Indonesia. *Microbiology Resource Announcements, 8*. https://doi.org/10.1128/MRA.01605-18

Wang, L., Tan, H., Medina-Puche, L., Wu, M., Garnelo Gomez, B., Gao, M., Shi, C., Jimenez-Gongora, T., Fan, P., Ding, X., & Zhang, D. (2022). Combinatorial interactions between viral proteins expand the potential functional landscape of the tomato yellow leaf curl virus proteome. *PLoS pathogens, 18*(10), e1010909.

Wang, X., Kong, L., Zhi, P., & Chang, C. (2020). Update on cuticular wax biosynthesis and its roles in plant disease resistance. *International Journal of Molecular Sciences, 21*(15), 5514.

Zaidi, S., Martin, D., Amin, I., Farooq, M., & Mansoor, S. (2017). Tomato leaf curl New Delhi virus: A widespread bipartite begomovirus in the territory of monopartite begomoviruses. *Molecular Plant Pathology, 18*(7), 901–911. https://doi.org/10.1111/mpp.12481

Zhou, M., Palanca, A. M. S., & Law, J. A. (2018). Locus-specific control of the de novo DNA methylation pathway in Arabidopsis by the CLASSY family. *Nature genetics, 50*(6), 865–873.

Zhao, S., & Li, Y. (2021). Current understanding of the interplays between host hormones and plant viral infections. *PLoS Pathogens, 17*(2), e1009242.

4 Mechanistic Understanding of Vector Dynamics and Tomato Leaf Curl New Delhi Virus-Vector Relationship

Bimal Kumar Sahoo, Ipsita Samal, Ansh Raj, Tanmaya Kumar Bhoi, and Deepak Kumar Mahanta

Abstract

Tomato leaf curl New Delhi virus (ToLCNDV), a member of the *Begomovirus* genus, is a highly devastating pathogen accountable for significant yield losses in tomatoes, cucurbits, and other horticultural crops. This virus is primarily transmitted by the whitefly (*Bemisia tabaci*), an invasive, polyphagous, and cosmopolitan insect with a broad host range. Epidemiology and rapid spread of ToLCNDV depend on the Genetic variability, biotype distribution, reproductive efficiency, and feeding habit of the whitefly. The intricate mechanism of virus-vector interaction involves the binding of viral coat proteins with the apical membrane receptors of the salivary gland and midgut, facilitating the circulative and persistent transmission. The population dynamics of the vector and the systemic spread of the virus also depend on the bioavailability of host plants, alternate hosts, as well as abiotic factors like relative humidity, temperature, and rainfall. ToLCNDV and *B. tabaci* can be adapted to each other's biology and environment as a result of co-evolution to enhance the ability of the virus to thrive inside the vector and the vector's ability to transmit the virus. A thorough

B. K. Sahoo (✉)
Department of Agricultural Entomology, Tamil Nadu Agricultural University, Coimbatore, Tamil Nadu, India

I. Samal · A. Raj
ICAR-National Research Centre on Litchi, Muzaffarpur, Bihar, India

T. K. Bhoi
Forest Protection Division, ICFRE-Arid Forest Research Institute (ICFRE-AFRI), Jodhpur, Rajasthan, India

D. K. Mahanta
Forest Entomology Discipline, Forest Protection Division, Indian Council of Forestry Research and Education (ICFRE)-Forest Research Institute (ICFRE-FRI), Dehradun, Uttarakhand, India

R. Kumar et al. (eds.), *Tomato Leaf Curl New Delhi Virus (ToLCNDV)*,
https://doi.org/10.1007/978-981-95-2852-3_4

knowledge of the intricate process of how viruses and vectors interact in a way that leads to virus spread can help in developing effective management practices for ToLCNDV sustainably to curb the deleterious impacts of this virus on the global economy.

Keywords

Tomato leaf curl New Delhi virus · Begomovirus · *Bemisia tabaci* · Coat protein · Polyphagous pest

1 Introduction

Tomato leaf curl New Delhi virus belongs to the family Geminiviridae and genus Begomovirus, which causes a pathogenic threat in tomato production within India. Global incidences of infection by this virus are mainly on vegetable crops belonging to the Cucurbitaceae and Solanaceae families (Tolin & Fayad, 2016). The first report of ToLCNDV infecting tomato was documented by Padidam et al. (1995). Later on, tomatoes were severely damaged by ToLCNDV in Pakistan as well as Northern India, resulting in drastic yield losses (Akhtar et al., 2019; Aslam et al., 2017). Asia is the leading geographical hotspot for accommodating the ToLCNDV invasion throughout the world. Since 2012, several countries, including Algeria, Spain, Morocco, Tunisia, Italy, Greece, and Iran, have again reported cucurbits' pathogenic struggle with ToLCNDV (Yazdani-Khameneh et al., 2016; Kheireddine et al., 2019; Orfanidou et al., 2019). Nations like Cambodia, China, Laos, Malaysia, and Indonesia have recently reported the devastating impact of ToLCNDV on cucurbit crops (Cai et al., 2023; Neoh et al., 2023; Desbiez et al., 2021; Vignesh et al., 2023).

Both variants belong to strains of ToLCNDV: ToLCNDV-ES and ToLCNDV-In, which originated from the Mediterranean Basin and the Indian subcontinent, respectively (Gautam et al., 2022; Zaidi et al., 2017; Fortes et al., 2016). The year 2012 marked the emergence of a new strain of ToLCNDV in Spain called ToLCNDV-ES, entirely destructive for zucchini (*Cucurbita pepo*) production (Janssen et al., 2022). Later on, it spread to 16 different countries, infecting more than 11 cucurbit crops and resulting in significant crop loss (Cai et al., 2023). There have been several reports of production loss of zucchini by ToLCNDV infection, like20% loss in Spain and 22% loss in Almeria (Crespo et al., 2020). Similarly, other cucurbit crops, such as *C. moschata* and *Sechium edule*, experienced a harvest loss of up to 100% in Italy and India. ToLCNDV-ES has the ability not only to affect cucurbit crops but also to reduce tomato yield up to 100% even in complete drop failure in severe pathogenic circumstances (Van Vu et al., 2013). ToLCNDV-In and ToLCNDV-ES attack the solanaceous crops and the cucurbits, respectively. In Italy, two evolutionary subgroups of this strain have been identified, namely subgroups I and II. The first one doesn't have the potential to infect tomato plants (Troiano & Parrella, 2023; Vo et al., 2023), whereas the second one shows a lesser incidence and mild symptoms in tomato under field conditions (Yamamoto et al., 2021; Ruiz et al., 2017).

Studying the mechanism behind vector behavior enables scientists to determine how ToLCNDV survives and transmits among diversified host plants. Whitefly (*Bemisia tabaci*) plays a significant role in transmitting the ToLCNDV in a circulative and persistent manner (Moriones et al., 2017, Naga et al., 2021a, b). The success of virus transmission is dependent on the perfect interaction of whitefly midgut receptors with the viral proteins (Pan et al., 2017). Interaction of viral proteins with midgut receptors results in the endocytosis of the virus and subsequently systemic movement through insect hemolymph to get into salivary glands (Xia et al., 2018). Climatic conditions, feeding habits, vector competence, and virus persistence in the vector body affect the epidemics and spread of ToLCNDV (Pakkianathan et al., 2015). The controlling mechanisms of diseases fail because viral proteins strongly interact with midgut proteins of the vector, which subsequently enables them to multiply and disseminate infection (Rana et al., 2016; Zhao et al., 2020). Therefore, a healthy comprehension of the virus-vector association is far more essential for formulating appropriate management practices to control this virus because of its destructive nature (Modgill et al., 2025). This chapter provides a comprehensive and elaborate explanation of the complicated process involved in the transmission of ToLCNDV Virus, including vector biology and molecular determinants regulating interaction.

2 Genomic Structure of Begomovirus ToLCNDV

Begomoviruses are of two different types, Monopartite and bipartite (Akhtar et al., 2019). In the case of a monopartite virus, only one genomic DNA is present, whereas a bipartite virus has two DNA components, named DNA A and B, on a single-stranded circular DNA (Bibi et al., 2024). Usually, monopartite viruses are smaller than those viruses, and they are more or less functionally equivalent. In the monopartite type, viral movement is controlled by either the coat protein (V1 gene) or the V2 ORF (open reading frame), where the V2 encodes the movement-associated proteins (Barboza et al., 2019). In case of a bipartite virus, the DNA A component comprises six ORFs, AC1, AC2, AC3, AC4, AV1 and AV2, whereas DNA B regulates the activities of BV1 and BV2 proteins. Six ORFs of DNA A are involved in different viral functions described below (Jeevalatha et al., 2021, 2023). AC1 is responsible for encoding the replication initiator protein (Rep), helping in DNA replication through ligation and nicking (Chellappan et al., 2005; Kumar et al., 2021, 2023). Other functions like RNA silencing, transcription regulation, and pathogenicity are regulated by transcriptional activator protein (TrAP), which is encoded by AC2 (Voinnet et al., 1999). Similarly, AC3 encodes for the replication enhancer protein (Ren), which enhances the viral replication ability (Phaneendra et al., 2012; Mazumder et al., 2024). AC4 works opposite to the AC2 encoding for RNA-silencing suppressor. The coat protein also plays a crucial role in the movement and assembly of the virus inside the vector, which AV1 encodes. The overlapping of AV2 with AV1 codes for pre-coat protein (PCP), causing viral movement as well as RSS activities (Sarkar et al., 2024; Cai et al., 2023).

The coat protein is not even necessary in ToLCNDV for symptom expression and systemic movement. These processes are controlled by DNA B (Padidam et al., 1995). BC1 has the protein responsible for viral symptoms formation and systemic spreading in DNA B (Padidam et al., 1995; Von Arnim & Stanley, 1992). The BV1 gene codes for a nuclear shuttle protein (NSP) in the sense orientation, whereas in the complementary sense orientation, the BV1 gene codes for a movement protein (MP). The common region (CR) is the analogous sequence of both DNA A and B, which is involved in DNA B replication by the rep encoded by DNA A.

3 ToLCNDV and Its Host Range

ToLCNDV has a wide range of hosts covering more than 60 crop species, including vegetable and ornamental plants under the families like solanaceae, malvaceae and cucurbitaceae (Anwar et al., 2020; Ashwathappa et al., 2020; Pant et al., 2018). The ToLCNDV-ES strain mainly infects cucurbit crops such as *Cucurbita melo* var. *flexuosus*, *C. pepo*, *C. melo*, *Citrullus lanatus*, and *Cucumis sativus*, whereas ToLCNDV-In mainly infects Solanaceous crops including *Solanum tuberosum*, *Solanum melongena*, *Capsicum annuum*, and *C. frutescens* (Siskos et al., 2022; Fortes et al., 2016; Parrella et al., 2020; Yamamoto et al., 2021; Juarez et al., 2019).

ToLCNDV also has the potential to inflict severe damage in other crops like *Glycine max*, *Carica papaya*, *Hibiscus cannabinus*, *Vigna radiata*, *Daucus carota*, and *Abelmoschus esculentus* (Pant et al., 2018; Moriones et al., 2017; Jamil et al., 2017; Venkataravanappa et al., 2018). Apart from the major hosts, ToLCNDV can perpetuate well in their alternate hosts, including some weed crop plants like *Ecballium elaterium*, *Sonchus oleraceus*, *Solanum nigrum*, and *Datura stramonium* (Kassem et al., 2013). These weed plants supplement nutrition to the growth and development of the virus and also act as a reservoir for virus source inoculum (Ferro et al., 2017; Juarez et al., 2019). The weed crops are more resilient to the virus due to their better genetic variability and heterogeneity, which can improve the spatio-temporal pattern of crop-enduring viruses (Kassem et al., 2013).

4 Typical Symptoms of ToLCNDV Infection

Stunting of plant height, reduction in the number of leaves, leaf curling, and thick leaf veins (Kiran et al., 2021; Aslam et al., 2017; Kumar et al., 2021), Yellow mosaic patterns and dwarfing, shortening of the internodes (Sarkar et al., 2024; Zaidi et al., 2017), fruits with longitudinal cracking and rough skin (Siskos et al., 2022; Yamamoto et al., 2021; Bragard et al., 2020), puckering, purpling or darkening of leaf margins, and a reduction in leaf area (Moriones et al., 2017) are the main symptoms displayed by the ToLCNDV.

5 Pathogenesis of ToLCNDV

Pathogenicity of ToLCNDV is defined by its coat protein, whose primary function is viral propagation rather than enabling movement from one cell to another. NSP performs a significant role in the transportation of virus between cells, enabling systemic infection in various bipartite Begomoviruses (Gafni & Epel, 2002). Clones infected with single amino acid alterations using functional analyses have reported that the residue at position 143 in the coat protein is crucial for ToLCNDV infection in tomato plants (Vo et al., 2023). For unknown reasons, ToLCNDV-ES CP was blocked in the early phases of viral production, which probably prevented infection and prevented NSP from continuing to cause systemic infection. When the CP from the ToLCNDV-In was integrated into the ToLCNDV-ES genomic backbone, it promoted the movement of viral proteins that led to the systemic infection, indicating that CP may be an essential factor in the pathogenicity of ToLCNDV in tomatoes in addition to NSP (Vo et al., 2023). However, compared to the ToLCNDV-In DNA-B component, its interaction with the ToLCNDV-ES movement proteins was less pronounced, which led to milder symptoms in ToLCNDV-ES-infected plants as opposed to ToLCNDV-In. MiRNAs also regulate the leaf curl symptoms in tomato cultivars infected by ToLCNDV. A report by Naqvi et al. (2010) revealed that down-regulation of multiple miRNAs, including 13 conserved miRNAs like miR319 and miR172, occurred due to infection by ToLCNDV. At the same time, the increased accretion of miR159/319 and miR172 was recorded as the days post-inoculation progressed.

A key factor in determining the virulence of the ToLCNDV is protein-protein interactions. To affect viral replication, systemic mobility, and symptom manifestation, the viral coat protein (CP) interacts with several protein molecules of host plants during the infection process. Five host proteins, viz., Protein LOW PSII ACCUMULATION 1, Cathepsin B-like protease 3, Protein phosphatase 2C (DIG3), NAD(P)H-quinone oxidoreductase subunit M and Ribulose bisphosphate carboxylase small chain 3B (RBCS4), interact with the coat protein of ToLCNDV-In for symptom development. In contrast, a different nature of proteins like NAD(P)H-quinone oxidoreductase subunit M protein, GAGA-binding transcriptional activator, protein CROWDED NUCLEI 4, 1-D-deoxyxylulose 5-phosphate synthase, RING finger protein 44-like and thaumatin-like protein 1b are involved in the pathogenicity of ToLCNDV-ES CP when interacting with coat protein (Vo et al., 2023).

6 Suppression of Plant Defenses by Viruses

Viruses have developed the ability to encode for several proteins with multifunctional skills that are essential for controlling their life cycle and enabling them to evade host defense mechanisms. When a virus infects a host, specific defense mechanisms are triggered. Two well-known defense mechanisms in plants against viral pathogen infection are resistance genes (R-genes) and RNA interference (RNAi) (also known as RNA, 264 silence). The role of viral proteins encoded by ORFs AC2,

AC4, AV2, and BV1 is to determine pathogenicity, most likely by controlling RNAi, a host basal defense mechanism. To reduce losses from ToLCNDV, Ty-resistant loci driven genetic resistance for managing viruses linked to TYLCD has been used in tomatoes because of their functional characterization points to their potential contribution to RNA interference-mediated host defense (Prasanna et al., 2015; Butterbach et al., 2014).

7 Hormonal Regulation Pathway Modification by a Virus

Monopartite geminiviruses encode the protein C4, which functions similarly to AC4, which interacts with a set of SHAGGY-like kinases that influence the brassinosteroid pathway (Fondong, 2022), promotes cell proliferation, and causes distinctive symptoms. Similar to a viral "oncogene," C4/AC4 causes hyperplasia and tumorigenic development in the infected plant tissues (Fondong, 2013). Monopartite geminiviruses encode the protein C4, which interacts with a set of SHAGGY-like kinases that influence the brassinosteroid pathway (Piroux et al., 2007), promotes cell proliferation, and causes distinctive symptoms. Like a viral "oncogene," C4/AC4 provokes hyperplasia and tumorous growth in the infected plant tissues (Fondong, 2013). As a consequence of the overexpression of AC4 in tomato plants, several phenotypic anomalies such as stunting, smaller leaf margin area, leaf curling, and abnormal root growth are apparent under TOLCNDV infection (Vinutha et al., 2018; 2020). These symptoms probably develop due to the interaction of host proteins associated with auxin biosynthesis with the AC4 protein of the virus, leading to increased auxin production. This increase in auxin synthesis further signals the viral replication and development of symptoms. Other host proteins, such as TAR1 (tryptophan aminotransferase 1), CYP450 monooxygenase, indole acetic acid biosynthase, and senescence-associated genes governed by the jasmonic acid pathway, are also engaged in interaction with ToLCNDV-AC4. Similarly, typical symptoms of TOLCNDV can also be resolved by exogenous application of IAA (Vinutha et al., 2020).

Apart from the decrease in endogenous IAA levels, ToLCNDV infection also results in the overexpression of host miRNAs, specifically miR164, miR167, miR319and miR393, implicated in auxin signaling and leaf morphogenesis (Vinutha et al., 2020). The disruption of auxin production and other transcriptome characteristics of the infected leaves were replicated by ectopic overexpression of ToLCNDV-derived AC4 in tomatoes. In auxin signaling, host miRNAs miR164 and miR167 are essential for the development of different plant organs, leaf shape, and root initiation, apical dominance, and vascular tissue differentiation, among other plant growth and developmental processes (Chandler, 2016; Luo et al., 2022; Li et al., 2016). By controlling host transcripts TCP and LoxD, miR319 also regulates the JA pathway (Schommer et al., 2008). ToLCNDV infection causes target transcripts TCP1 and LoxD to be downregulated and miR319 to be upregulated, indicating that auxin and JA signaling molecular crosstalk is involved in viral infection.

8 Role of Satellites in ToLCNDV Pathogenicity

Satellites are the additional circular single-stranded DNA molecules often linked with monopartite begomoviruses. These satellites can be categorized into several types, such as alpha, beta, and delta. These satellite molecules play an essential part in controlling viral pathogenesis and symptom severity. The *B. tabaci* facilitates the transmission of begomoviruses along with their associated satellites from infected plants to healthy hosts (Bibi et al., 2024). Alpha satellites can self-replicate due to the presence of a single Rep protein. Meanwhile, Beta satellites cannot self-replicate, but they express the crucial pathogenicity and symptom determinant βC1 protein. However, both alpha and beta satellites are exclusively reliant on helper virus (DNA A) for replication, mobility, encapsidation, and viral transmission through vector (Briddon et al., 2003).

In general, a satellite molecule is not essential for the pathogenicity of ToLCNDV. ToLCNDV acts as a helper molecule in encapsidating and transreplicating the satellites, resulting in complicated infections, even while no satellites are expressly linked to the virus. As evidence, some reports revealed that when tomatoes were coinfected with two different viruses, like ToLCNDV and the cotton leaf curl Multan beta satellite (CLCuMB), for severe symptom development, the ToLCNDV DNA A and DNA B must interact with CLCuMB, confirming no role of satellites in the infection process (Sivalingam & Varma, 2012; Saeed, 2010).

However, contrasting findings were reported by Agnihotri et al. (2018) and Akhter et al. (2014), who first documented the association of a chilli leaf curl beta satellite (ChLCB) with ToLCNDV in a synergistic manner, where systemic translocation of beta satellite along with DNA B to the upper part of the leaf was observed, leading to leaf curling and yellow mosaic in *Nicotiana benthamiana* following artificial biolistic inoculation with ToLCNDV DNA A, DNA B, and ChLCB. In addition, there have been reports of intricate viral infections in cucurbit crops that include many satellite molecules. ToLCNDV was reported to co-infect *Cucurbita pepo* from Pakistan with the cucurbit yellow mosaic alpha satellite and Papaya leaf curl betasatellite (Anwar, 2017). In another study conducted by Anwar et al., 2020, it was revealed that ToLCNDV coexisted with CLCuMB and *Ageratum conyzoides* alpha satellite in *Luffa cylindrica*.

Ranjan et al. (2014) stated that DNA A participates in trans-replication. To understand the interaction between the virus and vector in tomato and *N. benthamiana* plants, different kinds of beta satellites are needed. The role of beta satellites in the ToLCNDV infection process was further emphasized by the findings of Singh et al. (2016), who revealed that integration of Tomato leaf curl Gujarat virus (ToLCGV) and Tobacco leaf curl beta satellite with natural ToLCNDV infection further amplified the symptoms of disease in resistant chilli varieties. This proved the greater potential of DNA B and beta satellite to cause infections on a larger scale. Furthermore, in the absence of DNA B, the movement of the beta satellite might be partially compensated, highlighting the importance of DNA B in the infection process.

9 Transmission of ToLCNDV

ToLCNDV is transmitted by the whitefly *B. tabaci* in a circulative and persistent fashion (Janssen et al., 2022), which means the whitefly remains viruliferous after obtaining the virus from the infected ingested host plant phloem sap. The acquired virus remains active in the vector body for a long period, so that further feeding by the insect vector on the healthy plant results in successful virus transmission, leading to secondary infection.

ToLCNDV is made up of twin-shaped (geminate) icosahedral particles that are encased in circular single-stranded DNA (ssDNA) (Jeske, 2009). The viral single-stranded DNA is liberated from the viral capsid and transformed into the replicative form of double-stranded DNA (dsDNA) following transmission of the begomovirus to the plant. Then, with the aid of the viral replication-associated protein (Rep) and the plant enzyme machinery, complementary-sense (CS) and virion-sense (VS) DNA strands are created by rolling circle replication (RCR) (Hanley-Bowdoin et al., 2013). Therefore, the occurrence of viral replication can be supported by the CS DNA synthesis. After that, ssDNA and CP combine to create the ssDNA: CP complex, which enters the host cell nucleus to replicate and transcribe viral genes. The components required for this viral transcription and replication process are found in the host cell nucleus. TYLCV can fully enter the whole plant, damaging all parts and sections, after staying 1–2 weeks inside healthy leaves (Wege, 2007).

When *B. tabaci* feeds on the phloem of virus-infected plants, the virions of ToLCNDV move in the stylet through the food canal and enter the esophagus and the gut. From there, they travel through the midgut to the hemolymph and accumulate in the salivary glands and are subsequently transferred to healthy plants via saliva during subsequent feeding (Whitfield et al., 2015). An effective transmission of ToLCNDV requires a minimum acquisition access period (AAP) of 30 min and an inoculation access period (IAP) of 10 min (Rajeshwari & Reddy, 2014). Janssen et al. (2022) reported that the highest recorded retention duration of the ToLCNDV-ES virus in the *B. tabaci* was 16 days. An acquisition feeding hour of 5 min was not sufficient for the disease development of zucchini, as they need a minimum IAP of 15 min to infect 20–60% plants. However, after 5 min of feeding with viruliferous *B. tabaci*, 53 and 43% of zucchini plants and tomato plants were infected, respectively.

While feeding on diseased plants, *B. tabaci* acquire ToLCNDV, which they can then disseminate to healthy plants. The viral load of the insect body and the plant source are two main factors that affect transmission efficiency. Newly emerging whiteflies from infected plants often have lower levels of ToLCNDV than the plants themselves, so whiteflies that emerged from pupae on dried infected tomato leaves were incapable of infecting healthy plants due to lower viral loads (Janssen et al., 2022).

10 Interaction of ToLCNDV with Midgut and Salivary Glands

The life cycle of ToLCNDV starts when whiteflies acquire viruses through phloem sap ingestion from a virus-infected plant. A sequential path is followed for the transmission of ToLCNDV within its insect vectors traveling through the stylet, midgut, hemolymph, and salivary glands before being introduced into plants via saliva secretion (Hogenhout et al., 2008). During this sequential pathway, our primary barriers were encountered by the viruses, viz., midgut invasion, midgut escape, salivary gland invasion, and salivary gland escape (Fig. 4.1) (Gray et al., 2014). Among these, the midgut cells serve as the initial cellular barrier.

Virus binding to specific cell surface molecules on the apical membrane of the midgut is facilitated by the interaction of virions and vector midgut cells (Zhao et al., 2020). The movement of ToLCNDV from the digestive tract into the hemolymph plays a crucial role through the specific midgut protein (Rana et al., 2016). The association of ToLCNDV with the midgut and salivary glands is not yet understood. In contrast, the mechanism of Tomato yellow leaf curl virus (TYLCV) in the midgut as well as in the salivary glands was thoroughly studied.

Viral endocytosis generally occurs on the brush border epithelium located in the apical membrane of midgut epithelial cells (Brown & Czosnek, 2002; Czosnek et al., 2002; Mercer et al., 2010). The important endocytic structures called clathrin-coated pits are involved in viral endocytosis, through which the virus enters through specific receptors (Mercer et al., 2010). In the first step, the TOLCNDV enters the hemolymph and reaches the apical cells of primary salivary glands. On the basal lamina, the virus interacts with the membrane receptors and enters the apical cells through clathrin-mediated endocytosis. The fusion of these endocytic vesicles with early endosomes further transfers the virus to the apical membrane and then releases

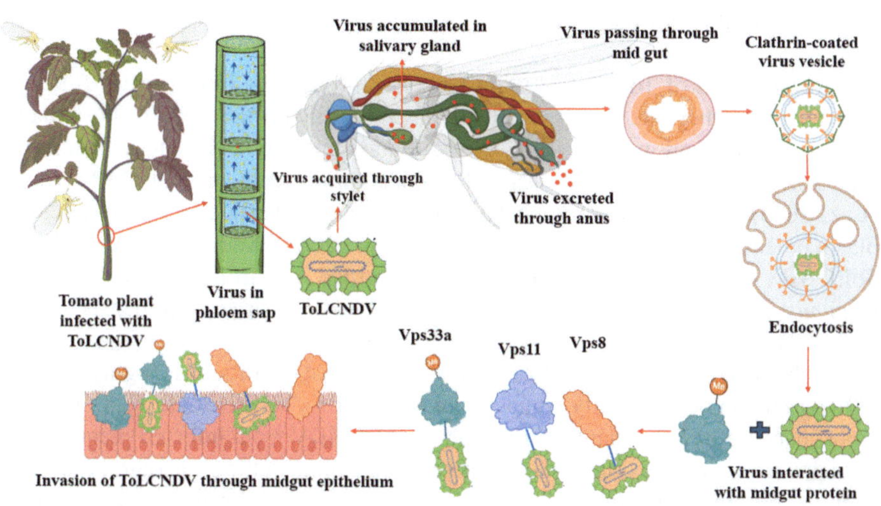

Fig. 4.1 Mechanism of ToLCNDV transmission and midgut invasion in whiteflies

it into the central lumen of the salivary gland. TYLCV trafficking in the midgut wall and primary salivary gland of white fly is facilitated by the early endosomes and clathrin-mediated endocytosis (Pan et al., 2017; Xia et al., 2018). There is a wide range of regulatory proteins involved in facilitating the passage of TYLCV in the midgut cells of whiteflies, which includes the Arp2/3 complex, the sorting nexin Snx12 and the CORVET complex subunits Vps8, Vps11, and Vps33a. However, these subunits do not play a substantial role in the passage of TYLCV to primary salivary glands (Guo et al., 2023). It has been reported that two white fly proteins, viz., BtCUBN (B. tabaci cubilin) and BtAMN (B. tabaci amnionless), are involved in the formation of the receptor complex as BtCubam. In this complex, the BtCUBN functions as a viral binding region while BtAMN helps in interaction during clathrin-dependent endocytosis for membrane anchorage (Zhao et al., 2020). The BtCubam receptor complex then separates from virions in early endosomes and recycles to return to the apical membrane.

Vitellogenin (Vg), a vital insect reproductive protein, is considered to be produced by the fat body and then internalised by growing oocytes by endocytosis mediated by the Vg receptor (VgR) (Snigirevskaya & Raikhel, 2005). In the midgut of whitefly, vitellogenin interacts with the coat protein of TYLCV to form the TYLCV-Vg complex and further moves into the epithelial cells via endocytosis. The TYLCV-Vg complexes are internalized and transported through early endosomes, indicating a stepwise intracellular trafficking pathway (Wei et al., 2017). TYLCV was observed to colocalize with Vg at multiple locations on the lumen side of the microvillar membrane, within the membrane itself, in the cytoplasm, and near the basal membrane, suggesting that the virus binds to Vg at the microvillar surface and is subsequently transported as a complex toward the basal membrane for systemic movement.

11 Transmission Route

The transmission of ToLCNDV mainly occurs through the insect vector *B. tabaci* naturally, and they are not transmitted mechanically to the healthy host plants. However, an exception is the ToLCNDV oriental melon (OM) isolate, which can be mechanically enabled by the DNA-B encoded movement protein (MP) (Chang et al., 2023). Transmission by mechanical means was also successful in the germplasm breeding program of cucurbits (Lopez et al., 2015). Additionally, three Italian ToLCNDV isolates were successfully transmitted by artificial introduction to *Cucumis melo*, *Cucurbita pepo* and *Cucumis melo* var. cantalupensis (Panno et al., 2019).

Mechanical transmission of ToLCNDV in real field conditions is highly unlikely, although it can be efficiently achieved under experimental settings. There is a possible chance of mechanical transmission of ToLCNDV from infected plants during field practices, such as pruning, vegetative propagation, cuttings, and grafting; however, the possibility is unlikely. Although there is ongoing discussion regarding the transmission of begomoviruses through seeds (Kothandaraman et al., 2016), seeds of zucchini squash were able to transmit the ToLCNDV in Italy, which was a different isolate of ToLCNDV infecting chayote (*Sechium edule* L.) in India and Italy (Sangeetha et al., 2018; Kil et al., 2020). The possible explanation for the seed transmission of this virus might be that infected seeds from fallen fruits can germinate into virus-infected seedlings, indicating a potential route for vertical transmission. Cucurbit seedlings are exclusively grown in plant nurseries in the southern part of Spain; therefore, infected seedlings were another way that ToLCNDV spread throughout Spain. According to Juarez et al. (2019), the sale of infected seedlings through trade practices may be the cause of the establishment and predominance of ToLCNDV populations in central Spain.

12 Characteristics of *B. tabaci* to Be a Potential ToLCNDV Vector

Begomoviruses are transmitted by the efficient vector *B. tabaci* due to their high reproductive potential, rapid dispersal ability, and adaptability to diverse agro-ecological zones worldwide (Seal et al., 2006). Being a polyphagous pest, it can feed on a diverse range of plants by using its piercing-sucking mouthparts (Shah et al., 2015). Due to its capability of detoxifying and metabolizing plant compounds, *B. tabaci* aids in its adaptation to various hosts (Lin et al., 2017). Small RNAs help in modifying the host plant defense mechanism, which increases their survival and virus transmission potential (Islam et al., 2017). Being a phloem-feeding insect, *B. tabaci* helps in the rapid and efficient transmission of begomoviruses (Navas-Castillo et al., 2011).

13 Role of *B. tabaci* in Global Dispersal of ToLCNDV

The persistent nature of transport of ToLCNDV enables them to colonize in all life stages of *B. tabaci*, such as juvenile, pupal, and adult stages (Islam et al., 2017). The global spread of ToLCNDV is primarily transmitted by its insect vector, *B. tabaci*, along with human-mediated factors such as the exchange and trade of infested plant materials. The transport of infected plant materials not only contributes to virus transmission but also helps in the diversification of *B. tabaci* populations in the newly introduced areas (Polston & Capobianco, 2013). A recent example is the introduction of ToLCNDV from India to southern Spain, likely facilitated by the movement of *B. tabaci* through trade practices (Juarez et al., 2014). Since *B. tabaci* can acquire multiple virus types from various hosts and transfer them to new plants, it enables the long-distance transportation of begomoviruses. Sometimes, infected plants escape quarantine checks due to the absence of symptoms, which later acts as a key factor in the disease dissemination process unknowingly (Melgarejo et al., 2013). So, in areas where these plant trade activities are concentrated, the rapid acquisition and diffusion of begomoviruses by *B. tabaci* can result in the abrupt advent of new viral strains.

14 Factors Affecting Whitefly Population: Emergence and Evolution

The insect vector, *B. tabaci*, due to its polyphagous nature, particularly two strains, such as Mediterranean and Middle East-Asia Minor-1, is the reason for the rapid spread of ToLCNDV, which is now widely distributed across the globe (Lee et al., 2013; Boykin et al., 2012). The development and establishment of *B. tabaci* populations are influenced by several environmental factors, including high temperatures, varying humidity, and intensive land use activities such as irrigation in arid locations, as well as the homogeneous distribution of uniform horticultural and crop types. These factors lead to the dominance of the vector B. tabaci in transmitting ToLCNDV (Barboza et al., 2019; Zerbini et al., 2017; Bragard et al., 2020). The diversified landscape can influence the spreading rate of the begomovirus. According to the report by Rodelo-Urrego et al. (2013), the change in the land usage pattern, as well as the wild hosts inhabiting the land, can impact the population structure of begomovirus in *Capsicum species* in Mexico. Weed plants can also act as a driving force in the emergence of novel viral strains and outbreaks (Juarez et al., 2019).

Major factors contributing to the spread and efficacy of begomovirus include accessibility of a broad host range to the virus, number of efficient virus-transmitting vectors, dissemination by human activity, and chances of emerging new viral strains as a result of recombination (Varma & Malathi, 2003). ToLCNDV genomes are highly susceptible to recombination and mutation, allowing them to rapidly accumulate genetic diversity through rapid evolution (Sanchez-Campos et al., 2018; Ge et al., 2007). However, introgression or gene flow due to new host adaptation, agro-ecological variables, or even genetic drift through vector transmission are the reasons for the alteration of the genomic composition of ToLCNDV (Cai et al., 2023).

15 Disease Tolerance Mechanism of the Host Plant Against ToLCNDV

Plants generally employ the small RNA-based defense mechanisms, which have the potential to counteract the DNA viral transcripts and RNA viruses (Muthamilarasan & Prasad, 2013). This type of defense system generally consists of two processes, Transcriptional gene silencing (TGS) and post-transcriptional gene silencing (PTGS) to form small interfering RNAs (siRNAs) (Bian et al., 2006; Muthamilarasan & Prasad, 2013). According to the reports by higher accumulation of 24-nt IR-specific siRNAs and 21-nt AC1-specific siRNAs, as well as increased percentage of methylation in IR and AC1 regions resulted in nuclear protein binding and AC1, AC2, and AC3 gene expression, was observed in tolerant cultivars compared to susceptible ones when examined through mechanism involving post-transcriptional gene silencing and DNA methylation-specific RNA silencing, suggesting their substantial role in disease tolerance.

16 Conclusion and Prospects

An effective management strategy to suppress ToLCNDV infectivity needs a proper understanding of the complex process of virus vector interaction. To understand this virus vector interaction, many challenges need to be overcome despite outstanding progress. The challenges are gigantic genetic variation of ToLCNDV, enhanced virus adaptability and spread, and more complex and challenging management practices. In addition, ignorance about virus-vector co-evolution, intracellular trafficking and receptor-mediated viral acquisition highlights the call for additional research. Increased effort must be allocated to integrating molecular virology, vector ecology and omics-based technologies in attempting to unravel the multifaceted virus vector interactions. Progress in crop modelling, RNAi, and CRISPR-based gene editing technology may render targeted treatment for vector transmission possible. The production of resistant crop varieties through transgenic and gene-editing technologies is a feasible option to control the disease in the long term. The development of sustainable disease management systems to downregulate the ToLCNDV infection globally will need a multidisciplinary approach that combines ecological, proteomic, and genomic studies.

References

Agnihotri, A. K., Mishra, S. P., Tripathi, R. C., Ansar, M., Srivastava, A., & Tripathi, I. P. (2018). First natural co-occurrence of tomato leaf curl New Delhi virus DNA-A and chili leaf curl betasatellite on tomato plants (*Solanum lycopersicum* L.) in India. *Journal of General Plant Pathology, 84*(6), 414–417.

Akhtar, K. P., Akram, A., Ullah, N., Saleem, M. Y., & Saeed, M. (2019). Evaluation of Solanum species for resistance to tomato leaf curl New Delhi virus using chip grafting assay. *Scientia Horticulturae, 256*, 108646.

Akhter, A., Akhtar, S., Saeed, M., & Mansoor, S. (2014). Chili leaf curl betasatellite enhances symptoms induced by tomato leaf curl New Delhi virus a bipartite begomovirus. *International Journal of Agriculture and Biology, 16*(6), 1225–1228.

Anwar, S. (2017). Distinct association of an alphasatellite and a betasatellite with tomato leaf curl New Delhi virus in field-infected cucurbit. *Journal of General Plant Pathology, 83*, 185–188.

Anwar, I., Bukhari, H. A., Nahid, N., Rashid, K., Amin, I., Shaheen, S., et al. (2020). Association of cotton leaf curl Multan betasatellite and *Ageratum conyzoides* symptomless alphasatellite with tomato leaf curl New Delhi virus in Luffa cylindrica in Pakistan. *Australasian Plant Pathology, 49*, 25–29.

Ashwathappa, K. V., Venkataravanappa, V., Reddy, C. L., & Reddy, M. K. (2020). Association of Tomato leaf curl New Delhi virus with mosaic and leaf curl disease of Chrysanthemum and its whitefly cryptic species. *Indian Phytopathology, 73*, 533–542.

Aslam, M. Q., Akhtar, K. P., Akram, M. U. D., Saleem, M. Y., Sarwar, N., & Yousaf, S. (2017). Association of chili leaf curl betasatellite with tomato leaf curl disease. *Journal of General Plant Pathology, 83*, 402–405.

Barboza, N., Hernåndez, E., Inoue-Nagata, A. K., Moriones, E., & Hilje, L. (2019). Achievements in the epidemiology of begomoviruses and their vector *Bemisia tabaci* in Costa Rica. *Revista de Biología Tropical, 67*(3), 419–453.

Bian, X. Y., Rasheed, M. S., Seemanpillai, M. J., & Rezaian, M. A. (2006). Analysis of silencing escape of tomato leaf curl virus: An evaluation of the role of DNA methylation. *Molecular Plant-Microbe Interactions, 19*(6), 614–624.

Bibi, C., Khan, R. S., Rahat, M. A., Zahoor, M., Sher, H., Hussain, K., et al. (2024). First report of tomato leaf curl New Delhi virus (a bipartite begomovirus) from round gourd plant in swat region of Pakistan: Its phylogenetic and biogeographic analysis. *Advancements in Life Sciences, 11*(2), 398–405.

Boykin, L. M., Armstrong, K. F., Kubatko, L., & De Barro, P. (2012). Species delimitation and global biosecurity. *Evolutionary Bioinformatics, 8*, EBO-S8532.

Bragard, C., Dehnen-Schmutz, K., Di Serio, F., Gonthier, P., Jacques, M. A., et al. (2020). Pest categorisation of tomato leaf curl New Delhi virus. *EFSA Journal, 18*(7), e06179.

Briddon, R. W., Bull, S. E., Amin, I., Idris, A. M., Mansoor, S., Bedford, I. D., et al. (2003). Diversity of DNA β, a satellite molecule associated with some monopartite begomoviruses. *Virology, 312*(1), 106–121.

Brown, J. K., & Czosnek, H. (2002). Whitefly transmission of plant viruses.

Butterbach, P., Verlaan, M. G., Dullemans, A., Lohuis, D., Visser, R. G., Bai, Y., & Kormelink, R. (2014). Tomato yellow leaf curl virus resistance by ty-1 involves increased cytosine methylation of viral genomes and is compromised by cucumber mosaic virus infection. *Proceedings of the National Academy of Sciences, 111*(35), 12942–12947.

Cai, L., Mei, Y., Ye, R., Deng, Y., Zhang, X., Hu, Z., et al. (2023). Tomato leaf curl New Delhi virus: An emerging plant begomovirus threatening cucurbit production. *aBIOTECH, 4*(3), 257–266.

Chandler, J. W. (2016). Auxin response factors. *Plant, Cell & Environment, 39*(5), 1014–1028.

Chang, H. H., Gustian, D., Chang, C. J., & Jan, F. J. (2023). Virus-virus interactions alter the mechanical transmissibility and host range of begomoviruses. *Frontiers in Plant Science, 14*, 1092998.

Chellappan, P., Vanitharani, R., & Fauquet, C. M. (2005). MicroRNA-binding viral protein interferes with Arabidopsis development. *Proceedings of the National Academy of Sciences, 102*(29), 10381–10386.

Crespo, O., Robles, C., Ruiz, L., & Janssen, D. (2020). Antagonism of cucumber green mottle mosaic virus against tomato leaf curl New Delhi virus in zucchini and cucumber. *Annals of Applied Biology, 176*(2), 147–157.

Czosnek, H., Ghanim, M., & Ghanim, M. (2002). The circulative pathway of begomoviruses in the whitefly vector *Bemisia tabaci*—Insights from studies with tomato yellow leaf curl virus. *Annals of Applied Biology, 140*(3), 215–231.

Desbiez, C., Gentit, P., Cousseau-Suhard, P., Renaudin, I., & Verdin, E. (2021). First report of tomato leaf curl New Delhi virus infecting courgette in France. *New Disease Reports, 43*(2), e12006.

Ferro, C. G., Silva, J. P., Xavier, C. A. D., Godinho, M. T., Lima, A. T. M., Mar, T. B., et al. (2017). The ever increasing diversity of begomoviruses infecting non-cultivated hosts: New species from *Sida* spp. and *Leonurus sibiricus*, plus two New World alphasatellites. *Annals of Applied Biology, 170*(2), 204–218.

Fondong, V. N. (2013). Geminivirus protein structure and function. *Molecular Plant Pathology, 14*(6), 635–649.

Fondong, V. N. (2022). The diverse roles of the multifunctional C4/AC4 protein in geminivirus infection. In *Geminivirus: Detection, diagnosis and management* (pp. 309–322). Academic Press.

Fortes, I. M., Sánchez-Campos, S., Fiallo-Olivé, E., Díaz-Pendón, J. A., Navas-Castillo, J., & Moriones, E. (2016). A novel strain of tomato leaf curl New Delhi virus has spread to the Mediterranean basin. *Viruses, 8*(11), 307.

Gafni, Y., & Epel, B. L. (2002). The role of host and viral proteins in intra-and inter-cellular trafficking of geminiviruses. *Physiological and Molecular Plant Pathology, 60*(5), 231–241.

Gautam, S., Mugerwa, H., Buck, J. W., Dutta, B., Coolong, T., Adkins, S., & Srinivasan, R. (2022). Differential transmission of old and new world begomoviruses by middle East-Asia minor 1 (MEAM1) and Mediterranean (MED) cryptic species of *Bemisia tabaci*. *Viruses, 14*(5), 1104.

Ge, L., Zhang, J., Zhou, X., & Li, H. (2007). Genetic structure and population variability of tomato yellow leaf curl China virus. *Journal of Virology, 81*(11), 5902–5907.

Gray, S., Cilia, M., & Ghanim, M. (2014). Circulative, "nonpropagative" virus transmission: An orchestra of virus-, insect-, and plant-derived instruments. In *Advances in virus research* (Vol. 89, pp. 141–199). Academic Press.

Guo, Q., Ban, F. X., Xia, W. Q., Shu, Y. N., Liu, Y. Q., Liu, S. S., et al. (2023). The essential role of clathrin-mediated endocytosis and early endosomes in the trafficking of begomoviruses through the primary salivary glands of their whitefly vectors. *Journal of Virology, 97*(11), e01067–e01023.

Hanley-Bowdoin, L., Bejarano, E. R., Robertson, D., & Mansoor, S. (2013). Geminiviruses: Masters at redirecting and reprogramming plant processes. *Nature Reviews Microbiology, 11*(11), 777–788.

Hogenhout, S. A., Ammar, E. D., Whitfield, A. E., & Redinbaugh, M. G. (2008). Insect vector interactions with persistently transmitted viruses. *Annual Review of Phytopathology, 46*(1), 327–359.

Islam, W., Islam, S. U., Qasim, M., & Wang, L. (2017). Host-pathogen interactions modulated by small RNAs. *RNA Biology, 14*(7), 891–904.

Jamil, N., Rehman, A., Hamza, M., Hafeez, A., Ismail, H., Zubair, M., et al. (2017). First report of tomato leaf curl New Delhi virus, a bipartite begomovirus, infecting soybean (Glycine max). *Plant Disease, 101*(5), 845–845.

Janssen, D., Simón, A., Boulares, M., & Ruiz, L. (2022). Host species-dependent transmission of tomato leaf curl New Delhi virus-ES by *Bemisia tabaci*. *Plants, 11*(3), 390.

Jeevalatha, A., Vanishree, G., Siddappa, S., Kumar, R., Kaundal, P., Kumar, A., & Chakrabarti, S. K. (2021). Molecular characterization and infectivity analysis of tomato leaf curl New Delhi virus isolates infecting potato. *3 Biotech, 11*, 203. https://doi.org/10.1007/s13205-021-02752-5

Jeevalatha, A., Siddappa, S., Kumar, R., Tiwari, R. K., Lal, M. K., Sharma, S., Chakrabarty, S. K., & Singh, B. P. (2023). RNA-seq analysis reveals an early defense response to tomato leaf curl New Delhi virus in potato cultivar Kufri Bahar. *Functional & Integrative Genomics, 23*, 215. https://doi.org/10.1007/s10142-023-01138-5

Jeske, H. (2009). Geminiviruses. In *TT viruses: the still elusive human pathogens* (pp. 185–226). Springer.

Juarez, M., Tovar, R., Fiallo-Olivé, E., Aranda, M. A., Gosálvez, B., Castillo, P., et al. (2014). First detection of tomato leaf curl New Delhi virus infecting zucchini in Spain. *Plant Disease, 98*(6), 857–857.

Juarez, M., Rabadán, M. P., Martínez, L. D., Tayahi, M., Grande-Pérez, A., & Gómez, P. (2019). Natural hosts and genetic diversity of the emerging tomato leaf curl New Delhi virus in Spain. *Frontiers in Microbiology, 10*, 140.

Kassem, M. A., Juarez, M., Gómez, P., Mengual, C. M., Sempere, R. N., Plaza, M., et al. (2013). Genetic diversity and potential vectors and reservoirs of cucurbit aphid-borne yellows virus in southeastern Spain. *Phytopathology, 103*(11), 1188–1197.

Kheireddine, A., Sifres, A., Sáez, C., Picó, B., & López, C. (2019). First report of tomato leaf curl New Delhi virus infecting cucurbit plants in Algeria. *Plant Disease, 103*(12), 3291–3291.

Kil, E. J., Vo, T. T. B., Fadhila, C., Ho, P. T., Lal, A., Troiano, E., et al. (2020). Seed transmission of tomato leaf curl New Delhi virus from zucchini squash in Italy. *Plants, 9*(5), 563.

Kiran, G. V. M., Nagaraju, N., Girish, T. R., & Ashwini, B. N. (2021). Molecular investigations reveal bitter gourd crop is more susceptible to tomato leaf curl New Delhi virus infection in diverse crop cultivation practices. *3 Biotech, 11*, 1–9.

Kothandaraman, S. V., Devadason, A., & Ganesan, M. V. (2016). Seed-borne nature of a begomovirus, mung bean yellow mosaic virus in black gram. *Applied Microbiology and Biotechnology, 100*, 1925–1933.

Kumar, R., Tiwari, R. K., Jeevalatha, A., Sundaresha, S., Shah, M. A., Sharma, S., Sagar, V., Kumar, M., & Chakrabarti, S. K. (2021). Potato apical leaf curl disease: Current status and perspectives on a disease caused by tomato leaf curl New Delhi virus. *Journal of Plant Diseases and Protection, 128*, 897–911. https://doi.org/10.1007/s41348-021-00463-w

Kumar, R., Lal, M. K., Tiwari, R. K., Chourasia, K. N., Kumar, A., Kumar, R., Sharma, S., & Singh, B. (2023). Investigating the interplay between tomato leaf curl New Delhi virus infection, starch metabolism and antioxidant defence system in potato (*Solanum tuberosum* L.). *Antioxidants, 12*, 1447. https://doi.org/10.3390/antiox12071447

Lee, W., Park, J., Lee, G. S., Lee, S., & Akimoto, S. I. (2013). Taxonomic status of the Bemisia tabaci complex (Hemiptera: Aleyrodidae) and reassessment of the number of its constituent species. *PLoS One, 8*(5), e63817.

Li, S. B., Xie, Z. Z., Hu, C. G., & Zhang, J. Z. (2016). A review of auxin response factors (ARFs) in plants. *Frontiers in Plant Science, 7*, 47.

Lin, Y., Qasim, M., Hussain, M., Akutse, K. S., Avery, P. B., Dash, C. K., & Wang, L. (2017). The herbivore-induced plant volatiles methyl salicylate and menthol positively affect growth and pathogenicity of entomopathogenic fungi. *Scientific Reports, 7*(1), 40494.

Lopez, C., Ferriol, M., & Picó, M. B. (2015). Mechanical transmission of tomato leaf curl New Delhi virus to cucurbit germplasm: Selection of tolerance sources in Cucumis melo. *Euphytica, 204*, 679–691.

Luo, P., Di, D., Wu, L., Yang, J., Lu, Y., & Shi, W. (2022). MicroRNAs are involved in regulating plant development and stress response through fine-tuning of TIR1/AFB-dependent auxin signaling. *International Journal of Molecular Sciences, 23*(1), 510.

Mazumder, P., Mondal, F., Sarkar, M., Majumdar, A., Biswas, K. K., Sharma, S. K., Lal, M. K., Tiwari, R. K., Kumar, R., & Roy, A. (2024). Biological and molecular characterization of a new isolate of tomato mottle mosaic virus causing severe shoestring and fruit deformities in tomato plants in India. *Plants, 13*, 2811. https://doi.org/10.3390/plants13192811

Melgarejo, T. A., Kon, T., Rojas, M. R., Paz-Carrasco, L., Zerbini, F. M., & Gilbertson, R. L. (2013). Characterization of a new world monopartite begomovirus causing leaf curl disease of tomato in Ecuador and Peru reveals a new direction in geminivirus evolution. *Journal of Virology, 87*(10), 5397–5413.

Mercer, J., Schelhaas, M., & Helenius, A. (2010). Virus entry by endocytosis. *Annual Review of Biochemistry, 79*(1), 803–833.

Modgill, N., Naga, K. C., Firake, D. M., Tiwari, R. K., Shivaramu, S., Kumar, R., Bhatnagar, A., Buckseth, T., Bairwa, A., Maharana, C., Sharma, S., & Singh, B. (2025). Flupyradifurone application: A promising approach to curtail tomato leaf curl New Delhi virus (ToLCNDV) transmission vectored by *Bemisia tabaci* in potato farming. *Crop Protection, 187*, 106951.

Moriones, E., Praveen, S., & Chakraborty, S. (2017). Tomato leaf curl New Delhi virus: An emerging virus complex threatening vegetable and fiber crops. *Viruses, 9*(10), 264.

Muthamilarasan, M., & Prasad, M. (2013). Plant innate immunity: An updated insight into defense mechanism. *Journal of Biosciences, 38*, 433–449.

Naga, K. C., Subhash, S., Bhatnagar, A., Tiwari, R. K., Kumar, R., Verma, G., Sharma, S., & Chakrabarti, S. K. (2021a). Influence of host plants on virus acquisition and endosymbionts of whitefly, *Bemisia tabaci* (Gennadius). *Indian Journal of Entomology, 83*, 371–375. https://doi.org/10.5958/0974-8172.2020.00206.0

Naga, K. C., Sundaresha, S., Kumar, R., Tiwari, R. K., Subhash, S., Verma, G., Buckseth, T., Bairwa, A., Sharma, S., Katare, S., & Srivastava, R. M. (2021b). A new record of Asia II 5 genetic group of *Bemisia tabaci* (Gennadius) in the major potato growing areas of India and its relationship with tomato leaf curl New Delhi virus infecting potato. *3 Biotech, 11*, 421. https://doi.org/10.1007/s13205-021-02966-7

Naqvi, A. R., Haq, Q. M., & Mukherjee, S. K. (2010). MicroRNA profiling of tomato leaf curl New Delhi virus (ToLCNDV) infected tomato leaves indicates that deregulation of mir159/319 and mir172 might be linked with leaf curl disease. *Virology Journal, 7*, 1–16.

Navas-Castillo, J., Fiallo-Olivé, E., & Sánchez-Campos, S. (2011). Emerging virus diseases transmitted by whiteflies. *Annual Review of Phytopathology, 49*(1), 219–248.

Neoh, Z. Y., Lai, H. C., Lin, C. C., Suwor, P., & Tsai, W. S. (2023). Genetic diversity and geographic distribution of cucurbit-infecting begomoviruses in The Philippines. *Plants, 12*(2), 272.

Orfanidou, C. G., Malandraki, I., Beris, D., Kektsidou, O., Vassilakos, N., Varveri, C., et al. (2019). First report of tomato leaf curl New Delhi virus in zucchini crops in Greece. *Journal of Plant Pathology, 101*, 799–799.

Padidam, M., Beachy, R. N., & Fauquet, C. M. (1995). Tomato leaf curl geminivirus from India has a bipartite genome and coat protein is not essential for infectivity. *Journal of General Virology, 76*(1), 25–35.

Pakkianathan, B. C., Kontsedalov, S., Lebedev, G., Mahadav, A., Zeidan, M., Czosnek, H., & Ghanim, M. (2015). Replication of tomato yellow leaf curl virus in its whitefly vector, *Bemisia tabaci*. *Journal of Virology, 89*(19), 9791–9803.

Pan, L. L., Chen, Q. F., Zhao, J. J., Guo, T., Wang, X. W., Hariton-Shalev, A., et al. (2017). Clathrin-mediated endocytosis is involved in tomato yellow leaf curl virus transport across the midgut barrier of its whitefly vector. *Virology, 502*, 152–159.

Panno, S., Caruso, A. G., Troiano, E., Luigi, M., Manglli, A., Vatrano, T., et al. (2019). Emergence of tomato leaf curl New Delhi virus in Italy: Estimation of incidence and genetic diversity. *Plant Pathology, 68*(3), 601–608.

Pant, R. P., Bhatnagar, A., & Lal, M. (2018). Role of alternate host plants in the transmission of apical leaf curl disease of potato caused by tomato leaf curl New Delhi virus-potato (ToLCNDV-pot.) in northern India. *Indian Journal of Agricultural Sciences, 88*(8), 1258–1262.

Parrella, G., Troiano, E., Lee, S., & Kil, E. J. (2020). Tomato leaf curl New Delhi virus found associated with eggplant yellowing disease in Italy. *Plant Disease, 104*(2034), 10–1094.

Phaneendra, C., Rao, K. S., Jain, R. K., & Mandal, B. (2012). Tomato leaf curl New Delhi virus is associated with pumpkin leaf curl: A new disease in northern India. *Indian Journal of Virology, 23*, 42–45.

Piroux, N., Saunders, K., Page, A., & Stanley, J. (2007). Geminivirus pathogenicity protein C4 interacts with Arabidopsis thaliana shaggy-related protein kinase AtSKη, a component of the brassinosteroid signalling pathway. *Virology, 362*(2), 428–440.

Polston, J. E., & Capobianco, H. (2013). Transmitting plant viruses using whiteflies. *Journal of Visualized Experiments*, (81), e4332.

Prasanna, H. C., Sinha, D. P., Rai, G. K., Krishna, R., Kashyap, S. P., Singh, N. K., et al. (2015). Pyramiding T y-2 and T y-3 genes for resistance to monopartite and bipartite tomato leaf curl viruses of India. *Plant Pathology, 64*(2), 256–264.

Rajeshwari, R., & Reddy, M. K. (2014). Biological characterisation of tomato leaf curl New Delhi virus infecting bottle gourd (*Lagenaria siceraria*) from Karnataka. *Mysore Journal of Agricultural Sciences, 48*(3), 387–393.

Rana, V. S., Popli, S., Saurav, G. K., Raina, H. S., Chaubey, R., Ramamurthy, V. V., & Rajagopal, R. (2016). A *Bemisia tabaci* midgut protein interacts with begomoviruses and plays a role in virus transmission. *Cellular Microbiology, 18*(5), 663–678.

Ranjan, P., Singh, A. K., Kumar, R. V., Basu, S., & Chakraborty, S. (2014). Host-specific adaptation of diverse betasatellites associated with distinct Indian tomato-infecting begomoviruses. *Virus Genes, 48*, 334–342.

Rodelo-Urrego, M., Pagán, I., González-Jara, P., Betancourt, M., Moreno-Letelier, A., Ayllón, M. A., et al. (2013). Landscape heterogeneity shapes host-parasite interactions and results in apparent plant–virus codivergence. *Molecular Ecology, 22*(8), 2325–2340.

Ruiz, L., Simon, A., Velasco, L., & Janssen, D. (2017). Biological characterization of tomato leaf curl New Delhi virus from Spain. *Plant Pathology, 66*(3), 376–382.

Saeed, M. (2010). Tomato leaf curl New Delhi virus DNA A component and cotton leaf curl Multan betasatellite can cause mild transient symptoms in cotton. *Acta Virologica, 54*(4), 317.

Sanchez-Campos, S., Domínguez-Huerta, G., Díaz-Martínez, L., Tomás, D. M., Navas-Castillo, J., Moriones, E., & Grande-Pérez, A. (2018). Differential shape of geminivirus mutant spectra across cultivated and wild hosts with invariant viral consensus sequences. *Frontiers in Plant Science, 9*, 932.

Sangeetha, B., Malathi, V. G., Alice, D., Suganthy, M., & Renukadevi, P. (2018). A distinct seed-transmissible strain of tomato leaf curl New Delhi virus infecting Chayote in India. *Virus Research, 258*, 81–91.

Sarkar, M., Gupta, D., Singh, O. W., Paul, S., Kumar, R., Mandal, B., & Roy, A. (2024). Identification of pathogenicity determinants in ToLCNDV and their RNAi-based knockdown for disease management in *Nicotiana benthamiana* and tomato plants. *Frontiers in Microbiology, 15*, 1481523. https://doi.org/10.3389/fmicb.2024.1481523

Schommer, C., Palatnik, J. F., Aggarwal, P., Chételat, A., Cubas, P., Farmer, E. E., et al. (2008). Control of jasmonate biosynthesis and senescence by miR319 targets. *PLoS Biology, 6*(9), e230.

Seal, S. E., VandenBosch, F., & Jeger, M. J. (2006). Factors influencing begomovirus evolution and their increasing global significance: Implications for sustainable control. *Critical Reviews in Plant Sciences, 25*(1), 23–46.

Shah, M. M. R., Zhang, S., & Liu, T. (2015). Whitefly, host plant and parasitoid: A review on their interactions. *Asian Journal of Applied Science and Engineering, 4*(1), 48–61.

Singh, A. K., Kushwaha, N., & Chakraborty, S. (2016). Synergistic interaction among begomoviruses leads to the suppression of host defense-related gene expression and breakdown of resistance in chilli. *Applied Microbiology and Biotechnology, 100*, 4035–4049.

Siskos, L., Cui, L., Wang, C., Visser, R. G., Bai, Y., & Schouten, H. J. (2022). A new challenge in melon resistance breeding: The ToLCNDV case. *Euphytica, 218*(9), 129.

Sivalingam, P. N., & Varma, A. (2012). Role of betasatellite in the pathogenesis of a bipartite begomovirus affecting tomato in India. *Archives of Virology, 157*, 1081–1092.

Snigirevskaya, E. S., & Raikhel, A. S. (2005). Receptor-mediated endocytosis of yolk proteins in insect oocytes. In *Reproductive biology of invertebrates* (Vol. 12, Part B, pp. 215–244). CRC Press.

Tolin, S. A., & Fayad, A. (2016). Virus diseases of tropical vegetable crops and their management. In *Integrated pest management of tropical vegetable crops* (pp. 41–76). Springer.

Troiano, E., & Parrella, G. (2023). First report of tomato leaf curl New Delhi virus in *Lagenaria siceraria* var. longissima in Italy. *Phytopathologia Mediterranea, 62*(1), 17–24.

Van Vu, T., Choudhury, N. R., & Mukherjee, S. K. (2013). Transgenic tomato plants expressing artificial microRNAs for silencing the pre-coat and coat proteins of a begomovirus, tomato leaf curl New Delhi virus, show tolerance to virus infection. *Virus Research, 172*(1–2), 35–45.

Varma, A., & Malathi, V. G. (2003). Emerging geminivirus problems: A serious threat to crop production. *Annals of Applied Biology, 142*(2), 145–164.

Venkataravanappa, V., Reddy, C. L., Saha, S., & Reddy, M. K. (2018). Recombinant tomato leaf curl New Delhi virus is associated with yellow vein mosaic disease of okra in India. *Physiological and Molecular Plant Pathology, 104*, 108–118.

Vignesh, S., Renukadevi, P., Nagendran, K., Senthil, N., Kumar, R. V., SwarnaPriya, R., et al. (2023). A distinct strain of tomato leaf curl New Delhi virus that causes mosaic disease in ash gourd and other cucurbitaceous crops. *Frontiers in Microbiology, 14*, 1268333.

Vinutha, T., Kumar, G., Garg, V., Canto, T., Palukaitis, P., Ramesh, S. V., & Praveen, S. (2018). Tomato geminivirus encoded RNAi suppressor protein, AC4 interacts with host AGO4 and precludes viral DNA methylation. *Gene, 678*, 184–195.

Vinutha, T., Vanchinathan, S., Bansal, N., Kumar, G., Permar, V., Watts, A., et al. (2020). Tomato auxin biosynthesis/signaling is reprogrammed by the geminivirus to enhance its pathogenicity. *Planta, 252*, 1–14.

Vo, T. T., Lal, A., Nattanong, B., Tabassum, M., Qureshi, M. A., Troiano, E., et al. (2023). Coat protein is responsible for tomato leaf curl New Delhi virus pathogenicity in tomato. *Frontiers in Plant Science, 14*, 1206255.

Voinnet, O., Pinto, Y. M., & Baulcombe, D. C. (1999). Suppression of gene silencing: A general strategy used by diverse DNA and RNA viruses of plants. *Proceedings of the National Academy of Sciences, 96*(24), 14147–14152.

Von Arnim, A., & Stanley, J. (1992). Determinants of tomato golden mosaic virus symptom development located on DNA B. *Virology, 186*(1), 286–293.

Wege, C. (2007). Movement and localization of tomato yellow leaf curl viruses in the infected plant. In *Tomato yellow leaf curl virus disease: Management, molecular biology, breeding for resistance* (pp. 185–206). Springer Netherlands.

Wei, J., He, Y. Z., Guo, Q., Guo, T., Liu, Y. Q., Zhou, X. P., et al. (2017). Vector development and vitellogenin determine the transovarial transmission of begomoviruses. *Proceedings of the National Academy of Sciences, 114*(26), 6746–6751.

Whitfield, A. E., Falk, B. W., & Rotenberg, D. (2015). Insect vector-mediated transmission of plant viruses. *Virology, 479*, 278–289.

Xia, W. Q., Liang, Y., Chi, Y., Pan, L. L., Zhao, J., Liu, S. S., & Wang, X. W. (2018). Intracellular trafficking of begomoviruses in the midgut cells of their insect vector. *PLoS Pathogens, 14*(1), e1006866.

Yamamoto, H., Wakita, Y., Kitaoka, T., Fujishiro, K., Kesumawati, E., & Koeda, S. (2021). Southeast Asian isolate of the tomato leaf curl New Delhi virus shows higher pathogenicity against tomato and cucurbit crops compared to that of the Mediterranean isolate. *The Horticulture Journal, 90*(3), 314–325.

Yazdani-Khameneh, S., Aboutorabi, S., Shoori, M., Aghazadeh, A., Jahanshahi, P., Golnaraghi, A., & Maleki, M. (2016). Natural occurrence of tomato leaf curl New Delhi virus in Iranian cucurbit crops. *The Plant Pathology Journal, 32*(3), 201.

Zaidi, S. S. E. A., Martin, D. P., Amin, I., Farooq, M., & Mansoor, S. (2017). Tomato leaf curl New Delhi virus: A widespread bipartite begomovirus in the territory of monopartite begomoviruses. *Molecular Plant Pathology, 18*(7), 901–911.

Zerbini, F. M., Briddon, R. W., Idris, A., Martin, D. P., Moriones, E., Navas-Castillo, J., et al. (2017). ICTV virus taxonomy profile: Geminiviridae. *Journal of General Virology, 98*(2), 131–133.

Zhao, J., Lei, T., Zhang, X. J., Yin, T. Y., Wang, X. W., & Liu, S. S. (2020). A vector whitefly endocytic receptor facilitates the entry of begomoviruses into its midgut cells via binding to virion capsid proteins. *PLoS Pathogens, 16*(12), e1009053.

Seed Transmission: A New Tool in the Weaponry of Tomato Leaf Curl New Delhi Virus

5

P. Renukadevi, S. Kousalya, M. Gayathri, R. Sharanya, A. Kokila, B. Sangeetha, and V. G. Malathi

Abstract

Whitefly-transmitted begomoviruses, the largest group of plant viruses, comprising about 445 members, affect the productivity of a wide range of crops. In the last decade, the most essential begomoviruses of considerable economic importance were the African cassava mosaic virus and tomato yellow leaf curl virus due to the losses they caused. However, in recent years, due to transboundary movement to Mediterranean and Southeast Asian regions from India and the expanding host range, tomato leaf curl New Delhi virus (ToLCNDV) has emerged as the most devastating pathogen. The active spread and increased virulence of the begomoviruses are generally attributed to their genetic propensity for recombination and mutation, their ability to capture satellite DNAs and their survival in alternative hosts. Another unexpected new dimension, for enhanced pathogenic potential, is the seed transmission witnessed in selected begomoviruses, including ToLCNDV. Though it is still debated and the mechanism of seed transmission in phloem-bound DNA viruses is not yet resolved, seed transmission of ToLCNDV in cucurbitaceous hosts has been well established. The term seed transmission is used to describe a situation where the embryo is infected, and the germinating seedlings contain the virus. It is essential to differentiate this from seed-borne nature, where the virus is present in the seed coat or other maternal tissue, and contamination of the seedling from the externally borne virus inoculum results in infections. In the case of begomoviruses, there have been many

P. Renukadevi (✉) · S. Kousalya · M. Gayathri · R. Sharanya · A. Kokila · B. Sangeetha
Department of Plant Pathology, Centre for Plant Protection Studies, Tamil Nadu Agricultural University, Coimbatore, Tamil Nadu, India
e-mail: renukadevi.p@tnau.ac.in

V. G. Malathi
Retired Scientist, Division of Plant Pathology, ICAR-Indian Agricultural Research Institute, New Delhi, Delhi, India

R. Kumar et al. (eds.), *Tomato Leaf Curl New Delhi Virus (ToLCNDV)*,
https://doi.org/10.1007/978-981-95-2852-3_5

controversies about seed-borne viruses being mistaken for seed-transmitted ones. However, in the case of ToLCNDV in cucurbitaceous hosts, it has been proven beyond doubt that the virus is seed-transmitted. The seed transmission is of any consequence only when the seedlings raised from infected seeds serve as inoculum and contribute to the active buildup of the virus population. These aspects of seed-borne inoculum contributing to ToLCNDV infection establishment are discussed in this chapter.

Keywords

Begomoviruses · ToLCNDV · Seed transmission · Cucurbitaceous · Epidemiology · Whitefly

1 Introduction

The emergence of plant virus diseases at epidemic levels poses a significant challenge to global agricultural productivity. Viral infections affect major food and feed crops, as well as fiber and ornamental plants, leading to substantial yield losses and quality reduction. Among these, seed transmission, also known as vertical transmission, plays a critical role in the persistence and global spread of plant viruses. This process involves the passage of viruses from infected parent plants to their progeny via seeds (Hull, 2013). Seed transmission has been documented in approximately 18% of plant RNA viruses, including members of the families *Potyviridae*, *Tobamoviridae*, *Comoviridae*, *Flexiviridae*, *Partitiviridae*, and viroids. Several economically important viruses, such as lettuce mosaic virus (LMV), pea seed-borne mosaic virus (PSbMV), barley stripe mosaic virus (BSMV), potato virus Y (PVY), plum pox virus (PPV), and wheat streak mosaic virus (WSMV), are known to be seed transmitted (Stace Smith & Hamilton, 1988; Cambra et al., 2006; Valkonen, 2007; Shanmugam et al., 2024). These viruses can infect embryos during their formation and persist through germination, allowing infected seedlings to emerge and act as sources of inoculum for vectors. While historically, seed transmission is associated with RNA viruses, seed transmission among DNA viruses is an emerging issue. The seed-borne nature of cocoa swollen shoot virus was long suspected and later confirmed (Quainoo et al., 2008). Similarly, early suspicions of geminiviruses being seed transmitted dating back to the 1930s (Keur, 1934; Bennett & Esau, 1936) have now been substantiated by molecular evidence showing their presence in seed tissues (Kim et al., 2015). This has prompted new interest in the seedborne virus as an inoculum in the epidemiological spread of DNA viruses. The economic implications of seed-transmitted viruses are far-reaching. Globally, plant virus-related losses are estimated at over US$30 billion annually (Jones & Naidu, 2019). Seed transmission facilitates early-stage infection, often resulting in systemic disease and yield losses ranging from 35% to 98% for some viruses like bean common mosaic virus and bean yellow mosaic virus (Albrechtsen, 2006). Moreover, virus-laden seeds serve as reservoirs of inoculum between growing seasons, and their distribution through international trade enhances the risk of introducing pathogens into new regions. This underlines the critical importance of seed health testing, molecular diagnostics, certification schemes, and quarantine enforcement to curb cross-border virus dissemination. For example, the detection of mungbean yellow mosaic India virus (MYMIV), a begomovirus in yardlong beans in Indonesia, highlights the risk of cross-border movement via seeds (Sutrawati & Aulia, 2024). One of the key biological barriers to seed transmission in the case of begomoviruses is its confined presence only in phloem tissue. Many viruses are phloem-limited and require specific adaptations to move into reproductive tissues and invade seeds. This limitation was previously believed to reduce the likelihood of seed transmission; however, emerging data challenge this assumption, especially for viruses that can overcome this barrier and systemically infect seeds. Among the most impactful DNA viruses, the group that shows seed transmission is the begomoviruses. Members of the genus *Begomovirus* belong to the family *Geminiviridae*. With over 445 species (Current ICTV Taxonomy Release | ICTV), begomoviruses are characterized by their twinned icosahedral particles encapsidating circular, single-stranded DNA genomes (Fig. 5.1). They are categorized as monopartite or bipartite, with the latter comprising DNA A encoding genes required for replication, encapsidation and gene expression and DNA B encoding genes for intra- and inter-cellular movement.

An important example is tomato leaf curl New Delhi virus (ToLCNDV), a bipartite begomovirus that affects solanaceous and cucurbitaceous crops across tropical and subtropical regions. The recent confirmation of seed-borne transmission in begomoviruses like ToLCNDV has raised alarm due to its implications for long-distance movement and management challenges. This chapter explores the mechanisms, evidence, and epidemiological consequences of seed transmission with a particular focus on ToLCNDV, an important begomovirus. It highlights the significance of seed transmission as a novel and significant tool in the virus weaponry,

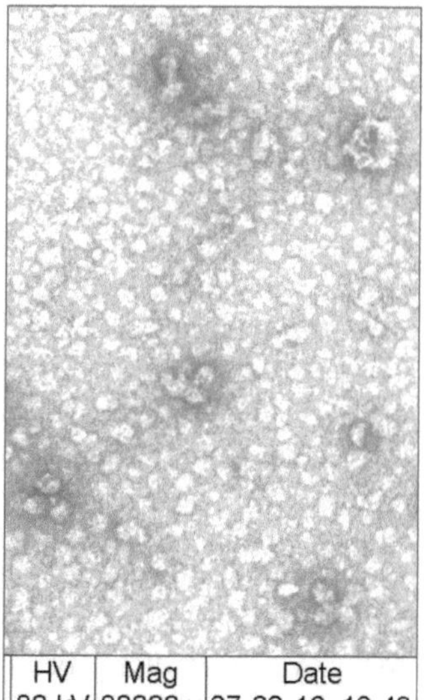

Fig. 5.1 Electron micrograph of ToLCNDV particle isolated from infected chayote

allowing for its persistence and spread across borders. The chapter discusses the molecular evidence supporting seed transmission of ToLCNDV, recent cases of its spread through international seed trade, and management strategies aimed at preventing further dissemination. With an emphasis on molecular detection techniques and quarantine measures, this chapter sheds light on the growing challenge posed by ToLCNDV and its impact on agricultural biosecurity.

2 Emergence of Seed Transmission in Begomoviruses

2.1 Early Views and Historical Evidence

For decades, seed transmission of begomoviruses, a genus within the family *Geminiviridae*, was considered implausible. This belief stemmed from the known tissue tropism of these viruses, which are restricted to phloem-associated cells such as companion cells and phloem parenchyma. Because seeds were thought to lack direct vascular connections and plasmodesmatal continuity with the rest of the plant, the possibility of virus movement into embryonic tissues seemed unlikely (Renukadevi et al., 2022).

2.2 Recent Discoveries and Expanding Host Range

This perspective began to change with breakthroughs in the mid-2010s. In 2015, Kim et al. provided the first clear evidence of seed transmission in a begomovirus, demonstrating that sweet potato leaf curl virus (SPLCV) could be transmitted through seeds in *Ipomoea batatas*. This was a pivotal moment that challenged existing dogma. The following year, Kil et al. (2016) reported seed transmission of tomato yellow leaf curl virus (TYLCV) in tomato, and subsequently in soybean (*Glycine max*) and sweet pepper (*Capsicum annuum*) (Kil et al., 2017, 2018). Parallel studies have expanded the list of seed transmissible begomoviruses. Mungbean yellow mosaic virus (MYMV) was detected in seed parts of urdbean (*Vigna mungo*) by Kothandaraman et al. (2016). In 2018, Suruthi et al. demonstrated seed transmission of dolichos yellow mosaic virus (DoYMV) in lablab bean (*Lablab purpureus*), while bitter gourd yellow mosaic virus (BgYMV) showed a seed infectivity rate of 79.16% and a seedling transmission rate of 32.05% in bitter gourd (*Momordica charantia*) (Manivannan et al., 2019). In 2018, ToLCNDV was reported to be seed transmissible in chayote (*Sechium edule*) by Sangeetha et al. (2018) and again in bitter gourd by Gomathi Devi et al. (2023). The list continued to grow with pepper yellow leaf curl Indonesia virus (PepYLCIV) in pepper (*Capsicum annuum*) (Fadhila et al., 2020), sweet potato symptomless virus 1 (SPSMV-1) in sweet potato (Qiao et al., 2020), and ToLCNDV in zucchini squash (*Cucurbita pepo*) (Kil et al., 2020) and mungbean yellow mosaic India virus (MYMIV) in yardlong bean (*Vigna unguiculata subsp. sesquipedalis*) (Mulyadi et al., 2021).

3 ToLCNDV

ToLCNDV is primarily transmitted by the whitefly, *Bemisia tabaci*, in a persistent circulative non-propagative manner (Sarkar et al., 2024). While mechanical and graft transmission have been demonstrated in controlled conditions, whitefly transmission remains the primary natural route, enabling the virus to spread rapidly and cause severe outbreaks in both open field and greenhouse systems (Kumar et al., 2021; Naga et al., 2021a). Seed transmission of ToLCNDV has been studied in various crops, particularly cucurbits such as bitter gourd, chayote, melon, potato, and zucchini squash. These studies have revealed varying levels of seed transmission across host species, indicating the potential for the virus to be seed-borne and seed-transmitted. Molecular investigations have led to the cloning, sequencing, and identification of the virus genome, confirming that the virus was cloned from the embryo belonging to the ToLCNDV (Jeevalatha et al., 2021, 2023).

3.1 Chayote (*Sechium edule*)

Chayote (*Sechium edule* (Jacq.) Sw.) is an important cucurbitaceous vegetable crop, primarily cultivated for its edible fruits, which are consumed globally. During 2015–2016, reports from the hilly regions of Tamil Nadu, India, highlighted a severe outbreak of viral disease in chayote plants. The plants exhibited typical symptoms of mosaic, yellow spots, yellowing, leaf curling, puckering, and enations (Fig. 5.2). Additionally, the fruits of the infected plants were severely malformed, rendering them unmarketable. Disease incidence varied widely across different regions, with reported rates ranging from 50.3% to 100%, and the overall yield loss reached an alarming 69.9% in Dindigul district (Sangeetha et al., 2018).

Interestingly, the chayote isolates of ToLCNDV DNA A exhibited only 91–92% sequence identity with isolates from ridge gourd in South India, suggesting that the chayote variant represents a distinct strain of ToLCNDV. Phylogenetic analysis, which compared the DNA A component of the virus, revealed that the chayote, ridge gourd, and ash gourd isolates from India shared a common origin with Spain

Fig. 5.2 Different symptoms of ToLCNDV infection in Chayote. (1) Severe mosaic in young seedling; (2) blistering; (3) enations; (4) malformed fruit

(ToLCNDV-ES) isolates of ToLCNDV. However, these Indian isolates diverged significantly from isolates found in other countries, highlighting regional variation in the virus (Sangeetha et al., 2018). Further studies on the vector transmission of the chayote ToLCNDV showed that *B. tabaci* could effectively transmit ToLCNDV to bottle gourd within a 4-h acquisition access period (AAP) followed by a 24-h inoculation feeding period (IFP). This confirmed the whitefly's role as a vector for ToLCNDV in cucurbit crops. Additionally, an increased greenhouse whitefly population (*Trialeurodes vaporariorum*) was found on infected chayote plants in Kodaikanal, prompting studies into the vector capacity of this species. Using PCR assays, the field population of *T. vaporariorum* was confirmed to be viruliferous, and when reared in controlled greenhouse conditions, these whiteflies transmitted the virus efficiently, similar to *B. tabaci*, further establishing their role as vectors of ToLCNDV (Sangeetha et al., 2018).

3.1.1 Evidence of Seed Transmission and Confirmation by Molecular Tools

One of the most significant findings of the study conducted by Sangeetha et al. (2018) was the first-ever report of the seed-borne nature of ToLCNDV in chayote. This discovery has important implications for the management of the virus in agricultural practices, particularly in regions where chayote is propagated through fruits (in situ germination of the embryo while within the fruit) (Fig. 5.3a). To investigate the potential for seed transmission, the researchers collected fruits from infected plants. They dissected them into fruit parts, including pericarp, mesocarp, seed coat, endosperm, and embryo, from different locations, including Thandikudi, Adalur, Pudhur, and Palaiyur (Fig. 5.3b). PCR analysis, using both Rojas degenerate primers (PAR1v 772 and PAL1c) and ToLCNDV specific primers, revealed that the virus was present in the seeds, particularly in the embryos, in 23 of the 30 fruit samples analyzed (Fig. 5.3c). In contrast, no viral DNA was detected in the pericarp and mesocarp tissues of samples collected from Kodaikanal, indicating that the presence within fruit tissue may vary from one location to another (Tables 5.1 and 5.3). Furthermore, no viral presence was detected in symptom-free chayote plants from Kotagiri, highlighting the importance of field surveillance in identifying infected plants (Sangeetha et al., 2018).

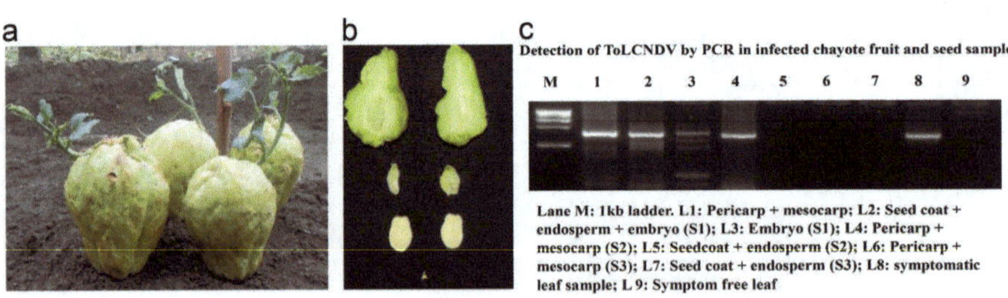

Fig. 5.3 (**a**–**c**) Detection of ToLCNDV in different fruit and seed parts. (**a**) In situ germination of chayote fruit; (**b**) dissection of fruit and seed parts; (**c**) confirmation of ToLCNDV in seed parts

Table 5.1 Evidence of seed transmission of tomato leaf curl New Delhi virus (ToLCNDV) across different host plants

Host plant	Seed-borne	Seed transmitted	Evidence type	Transmission rate (%)	Key references
Chayote (*Sechium edule*)	Yes	Yes (limited)	PCR, RCA, grow-out	25	Sangeetha et al. (2018)
Zucchini squash (*Cucurbita pepo*)	Yes	Yes (strong)	PCR, grow-out, sap inoculation	61.36	Kil et al. (2020)
Bitter gourd (*Momordica charantia*)	Yes	Yes (strong)	PCR, DAS-ELISA, grow-out, vector study	Up to 43.2	Gomathi Devi et al. (2023)

Table 5.2 Evidence of only seed association of ToLCNDV in cucurbitaceous hosts

Host plant	Seed-borne	Seed transmitted	Evidence type	Transmission rate (%)	Key references
Melon (*Cucumis melo*)	Yes	No	PCR, qPCR, seed dissection, grow-out	0	Fortes et al. (2023)
Cucumber, Winter squash	Yes	Unclear	PCR, qPCR, grow-out	Not quantified	Sáez et al. (2023)
Watermelon (*Citrullus lanatus*)	No	No	PCR	0	Sáez et al. (2023)

Table 5.3 Summary of experimental methods and outcome of seed transmission studies of ToLCNDV

Study (author, year)	Host/sample type	Detection methods	Systemic infection	Replicative DNA forms	Interpretation
Sangeetha et al. (2018)	Chayote embryo, seed coat	PCR, RCA	Yes (25%)	Yes (RCA)	Confirmed transmission
Kil et al. (2020)	Zucchini squash Seedling, embryo	PCR, sap inoculation	Yes (61.36%)	Not done	Likely transmission
Gomathi Devi et al. (2023)	Bitter gourd Seedling, embryo	PCR, DAS-ELISA, grow-out	Yes (up to 43.2%)	Not confirmed	Strong evidence
Sáez et al. (2023)	Water melon Seed coat, embryo	qPCR, RCA, Southern blot	No	No	Not seed transmitted
Fortes et al. (2023)	Melon Dissected seed	PCR, qPCR, histological assay	No	No	Contamination only

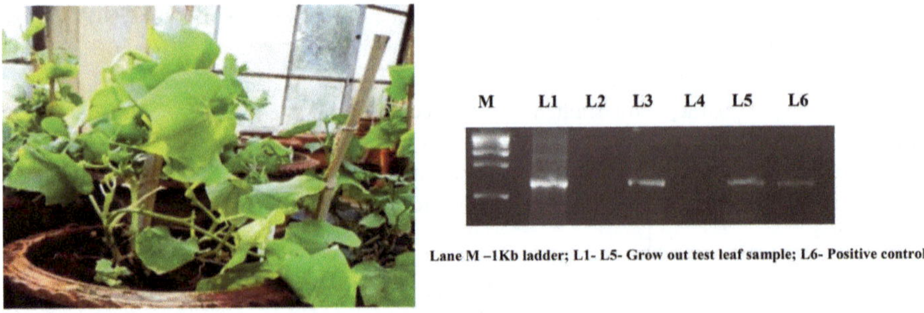

a. Grow test plant (asymptomatic); b. Molecular confirmation of ToLCNDV in grow out test leaf sample

Fig. 5.4 Asymptomatic PCR positive (ToLCNDV) grow-out test chayote plant. (**a**) Grow test plant (asymptomatic); (**b**) molecular confirmation of ToLCNDV in the grow-out test leaf sample

In addition, the study extended its investigation into the seed markets of Tamil Nadu, where 20 samples of fruits with in situ germinating seeds were collected from regions such as Sirumalai and Pannaikaadu. Remarkably, 100% of the seed samples tested positive for ToLCNDV, underlining the risk of virus transmission through infected seeds being sold in the market (Sangeetha et al., 2018). To assess the potential for seed transmission of ToLCNDV, 25 infected seeds from different regions were germinated in an insect-proof net house to prevent vector-mediated transmission. Of the 25 seeds planted, 12 germinated and showed good crop growth. However, they exhibited dark green, leathery foliage, which was distinct from the normal pale green leaves typically seen in healthy chayote plants from non-infected regions. Although these seedlings appeared symptom-free, PCR tests conducted 60 days after sowing revealed that the virus was present in 3 of the 12 plants, confirming a 25% seed transmission rate. These findings suggest that even symptom-free plants may harbor the virus, which poses a significant challenge for virus management strategies in seed production (Sangeetha et al., 2018) (Fig. 5.4).

Furthermore, to understand whether the viral genome in seed tissues was intact, the researchers performed rolling circle amplification (RCA) on DNA extracted from the embryos of infected seeds from the Adalur region. The subsequent cloning and sequencing of the DNA-A component showed 95% sequence identity with viral isolates derived from leaf samples of infected chayote plants, confirming that the full-length viral genome was present in the seed tissues. This molecular evidence firmly establishes the seed transmission of ToLCNDV in chayote as a critical route for virus spread (Sangeetha et al., 2018).

3.2 Zucchini Squash (*Cucurbita pepo* L.)

Kil et al. (2020) reported that ToLCNDV has emerged as a major threat to zucchini squash (*Cucurbita pepo*) cultivation in Italy. Initially restricted to the Indian subcontinent, ToLCNDV was first detected in Spain in 2016 and has since spread to other Mediterranean countries, including Tunisia and Italy, where it causes significant yield and quality losses (Panno et al., 2016). In surveys conducted in the Campania region of southern Italy, symptomatic zucchini plants were found with severe leaf curling, yellow mosaic, vein swelling, shortened internodes, and fruit deformation, which are hallmark symptoms of ToLCNDV infection. Molecular analysis of symptomatic plants using PCR with ToLCNDV-specific primers targeting the DNA A component yielded amplicons that matched the Mediterranean strain ToLCNDV ES (GenBank: MF688670), which shares ~90% nucleotide identity with Asian strains, indicating regional evolution. In open fields, Kil et al. (2020) also observed young seedlings germinating from the fallen fruits of ToLCNDV-infected plants from the previous season. All five such self-sown seedlings tested were PCR positive for ToLCNDV. However, due to the natural infestation of whiteflies (*Bemisia tabaci*) in these fields, it remained unclear whether infection occurred post-emergence via vectors or through vertical transmission from the seeds themselves. This prompted further controlled studies under insect-proof conditions to examine the possibility of accurate seed transmission.

3.2.1 Confirmation of Seed-Borne Nature and Infectivity in Zucchini Squash

To verify the seed-borne nature of ToLCNDV, Kil et al. (2020) collected seeds from infected fruits of two zucchini squash cultivars, Milos and Ortano. Thirty seeds from Milos and ten from Ortano were surface sterilized using ethanol and sodium hypochlorite to eliminate any external contamination. DNA extracted from the sterilized seeds tested 100% positive for ToLCNDV by PCR, strongly suggesting internal viral presence. To confirm vertical transmission, 44 seeds (28 from Milos and 16 from Ortano) were germinated in insect-proof chambers. Seedlings were sampled between 1 and 3 weeks after germination, and PCR results showed that 16 of 28 Milos seedlings (57.14%) and 11 of 16 Ortano seedlings (68.75%) were infected, indicating a substantial overall seed transmission rate of 61.36%. Both seed coats and seedling tissues were negative, eliminating the possibility of surface contamination and confirming systemic viral infection. Further, sap from the PCR-positive seedlings was mechanically inoculated into healthy zucchini squash plants. Within 3 weeks, all recipient plants (five from Ortano and four from Milos) developed characteristic ToLCNDV symptoms and tested positive by PCR. This experiment confirmed that not only is ToLCNDV seed-borne, but the virus also remains infective and transmissible, initiating new infections in healthy hosts. These findings provide strong molecular evidence that ToLCNDV can be vertically transmitted via seeds in zucchini squash and subsequently act as a primary source of inoculum, especially in vector-free or early-season conditions. Such transmission dynamics highlight the importance of incorporating seed health testing and certification protocols into integrated management strategies for begomoviruses in cucurbits (Tables 5.1 and 5.3).

3.3 Bitter Gourd (*Momordica charantia*)

Bitter gourd (*Momordica charantia* L.), a nutritionally and medicinally important vegetable crop, is increasingly threatened by yellow mosaic disease with severe mosaic, leaf curling and bushiness caused by two major begomoviruses: ToLCNDV and bitter gourd yellow mosaic virus (BgYMV) (Fig. 5.5).

In a study by Gomathi Devi et al. (2023), seeds from both commercial markets and infected farmer fields were evaluated to investigate the possibility of seedborne viral transmission. Using DAS-ELISA, a high percentage of embryo infection was recorded up to 62.96% in hybrid H2, followed by H1 (26.6%), H3 (20%), and H4 (10%). PCR analyses using virus-specific primers further confirmed the presence of ToLCNDV alone or in combination with BgYMV in these embryos. Interestingly, commercial seed lots exhibited a higher viral load than field-collected samples, indicating a potential risk in seed market practices. The viral DNA from the embryo was amplified using a CP-specific primer, and the amplicons were then sequenced. This strengthened the molecular evidence of embryo-borne infection. These findings provide compelling confirmation of vertical transmission of begomoviruses via seeds in bitter gourd (Gomathi Devi et al., 2023) (Fig. 5.6).

a. Mosaic symptom; b. Leaf curling; c. Stunted and bushy plant; d. Malformed fruits

Fig. 5.5 Different patterns of symptoms caused by ToLCNDV in bitter gourd. (**a**) Mosaic symptom; (**b**) leaf curling; (**c**) stunted and bushy plant; (**d**) malformed fruits

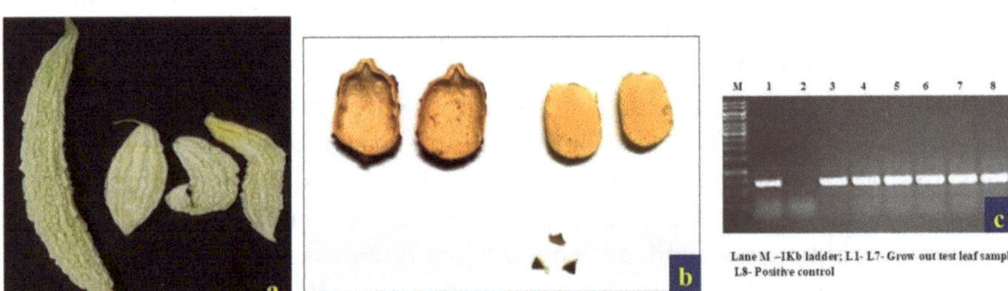

a. Fruits collected from infected plant; b. Dissection of seed into different parts; c. Molecular confirmation of ToLCNDV in embryo

Fig. 5.6 Confirmation of seed-borne nature of ToLCNDV in bitter gourd. (**a**) Fruits collected from infected plant; (**b**) dissection of seed into different parts; (**c**) molecular confirmation of ToLCNDV in embryo

3.4 Evidence of Active Seed-to-Seedling Transmission and Vector-Mediated Spread

The study by Gomathi Devi et al. (2023) also established that begomoviruses present in seeds can actively transmit to seedlings and initiate systemic infections. This was demonstrated through insect-proof grow-out tests using hybrids H1, H2, and H3, where ToLCNDV was detected in 3–5% of the emerging seedlings, while bitter gourd yellow mosaic virus (BgYMV) was absent. In a more extensive microplot experiment, ToLCNDV seed transmission reached as high as 43.2%, with both symptomatic (4.8%) and asymptomatic (38.4%) plants testing PCR positive (Fig. 5.7). The significance of this seedborne inoculum was further highlighted by a whitefly transmission study using *Bemisia tabaci* (Asia I genotype), viruliferous whiteflies acquired the virus from infected seedlings became viruliferous and efficiently transmitted it to healthy plants. The study reported that symptom expression occurred 20 days post-inoculation, with an up to 72% infection rate in plants after whitefly release, compared to 43.3% before release. These findings clearly illustrate that seed-derived inoculum serves as a potent and initial source for vector-mediated disease outbreaks in bitter gourd cultivation (Gomathi Devi et al., 2023) (Fig. 5.8).

Sáez et al. (2023) undertook a detailed study to evaluate the potential for seedborne and seed-transmitted spread of the ToLCNDV-ES isolate from Spain in cucurbits, including cucumber, musk melon, winter squash, zucchini, and watermelon. In *Cucumis melo* (melon), 24 genotypes from the *inodorus*, *cantalupensis*, and *flexuosus* groups were evaluated, with all showing ToLCNDV infection in leaves, flowers, and seed coats. Some seeds also had detectable virus in the endosperm and embryos. However, among 255 seedlings, only a small number showed transient PCR positivity, and no replicative viral forms were found, indicating the absence of vertical transmission. In *Cucurbita moschata* (pumpkin), including Nigerian Local and two derived lines, plants were asymptomatic or mildly symptomatic. Viral DNA was found in seed coats but not in endosperms or embryos. Though 7 and 19 of 45 seedlings tested positive at 30 and 60 days post germination, respectively, only four were positive at both times, and no replicative forms were found, again confirming the lack of vertical transmission. In *Cucurbita pepo* (zucchini), severe infection prevented fruit and seed development under experimental

Fig. 5.7 Confirmation of seed transmission in bitter gourd through the grow-out test in microplot. (**a**) Microplot study; (**b, c**) mosaic pattern of symptom in grow-out plants; (**d**) molecular confirmation of ToLCNDV in grow-out test leaves

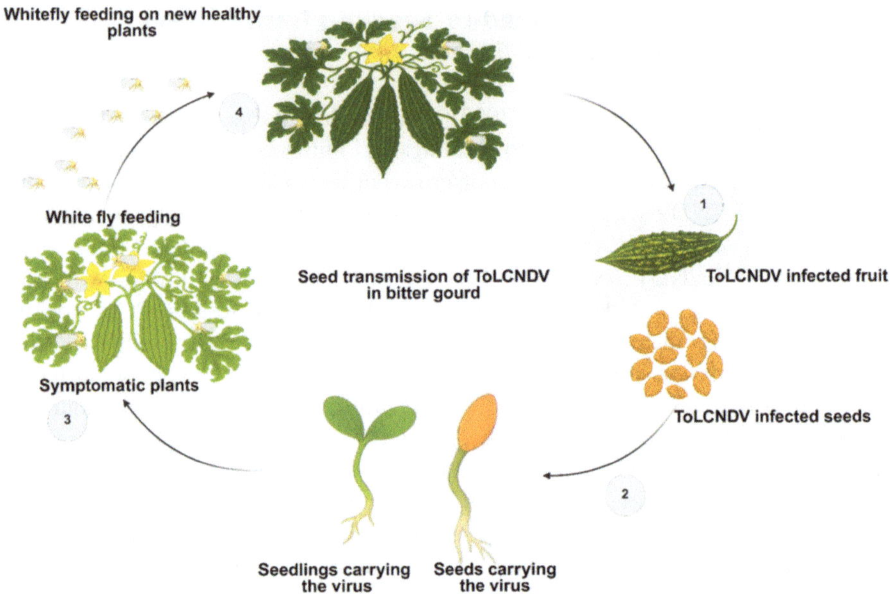

Fig. 5.8 Graphical representation of seed transmission of ToLCNDV in bitter gourd

conditions, but commercial seeds tested positive in one instance; however, resulting seedlings showed no symptoms or disease. For *Cucumis sativus* (cucumber), viral DNA was detected in one commercial seed, but seedlings were virus-free, with no evidence of replication. Lastly, *Citrullus lanatus* (watermelon) seeds tested entirely negative for ToLCNDV, confirming it is unlikely to act as a seedborne reservoir (Tables 5.1 and 5.3).

4 Contradictions in the Seed-Borne and Seed-Transmitted Nature of ToLCNDV

4.1 Melon (*Cucumis melo*): Seed Association, Localization, and Stability of ToLCNDV in Melon

Melon (*Cucumis melo* L.), a widely cultivated cucurbitaceous crop, has been confirmed as a host for ToLCNDV. In a comprehensive study by Fortes et al. (2023), the potential seed association of the "Spain" strain (ToLCNDV-ES) of ToLCNDV was examined using three susceptible F1 hybrid melon cultivars Brimos (Galia type), Mayor, and Nesta (Yellow types). Plants were experimentally inoculated at the two to three-leaf stage using infectious constructs in *Agrobacterium tumefaciens* by the stem puncture method. At 32 days post-inoculation, both male and hermaphrodite flowers were sampled and dissected into sepals, petals, stamens, and pistils. ToLCNDV detection was performed using conventional PCR targeting total DNA from dissected floral tissues, with the COX gene as an internal control and leaf DNA as a positive control. A total of ten flowers (four male, six hermaphrodite) from infected plants were analyzed, and all tissue samples tested positive, while tissues from virus-free plants showed no amplification. Mature seeds harvested from infected plants were analyzed using PCR and qPCR, revealing the presence of ToLCNDV DNA in bulked seed samples of all three cultivars. To evaluate whether the virus was internally localized or externally seed coat-borne, contaminating the seeds, different treatments were applied. Seeds were either left untreated, washed with water, or disinfected using 1.5% sodium hypochlorite. A significant reduction in viral DNA was observed following surface disinfection, suggesting predominant localization of ToLCNDV on the seed coat (Table 5.2).

To further determine the precise location of the virus, both treated and untreated seeds were dissected into seed coat and internal fractions (cotyledons and embryo). In untreated seeds, both the internal and external tissues exhibited similar levels of viral DNA, likely due to surface contamination or potential cross-contamination during dissection. In contrast, disinfected seeds showed a significantly higher viral load in the seed coat compared to the embryo, indicating that the viral presence in internal tissues was minimal or incidental. Control seeds from virus-free plants consistently tested negative, validating the reliability of the assays. These results firmly established that ToLCNDV is seed-borne in melon, with primary association on the outer seed surface and minimal presence in internal seed tissues. Although detection in embryos was occasionally observed, its significance remains uncertain, and the potential for systemic infection from such an association appeared limited under the experimental conditions. The findings highlight the importance of effective seed sanitation practices in melon seed production and stress the need for sensitive diagnostic protocols to monitor seedborne viral contamination.

4.2 Lack of Seed Transmission in Melon

Further, Fortes et al. (2023) conducted detailed growth-out experiments under controlled, insect-proof environments to confirm transmission of ToLCNDV from seed to progeny plants. Seeds obtained from ToLCNDV-infected Brimos, Mayor, and Nesta melon plants were used to germinate progeny, with evaluations conducted at both early (15 days post sowing) and late (60 days post sowing) growth stages. In early-stage experiments, seeds were germinated in moistened filter paper within sterile plastic boxes, and tissues such as seed coats, cotyledons, and first true leaves were collected for qPCR analysis. Although viral DNA was detected in the seed coats of seedlings, none of the internal seedling tissues, cotyledons or first leaves tested positive for ToLCNDV, indicating a lack of infection of embryonic tissue.

Large-scale late-stage evaluations involved a total of 775 progeny plants across three independent trials at two locations. These plants were monitored for symptom development and tested for viral presence at the 20-leaf and 60-day stages using tissue blot hybridization, conventional PCR, and qPCR. None of the progeny plants derived from infected seed showed ToLCNDV infection by any method, and no visual symptoms characteristic of begomovirus infection were observed. Positive control plants (experimentally inoculated with ToLCNDV) consistently showed strong viral detection and symptom expression, while negative controls from mock-inoculated plants remained virus-free. The complete absence of ToLCNDV in all progeny plants demonstrated unequivocally that, although ToLCNDV is seedborne in melon, it is not seed-transmitted under the tested conditions. These results suggest that while melon seeds may harbor detectable viral DNA, this does not translate to infection in emerging seedlings, thus eliminating the risk of virus spread through seed in this crop (Tables 5.2 and 5.3).

Several studies have presented evidence supporting the detection of ToLCNDV in seed tissues. However, these findings have been contradicted by more recent and comprehensive studies. Notably, Sáez et al. (2023) conducted a rigorous molecular investigation using quantitative PCR, rolling circle amplification (RCA), and Southern blotting across multiple cucurbit crops (melon, zucchini, cucumber) and found that while viral DNA could be detected in seed coats and floral tissues, no systemic infection or replicative forms of ToLCNDV were observed in seedlings, even those that tested were only weakly PCR positive. The absence of full-length viral genomes and lack of symptom development led the authors to conclude that the virus may contaminate seeds but is not truly seed-transmitted. Recent investigations by Fortes et al. (2023) and Sáez et al. (2023) in cucurbits like melon (*Cucumis melo*), watermelon (*Citrullus lanatus*), and pumpkin (*Cucurbita moschata*) revealed that although ToLCNDV could be detected in seed tissues, it did not result in infected progeny, indicating that the virus is seed-borne but not truly seed transmitted. These recent studies have redefined the role of seeds in the epidemiology and long-distance spread of begomoviruses. Recognizing seed transmission as a viable pathway urges the need for stricter seed health regulations, deeper investigation into transmission mechanisms, and more robust virus management strategies.

These conflicting reports underline a critical need to distinguish between seed-borne contamination (presence of viral DNA) and accurate seed transmission (systemic infection and epidemiological relevance). While the presence of ToLCNDV in seed tissues is well documented, consistent field-level transmission and outbreak initiation from seeds remain unproven, reinforcing the current consensus that whitefly (*Bemisia tabaci*) remains the predominant and epidemiologically significant vector for ToLCNDV spread.

5 Epidemiological Implications of Seed Transmission in Tomato Leaf Curl New Delhi Virus

Seed transmission represents a critical, yet often underestimated, component in the epidemiology of plant viruses, including begomoviruses like ToLCNDV. This route allows the virus to persist under unfavorable environmental conditions and spread across geographies, even in the absence of insect vectors. In begomoviruses, including ToLCNDV, seed-mediated dissemination poses a major threat due to the global movement of seeds, which are integral to agriculture, germplasm exchange, and trade. If infected, seeds become silent carriers, enabling transboundary introduction of the virus and triggering outbreaks in distant and diverse agroecological zones (Ali & Kobayashi, 2010; Pagán, 2022; Mink, 1993; Sastry, 2013; Chang et al., 2023).

5.1 Seed Transmission as a Vehicle for Virus Circulation

Recent studies have confirmed seed transmission of ToLCNDV in several hosts. Infected seedlings serve as critical foci for initial virus acquisition by whitefly vectors (*Bemisia tabaci*), from where secondary spread occurs rapidly within and between fields. Even low percentages of seed-borne infection have been shown to lead to significant epidemics, particularly when vector populations surge under favorable conditions.

Moreover, the virus's ability to persist in alternate hosts and weeds via seed transmission further amplifies its epidemiological footprint. Asymptomatic wild hosts can maintain viral populations between cropping seasons or in vector-free periods, ensuring continuity and potential resurgence of infection (Regassa et al., 2021). The colonization of non-traditional hosts, including monocots such as maize and onion (Fiallo-Olivé & Navas-Castillo, 2023), underscores the potential for seed transmission to assist in host range expansion and evolution of novel viral strains.

5.2 Role in Outbreak Initiation

Although vector transmission is the primary mechanism of ToLCNDV dissemination, seed transmission plays a key role in epidemic initiation. In regions where whiteflies are seasonal, infected seeds may initiate low-level infections that later intensify due to secondary spread (Naga et al., 2021b; Modgill et al., 2025). In bitter gourd, up to 43% of early infections have been traced to seed transmission, with the total infection rate rising to 70% after whitefly activity commenced (Gomathi Devi et al., 2023).

5.3 Need for Risk Assessment and Predictive Modeling

Addressing the risks posed by ToLCNDV seed transmission requires a multidimensional risk assessment framework. This involves understanding how infected seeds influence Seed germination and seedling vigor, Initial infection establishment, subsequent disease development, yield and quality losses. Jones (2000) highlighted the importance of establishing threshold levels of acceptable seed infection, crucial for defining quarantine regulations and seed certification standards. In addition, predictive epidemiological models that simulate seed transmission, vector migration, host response, and climatic variables would be invaluable tools for forecasting future ToLCNDV outbreaks. The integration of molecular detection techniques with geographic information systems (GIS) and weather-based modeling could significantly enhance early warning systems and help to mitigate the impact of ToLCNDV on food security and seed trade.

6 Phytosanitary Concerns and Regulatory Status of ToLCNDV

ToLCNDV, a highly adaptable begomovirus, has become a critical phytosanitary threat due to its rapid spread, extensive host range, and emerging evidence of seed transmission. Since its first detection in India in 1948 and characterization in the 1990s, ToLCNDV has been reported in over 15 countries, including Italy, Spain, Tunisia, Turkey, Pakistan, and Thailand (EPPO Global Database, 2024). In Europe, yield losses up to 70% have been reported in zucchini and melon crops due to ToLCNDV outbreaks (Juárez et al., 2014; Mnari-Hattab et al., 2015). Its detection in seed lots of cucurbits and tomatoes has raised serious concerns, as even seed infection rates as low as 0.01% can initiate epidemics under favorable vector conditions (Pagán, 2022; Gomathi Devi et al., 2023) (Fig. 5.9). The European and Mediterranean Plant Protection Organization (EPPO) added ToLCNDV to its alert list in 2012 and currently recommends rigorous phytosanitary inspections and seed health testing. According to EFSA (2020), ToLCNDV meets the criteria for a Union quarantine pest, and import regulations have been tightened for solanaceous and cucurbitaceous crops. EPPO standards now emphasize the need for zero-tolerance policies in the certified seed trade, especially for crops exported to virus-free zones. Surveillance programs in Spain and Italy reported ToLCNDV incidence in over 60% of surveyed cucurbit fields, further underlining the importance of strict quarantine and certification procedures (EFSA, 2020). A proactive phytosanitary strategy involving molecular diagnostics, seed testing, vector monitoring, and quarantine enforcement is crucial to contain ToLCNDV and prevent global dissemination

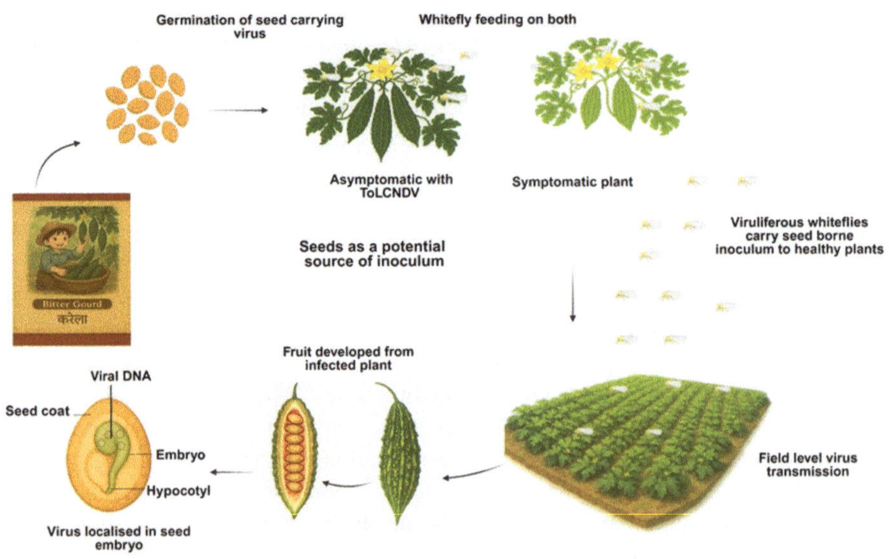

Fig. 5.9 Graphic representation of seed as a potential source of inoculum for the rapid spread by the vector

through seed trade (Kumar et al., 2022; Raigond et al., 2022). Harmonization of international phytosanitary standards is urgently needed as climate change and globalization continue to alter vector dynamics and increase the risk of virus introduction into new territories.

7 Challenges and Future Perspectives

Several challenges remain in understanding and managing ToLCNDV seed transmission. One of the primary gaps is the need for conclusive evidence on whether seedborne viral particles can establish systemic infections in emerging seedlings under natural field conditions (Fortes et al., 2023; Sangeetha et al., 2018). To combat these challenges, integrating advanced molecular breeding techniques, such as CRISPR/Cas9 genome editing, could offer a promising approach to develop virus-resistant cultivars by targeting host susceptibility factors. Additionally, strengthening phytosanitary regulations and implementing stricter seed health certification programs will be essential in limiting the risk of seed-associated transmission.

8 Conclusion

The potential role of seeds in ToLCNDV transmission presents a significant challenge for disease management and global trade regulations. While whiteflies remain the primary vectors, the detection of viral genomes in seeds underscores the need for further investigation into their epidemiological significance. Advances in molecular diagnostics, seed treatment strategies, and the development of resistant cultivars will be critical in mitigating the risks associated with ToLCNDV. Future research should emphasize interdisciplinary collaboration among plant pathologists, breeders, and regulatory bodies to develop integrated management strategies that ensure sustainable agricultural practices.

References

Albrechtsen, S. E. (2006). Seed transmission of viruses. In *Test methods seed-transmitted viruses: principles and protocols* (pp. 13–26). CABI. https://doi.org/10.1079/9780851990163.0013

Ali, A., & Kobayashi, M. (2010). Seed transmission of Cucumber mosaic virus in pepper. *Journal of Virological Methods, 163*, 234–237. https://doi.org/10.1016/J.JVIROMET.2009.09.026

Bennett, C., & Esau, K. (1936). Further studies on the relation of the curly top virus to plant tissues. *Journal of Agricultural Research, 53*, 595–620. https://doi.org/10.5555/19371100529

Cambra, M., Capote, N., Myrta, A., & Llácer, G. (2006). Plum pox virus and the estimated costs associated with sharka disease. *Wiley Online Library, 36*, 202–204. https://doi.org/10.1111/J.1365-2338.2006.01027.X

Chang, H. H., Gustian, D., Chang, C. J., & Jan, F. J. (2023). Seed and pollen transmission of tomato leaf curl New Delhi virus, tomato leaf curl Taiwan virus, and tomato yellow leaf curl Thailand virus in cucumbers and tomatoes. *Plant Disease, 107*, 2002–2008.

EFSA Panel on Plant Health (PLH), Bragard, C., Dehnen-Schmutz, K., Di Serio, F., et al. (2020). Pest categorisation of tomato leaf curl New Delhi virus. *EFSA Journal, 18*(7), e06179.

EPPO Global Database (2024). European and Mediterranean Plant Protection Organization (EPPO), Paris, France. https://gd.eppo.int

Fadhila, C., Lal, A., Vo, T. T. B., et al. (2020). The threat of seed-transmissible pepper yellow leaf curl Indonesia virus in chili pepper. *Microbial Pathogenesis, 143*, 104132. https://doi.org/10.1016/J.MICPATH.2020.104132

Fiallo-Olivé, E., & Navas-Castillo, J. (2023). Begomoviruses: What is the secret(s) of their success? *Trends in Plant Science, 28*, 715–727. https://doi.org/10.1016/J.TPLANTS.2023.01.012

Fortes, I. M., Pérez-Padilla, V., Romero-Rodriguez, B., et al. (2023). Begomovirus tomato leaf curl New Delhi virus is seedborne but not seed transmitted in melon. *Plant Disease, 107*, 473–479.

Gomathi Devi, R., Jothika, C., Sankari, A., et al. (2023). Seed transmission of begomoviruses: A potential threat for bitter gourd cultivation. *Plants, 12*, 1396. https://doi.org/10.3390/PLANTS12061396/S1

Hull, R. (2013). *Plant virology* (5th ed., pp. 1–1104). https://doi.org/10.1016/C2010-0-64974-1

ICTV. *Current ICTV taxonomy release*. https://ictv.global/taxonomy. Accessed 14 May 2024.

Jeevalatha, A., Vanishree, G., Siddappa, S., Kumar, R., Kaundal, P., Kumar, A., & Chakrabarti, S. K. (2021). Molecular characterization and infectivity analysis of tomato leaf curl New Delhi virus isolates infecting potato. *3 Biotech, 11*, 203. https://doi.org/10.1007/s13205-021-02752-5

Jeevalatha, A., Siddappa, S., Kumar, R., Tiwari, R. K., Lal, M. K., Sharma, S., Chakrabarty, S. K., & Singh, B. P. (2023). RNA-seq analysis reveals an early defense response to tomato leaf curl New Delhi virus in potato cultivar Kufri Bahar. *Functional & Integrative Genomics, 23*, 215. https://doi.org/10.1007/s10142-023-01138-5

Jones, R. A. C. (2000). Determining "threshold" levels for seed-borne virus infection in seed stocks. *Virus Research, 71*, 171–183. https://doi.org/10.1016/S0168-1702(00)00197-0

Jones, R. A. C., & Naidu, R. A. (2019). Global dimensions of plant virus diseases: Current status and future perspectives. *Annual Review of Virology, 6*, 387–409. https://doi.org/10.1146/annurev-virology-092818-015606/1

Juárez, M., Tovar, R., Fiallo-Olivé, E., et al. (2014). First detection of tomato leaf curl New Delhi virus infecting zucchini in Spain. *Plant Disease, 98*, 857. https://doi.org/10.1094/PDIS-10-13-1050-PDN

Keur, J. Y. (1934). Studies of the occurrence and transmission of virus diseases in the genus abutilon. *Bulletin of the Torrey Botanical Club, 61*, 53. https://doi.org/10.2307/2480784

Kil, E. J., Kim, S., Lee, Y. J., et al. (2016). Tomato yellow leaf curl virus (TYLCV-IL): A seed-transmissible geminivirus in tomatoes. *Scientific Reports, 6*, 1–10.

Kil, E. J., Park, J., Choi, H. S., et al. (2017). Seed transmission of tomato yellow leaf curl virus in white soybean (Glycine max). *Plant Pathology Journal, 33*, 424–428. https://doi.org/10.5423/PPJ.NT.02.2017.0043

Kil, E. J., Park, J., Choi, E. Y., et al. (2018). Seed transmission of tomato yellow leaf curl virus in sweet pepper (Capsicum annuum). *European Journal of Plant Pathology, 150*, 759–764. https://doi.org/10.1007/S10658-017-1304-8/METRICS

Kil, E. J., Bich Vo, T. T., Fadhila, C., et al. (2020). Seed transmission of tomato leaf curl New Delhi virus from zucchini squash in Italy. *Plants, 9*, 563. https://doi.org/10.3390/PLANTS9050563

Kim, J., Kil, E. J., Kim, S., et al. (2015). Seed transmission of sweet potato leaf curl virus in sweet potato (Ipomoea batatas). *Plant Pathology, 64*, 1284–1291.

Kothandaraman, S. V., Devadason, A., & Ganesan, M. V. (2016). Seed-borne nature of a begomovirus, Mung bean yellow mosaic virus in black gram. *Applied Microbiology and Biotechnology, 100*, 1925–1933. https://doi.org/10.1007/S00253-015-7188-7/METRICS

Kumar, R., Tiwari, R. K., Jeevalatha, A., Sundaresha, S., Shah, M. A., Sharma, S., Sagar, V., Kumar, M., & Chakrabarti, S. K. (2021). Potato apical leaf curl disease: Current status and perspectives on a disease caused by tomato leaf curl New Delhi virus. *Journal of Plant Diseases and Protection, 128*, 897–911. https://doi.org/10.1007/s41348-021-00463-w

Kumar, R., Tiwari, R. K., Sundaresha, S., Kaundal, P., & Raigond, B. (2022). Potato viruses and their management. In S. K. Chakrabarti, S. Sharma, & M. A. Shah (Eds.), *Sustainable management of potato pests and diseases*. Springer. https://doi.org/10.1007/978-981-16-7695-6_12

Manivannan, K., Renukadevi, P., Malathi, V. G., et al. (2019). A new seed-transmissible begomovirus in bitter gourd (Momordica charantia L.). *Microbial Pathogenesis, 128*, 82–89. https://doi.org/10.1016/J.MICPATH.2018.12.036

Mink, G. I. (1993). Pollen- and seed-transmitted viruses and viroids. *Annual Review of Phytopathology, 31*, 375–402. https://doi.org/10.1146/ANNUREV.PY.31.090193.002111/CITE/REFWORKS

Mnari-Hattab, M., Zammouri, S., Belkadhi, M. S., et al. (2015). First report of tomato leaf curl New Delhi virus infecting cucurbits in Tunisia. *New Disease Reports, 31*, 21–21. https://doi.org/10.5197/J.2044-0588.2015.031.021

Modgill, N., Naga, K. C., Firake, D. M., Tiwari, R. K., Shivaramu, S., Kumar, R., Bhatnagar, A., Buckseth, T., Bairwa, A., Maharana, C., Sharma, S., & Singh, B. (2025). Flupyradifurone application: A promising approach to curtail tomato leaf curl New Delhi virus (ToLCNDV) transmission vectored by Bemisia tabaci in potato farming. *Crop Protection, 187*, 106951.

Mulyadi, D., Sulandari, S., Hartono, S., & Somowiyarjo, S. (2021). Distribution, host range and detection of seed-borne yellow mosaic disease on yardlong beans (*Vigna unguiculata* subsp. *sesquipedalis* L.) in the special region of Yogyakarta, Indonesia. *Biodiversitas, 22*, 3949–3957. https://doi.org/10.13057/BIODIV/D220942

Naga, K. C., Subhash, S., Bhatnagar, A., Tiwari, R. K., Kumar, R., Verma, G., Sharma, S., & Chakrabarti, S. K. (2021a). Influence of host plants on virus acquisition and endosymbionts of whitefly, Bemisia tabaci (Gennadius). *Indian Journal of Entomology, 83*, 371–375. https://doi.org/10.5958/0974-8172.2020.00206.0

Naga, K. C., Sundaresha, S., Kumar, R., Tiwari, R. K., Subhash, S., Verma, G., Buckseth, T., Bairwa, A., Sharma, S., Katare, S., & Srivastava, R. M. (2021b). A new record of Asia II 5 genetic group of Bemisia tabaci (Gennadius) in the major potato growing areas of India and its relationship with tomato leaf curl New Delhi virus infecting potato. *3 Biotech, 11*, 421. https://doi.org/10.1007/s13205-021-02966-7

Pagán, I. (2022). Transmission through seeds: The unknown life of plant viruses. *PLoS Pathogens, 18*, e1010707. https://doi.org/10.1371/JOURNAL.PPAT.1010707

Panno, S., Iacono, G., Davino, M., et al. (2016). First report of tomato leaf curl New Delhi virus affecting zucchini squash in an important horticultural area of southern Italy. *New Disease Reports, 33*, 6. https://doi.org/10.5197/j.2044-0588.2016.033.006

Qiao, Q., Zhang, Z. C., Zhao, X. L., et al. (2020). Evidence for seed transmission of sweet potato symptomless virus 1 in sweet potato (*Ipomoea batatas*). *Journal of Plant Pathology, 102*, 299–303. https://doi.org/10.1007/S42161-019-00427-Y/METRICS

Quainoo, A. K., Wetten, A. C., & Allainguillaume, J. (2008). Transmission of cocoa swollen shoot virus by seeds. *Journal of Virological Methods, 150*, 45–49. https://doi.org/10.1016/J.JVIROMET.2008.03.009

Raigond, B., Verma, G., Kumar, R., & Tiwari, R. K. (2022). Serological and molecular diagnosis of potato viruses: An overview. In S. K. Chakrabarti, S. Sharma, & M. A. Shah (Eds.), *Sustainable management of potato pests and diseases* (pp. 337–352). Springer. https://doi.org/10.1007/978-981-16-7695-6_13

Regassa, B., Abraham, A., Fininsa, C., & Wegary, D. (2021). Alternate hosts and seed transmission of maize lethal necrosis in Ethiopia. *Journal of Phytopathology, 169*, 303–315. https://doi.org/10.1111/JPH.12986;wgroup:string:publication

Renukadevi, P., Sangeetha, B., Malathi, V. G., et al. (2022). Enigmatic emergence of seed transmission of geminiviruses. In *Geminivirus: Detection, diagnosis and management* (pp. 285–306). Academic Press. https://doi.org/10.1016/B978-0-323-90587-9.00003-1

Sáez, C., Kheireddine, A., García, A., et al. (2023). Further molecular diagnosis determines lack of evidence for real seed transmission of tomato leaf curl New Delhi virus in cucurbits. *Plants, 12*, 3773. https://doi.org/10.3390/plants12213773

Sangeetha, B., Malathi, V. G., Alice, D., et al. (2018). A distinct seed-transmissible strain of tomato leaf curl New Delhi virus infecting Chayote in India. *Virus Research, 258*, 81–91. https://doi.org/10.1016/j.virusres.2018.10.009

Sarkar, M., Gupta, D., Singh, O. W., Paul, S., Kumar, R., Mandal, B., & Roy, A. (2024). Identification of pathogenicity determinants in ToLCNDV and their RNAi-based knock-down for disease management in Nicotiana benthamiana and tomato plants. *Frontiers in Microbiology, 15*, 1481523. https://doi.org/10.3389/fmicb.2024.1481523

Sastry, K. S. (2013). *Seed-borne plant virus diseases*. Springer. https://doi.org/10.1007/978-81-322-0813-6

Shanmugam, K., Perumal, R., Oliva, R., et al. (2024). Seed transmission of potyviruses: A threat to crop health. *Notulae Botanicae Horti Agrobotanici Cluj-Napoca, 52*(4), 13698. https://doi.org/10.15835/nbha52413698

Stace Smith, R., & Hamilton, R. (1988). Inoculum thresholds of seedborne pathogens. *Viruses. Phytopathology, 78*, 875–880.

Suruthi, V., Nakkeeran, S., Renukadevi, P., et al. (2018). Evidence of seed transmission of dolichos yellow mosaic virus, a begomovirus infecting lablab-bean in India. *Virus, 29*, 506–512.

Sutrawati, M., & Aulia, E. (2024). Virus identification in yard long bean plants with yellow mosaic symptoms. *AGRITROPICA: Journal of Agricultural Sciences, 7*, 116–121. https://doi.org/10.31186/J.Agritropica.7.2.116-121

Valkonen, J. P. T. (2007). Viruses: Economical losses and biotechnological potential. In *Potato biology and biotechnology: Advances and perspectives* (pp. 619–641). Elsevier.

Epidemiology and Pathophysiology of Tomato Leaf Curl New Delhi Virus Infection in Horticultural Crops

6

Varshini Pillai and A. Swapna Geetanjali

Abstract

Tomato leaf curl New Delhi virus (ToLCNDV) is a bipartite begomovirus that has been a serious threat to horticultural crops grown worldwide. The virus primarily affects economically important crops of solanaceous and cucurbitaceous families, such as tomato, peppers, cucurbits, etc., causing severe yield losses. ToLCNDV is transmitted efficiently by the whitefly, *Bemisia tabaci*, which has been a reported vector globally. Factors that contribute to the disease spread include global distribution of the vector population, temperature, humidity, and seasonal changes, which also affect the vector dynamics. First reported in India, the virus is mainly present in Asia, Europe, and the Middle East, causing sporadic economic losses. The infection involves virus and host plant interactions, where the bipartite genome of the virus, upon entering the host, encodes several essential proteins for replication, movement, and suppression of the host's defense mechanisms. The host plant develops characteristic symptoms, including leaf curling, vein thickening, chlorosis, and stunted growth. These symptoms often lead to reduced or complete loss of quality and yield loss. To mitigate the virus and overcome the impact on the global horticulture production, it is essential to understand its molecular characteristics, functions, and interactions with the host. This chapter discusses the epidemiological and pathophysiological aspects of ToLCNDV infection in horticultural crops, focusing on the interaction between the virus, vector, and host at the cellular and molecular levels.

V. Pillai · A. Swapna Geetanjali (✉)
Department of Genetic Engineering, SRM Institute of Science and Technology, Kattankulathur, Tamil Nadu, India

R. Kumar et al. (eds.), *Tomato Leaf Curl New Delhi Virus (ToLCNDV)*,
https://doi.org/10.1007/978-981-95-2852-3_6

Keywords

Begomovirus · Epidemiology · Pathophysiology · Tomato leaf curl New Delhi virus

1 Introduction

Tomato leaf curl New Delhi virus (ToLCNDV) is one of the "Old World" begomoviruses belonging to the *Geminiviridae* family (Fortes et al., 2016). It is a bipartite virus consisting of two genomes—DNA-A and DNA-B. DNA-A encodes proteins essential for viral replication (Rep), transcriptional activation (TrAP), capsid formation (CP), and suppression of host defense responses (C4). On the other hand, DNA-B plays a vital role in the movement of the virus within plant tissues (MP). In most infections, ToLCNDV is accompanied by betasatellites, which are approximately 1.4 kb in size (Jyothsna et al., 2013; Jeevalatha et al., 2018). ToLCNDV has emerged as a major threat to horticultural crops, mainly in tropical and subtropical regions, including India, Pakistan, Southeast Asia and parts of Europe and the Middle East (Moriones et al., 2017; Kumar et al., 2022, 2023). Plant families such as *Solanaceae* (e.g., tomato, chili, potato), *Cucurbitaceae* (e.g., zucchini, melon, cucumber, squash, gourd, chayote), and *Malvaceae* (e.g., okra and cotton) have been reported hosts of ToLCNDV (Cai et al., 2023; Kumar et al., 2023). Infected plants typically exhibit symptoms such as leaf curling, chlorosis, vein clearing, stunted growth, and deformed fruits that significantly impact crop yield (Moriones et al., 2017).

The silverleaf whitefly transmits ToLCNDV, *Bemisia tabaci* (Order: *Hemiptera*), a common vector of begomoviruses. The vector acquires the virus during phloem feeding using its stylet and transmits it in a circulative, persistent manner (Navas-Castillo et al., 2011). Once the virus is transmitted to the host plant, it takes over the plant's vascular system and starts replicating, thus spreading the infection. The key factor in managing ToLCNDV is controlling the vector. Unfortunately, this remains one of the biggest challenges due to the whitefly's free movement in open fields and its high reproductive capacity. While chemical control methods such as insecticides have been used to suppress whitefly populations, their efficacy is often undermined by the development of insecticide resistance (Lapidot et al., 2014). Apart from the common method of vector transmission, the virus may also spread through infected plant materials, seeds, or mechanical means (Rosen et al., 2015). Thus, having a more profound knowledge of both the spread patterns and the host-virus interactions is critical for designing effective control strategies. From what we currently know, an effective management of ToLCNDV demands an integrated approach that considers both vector control and disease spread mitigation.

2 Epidemiology

2.1 Geographical Distribution

In the winter of 1947, at New Delhi, 90% of tomato fields were showing early symptoms of the leaf curl disease such as leaf puckering, vein clearing, dwarfing and stunted growth (Vasudeva & Sam Raj, 1948). A few decades later, in 1995, Padidam et al., 1995 gave the toponymic name ToLCNDV to the virus. Field surveys still show that the pathogen is more prevalent in northern regions of India than in the southern states (Fortes et al., 2016). Several reports now trickle in from Bangladesh and Pakistan and, more recently, from Thailand, Indonesia, and several provinces in China (see Table 6.1). All reports state that ToLCNDV has been confirmed in almost 43 plant species (Moriones et al., 2017), which is comparatively higher than most host-specific viruses.

In recent years, ToLCNDV has disseminated across Central Asia, like Iran, to the areas of the Mediterranean, such as Tunisia and Spain (Ruiz et al., 2015), Greece, Italy (Orfanidou et al., 2019), and North African nations, Algeria, and Morocco (Radouane et al., 2022) (Fig. 6.1; Table 6.1). There is minimal evidence of the existence of ToLCNDV infesting horticultural crops in the Americas; nevertheless, the virus has caused devastating economic losses around the globe.

2.2 Host Range and Susceptibility

Host range depends on climatic conditions and environmental factors, as both the vector, *Bemisia tabaci*, and the host plants need suitable temperatures for thriving (Berlinger et al., 1996). Though ToLCNDV was first reported in its natural and primary host, tomato (*Solanum lycopersicum*), the virus has extended its range exceptionally broadly in recent years. Other Solanaceous crops, chilli (*Capsicum* spp.) (Khan et al., 2006), eggplant (*Solanum melongena*) (Pratap et al., 2011), and potato (*Solanum tuberosum*) (Usharani et al., 2004) have also been reported to be infected with ToLCNDV in India. Potato cultivation has faced significant challenges in recent years due to ToLCNDV infections. Affected plants exhibit symptoms such as leaf curling, crinkling, mosaic patterns, and, in severe cases, complete stunting of growth (Kumar et al., 2023; Raigond et al., 2022; Jeevalatha et al., 2023). Co-cultivation with infected tomato and sponge gourd resulted in the virus transmitting to potato, as reported by Saha et al. (2014) and Sohrab et al. (2013). Top potato-producing states in India have been found with 40–100% of recorded ToLCNDV infections and have high yield loss of up to 50% (Jeevalatha et al., 2018). Cucurbitaceous vegetables like cucumber (*Cucumis sativus*) (Venkataravanappa et al., 2018), pumpkin (*Cucurbita moschata*) (Phaneendra et al., 2012; Parrella et al., 2017), chayote (Nagendran et al., 2017), and squash (Panno et al., 2019) have been infected by ToLCNDV in some parts of India and Italy. In addition to horticultural crops, weeds and other flowering crops were also reported to act as reservoirs for ToLCNDV over non-cropping seasons. For example, in the Sindh and Punjab

Table 6.1 List of horticultural crops infected by ToLCNDV worldwide

Country	Horticultural crop	Genome	Reference
India	Tomato	DNA A + DNA B	Padidam et al. (1995)
	Chilli	DNA A	Khan et al. (2006)
	Potato	DNA A + DNA B	Usharani et al. (2004)
	Chayote		Nagendran et al. (2017)
	Cucumber		Venkataravanappa et al. (2018)
	Eggplant		Pratap et al. (2011)
	Pumpkin	DNA A	Phaneendra et al. (2012)
	Gourd	DNA A + DNA B + Alphasatellite and Betasatellite	Venkataravanappa et al. (2018)
Pakistan	Potato	DNA A	Hameed et al. (2017)
	Cotton	DNA A + DNA B	Zaidi et al. (2016)
	Chilli		
Italy	Melon	DNA A	Panno et al. (2019)
	Pumpkin	DNA A + DNA B	Parrella et al. (2017)
	Zucchini		Mastrochirico et al. (2023)
Bangladesh	Tomato	DNA A + DNA B	Maruthi et al. (2007)
Thailand	Cucumber		Ito et al. (2008)
	Bottle gourd	DNA A	
	Muskmelon		
Indonesia	Cucumber	DNA A + DNA B	Mizutani et al. (2011)
	Melon		Wilisiani et al. (2019)
China	Muskmelon	DNA A + DNA B	Gu et al. (2023)
	Oriental melon		
	Pumpkin		
	Tomato		Li et al. (2023)
	Cucumber		Zeng et al. (2023)
	Luffa		
Tunisia	Melon	DNA A + DNA B	Mnari-Hattab et al. (2015)
	Cucumber		
	Zucchini		
Spain	Melon	DNA A + DNA B	Juárez et al. (2019)
	Zucchini		
	Pumpkin		
	Tomato		
Morocco	Zucchini	DNA A + DNA B	Radouane et al. (2022)
Algeria	Cucumber	DNA A + DNA B	Kheireddine et al. (2019)
	Melon		
	Zucchini		

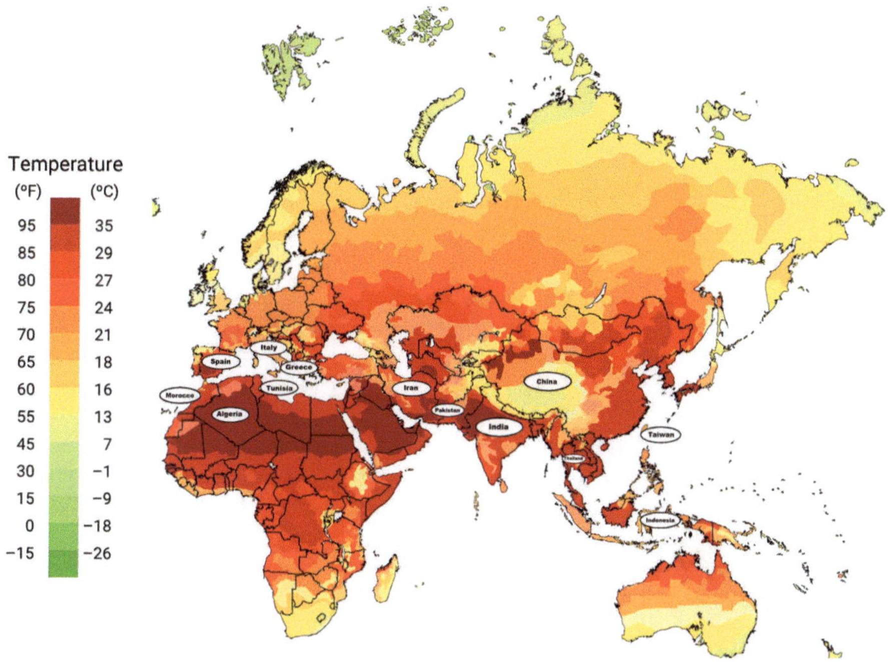

Fig. 6.1 Geographical distribution of ToLCNDV across various climatic regions of the world

provinces of Pakistan, ToLCNDV was found in *Calotropis procera*, which is popularly also known as the Rubber bush, serving as a reservoir host (Zaidi et al., 2017a, b). Similarly, in Northern India during the Rabi season, Nightshade (*Solanum nigrum*) acts as a reservoir during off-seasons (Ansar et al., 2021). The host range of the virus is also expanding due to mixed cropping in endemic regions such as India and Pakistan.

2.3 Modes of Transmission

ToLCNDV is transmitted by *Bemisia tabaci* (Order: *Hemiptera*), a common vector of begomoviruses (Gilbertson et al., 2015; Czosnek et al., 2017). It is said to be transmitted in a persistent circulative (non-propagative) manner (Fig. 6.2) (Renukadevi et al., 2022). *Bemisia tabaci*, also known as Silver Whitefly with its distinct silver wings, is a species present in the tropical and subtropical regions, with a less predominant presence in the temperate areas. *Bemisia tabaci* feeds primarily on members of the *Solanaceae* and cucurbit family (Kavallieratos et al., 2024), explaining the host range of ToLCNDV. Current reported hosts are horticultural crops that grow between the temperatures of 25 and 35 °C (Chaudhary et al., 2022). The whitefly acquires the virus through its stylet and circulates between the gut and saliva before disseminating the virus to other plants (Rosen et al., 2015).

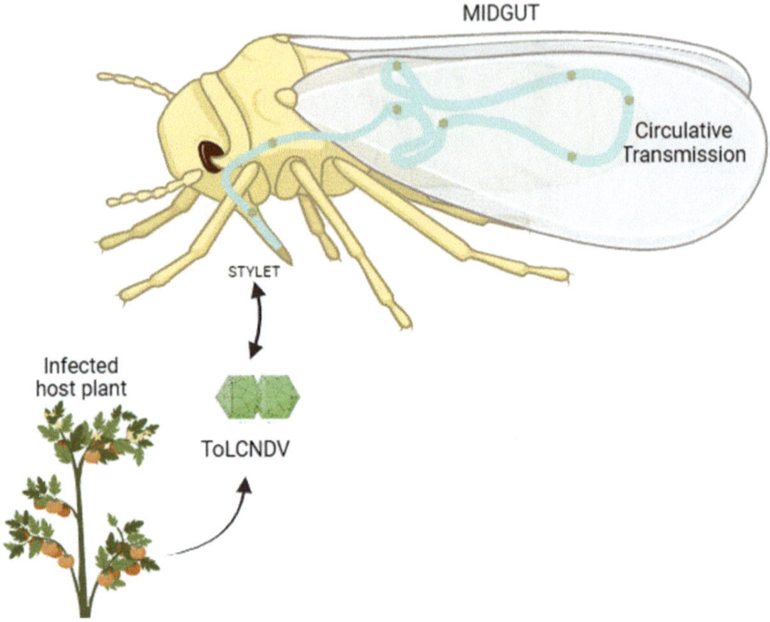

Fig. 6.2 Mode of transmission of ToLCNDV

After approximately 8 h of incubation period, the vector can now transmit the virus for the rest of its lifespan.

For successful transmission of ToLCNDV, the whitefly requires a defined 15–30 min of feeding known as the acquisition access period (AAP) and followed by a 10–60 min of feeding time known as the inoculation access period (IAP) on a virus-free plant. While the primary mode of spread remains vector-mediated, mechanical transmission via infected sap has been demonstrated under controlled experimental conditions, though its role in field transmission is limited (Janssen et al., 2022). Recent findings have also suggested the possibility of seed-borne transmission of ToLCNDV in certain horticultural crops. Notably, seed transmission has been reported in chayote grown in India and in seeds of squash plants cultivated in Spain (Fortes et al., 2016; Sandra &Mandal, 2024). Although such reports are relatively new and require solid validation, they raise important considerations for seed hygiene and biosecurity in virus management. It has also become increasingly more complex to control the whitefly population in open fields. The hot and dry climates of tropical and subtropical regions, along with the increasing temperature worldwide due to global warming, favor the reproductive efficiency of whiteflies. Monoculture and highly susceptible host plants also increase the rate of infectivity of ToLCNDV. Accordingly, integrated management strategies comprising vector control, development of resistant cultivars, crop rotation, and maintenance of field sanitation have been essential to reduce the spread of the virus.

2.4 Outbreak Dynamics

The dynamics of ToLCNDV outbreaks depend on several factors, including transmission mode, environmental factors, climate, and vector activity, which control the spatial and temporal dissemination patterns of the virus as well as endemic hotspots. Initially, the outbreak begins in fields with a higher prevalence of infected plants or seeds. The vectors then facilitate the localized spread of the virus by feeding on infected crops, moving across and adjacent plants (Fig. 6.3). Long-distance dispersal of infection occurs when whiteflies migrate between fields, dispersing over 5 km (Byrne, 1999) due to environmental changes, resource depletion, or non-cropping season. Infected adult whiteflies can introduce the virus to previously uninfected areas. The spread of ToLCNDV is seasonal, with outbreaks peaking during warm and humid periods that favor whitefly population and host plants cultivation period (Chaudhary et al., 2022). High temperatures are preferred by *Bemisia tabaci*; thus, hot summers, warm springs, fall, and mild winters permit the spread of ToLCNDV (Naranjo et al., 2010). The incidence is more often during the early cropping season as young plants are more susceptible to infection (Padidam et al., 1995). Continuous planting of susceptible crops (e.g., tomatoes, cucurbits) acts as a corridor, enhancing the activity of whitefly movement and virus transmission. Areas with a high density of horticultural cropping, especially of tomatoes and cucurbits, serve as hotspots due to the availability of a continuous host reservoir. Regions near international borders, such as Punjab (Gaikwad et al., 2011; Zaidi et al., 2016), where agricultural goods and plant materials are traded, are particularly vulnerable to ToLCNDV introduction and spread.

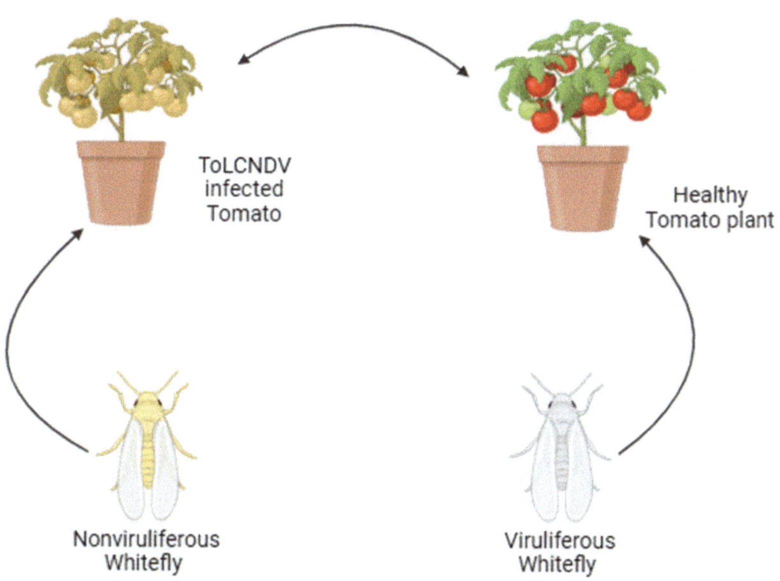

Fig. 6.3 Dissemination of ToLCNDV across host plants

3 Pathophysiology

3.1 Mechanism of Infection

Entry and initial infection begin with the introduction of ToLCNDV into the host plant during feeding by viruliferous whiteflies. When the whitefly feeds on the plant, the virus gets transferred through saliva into the phloem tissues of the host, where it primarily infects and begins its replication process (Fig. 6.4) (Cai et al., 2023; Moriones et al., 2017). Several key viral proteins mediate the replication and systemic spread of ToLCNDV within host plants. Among these, the nuclear shuttle protein (NSP) plays a critical role in facilitating the transport of viral DNA since nuclear localization is essential for viral replication. Here, the virus replicates by rolling circle mechanism via the Rep protein by binding to the origin of replication (ori) and subsequently synthesizing the new viral DNA using host polymerases. Functional studies conducted on *N. tabacum*, *N. benthamiana*, and *L. esculentum* have consistently solidified that the NSP functions as a central pathogenicity determinant. It was demonstrated that NSP plays a key role in the early stages of ToLCNDV infection by transporting the viral DNA into the host cell nucleus (Hussain et al., 2004). Alongside NSP, the coat protein (CP) is another major determinant of viral infectivity and pathogenicity. CP is involved in viral encapsidation, mediating host–virus interactions, and facilitating whitefly transmission. Recent studies using chimeric clones have shown that sequence variations in the CP gene can significantly influence both symptom expression and host range (Vo et al., 2023). Further, Kaur et al. (2015) investigated gene expression changes in *Solanum*

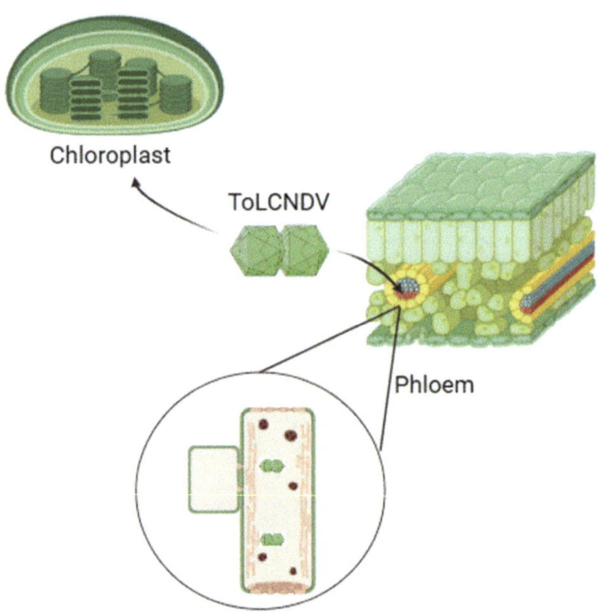

Fig. 6.4 Interaction of ToLCNDV with host cells

lycopersicum following viral infection and reported several alterations in host metabolic pathways. Notably, they observed the activation of stress-related genes and suppression of photosynthetic gene expression, suggesting a metabolic shift that favors viral replication over plant growth and productivity.

3.2 Host-Virus Interactions

ToLCNDV, like any plant virus, sets off a range of responses in the host plant. It's been shown to modulate key pathways involving auxin, jasmonic acid (JA), and salicylic acid (SA), all of which are central to plant growth and defense (Kushwaha et al., 2015; Sarkar et al., 2024). This suggests that ToLCNDV not only avoids or circumvents the plant's immune system but also actively modulates it to support infection and movement. One of the main antiviral defense tools the plant uses is RNA interference (RNAi). Core proteins such as Dicer-like (DCL) and Argonaute (AGO) detect double-stranded viral RNA, chop it into small interfering RNAs (siRNAs), and guide the degradation of viral transcripts (Ho et al., 2010). This mechanism also triggers other immune pathways, activating genes for pathogenesis-related (PR) proteins, ROS-detoxifying enzymes, and elements of the ubiquitin–proteasome system (UPS). Alongside this, SA- and JA-related defense genes are also upregulated, highlighting how the plant tries to overcome or fight the infection in a multi-layered defense (Sahu et al., 2010). The TRN1 gene is another key part of the plant's response, which helps maintain the structure and function of vascular tissues, particularly the phloem. When TRN1 is downregulated, phloem integrity breaks down and auxin transport is disrupted, giving the virus easier access to spread systemically (Mandal et al., 2015). A study done in melon (*Cucumis melo*) recently identified a major QTL on chromosome 11 that confers strong resistance to ToLCNDV. Resistant lines show lower viral titers, milder symptoms, and higher expression of defense-related genes (Sáez et al., 2017). Similar progress has been seen in tomato, where genotypes like H-88-78-1 and LA1777 have shown reduced susceptibility to the virus (Sahu et al., 2010; Sadashiva et al., 2017). Another study compared the transcriptomic data between the resistant tomato cultivar H-88-78-1 and the susceptible "Punjab Chhuhara" variety where they identified SlATG18f, a gene involved in autophagy, as a critical factor in viral defense (Prasad et al., 2021). Functional characterization suggests that SlATG18f contributes to antiviral autophagy, a conserved cellular degradation pathway that targets viral components and limits systemic viral spread within plant tissues.

Similar trends have been observed in potato, where Kufri Bahar demonstrated the highest degree of resistance to ToLCNDV. Kufri Bahar showed upregulation of genes involved in RNA silencing, hypersensitivity, and secondary metabolite production, highlighting an enhanced antiviral defense (Jeevalatha et al., 2014, 2023). In contrast, the susceptible variety Kufri Pukhraj exhibited elevated expression of genes related to viral replication and movement, along with suppressed expression of immune-associated pathways.

3.3 Symptoms in Horticultural Crops

Infected plants often show upward or downward curling and crinkling of leaves. These symptoms are prevalent in tomatoes, where leaves may also show puckering and blistering. Impaired fruits and stunted growth are observed at the late stage of infection (Moriones et al., 2017). In potatoes, typical symptoms of diseases, such as apical leaf curl, crinkled leaves, and mosaic patterns, were observed (Kumar et al., 2021). Similarly, in okra, yellow mosaic was apparent on the leaves, indicating disruption of chlorophyll. Thickening of veins in leaves, vein twisting, petiole bending, upward leaf curling and irregular plant growth were also noticed in infected plants (Venkataravanappa et al., 2018; Kaur et al., 2015). Yellow mottling or mosaic spots on leaves is a common symptom reported in cucurbit crops, such as zucchini, cucumber, and melon (Li et al., 2023). Juárez et al. noted that in 2012, zucchini (*Cucurbita pepo*) plants in Murcia Province, south-eastern Spain, showed leaf curling, swelling of leaf venation, typical mosaic patterns, shortened internodal lengths, and gritty leaf curling and fruits due to infection of ToLCNDV. ToLCNDV-infected eggplant shows leaf curling and vein clearing, leaf mottling and stunted growth (Pratap et al., 2011). In chilli plants, curling of the leaves is caused by ToLCNDV infection, and leaves curl upwards or downwards, usually with crinkling. Yellowing and mosaics are also observed with stunted growth (Khan et al., 2006). These symptoms need to be identified for early detection and ToLCNDV management in horticultural crops.

3.4 Impact on Plant Growth and Yield

The impact of plant viruses is mainly on the livelihood of the farmers and agriculturalists who depend on these crops for income and food security as much as it is on the crop itself. It has a direct and measurable impact on crop productivity and yield. With certain horticultural crops such as tomato, cucumber, melon and squash, the effects of ToLCNDV are particularly severe. The virus has now completely emerged as a significant pathogen, leading to substantial reductions in yield. Yield losses associated with ToLCNDV infection can range between 20% and 100%, depending on the host species, cultivar susceptibility, and the severity of disease (Cai et al., 2023). In highly susceptible tomato cultivars, losses of up to 90% have been documented in northern Indian states. The virus has become endemic, routinely causing severe outbreaks in tomato, cucurbits, potato, and other vegetable crops, which contribute to considerable economic setbacks in affected regions (Naga et al., 2021a, b; Modgill et al., 2025). Outside South Asia, ToLCNDV outbreaks have caused major disruptions in crop production across countries such as Spain and Italy. For example, in Almería, it was reported that yield losses reached up to 20% in melon and 22% in zucchini (Vo et al., 2023) and in southern Italy, infection rates in pumpkin fields have been alarmingly high, ranging between 80% and 100% (Fortes et al., 2016). A more recent outbreak in Shanghai, China, during the autumn of 2022, affected nearly 72.5% of the total cucurbit-growing area, leading to estimated losses of around USD 15 million (Zeng et al., 2023). These examples reflect the scale of economic impact ToLCNDV has when timely control measures are not in place. Beyond reducing the overall yield, ToLCNDV also impacts fruit quality, which is a critical factor for marketability. Infected crops often show deformities, uneven ripening, and discoloration, making the produce less appealing to consumers and harder to sell at a fair price.

4 Conclusion

Understanding the epidemiology and pathophysiology of ToLCNDV is essential for designing targeted, science-based control strategies. These studies help us trace how the virus spreads, identify outbreak patterns, and dissect its interactions with host plants and vectors at the molecular level. Despite these advances, ToLCNDV remains one of the most destructive plant viruses affecting solanaceous and cucurbitaceous crops due to its ability to move through the plant's vascular system, allowing it to disrupt key physiological functions and reduce overall yield. The visible signs of infection are just the surface, and internally, the plant's metabolism, hormone signaling, and defense mechanisms are being compromised in ways that directly impact both yield and quality. Over the past decade, the virus has shown remarkable adaptability by expanding its host range and geographic footprint across Asia, the Mediterranean, and parts of Europe. This growing distribution is complicated to manage in open-field systems, where controlling its vector, the silver whitefly's population, remains a persistent challenge. Managing ToLCNDV effectively requires an integrated approach. No single strategy is enough. Current methods include breeding resistant cultivars, deploying biological control agents, and improving field-level practices like crop rotation, timely planting, and sanitation to reduce whitefly pressure. On the research front, tools like molecular breeding, RNA interference, and CRISPR/Cas9 gene editing hold immense promise for creating crop varieties with long-lasting resistance. Ultimately, combining these advanced tools with well-planned field management can ease the burden on our farmers and make horticultural production systems more resilient not just to ToLCNDV but against future emerging plant viruses as well.

References

Ansar, M., Agnihotri, A., Ranjan, T., Karn, M., Kumar, R., Bhagat, A., Agni-hotri, A., Bertaccini, A., Mater Studiorum, A., Ansar, M., Kumar Agnihotri, A., Ranjan, T., Karn, M., Ranjan Kumar, R., & Prasad Bhagat, A. (2021). Nightshade (Solanum nigrum), an intermediate host between tomato and cucurbits of tomato leaf curl New Delhi virus. *Phytopathologia Mediterranea, 60*(3), 409–419. https://doi.org/10.36253/PHYTO-12745

Berlinger, M. J., Lehmann-Sigura, N., & Taylor, R. A. J. (1996). Survival of Bemisia tabaci adults under different climatic conditions. *Entomologia Experimentalis et Applicata, 80*(3), 511–519. https://doi.org/10.1111/J.1570-7458.1996.TB00967.X

Byrne, D. N. (1999). Migration and dispersal by the sweet potato whitefly, Bemisia tabaci. *Agricultural and Forest Meteorology, 97*(4), 309–316. https://doi.org/10.1016/S0168-1923(99)00074-X

Cai, L., Mei, Y., Ye, R., Deng, Y., Zhang, X., Hu, Z., Zhou, X., Zhang, M., & Yang, J. (2023). Tomato leaf curl New Delhi virus: An emerging plant begomovirus threatening cucurbit production. *aBIOTECH, 4*(3), 257–266. https://doi.org/10.1007/S42994-023-00118-4/FIGURES/3

Chaudhary, S., Devi, P., HanumanthaRao, B., Jha, U. C., Sharma, K. D., Prasad, P. V. V., Kumar, S., Siddique, K. H. M., & Nayyar, H. (2022). Physiological and molecular approaches for developing thermotolerance in vegetable crops: A growth, yield and sustenance perspective. *Frontiers in Plant Science, 13*, 878498. https://doi.org/10.3389/FPLS.2022.878498

Czosnek, H., Hariton-Shalev, A., Sobol, I., Gorovits, R., & Ghanim, M. (2017). The incredible journey of begomoviruses in their whitefly vector. *Viruses, 9*(10), 273. https://doi.org/10.3390/V9100273

Fortes, I. M., Sánchez-Campos, S., Fiallo-Olivé, E., Díaz-Pendón, J. A., Navas-Castillo, J., & Moriones, E. (2016). A novel strain of tomato leaf curl New Delhi virus has spread to the Mediterranean Basin. *Viruses, 8*(11), 307. https://doi.org/10.3390/V8110307

Gaikwad, K. A., Sharma, A., & Cheema, D. S. (2011). Molecular detection and characterization of leaf curl virus infecting tomato in Punjab, India. *Acta Horticulturae, 914*, 153–156. https://doi.org/10.17660/ACTAHORTIC.2011.914.27

Gilbertson, R. L., Batuman, O., Webster, C. G., & Adkins, S. (2015). Role of the insect supervectors Bemisia tabaci and Frankliniella occidentalis in the emergence and global spread of plant viruses. *Annual Review of Virology, 2*, 67–93. https://doi.org/10.1146/ANNUREV-VIROLOGY-031413-085410/CITE/REFWORKS

Gu, Q. S., Yan, L. Y., Liu, L. M., Bao, W. H., Fang, H., Xu, J. H., Li, J. F., Kang, B. S., Wu, H. J., Wang, K., Tao, X. R., & Peng, B. (2023). First report of tomato leaf curl New Delhi virus infecting several cucurbit plants in China. *Plant Disease, 107*(8), 2563. https://doi.org/10.1094/PDIS-01-23-0180-PDN

Hameed, A., Tahir, M. N., Amin, I., & Mansoor, S. (2017). First report of tomato leaf curl New Delhi virus and a tomato yellow leaf curl Thailand betasatellite causing severe leaf curl disease of potato in Pakistan. *Plant Disease, 101*(6), 1065. https://doi.org/10.1094/PDIS-09-16-1335-PDN

Ho, T., Wang, L., Huang, L., Li, Z., Pallett, D. W., Dalmay, T., Ohshima, K., Walsh, J. A., & Wang, H. (2010). Nucleotide bias of DCL and AGO in plant anti-virus gene silencing. *Protein & Cell, 1*(9), 847–858. https://doi.org/10.1007/s13238-010-0100-4

Hussain, M., Mansoor, S., Iram, S., Zafar, Y., & Briddon, R. W. (2004). First report of tomato leaf curl New Delhi virus affecting chilli pepper in Pakistan. *Plant Pathology, 53*(6), 794–794. https://doi.org/10.1111/J.1365-3059.2004.01073.X

Ito, T., Sharma, P., Kittipakorn, K., & Ikegami, M. (2008). Complete nucleotide sequence of a new isolate of tomato leaf curl New Delhi virus infecting cucumber, bottle gourd and muskmelon in Thailand. *Archives of Virology, 153*(3), 611–613. https://doi.org/10.1007/S00705-007-0029-Y/FIGURES/1

Janssen, D., Simón, A., Boulares, M., & Ruiz, L. (2022). Host species-dependent transmission of tomato leaf curl New Delhi virus-ES by Bemisia tabaci. *Plants, 11*(3), 390. https://doi.org/10.3390/PLANTS11030390/S1

Jeevalatha, A., Singh, B. P., Kaundal, P., Kumar, R., & Raigond, B. (2014). RCA-PCR: A robust technique for the detection of tomato leaf curl New Delhi virus-[potato] at ultra-low virus titre. *Potato Journal, 41*, 76–80.

Jeevalatha, A., Kaundal, P., Kumar, R., Raigond, B., Kumar, R., Sharma, S., & Chakrabarti, S. K. (2018). Optimized loop-mediated isothermal amplification assay for tomato leaf curl New Delhi virus-[potato] detection in potato leaves and tubers. *European Journal of Plant Pathology, 150*, 565–573. https://doi.org/10.1007/s10658-017-1300-z

Jeevalatha, A., Siddappa, S., Kumar, R., Tiwari, R. K., Lal, M. K., Sharma, S., Chakrabarty, S. K., & Singh, B. P. (2023). RNA-seq analysis reveals an early defense response to tomato leaf curl

New Delhi virus in potato cultivar Kufri Bahar. *Functional & Integrative Genomics, 23*, 215. https://doi.org/10.1007/s10142-023-01138-5

Juárez, M., Rabádan, M. P., Martínez, L. D., Tayahi, M., Grande-Pérez, A., & Gómez, P. (2019). Natural hosts and genetic diversity of the emerging tomato leaf curl New Delhi virus in Spain. Frontiers in Microbiology, 10(FEB), 432053. https://doi.org/10.3389/FMICB.2019.00140/BIBTEX

Jyothsna, P., Haq, Q. M. I., Singh, P., Sumiya, K. V., Praveen, S., Rawat, R., Briddon, R. W., & Malathi, V. G. (2013). Infection of tomato leaf curl New Delhi virus (ToLCNDV), a bipartite begomovirus with betasatellites, results in enhanced level of helper virus components and antagonistic interaction between DNA B and betasatellites. *Applied Microbiology and Biotechnology, 97*(12), 5457–5471. https://doi.org/10.1007/S00253-012-4685-9/METRICS

Kaur, H., Yadav, C. B., Alatar, A. A., Faisal, M., Jyothsna, P., Malathi, V. G., & Praveen, S. (2015). Gene expression changes in tomato during symptom development in response to leaf curl virus infection. *Journal of Plant Biochemistry and Biotechnology, 24*(3), 347–354. https://doi.org/10.1007/S13562-014-0280-8/FIGURES/3

Kavallieratos, N. G., Wakil, W., Eleftheriadou, N., Ghazanfar, M. U., El-Shafie, H., Simmons, A. M., Dimase, M., Smith, H. A., & Chandler, D. (2024). Integrated management system of the whitefly Bemisia tabaci: a review. Entomologia Generalis, 44(5), 1117–1133. https://doi.org/10.1127/ENTOMOLOGIA/2024/2764

Khan, M. S., Raj, S. K., & Singh, R. (2006). First report of tomato leaf curl New Delhi virus infecting chilli in India. *Plant Pathology, 55*(2), 289. https://doi.org/10.1111/J.1365-3059.2006.01324.X

Kheireddine, A., Sifres, A., Sáez, C., Picó, B., & López, C. (2019). First report of tomato leaf curl New Delhi virus infecting cucurbit plants in Algeria. *Plant Disease, 103*(12). https://doi.org/10.1094/PDIS-05-19-1118-PDN

Kumar, R., & Singh, B. P. (2023). Diagnostics for production of disease-free planting material of potato. *International Journal of Innovative Horticulture, 12*, 31–49. https://doi.org/10.5958/2582-2527.2023.00003.9

Kumar, R., Tiwari, R. K., Jeevalatha, A., Sundaresha, S., Shah, M. A., Sharma, S., Sagar, V., Kumar, M., & Chakrabarti, S. K. (2021). Potato apical leaf curl disease: Current status and perspectives on a disease caused by tomato leaf curl New Delhi virus. *Journal of Plant Diseases and Protection, 128*, 897–911. https://doi.org/10.1007/s41348-021-00463-w

Kumar, R., Tiwari, R. K., Sundaresha, S., Kaundal, P., & Raigond, B. (2022). Potato viruses and their management. In S. K. Chakrabarti, S. Sharma, & M. A. Shah (Eds.), *Sustainable management of potato pests and diseases*. Springer. https://doi.org/10.1007/978-981-16-7695-6_12

Kumar, R., Lal, M. K., Tiwari, R. K., Chourasia, K. N., Kumar, A., Kumar, R., Sharma, S., & Singh, B. (2023). Investigating the interplay between tomato leaf curl New Delhi virus infection, starch metabolism and antioxidant defence system in potato (Solanum tuberosum L.). *Antioxidants, 12*, 1447. https://doi.org/10.3390/antiox12071447

Kushwaha, N., Singh, A. K., Basu, S., & Chakraborty, S. (2015). Differential response of diverse solanaceous hosts to tomato leaf curl New Delhi virus infection indicates coordinated action of NBS-LRR and RNAi-mediated host defense. *Archives of Virology, 160*(6), 1499–1509. https://doi.org/10.1007/s00705-015-2399-x

Lapidot, M., Legg, J. P., Wintermantel, W. M., & Polston, J. E. (2014). Management of whitefly-transmitted viruses in open-field production systems. *Advances in Virus Research, 90*, 147–206. https://doi.org/10.1016/B978-0-12-801246-8.00003-2

Li, R., Liu, Y., Yin, C., Sun, K., & Zhang, P. (2023). Disease note diseases caused by viruses occurrence of tomato leaf curl New Delhi virus in tomato (Lycopersicon esculentum) in China. *Plant Disease, 107*(5). https://doi.org/10.1094/PDIS-06-22-1427-PDN

Mandal, A., Sarkar, D., Kundu, S., & Kundu, P. (2015). Mechanism of regulation of tomato TRN1 gene expression in late infection with tomato leaf curl New Delhi virus (ToLCNDV). Plant Science, 241, 221–237. https://doi.org/10.1016/j.plantsci.2015.10.008

Maruthi, M. N., Rekha, A. R., Cork, A., Colvin, J., Alam, S. N., & Kader, K. A. (2007). First report of tomato leaf curl New Delhi virus infecting tomato in Bangladesh. *Plant Disease, 89*(9), 1011. https://doi.org/10.1094/PD-89-1011C

Mastrochirico, M., Spanò, R., de Miccolis Angelini, R. M., & Mascia, T. (2023). Molecular characterization of a recombinant isolate of tomato leaf curl New Delhi virus associated with severe outbreaks in zucchini squash in southern Italy. *Plants, 12*(13), 2399. https://doi.org/10.3390/PLANTS12132399/S1

Mizutani, T., Daryono, B. S., Ikegami, M., & Natsuaki, K. T. (2011). First report of tomato leaf curl New Delhi virus infecting cucumber in Central Java, Indonesia. *Plant Disease, 95*(11), 1485. https://doi.org/10.1094/PDIS-03-11-0196

Mnari-Hattab, M., Zammouri, S., Belkadhi, M. S., Doña, D. B., ben Nahia, E., & Hajlaoui, M. R. (2015). First report of tomato leaf curl New Delhi virus infecting cucurbits in Tunisia. *New Disease Reports, 31*(1), 21–21. https://doi.org/10.5197/J.2044-0588.2015.031.021

Modgill, N., Naga, K. C., Firake, D. M., Tiwari, R. K., Shivaramu, S., Kumar, R., Bhatnagar, A., Buckseth, T., Bairwa, A., Maharana, C., Sharma, S., & Singh, B. (2025). Flupyradifurone application: A promising approach to curtail tomato leaf curl New Delhi virus (ToLCNDV) transmission vectored by Bemisia tabaci in potato farming. *Crop Protection, 187*, 106951.

Moriones, E., Praveen, S., & Chakraborty, S. (2017). Tomato leaf curl New Delhi virus: An emerging virus complex threatening vegetable and fiber crops. *Viruses, 9*(10), 264. https://doi.org/10.3390/V9100264

Naga, K. C., Subhash, S., Bhatnagar, A., Tiwari, R. K., Kumar, R., Verma, G., Sharma, S., & Chakrabarti, S. K. (2021a). Influence of host plants on virus acquisition and endosymbionts of whitefly, Bemisia tabaci (Gennadius). *Indian Journal of Entomology, 83*, 371–375. https://doi.org/10.5958/0974-8172.2020.00206.0

Naga, K. C., Sundaresha, S., Kumar, R., Tiwari, R. K., Subhash, S., Verma, G., Buckseth, T., Bairwa, A., Sharma, S., Katare, S., & Srivastava, R. M. (2021b). A new record of Asia II 5 genetic group of Bemisia tabaci (Gennadius) in the major potato growing areas of India and its relationship with tomato leaf curl New Delhi virus infecting potato. *3 Biotech, 11*, 421. https://doi.org/10.1007/s13205-021-02966-7

Nagendran, K., Mohankumar, S., Mohammed Faisal, P., Bagewadi, B., & Karthikeyan, G. (2017). Molecular evidence for the occurrence of tomato leaf curl New Delhi virus on chayote (Sechium edule) in southern India. *Virus Disease, 28*(4), 425–429. https://doi.org/10.1007/S13337-017-0403-7/FIGURES/3

Naranjo, S. E., Castle, S. J., de Barro, P. J., & Liu, S. S. (2010). Population dynamics, demography, dispersal and spread of Bemisia tabaci. In *Bemisia: Bionomics and management of a global pest* (pp. 185–226). Springer. https://doi.org/10.1007/978-90-481-2460-2_6/FIGURES/7

Navas-Castillo, J., Fiallo-Olivé, E., & Sánchez-Campos, S. (2011). Emerging virus diseases transmitted by whiteflies. *Annual Review of Phytopathology, 49*, 219–248. https://doi.org/10.1146/annurev-phyto-072910-095235

Orfanidou, C. G., Malandraki, I., Beris, D., Kektsidou, O., Vassilakos, N., Varveri, C., Katis, N. I., & Maliogka, V. I. (2019). First report of tomato leaf curl New Delhi virus in zucchini crops in Greece. *Journal of Plant Pathology, 101*(3), 799. https://doi.org/10.1007/S42161-019-00265-Y/METRICS

Padidam, R., Beachy, R. N., & Fauquet, C. M. (1995). Tomato leaf curl geminivirus from India has a bipartite genome and coat protein is not essential for infectivity. *The Journal of General Virology, 76*(1), 25–35. https://doi.org/10.1099/0022-1317-76-1-25

Panno, S., Caruso, A. G., Troiano, E., Luigi, M., Manglli, A., Vatrano, T., Iacono, G., Marchione, S., Bertin, S., Tomassoli, L., Parrella, G., & Davino, S. (2019). Emergence of tomato leaf curl New Delhi virus in Italy: Estimation of incidence and genetic diversity. *Plant Pathology, 68*(3), 601–608. https://doi.org/10.1111/PPA.12978

Parrella, G., Troiano, E., Formisano, G., Accotto, G. P., & Giorgini, M. (2017). First report of tomato leaf curl New Delhi virus associated with severe mosaic of pumpkin in Italy. *Plant Disease, 102*(2), 459. https://doi.org/10.1094/PDIS-07-17-0940-PDN

Phaneendra, C., Rao, K. R. S. S., Jain, R. K., & Mandal, B. (2012). Tomato leaf curl New Delhi virus is associated with pumpkin leaf curl: A new disease in northern India. *Indian Journal of Virology, 23*(1), 42–45. https://doi.org/10.1007/S13337-011-0054-Z/FIGURES/2

Prasad, A., Hari-Gowthem, G., Muthamilarasan, M., Hussain, Z., Yadav, P. K., Tripathi, S., & Prasad, M. (2021). Molecular characterization of SlATG18f in response to tomato leaf curl New Delhi virus infection in tomato and development of a CAPS marker for leaf curl disease tolerance. *Theoretical and Applied Genetics, 134*, 1463–1474. https://doi.org/10.1007/s00122-021-03783-5

Pratap, D., Kashikar, A. R., & Mukherjee, S. K. (2011). Molecular characterization and infectivity of a tomato leaf curl New Delhi virus variant associated with newly emerging yellow mosaic disease of eggplant in India. *Virology Journal, 8*, 305. http://www.virologyj.com/content/8/1/305

Radouane, N., Lahlali, R., Darif, M., Ezrari, S., Benjelloun, M., Belabess, Z., Barka, E. A., & Tahiri, A. (2022). Detection and molecular characterization of chickpea chlorotic dwarf virus and tomato leaf curl New Delhi virus in Morocco. *Horticulturae, 8*(10), 927. https://doi.org/10.3390/HORTICULTURAE8100927

Raigond, B., Verma, G., Kumar, R., & Tiwari, R. K. (2022). Serological and molecular diagnosis of potato viruses: An overview. In S. K. Chakrabarti, S. Sharma, & M. A. Shah (Eds.), *Sustainable management of potato pests and diseases* (pp. 337–352). Springer. https://doi.org/10.1007/978-981-16-7695-6_13

Renukadevi, P., Sangeetha, B., Malathi, V. G., Nakkeeran, S., & Satya, V. K. (2022). Enigmatic emergence of seed transmission of geminiviruses. In *Geminivirus: Detection, diagnosis and management* (pp. 285–306). Academic Press. https://doi.org/10.1016/B978-0-323-90587-9.00003-1

Rosen, R., Kanakala, S., Kliot, A., Cathrin Pakkianathan, B., Farich, B. A., Santana-Magal, N., Elimelech, M., Kontsedalov, S., Lebedev, G., Cilia, M., & Ghanim, M. (2015). Persistent, circulative transmission of begomoviruses by whitefly vectors. *Current Opinion in Virology, 15*, 1–8. https://doi.org/10.1016/J.COVIRO.2015.06.008

Ruiz, M. L., Simón, A., Velasco, L., García, M. C., & Janssen, D. (2015). First report of tomato leaf curl New Delhi virus infecting tomato in Spain. *Plant Disease, 99*(6), 894. https://doi.org/10.1094/PDIS-10-14-1072-PDN

Sadashiva, A. T., Hanson, P., Krishna Reddy, M., Ravishankar, K. V., Prasad, M., Prasanna, H. C., Madhavi Reddy, K., Singh, T. H., Saritha, R. K., Hussain, Z., Mythili, J. B., Shivashankara, K. S., Bhatt, R. M., Laxman, R. H., Tiwari, R. B., Sridhar, V., Sowmya, V., Kumar, N. P., Kumar, M., et al. (2017). Breeding tomato (Solanum lycopersicum L.) for resistance to biotic and abiotic stresses. *Journal of Horticultural Sciences, 12*(2), 91–105. https://doi.org/10.24154/jhs.v12i2.2

Sáez, C., Esteras, C., Martínez, C., Ferriol, M., Dhillon, N. P. S., López, C., & Picó, B. (2017). Resistance to tomato leaf curl New Delhi virus in melon is controlled by a major QTL located in chromosome 11. *Plant Cell Reports, 36*(10), 1571–1584. https://doi.org/10.1007/S00299-017-2175-3/FIGURES/6

Saha, A., Saha, B., & Saha, D. (2014). Molecular detection and partial characterization of a begomovirus causing leaf curl disease of potato in sub-Himalayan West Bengal, India. *Journal of Environmental Biology, 35*(3), 601–606. PMID: 24813019.

Sahu, P. P., Rai, N. K., Chakraborty, S., Singh, M., Chandrappa, P. H., Ramesh, B., Chattopadhyay, D., & Prasad, M. (2010). Tomato cultivar tolerant to tomato leaf curl New Delhi virus infection induces virus-specific short interfering RNA accumulation and defence-associated host gene expression. *Molecular Plant Pathology, 11*(4), 531–544. https://doi.org/10.1111/J.1364-3703.2010.00630.X

Sandra, N., & Mandal, B. (2024). Emerging evidence of seed transmission of begomoviruses: Implications in global circulation and disease outbreak. *Frontiers in Plant Science, 15*, 1376284. https://doi.org/10.3389/FPLS.2024.1376284/BIBTEX

Sarkar, M., Gupta, D., Singh, O. W., Paul, S., Kumar, R., Mandal, B., & Roy, A. (2024). Identification of pathogenicity determinants in ToLCNDV and their RNAi-based knockdown for disease management in *Nicotiana benthamiana* and tomato plants. *Frontiers in Microbiology, 15*, 1481523. https://doi.org/10.3389/fmicb.2024.1481523

Sohrab, S. S., Karim, S., Varma, A., Abuzenadah, A. M., Chaudhary, A. G., & Mandal, B. (2013). Role of sponge gourd in apical leaf curl disease of potato in northern India. *Phytoparasitica, 41*(4), 403–410. https://doi.org/10.1007/S12600-013-0301-5/FIGURES/5

Usharani, K. S., Surendranath, B., Paul-Khurana, S. M., Garg, I. D., & Malathi, V. G. (2004). Potato leaf curl – A new disease of potato in northern India caused by a strain of tomato leaf curl New Delhi virus. *Plant Pathology, 53*(2), 235–235. https://doi.org/10.1111/J.0032-0862.2004.00959.X

Vasudeva, R. S., & Sam Raj, J. (1948). A leaf-curl disease of tomato. *Phytopathology, 38*, 364–369. https://www.cabidigitallibrary.org/doi/full/10.5555/19481101313

Venkataravanappa, V., Lakshminarayana Reddy, C. N., Saha, S., & Krishna Reddy, M. (2018). Recombinant tomato leaf curl New Delhi virus is associated with yellow vein mosaic disease of okra in India. *Physiological and Molecular Plant Pathology, 104*, 108–118. https://doi.org/10.1016/J.PMPP.2018.10.004

Vo, T. T. B., Lal, A., Nattanong, B., Tabassum, M., Qureshi, M. A., Troiano, E., Parrella, G., Kil, E. J., & Lee, S. (2023). Coat protein is responsible for tomato leaf curl New Delhi virus pathogenicity in tomato. *Frontiers in Plant Science, 14*, 1206255. https://doi.org/10.3389/FPLS.2023.1206255/BIBTEX

Wilisiani, F., Mashiko, T., Wang, W. Q., Suzuki, T., Hartono, S., Neriya, Y., Nishigawa, H., & Natsuaki, T. (2019). New recombinant of tomato leaf curl New Delhi virus infecting melon in Indonesia. *Journal of General Plant Pathology, 85*(4), 306–310. https://doi.org/10.1007/S10327-019-00849-7/FIGURES/2

Zaidi, S. S. E. A., Shafiq, M., Amin, I., Scheffler, B. E., Scheffler, J. A., Briddon, R. W., & Mansoor, S. (2016). Frequent occurrence of tomato leaf curl New Delhi virus in cotton leaf curl disease affected cotton in Pakistan. *PLoS One, 11*(5), e0155520. https://doi.org/10.1371/JOURNAL.PONE.0155520

Zaidi, S. S., Shakir, S., Malik, H. J., Farooq, M., Amin, I., & Mansoor, S. (2017a). First report of tomato leaf curl New Delhi virus on Calotropis procera, a weed as potential reservoir begomovirus host in Pakistan. *Plant Disease, 101*(6), 1071. https://doi.org/10.1094/PDIS-10-16-1539-PDN

Zaidi, S. S. E. A., Martin, D. P., Amin, I., Farooq, M., & Mansoor, S. (2017b). Tomato leaf curl New Delhi virus: A widespread bipartite begomovirus in the territory of monopartite begomoviruses. *Molecular Plant Pathology, 18*(7), 901–911. https://doi.org/10.1111/MPP.12481

Zeng, R., Gu, H., Fan, J., Xu, L., Gao, S., Song, Z., Gao, P., Zhang, K., Zhang, C., Zhu, P., & Dai, F. (2023). Disease note diseases caused by viruses occurrence of tomato leaf curl New Delhi virus in cucurbit plants in China. *Plant Disease, 107*(8). https://doi.org/10.1094/PDIS-01-23-0059-PDN

Host-Pathogen Interactions: A Case Study on Tomato Leaf Curl New Delhi Virus- Tomato Pathosystem

7

Komal Devi, Tuyelee Das, Rohit Sharma, and Ashish Prasad

Abstract

Host-pathogen interactions are the fundamental basis of phytopathology, causing the suppression of plant immunity or interfering with normal cellular processes. Plants have developed multilayered defense systems to maintain resistance or induce susceptibility against these evolved pathogens. Tomato leaf curl New Delhi virus (ToLCNDV) offers a precise model for examining the molecular basics of plant-viral interactions in important crops. ToLCNDV is a bipartite begomovirus (family Geminiviridae) that poses a major threat to tomato production and has expanded its host range to various families of plants. Some hallmark disease symptoms are leaf curl upwards, puckering, vein clearing and distortion. The major cultivated host of ToLCNDV is tomato (*Solanum lycopersicum* L). However, the virus is known to infect several different plant species from various families, including Cucurbitaceae, Euphorbiaceae, Fabaceae, Malvaceae, and Solanaceae. ToLCNDV's success is likely due to its wide host range and extremely adaptable genomic structure, which allows its DNA-A component to interface and trans-replicate with a variety of DNA-B components and beta satellites. Knowing that ToLCNDV may infect multiple plant species and propagate over a vast and ecologically diverse geographic range may help to clarify the possible economic risks connected to similar begomoviral invasions. ToLCNDV infection in tolerant cultivars triggers the up-regulation of defense-associated host genes, suggesting a complex relationship between the virus and host defense mechanisms. This chapter examines the host's novel defense responses to

K. Devi · R. Sharma · A. Prasad (✉)
Department of Botany, Kurukshetra University, Kurukshetra, Haryana, India
e-mail: ashishprasad0411@kuk.ac.in

T. Das
University Centre for Research and Development, Chandigarh University, Mohali, Punjab, India

R. Kumar et al. (eds.), *Tomato Leaf Curl New Delhi Virus (ToLCNDV)*,
https://doi.org/10.1007/978-981-95-2852-3_7

ToLCNDV infection and how they affect tolerance or susceptibility. This chapter mainly discusses the molecular and genetic framework of ToLCNDV-tomato interaction, novel techniques and mechanisms for producing tolerant crops for quality food production, and suggests analytical insights to combat viral diseases in high-value crops.

Keywords

Host-pathogen interactions · Tomato Leaf Curl New Delhi Virus · ToLCNDV

1 Introduction

1.1 Decoding Host-Pathogen Dynamics

Viruses are known as obligatory intracellular parasites that rely on host machinery for their replication, viral migration, and pathogenesis at every phase of the infection cycle (Suman et al., 2023). The interactions of plant viruses with their hosts may trigger either host resistance or susceptibility. By interacting with several host variables, viruses in susceptible hosts manipulate complicated cellular events, which makes it easier for the virus to infect, replicate, and cause disease in host plants (Gaur et al., 2021). Plants possess a multilayered surveillance mechanism to protect themselves against the threat of various pathogens. Major antiviral responses include RNA silencing, innate immunological signaling, ubiquitination and autophagy-mediated destruction, and viral protein translation inhibition (Hoenen & Groseth, 2022). The viral infection process includes numerous protein-protein and protein-nucleic acid interactions with the host, resulting in a competitive molecular network. Overall, the interaction network includes processes by which the host plant activates defense mechanisms while the virus manipulates host components to increase replication and dodge immune responses (Wu et al., 2019; Sarkar et al., 2024). Exploring the interaction network of plant–virus is one of the major objectives of virology studies for better understanding the plant defense mechanism and for boosting crop health and yield. Crop productivity and sustainable agriculture are seriously threatened by plant viruses, which reduce crop output and quality (Jones, 2021). Despite inconsistent effects on host composition, viruses seem to converge, causing phenotypic changes in physiological and physical features like texture, color, volatile profiles, size, metabolite levels (primary and secondary), and metabolite distribution across tissues, defenses, and cuticular waxes (Mauck & Chesnais, 2020; Kumar et al., 2023). Viral infections are more complicated to control than vector-borne plant diseases. Plant viruses, including cucumber mosaic virus, tobacco mosaic virus, brome mosaic virus, potato virus Y, soybean mosaic virus, and plum pox virus, threaten the economic viability of many crops (Scholthof et al., 2011).

1.2 The ToLCNDV Crisis: A Global Challenge

The most prominent family of viruses that infect plants is the Geminiviridae, which contains over 450 known species. The family Geminiviridae consists of nine different genera that are grouped according to host range, genome structure, sequence association, and the transmission vector (such as aphids, leafhoppers, treehoppers, or whiteflies). Of these, *Begomovirus* comprises most of the viral species capable of infecting economically significant plants (Cana-Quijada et al., 2020). Among these species, the tomato leaf curl New Delhi virus (ToLCNDV) is one of the most prevalent and frequently encountered viruses that is spreading in the Indian subcontinent and leading to considerable crop damage in tomatoes in tropical and subtropical areas. It causes tomato leaf curl disease (ToLCD). ToLCD has been linked to at least 13 different viral species in the Indian subcontinent, with ToLCNDV being one of the most significant (Brustolini et al., 2015). This chapter focuses on research on the variation, symptoms, diagnosis, host range, potential management, issues related to agriculture, and dissemination of ToLCNDV affecting crops (Fig. 7.1). The impact of the active spread of the virus will also be briefly discussed. Control strategies include molecular diagnostics, whitefly population management, and genetic approaches, also mentioned in this chapter.

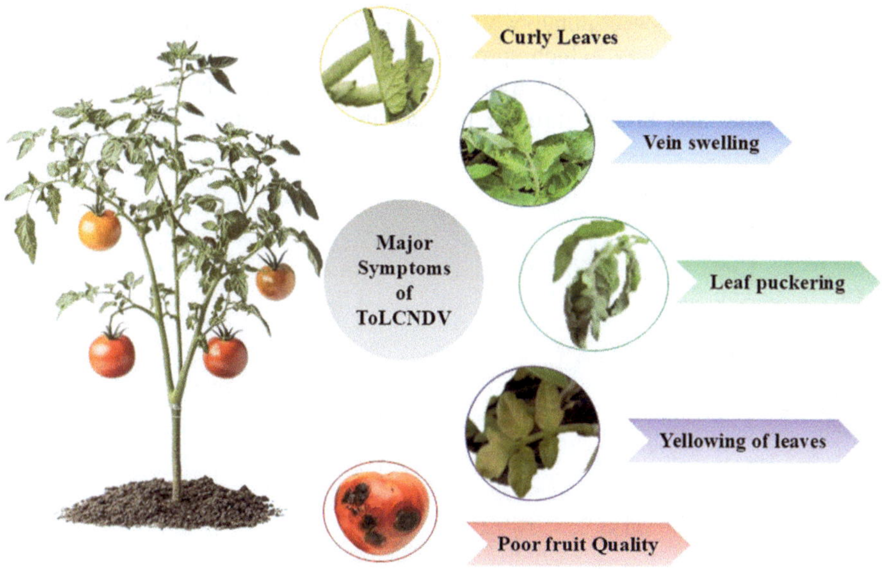

Fig. 7.1 Major symptoms of ToLCNDV in a host plant

2 ToLCNDV: Biology and Epidemiology

2.1 Virus Taxonomy and Genomic Organization

ToLCNDV, a bipartite (DNA-A and DNA-B), ssDNA *Begomovirus* from the Geminiviridae family. D-A and DNA-B components contain genes oriented in the virion-sense (AV1, AV2, AV3, and BV1; respectively) and complementary-sense (AC1, AC2, AC3, AC4, AC5, and *BC1*; respectively) (Padidam et al., 1995; Zaidi et al., 2017). Between DNA-A and DNA-B, a common region (CR) with conserved sequences is present, which is the part of the intergenic space that separates the genes found in complementary sense orientations on DNA-A and DNA-B. The main configuration of the CR is a hairpin structure that covers the replication origin of the virion strand and has a conserved nonanucleotide (TAATATT/AC). At 5′ of the hairpin, repeated sequences called iterons, which are roughly 5–7 nucleotides long, act as binding sites for the protein linked to virus duplication (Jeevalatha et al., 2021, 2023). Replication enhancer, precoat, replication initiator, transcriptional activator, coat, AV3, AC4 and AC5 proteins are all encoded by the ToLCNDV DNA-A. On the other hand, DNA-B genes encode a movement protein and a nuclear shuttle protein (NSP) (Fig. 7.2) (Gafni, 2003).

2.2 Transmission Dynamics

The *Bemisia tabaci* (sweet potato whitefly), a member of the Aleyrodidae family, is the carrier of ToLCNDV, which spreads the virus in a circulative and persistent manner in nature. Through the transmission channel, ToLCNDV interacts with insect proteins and breaks through barriers in the digestive, hemolymph, and

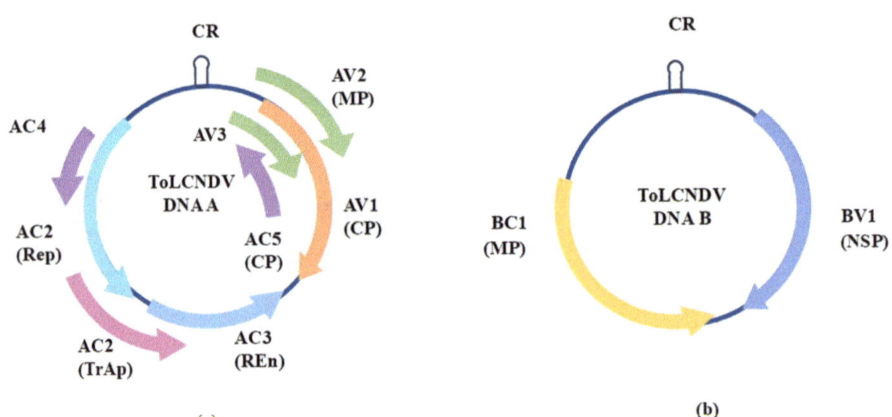

Fig. 7.2 Genome organization of bipartite two circular ssDNA of ToLCNDV (**a**) DNA-A (**b**) DNA-B (REn replication enhancer, CP coat protein, NSP nuclear shuttle protein, MP movement protein, CR common region, Rep replication protein, TrAp transcriptional activator protein)

salivary systems (Fiallo-Olivé et al., 2020). Whitefly with geminate particles, when uncoated, releases viral single-stranded (ssDNA) in the cytoplasm of infected plant cells, which is then transferred to the nucleus for viral DNA amplification. The host cell machinery finally transforms the ssDNA into double-stranded (dsDNA). Viral mRNAs can be translated into numerous viral proteins in the cytoplasm once the viral DNA interacts with nuclear proteins and transcribes them there. Viral proteins required for movement, replication or transcription of viral DNA enter the nucleus via nuclear pores (Fig. 7.3). Viral DNA can enter neighboring cells via plasmodesmatal channels with the assistance of the movement protein, resulting in virus transmission (Stanley, 1985). The effectiveness and specificity of the transmission are determined by these interactions as well as the tissue tropism in the vector body (Rosen et al., 2015). Whiteflies pose a serious threat to agriculture because they destroy their host plants by feeding on them and spreading plant diseases (Naveed et al., 2023). Nevertheless, the molecular basis of calculative transmission of *B. tabaci* virus group is still poorly understood (Zhao et al., 2022; Modgill et al., 2025). Further research is necessary to uncover the molecular connections between viruses and whiteflies, which will help identify the foundation for the species-specific differential transmission.

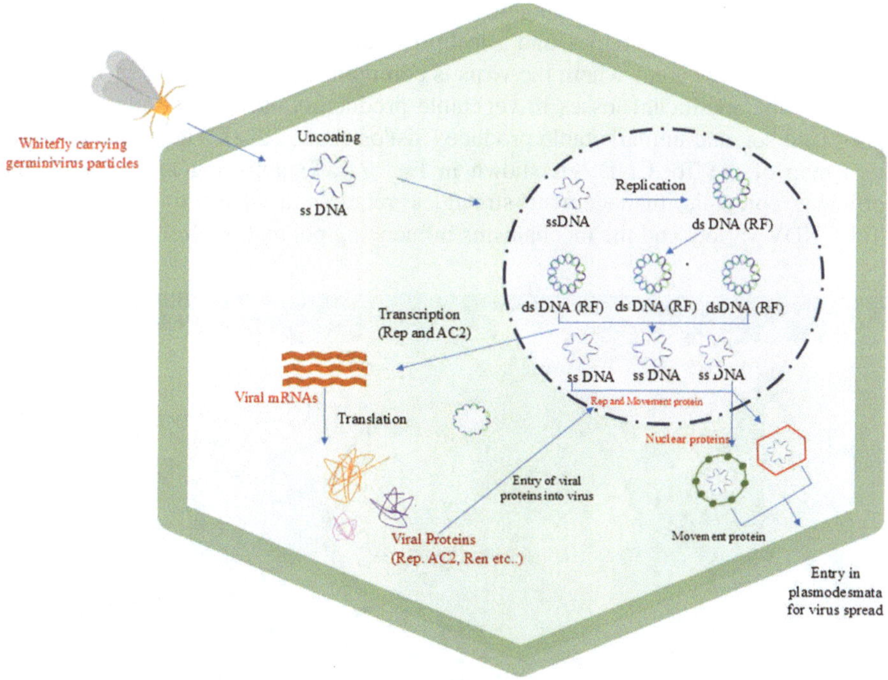

Fig. 7.3 Transmission of virus ssDNA to the cytoplasm of infected plant cells via whitefly (sRen replication enhancer, dsDNA double-stranded, RF replicative form, Rep replication protein)

2.3 Global Distribution and Economic Impact

In the Indian subcontinent, ToLCNDV is a major obstacle to tomato production since it produces widely occurring and commercially vital tomato diseases. Before recent times, ToLCNDV outbreaks were restricted to Asian nations; isolates were primarily detected in Bangladesh, India, the Indian subcontinent, and Pakistan, but ToLCNDV outbreaks were also discovered in Thailand, Indonesia, and Taiwan in Southeast and East Asia (Zaidi et al., 2016b; Chen et al., 2021; Kumar et al., 2021). Since 2012, this ToLCNDV has spread westward, with investigations coming from Iran in the Middle East and its first detection in the western Mediterranean Basin (Fortes et al., 2016). Severe disease outbreaks associated with ToLCNDV infections in greenhouses in Spain and open-field cucurbit crops have also been documented. ToLCNDV has spread its host spectrum and is going to new geographic regions in recent years, as evidenced by recent reports of isolates in Italy mainly close to Spain and in Tunisia that infect zucchini squash (*Cucurbita pepo*), melon (*Cucumis melo*), and cucurbits (*Cucumis sativus*) in Tunisia (Orfanidou et al., 2019; Chen et al., 2021). ToLCNDV is the only bipartite *Begomovirus* that brings diseases in several plant families, involving Cucurbitaceae, Euphorbiaceae, Fabaceae, Papaveraceae, Phyllanthaceae, Malvaceae, and Solanaceae. It affects papaya, cucumber, melon, pumpkin, cotton, opium poppy, sweet leaf, eggplant, pepper, potato, and tomato (Moriones et al., 2017; Chauhan et al., 2023). ToLCNDV has a major detrimental effect on the economy by reducing the yield of many crops, especially cucurbits like melons, cucumbers, potatoes and zucchini (Kumar et al., 2022, Raigond et al., 2022). Farmers in areas where the virus is common experience lower incomes as a result of the significant losses in vegetable production caused by stunted growth, poor fruit set, and unmarketable produce (Siskos et al., 2022). The global distribution map of the ToLCNDV is shown in Fig. 7.4. Establishing more robust and efficient epidemic management strategies requires a better understanding of ToLCNDV variety and the mechanisms influencing population evolution.

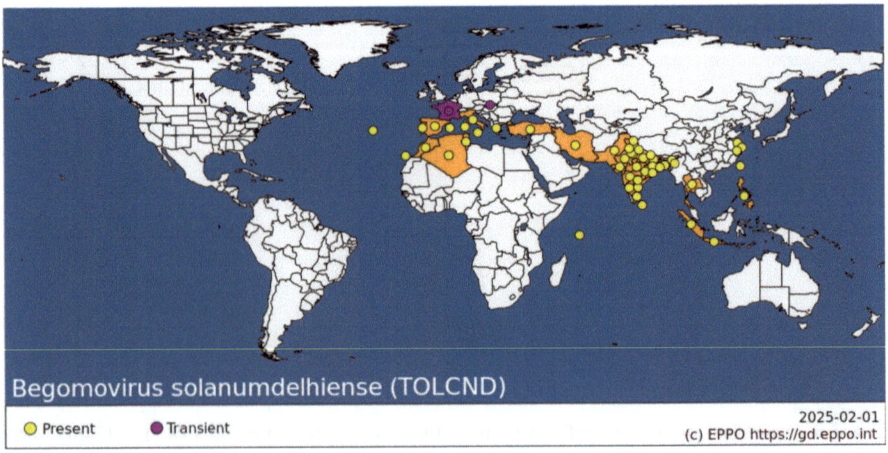

Fig. 7.4 ToLCND global distribution map. (Source: EPPO Global Database; last update 01/02/2025)

3 Tomato as a Host: Physiology and Defense Mechanisms

3.1 Plant Innate Immune Responses

Plants react very specifically to a wide range of microbial diseases through a series of phytohormone signaling cascades. The frequency and specificity of these defense reactions rely on the capabilities of the host (Hatsugai et al., 2017; Sarkar et al., 2024). Plants can recognize and respond to several variety of diseases, which triggers a strong defense response in the plant. Plant PR proteins are induced by these defense responses, which comprise resistance (R) and pathogenesis-related (PR) genes (Adss et al., 2024). Plants possess various disease resistance genes (R genes) that detect specific effector proteins, enabling them to initiate an effector-triggered immune response for hypersensitive resistance. Pathogen-associated molecular pattern (PAMP) is a conserved and distinct molecular structure that plant membrane pattern recognition receptors (PRRs) on the cell surface can recognize, activating signaling cascades that lead to PAMP-triggered immunity (PTI), a fundamental immune reaction. An ongoing arms race between hosts and pathogens leads to a hormone-mediated signaling pathway-based defensive and counter-defense strategy to exploit the host plant fully. Pathogens release virulence-boosting proteins. NSs or NSm code for non-structural effector proteins. Such action causes host plant cells to initiate effector-triggered immunity (ETI), which causes hypersensitive cell death. HR prevents pathogen spread in the host and is visible as necroses on plants at pathogen invasion sites.

3.2 Role of Resistance Genes

There are currently many R genes known to exist, including those involved in the direct detection of viruses, such as I locus for bean common mosaic virus, Sw-5 for tomato tospoviruses, and Rx2 for potato virus X. MiRNAs have been found to target R genes (mainly NBS-LRR) and thereby decreases cellular levels; for example, it was demonstrated that miR6019/6020 and miR9863 regulate the *SlN* gene (the R gene against tobacco mosaic virus) and *Mla* transcripts (the R gene against powdery mildew fungus). Studies have identified multiple mechanisms of resistance to ToLCNDV. Particular chromosomes of *C. melo* and *C. moschata* have been found to have significant resistance loci in cucurbits, where genes like *ARP4* and *SYTA* show differential expression between susceptible and resistant varieties (Román et al., 2023). The resistance of tomato to ToLCNDV has been linked to *Ty* genes, specifically *Ty-2* and *Ty-3* (Lata et al., 2019). In comparison to susceptible cultivars, the resistant tomato cultivar H-88-78-1 shows increased accumulation of virus-specific siRNAs and activation of defense genes (Sahu et al. 2010). Furthermore, the sly-miR159-SlMyb33 module controls the *Sw5a* gene expression, which has been found to confer resistance to ToLCNDV by identifying the viral AC4 effector protein and inducing a hypersensitive response, making it a possible target for creating resistant tomato cultivars (Sharma et al., 2021). To develop effective and long-lasting resistance methods, it is necessary to combine natural and genetically modified (GM) resistance measures, which are based on a better molecular knowledge of ToLCNDV's interactions with its plant hosts and vector.

3.3 Role of Ubiquitin and Ubiquitin-Like Proteins

The ubiquitin/26S proteasome pathway (UPS) is involved in various facets of the regulation of eukaryotic cells, as it swiftly eliminates intracellular proteins. Besides these functionalities, it is likewise linked to immune reactions against pathogen invasion. UPS components play a role, either directly or indirectly, in signaling and regulating non-host disease resistance, responses mediated by resistance genes, systemic acquired resistance and basal immunity. In addition to host cells, which deliver immunity and respond to biotic stress, pathogens, such as viruses, are also responsible for their benefit. Reports indicate that 26S proteasome (26SP) subunits autonomously target biological pathways and regulators, with these targets becoming more apparent. In addition to their role in protein deterioration and turnover, proteasomes also exhibit RNase activity. This action seems a crucial component of plant tolerance against viruses, as the RNase activity of 26SP subunits was emphasized during in vitro interactions with viral RNAs. The α5 subunit of the proteasome in *Arabidopsis thaliana* was recognized as a component that degrades RNAs derived from tobacco mosaic virus and lettuce mosaic virus in vitro. Viral proteins might interact and modify the catalytic activities of 20SP components. For instance, HcPro from potato virus Y engaged with the α1, β2, and β5 subunits of Arabidopsis thaliana 20SP and adjusted the RNase activity, resulting in weakened defense mechanisms against these viruses. Likewise, the 26S proteasomal subunit RPT4a (SlRPT4) gene in *Solanum lycopersicum* showed differential expression following ToLCNDV infection in the tolerant tomato cultivar H-88-78-1. Molecular studies indicated that the SlRPT4 protein exhibits active ATPase activity. SlRPT4 was able to interact specifically with the stem-loop structure found in the intergenic region (IR) of both the DNA-A and DNA-B components of the bipartite viral genome. The interaction of SlRPT4 with IR impeded the activity of RNA Polymerase-II and led to a decrease in the bi-directional transcription of the ToLCNDV genome (Sahu et al., 2016).

4 ToLCNDV Pathogenicity and Host Manipulation

4.1 Viral Proteins and Their Roles

ToLCNDV consists of two genomic parts, DNA-A and DNA-B. DNA-A comprises eight open reading frames (ORFs): AC1, AC2, AC3, AC4, AC5, AV1, AV2, and AV3. These ORFs produce eight proteins: AC1 (Replication initiator protein, Rep), AC2 (Transcriptional activator protein, TrAP), AC3 (Replication enhancer protein, Ren), AC4, AC5, AV1 (coat protein, CP), AV2 (precoat protein) and AV3. DNA-B possesses two ORFs, BC1 and BV1, with BC1 coding for a movement protein and

BV1 coding for a nuclear shuttle protein (NSP) (Moriones et al., 2017; Jeevalatha et al., 2018, 2021). Some of these proteins possess distinct kinase motifs, which suggests their participation in signal transduction pathways. The genomic structure and functional characteristics of the eight viral proteins in ToLCNDV indicate that the majority of these proteins perform multiple functions. The embedded and overlapping ORFs within the genome closely control the expression and functionality of adjacent proteins. Virus proteins encoded by ORFs AC2, AC4, AV2, and BV1 contribute to pathogenicity, disrupt RNA silencing mechanisms of plants, and frequently function as determinants for the symptom.

4.2 Host Manipulation Strategies

As nucleoprotein complexes, viruses are devoid of the fundamental molecular machinery needed for transcription, translation, and genome replication. Their hosts are their only source of support for performing these vital tasks, and they serve as vectors of transmission (Prasad et al., 2020). It has been found that AC2/C2 proteins influence several host immunological pathways to avoid host defense reactions (Guerrero et al., 2020). Apart from inhibiting the host's defense system, AC2/C2 is essential for controlling the coat protein's expression throughout the viral life cycle. These proteins engage with a range of host defensive mechanisms, including both inducible and basal defenses. These latter comprise the UPS, RNA silencing pathways, and the jasmonic acid (JA) and salicylic acid (SA) pathways. For example, SCF^{COI1} (coronatine insensitive 1) degrades the JAZ family of transcriptional regulators (TR) in a proteasome-dependent way in response to JA in plants. Accordingly, the AC2/C2 protein may modify JA retort by preventing JAZs from being targeted by ubiquitination and destruction through the 26S proteasome pathway, which would stop the gene expression that responds to JA (Lozano-Durán et al., 2011). Apart from this, AC2 was later discovered to inhibit the post-transcriptional gene silencing (PTGS) by multiple mechanisms. Remarkably, AC2 can prevent PTGS establishment and can stop the systemic spread of silencing (Jackel et al., 2015).

Various recent findings have also highlighted the significance of autophagy in plant-virus interactions. Prior research has demonstrated that ATG8 proteins can act as cargo receptors for virus proteins and facilitate the destruction of these proteins via the autophagy process. For example, the 2b Cucumber mosaic virus protein is degraded via autophagy (Shukla et al., 2022). A report has found that the transcriptional activator protein of ToLCNDV plays multiple roles in altering host cellular pathways and triggering viral transcription. It is also a well-known RNA silencing suppressor (Sarkar et al., 2021). ToLCNDV TrAP inhibits the host RNA silencing by interacting with AGO1. ToLCNDV TrAP and tomato SlATG8f proteins interact in the nucleus, produce granules outside the nucleus, and intervene in its breakdown by the autophagy pathway. These observations disclose that SlATG8f has a significant role in tomato defense against ToLCNDV (Prasad and Prasad, 2022). These investigations show that SlATG8f plays an important role in tomato protection against ToLCNDV. So, this can be used as antiviral defense strategy in the host, but more research needs to be done in this prospect. Another autophagy-related protein, ATG18f, is also related to defense against ToLCNDV in tomato, as its silencing by the virus-induced gene silencing approach results in susceptibility (Prasad et al., 2021). Some studies have also highlighted that miRNA-regulated defense or susceptibility against ToLCNDV (Naqvi et al., 2010). A study has shown that the miR166-Hd-ZIPIII module is involved in ToLCNDV infection. Silencing of the miR166-regulated Hd-ZIPIII gene leads to tolerance against the virus, suggesting that the gene might be a crucial proviral factor (Prasad et al., 2022).

5 Integrated Management Strategies

5.1 Vector Management

Vector management strategies that include host resistance, cultural control techniques, and chemical and biological management are crucial for effectively managing ToLCNDV while maintaining environmental and social accountability (Bhardwaj et al., 2019). Chemical management has mainly depended on insecticides to control *B. tabaci*, the effective vector for ToLCNDV (Palumbo et al., 2001). Nonetheless, dependence on pesticides can promote resistance and harm ecosystems, resulting in a growing interest in natural bio-stimulants and essential oils (EOs) as eco-friendly options, which have proven to be effective in boosting resistance in zucchini plants. Biological control strategies can help reduce the spread of ToLCNDV by employing predatory mites such as *Amblyseius swirskii* to lower *B. tabaci* populations, consequently decreasing secondary virus transmission in zucchini crops (Abubakar et al., 2022). Additionally, cultural practices play a significant role in virus management; utilizing healthy planting materials and enforcing stringent quarantine protocols helps prevent seed transmission, which is vital considering its prevalence in India and Italy (Naga et al., 2021a, b). Other techniques involve choosing specific crops, practicing intercropping, and managing weeds that act as alternative hosts for viruses, like *Datura stramonium* and *Solanum nigrum*. These combined methods improve the management of ToLCNDV and promote sustainable agricultural practices (Palumbo et al., 2001; Sani et al., 2020).

5.2 Biotechnological Approaches

Plant immune systems that protect against plant viruses at the molecular level can improve crop health, fertility, and yields, which in turn improves food security. Various methods have been used to enhance plant immunity by disrupting the transmission or life cycles of viruses, such as molecular breeding of resistance-associated quantitative trait loci (QTLs), mutagenesis, and introducing R genes or RNA interference (RNAi) constructs through transgenic methods (Hariharan et al., 2021). Gene silencing/RNA interference is the main defense against pathogen severity and reproduction. Synthetic viral gene suppression uses several RNAi mechanisms.

Spray, host, and virus-induced gene silencing are examples. RNAi deactivates viral suppressors on proteins to treat viral infection. An RNAi method is used to treat viral infections by disabling viral suppressors found on different proteins. These proteins consist of sugarcane mosaic virus, V2 from tomato yellow leaf curl virus, the helper component proteinase from potyviruses such as tobacco etch virus, P6 from olive mild mosaic virus, AC2 from *Begomoviruses*, *betasatellites* associated with tomato leaf curl China virus and turnip mosaic virus, P38 from turnip crinkle virus, C2 from Curtoviruses, and 2b from cucumber mosaic virus (Hariharan et al., 2021; Akbar et al., 2022). The most recent genome-editing methodologies include zinc finger nucleases (ZFNs), engineered homing endonucleases, transcription activator-like effector nucleases (TALENs) and meganucleases, clustered often interspersed short palindromic repeats (CRISPR), and CRISPR-associated protein 9 (Cas9) (CRISPR/Cas9) (Green and Hu, 2017; Jeyaraj et al., 2024). Successful resistance against plant viruses has been achieved utilizing CRISPR/Cas technology in a variety of ways, including plant resistance, altering host components essential for viral infection, and editing the virus genome. Instead of directly targeting and destroying viral DNA or RNA, the CRISPR/Cas system is used to change host components that are critical for viral infection in plant resistance mechanisms. This alteration is performed by reducing host susceptibility factors, such as plant susceptibility (S) genes, that prevent viral infection (Varanda et al., 2021). In contrast, virus-mediated resistance focuses on specifically modifying the virus genome, necessitating the use of CRISPR/Cas systems to detect and cut viral genetic material accurately. CRISPR/Cas9 gene editing can directly interfere with disease-causing genes known as "S-genes," resulting in disease-resistant crops. These methods demonstrate the great power of the CRISPR/Cas strategy for offering defense against viral infection in plants (Zaidi et al., 2016a). To address the need for tissue culture, modifications to molecular technology are still important for accurate editing and methods. Genome editing approaches that include editing to inadvertent sites and non-targets should be considered for better results.

6 Challenges and Future Directions

Although progress has been made in breeding crops resistant to ToLCNDV, genome editing techniques offer promising avenues for enhancing plant resistance, but key challenges associated with gene editing include effective target gene selection, reliable dsRNA delivery methods, and ensuring sufficient siRNA delivery to the right cells at the right time. Further study should be done to understand the pathogenesis mechanisms of ToLCNDV, its interactions with satellite DNAs, and its interactions with the whitefly vector *B. tabaci*. Therefore, applying multidisciplinary omics approaches, including genomics, proteomics, and metabolomics, is crucial for understanding virus-host interactions and developing effective antiviral strategies. These techniques provide insights into viral pathogenesis, host responses, and metabolic rewiring during infection. Metabolomics, in particular, offers a window into metabolic mechanisms and has identified common pathways utilized by viruses, potentially leading to broad-spectrum antiviral approaches.

7 Conclusion

In conclusion, determining the host and pathogen interactions, especially the ToLCNDV and its infection on tomato plants, reveals the complex interplay that occurs during viral invasion in important crops. The wide range of hosts and ToLCNDV genomic structure poses a significant economic threat to crop production in the Indian subcontinent and other regions. Understanding the molecular and genetic aspects of the association between the ToLCNDV and tomato is vital for developing novel perspectives towards growing tolerant plants and controlling viral infections. Combining these strategies with vector management, biotechnological approaches such as genome editing, and multidisciplinary omics technologies offers hope for developing plant resistance and an effective antiviral strategy. Further research is needed to understand the mechanisms of ToLCNDV pathology and pathogen-vector interactions, enabling the development of effective strategies to mitigate the impact of ToLCNDV on plant health and productivity.

References

Abubakar, M., Koul, B., Chandrashekar, K., Raut, A., & Yadav, D. (2022). Whitefly (*Bemisia tabaci*) Management (WFM) strategies for sustainable agriculture: A review. *Agriculture (Switzerland), 12*, 1317.

Adss, I. A., Al-Huqail, A. A., Khan, F., EL-Shamy, S. S., Amer, G. M., Hafez, E. E., Ibrahim, O. M., Sobhy, S. E., & Saleh, A. A. (2024). Physio-molecular responses of tomato cultivars to biotic stress: Exploring the interplay between *Alternaria alternata* OP881811 infection and plant defence mechanisms. *Plant Physiology and Biochemistry, 207*, 108421.

Akbar, S., Wei, Y., & Zhang, M. Q. (2022). RNA interference: Promising approach to combat plant viruses. *International Journal of Molecular Sciences, 23*, 5312.

Bhardwaj, V., Sood, S., Kumar, A., Vanishree, G., Sharma, S., Sundaresha, S., Raigond, B., Kumar, R., Bairwa, A., Lal, M., & Chakrabarti, S. K. (2019). Efficiency and reliability of marker-assisted selection for resistance to major biotic stresses in potato. *Potato Journal, 46*, 56–66.

Brustolini, O. J. B., Machado, J. P. B., Condori-Apfata, J. A., Coco, D., Deguchi, M., Loriato, V. A. P., Pereira, W. A., Alfenas-Zerbini, P., Zerbini, F. M., Inoue-Nagata, A. K., Santos, A. A., Chory, J., Silva, F. F., & Fontes, E. P. B. (2015). Sustained NIK-mediated antiviral signalling confers broad-spectrum tolerance to begomoviruses in cultivated plants. *Plant Biotechnology Journal, 13*, 1300–1311.

Cana-Quijada, P., Romero-Rodríguez, B., Vallejo, P. G., Castillo, A. G., & Bejarano, E. R. (2020). Cutting-edge technology to generate plant immunity against geminiviruses. *Current Opinion in Virology, 42*, 58–64.

Chauhan, P. K., Mehta, N., Chauhan, R. S., Kumar, A., Singh, H., Lal, M. K., Tiwari, R. K., & Kumar, R. (2023). Utilization of primary and secondary biochemical compounds in cotton as diagnostic markers for measuring resistance to cotton leaf curl virus (CLCuV). *Frontiers in Plant Science, 14*, 1185337. https://doi.org/10.3389/fpls.2023.1185337

Chen, Y. J., Lai, H. C., Lin, C. C., Neoh, Z. Y., & Tsai, W. S. (2021). Genetic diversity, pathogenicity and pseudorecombination of cucurbit-infecting begomoviruses in Malaysia. *Plants, 10*, 2396.

Fiallo-Olivé, E., Pan, L. L., Liu, S. S., & Navas-Castillo, J. (2020). Transmission of begomoviruses and other whitefly-borne viruses: Dependence on the vector species. *Phytopathology, 110*, 10–17.

Fortes, I. M., Sánchez-Campos, S., Fiallo-Olivé, E., Díaz-Pendón, J. A., Navas-Castillo, J., & Moriones, E. (2016). A novel strain of tomato leaf curl New Delhi virus has spread to the Mediterranean basin. *Viruses, 8*, 307.

Gafni, Y. (2003). Tomato yellow leaf curl virus, the intracellular dynamics of a plant DNA virus. *Molecular Plant Pathology, 4*, 9–15.

Gaur, R. K., Sharma, P., Khurana, S. M. P., & Hohn, T. (2021). *Plant virus-host interaction: Molecular approaches and viral evolution*. Academic Press.

Green, J. C., & Hu, J. S. (2017). Editing plants for virus resistance using CRISPR-Cas. *Acta Virologica, 61*, 138–142.

Guerrero, J., Regedanz, E., Lu, L., Ruan, J., Bisaro, D. M., & Sunter, G. (2020). Manipulation of the plant host by the Geminivirus AC2/C2 protein, a central player in the infection cycle. *Frontiers in Plant Science, 11*, 591.

Hariharan, G., Sivasubramaniam, N., & Prasannath, K. (2021). RNA interference as a promising strategy for plant disease management. In *Food security and plant disease management*. Woodhead Publishing.

Hatsugai, N., Igarashi, D., Mase, K., Lu, Y., Tsuda, Y., Chakravarthy, S., Wei, H., Foley, J. W., Collmer, A., Glazebrook, J., & Katagiri, F. (2017). A plant effector-triggered immunity signaling sector is inhibited by pattern-triggered immunity. *The EMBO Journal, 36*, 2758–2769.

Hoenen, T., & Groseth, A. (2022). Virus–host cell interactions. *Cells, 11*, 804.

Jackel, J. N., Buchmann, R. C., Singhal, U., & Bisaro, D. M. (2015). Analysis of geminivirus AL2 and L2 proteins reveals a novel AL2 silencing suppressor activity. *Journal of Virology, 89*, 3176–3187.

Jeevalatha, A., Kaundal, P., Kumar, R., Raigond, B., Kumar, R., Sharma, S., & Chakrabarti, S. K. (2018). Optimized loop-mediated isothermal amplification assay for tomato leaf curl New Delhi virus-[potato] detection in potato leaves and tubers. *European Journal of Plant Pathology, 150*, 565–573. https://doi.org/10.1007/s10658-017-1300-z

Jeevalatha, A., Vanishree, G., Siddappa, S., Kumar, R., Kaundal, P., Kumar, A., & Chakrabarti, S. K. (2021). Molecular characterization and infectivity analysis of tomato leaf curl New Delhi virus isolates infecting potato. *3 Biotech, 11*, 203. https://doi.org/10.1007/s13205-021-02752-5

Jeevalatha, A., Siddappa, S., Kumar, R., Tiwari, R. K., Lal, M. K., Sharma, S., Chakrabarty, S. K., & Singh, B. P. (2023). RNA-seq analysis reveals an early defense response to tomato leaf curl New Delhi virus in potato cultivar Kufri Bahar. *Functional & Integrative Genomics, 23*, 215. https://doi.org/10.1007/s10142-023-01138-5

Jeyaraj, G., Alphonse, V., Jayanthi, P., Angelin, F. N., Geetanjali, A. S., & Govindan, G. (2024). Harnessing the potential of CRISPR/Cas system for enhancing virus resistance in plants: Targets, strategies, and challenges. *Physiological and Molecular Plant Pathology, 129*, 102202.

Jones, R. A. C. (2021). Global plant virus disease pandemics and epidemics. *Plants, 10*, 233.

Kumar, R., Tiwari, R. K., Jeevalatha, A., Sundaresha, S., Shah, M. A., Sharma, S., Sagar, V., Kumar, M., & Chakrabarti, S. K. (2021). Potato apical leaf curl disease: Current status and perspectives on a disease caused by tomato leaf curl New Delhi virus. *Journal of Plant Diseases and Protection, 128*, 897–911. https://doi.org/10.1007/s41348-021-00463-w

Kumar, R., Tiwari, R. K., Sundaresha, S., Kaundal, P., & Raigond, B. (2022). Potato viruses and their management. In S. K. Chakrabarti, S. Sharma, & M. A. Shah (Eds.), *Sustainable management of potato pests and diseases*. Springer. https://doi.org/10.1007/978-981-16-7695-6_12

Kumar, R., Lal, M. K., Tiwari, R. K., Chourasia, K. N., Kumar, A., Kumar, R., Sharma, S., & Singh, B. (2023). Investigating the interplay between tomato leaf curl New Delhi virus infection, starch metabolism and antioxidant defence system in potato (*Solanum tuberosum* L.). *Antioxidants, 12*, 1447. https://doi.org/10.3390/antiox12071447

Lata, S., Hussain, Z., Mangal, M., Yadav, R. K., Vinutha, T., Jat, G. S., Gosavi, G., Kumar, P., Perveen, S., & Tomar, B. S. (2019). QPCR analysis of Ty-2 and Ty-3 gene pyramided lines of tomato for resistance to tomato leaf curl New Delhi virus (ToLCNDV). *Indian Journal of Agricultural Sciences, 89*, 1719–1722.

Lozano-Durán, R., Rosas-Díaz, T., Gusmaroli, G., Luna, A. P., Taconnat, L., Deng, X. W., & Bejarano, E. R. (2011). Geminiviruses subvert ubiquitination by altering CSN-mediated derubylation of SCF E3 ligase complexes and inhibit jasmonate signaling in *Arabidopsis thaliana. Plant Cell, 23*, 1014–1032.

Mauck, K. E., & Chesnais, Q. (2020). A synthesis of virus-vector associations reveals important deficiencies in studies on host and vector manipulation by plant viruses. *Virus Research, 285*, 197957.

Modgill, N., Naga, K. C., Firake, D. M., Tiwari, R. K., Shivaramu, S., Kumar, R., Bhatnagar, A., Buckseth, T., Bairwa, A., Maharana, C., Sharma, S., & Singh, B. (2025). Flupyradifurone application: A promising approach to curtail tomato leaf curl New Delhi virus (ToLCNDV) transmission vectored by Bemisia tabaci in potato farming. *Crop Protection, 187*, 106951.

Moriones, E., Praveen, S., & Chakraborty, S. (2017). Tomato leaf curl New Delhi virus: An emerging virus complex threatening vegetable and fiber crops. *Viruses, 9*, 264.

Naga, K. C., Subhash, S., Bhatnagar, A., Tiwari, R. K., Kumar, R., Verma, G., Sharma, S., & Chakrabarti, S. K. (2021a). Influence of host plants on virus acquisition and endosymbionts of whitefly, *Bemisia tabaci* (Gennadius). *Indian Journal of Entomology, 83*, 371–375. https://doi.org/10.5958/0974-8172.2020.00206.0

Naga, K. C., Sundaresha, S., Kumar, R., Tiwari, R. K., Subhash, S., Verma, G., Buckseth, T., Bairwa, A., Sharma, S., Katare, S., & Srivastava, R. M. (2021b). A new record of Asia II 5 genetic group of *Bemisia tabaci* (Gennadius) in the major potato growing areas of India and its relationship with tomato leaf curl New Delhi virus infecting potato. *3 Biotech, 11*, 421. https://doi.org/10.1007/s13205-021-02966-7

Naqvi, A. R., Haq, Q. M., & Mukherjee, S. K. (2010). MicroRNA profiling of tomato leaf curl New Delhi virus (tolcndv) infected tomato leaves indicates that deregulation of mir159/319 and mir172 might be linked with leaf curl disease. *Virology Journal, 7*, 281.

Naveed, H., Islam, W., Jafir, M., Andoh, V., Chen, L., & Chen, K. (2023). A review of interactions between plants and whitefly-transmitted Begomoviruses. *Plants, 12*, 3677.

Orfanidou, C. G., Malandraki, I., Beris, D., Kektsidou, O., Vassilakos, N., Varveri, C., Katis, N. I., & Maliogka, V. I. (2019). First report of tomato leaf curl New Delhi virus in zucchini crops in Greece. *Journal of Plant Pathology, 101*, 799.

Padidam, R., Beachy, R. N., & Fauquet, C. M. (1995). Tomato leaf curl geminivirus from India has a bipartite genome and coat protein is not essential for infectivity. *Journal of General Virology, 76*, 25–35.

Palumbo, J. C., Horowitz, A. R., & Prabhaker, N. (2001). Insecticidal control and resistance management for *Bemisia tabaci. Crop Protection, 20*, 739–765.

Prasad, A., & Prasad, M. (2022). Interaction of ToLCNDV TrAP with SlATG8f marks it susceptible to degradation by autophagy. *Cellular and Molecular Life Sciences, 79*, 241.

Prasad, A., Sharma, N., Hari-Gowthem, G., Muthamilarasan, M., & Prasad, M. (2020). Tomato yellow leaf curl virus: Impact, challenges, and management. *Trends in Plant Science, 25*, 897–911.

Prasad, A., Hari-Gowthem, G., Muthamilarasan, M., Hussain, Z., Yadav, P. K., Tripathi, S., & Prasad, M. (2021). Molecular characterization of SlATG18f in response to tomato leaf curl New Delhi virus infection in tomato and development of a CAPS marker for leaf curl disease tolerance. *Theoretical and Applied Genetics, 134*, 1463–1474.

Prasad, A., Sharma, N., Chirom, O., & Prasad, M. (2022). The sly-miR166-SlyHB module acts as a susceptibility factor during ToLCNDV infection. *Theoretical and Applied Genetics, 135*, 233–242.

Raigond, B., Verma, G., Kumar, R., & Tiwari, R. K. (2022). Serological and molecular diagnosis of potato viruses: An overview. In S. K. Chakrabarti, S. Sharma, & M. A. Shah (Eds.), *Sustainable management of potato pests and diseases* (pp. 337–352). Springer. https://doi.org/10.1007/978-981-16-7695-6_13

Román, B., Gómez, P., Janssen, D., & Ruiz, L. (2023). Insights into the key genes in *Cucumis melo* and *Cucurbita moschata* ToLCNDV resistance. *Horticulturae, 9*, 231.

Rosen, R., Kanakala, S., Kliot, A., Cathrin Pakkianathan, B., Farich, B. A., Santana-Magal, N., Elimelech, M., Kontsedalov, S., Lebedev, G., Cilia, M., & Ghanim, M. (2015). Persistent, circulative transmission of begomoviruses by whitefly vectors. *Current Opinion in Virology, 15*, 1–8.

Sahu, P. P., Rai, N. K., Chakraborty, S., Singh, M., Chandrappa, P. H., Ramesh, B., Chattopadhyay, D., & Prasad, M. (2010). Tomato cultivar tolerant to tomato leaf curl New Delhi virus infection induces virus-specific short interfering RNA accumulation and defence-associated host gene expression. *Molecular Plant Pathology, 11*, 531–544.

Sahu, P. P., Sharma, N., Puranik, S., Chakraborty, S., & Prasad, M. (2016). Tomato 26S proteasome subunit RPT4a regulates ToLCNDV transcription and activates hypersensitive response in tomato. *Scientific Reports, 6*, 27078.

Sani, I., Ismail, S. I., Abdullah, S., Jalinas, J., Jamian, S., & Saad, N. (2020). A review of the biology and control of whitefly, *Bemisia tabaci* (Hemiptera: Aleyrodidae), with special reference to biological control using entomopathogenic fungi. *Insects, 11*, 619.

Sarkar, M., Aggarwal, S., Mukherjee, S. K., Mandal, B., & Roy, A. (2021). Sub-cellular localization of suppressor proteins of tomato leaf curl New Delhi virus. *Virus Disease, 32*, 298–304.

Sarkar, M., Gupta, D., Singh, O. W., Paul, S., Kumar, R., Mandal, B., & Roy, A. (2024). Identification of pathogenicity determinants in ToLCNDV and their RNAi-based knockdown for disease management in *Nicotiana benthamiana* and tomato plants. *Frontiers in Microbiology, 15*, 1481523. https://doi.org/10.3389/fmicb.2024.1481523

Scholthof, K. B. G., Adkins, S., Czosnek, H., Palukaitis, P., Jacquot, E., Hohn, T., Hohn, B., Saunders, K., Candresse, T., Ahlquist, P., Hemenway, C., & Foster, G. D. (2011). Top 10 plant viruses in molecular plant pathology. *Molecular Plant Pathology, 12*, 938–954.

Sharma, N., Sahu, P. P., Prasad, A., Muthamilarasan, M., Waseem, M., Khan, Y., Thakur, J. K., Chakraborty, S., & Prasad, M. (2021). The Sw5a gene confers resistance to ToLCNDV and triggers an HR response after direct AC4 effector recognition. *Proceedings of the National Academy of Sciences of the United States of America, 118*, e2101833118.

Shukla, A., Hoffmann, G., Kushwaha, N. K., López-González, S., Hofius, D., & Hafrén, A. (2022). Salicylic acid and the viral virulence factor 2b regulate the divergent roles of autophagy during cucumber mosaic virus infection. *Autophagy, 18*, 1450–1462.

Siskos, L., Cui, L., Wang, C., Visser, R. G. F., Bai, Y., & Schouten, H. J. (2022). A new challenge in melon resistance breeding: The ToLCNDV case. *Euphytica, 218*, 129.

Stanley, J. (1985). The molecular biology of geminiviruses. *Advances in Virus Research, 30*, 139–177.

Suman, R., Rishi, N., & Chandel, V. (2023). Molecular approaches in plant host-virus interaction studies. In *Plant RNA viruses: Molecular pathogenesis and management*. Academic Press.

Varanda, C. M. R., Félix, M. D. R., Campos, M. D., Patanita, M., & Materatski, P. (2021). Plant viruses: From targets to tools for CRISPR. *Viruses, 13*, 141.

Wu, X., Valli, A., García, J. A., Zhou, X., & Cheng, X. (2019). The tug-of-war between plants and viruses: Great progress and many remaining questions. *Viruses, 11*, 203.

Zaidi, S. S. E. A., Tashkandi, M., Mansoor, S., & Mahfouz, M. M. (2016a). Engineering plant immunity: Using CRISPR/Cas9 to generate virus resistance. *Frontiers in Plant Science, 7*, 1673.

Zaidi, S. S. E. A., Shafiq, M., Amin, I., Scheffler, B. E., Scheffler, J. A., Briddon, R. W., & Mansoor, S. (2016b). Frequent occurrence of tomato leaf curl New Delhi virus in cotton leaf curl disease affected cotton in Pakistan. *PLoS One, 11*, e0155520.

Zaidi, S. S. E. A., Martin, D. P., Amin, I., Farooq, M., & Mansoor, S. (2017). Tomato leaf curl New Delhi virus: A widespread bipartite begomovirus in the territory of monopartite begomoviruses. *Molecular Plant Pathology, 18*, 901–911.

Zhao, K., Liu, S. S., Wang, X. W., Yang, J. G., & Pan, L. L. (2022). Manipulation of whitefly behavior by plant viruses. *Microorganisms, 10*, 2410.

Advancements in Diagnostic Tools for Rapid and Sensitive Detection of Tomato Leaf Curl New Delhi Virus in Diverse Crops

8

Bishal Roy, Namrata Verma, Anik Majumdar, Firoz Mondal, Nimmala Sree Valli, Mehulee Sarkar, Prantik Mazumder, Dharmappa Chavan, Ravinder Kumar, and Anirban Roy

Abstract

Tomato leaf curl New Delhi virus (ToLCNDV) is a serious threat to global crop productivity. Timely and accurate detection is essential for managing the disease effectively. While initial diagnostic efforts depended on visible symptoms, recent advances have led to more precise and reliable methods. Serological techniques such as ELISA—including DAC-ELISA, DAS-ELISA, and IC-ELISA—have proven effective in detecting viral antigens, especially when paired with molecular diagnostics. Among nucleic acid-based tools, conventional, nested, and multiplex PCR facilitate qualitative detection, while quantitative PCR (qPCR) enables early diagnosis through accurate viral load quantification. In the field, isothermal amplification methods like LAMP and RPA deliver rapid results without the need for advanced laboratory infrastructure. Newly emerging techniques such as CRISPR-LAMP and nanotechnology-based biosensors, using gold nanoparticles for scanometric DNA detection, are pushing the boundaries of sensitivity and specificity. Together, these diagnostic innovations-from traditional assays to cutting-edge technologies-offer powerful means to monitor and control ToLCNDV, safeguarding agricultural productivity.

Keywords

Detection · Diagnostics · ELISA · PCR · Molecular

B. Roy · N. Verma · A. Majumdar · F. Mondal · N. S. Valli · M. Sarkar · P. Mazumder · D. Chavan · R. Kumar · A. Roy (✉)
Division of Plant Pathology, ICAR-Indian Agricultural Research Institute, New Delhi, India

R. Kumar et al. (eds.), *Tomato Leaf Curl New Delhi Virus (ToLCNDV)*,
https://doi.org/10.1007/978-981-95-2852-3_8

1 Introduction

The tomato leaf curl New Delhi virus (ToLCNDV) is a highly destructive plant pathogen that severely impacts crop productivity worldwide, particularly in subtropical, tropical, and temperate regions. It emerged as a notable plant pathogen in the mid-1990s in northern India, where it was identified as the cause of tomato leaf curl disease in tomato crops (Solanum lycopersicum) (Padidam et al., 1995a, b). It primarily affects tomatoes (*Solanum lycopersicum*). Still, it can also infect other crops, including cucumbers (*Cucumis sativus*), zucchini (*Cucurbita pepo*), squash (*Cucurbita moschata*), bitter gourd (*Momordica charantia*), watermelons (*Cucumis melo*), and bottle gourd (*Lagenaria siceraria*), spanning families such as Cucurbitaceae, Euphorbiaceae, and Fabaceae. ToLCNDV poses a significant challenge to both large-scale farming and smallholder agriculture due to its rapid spread and devastating effects (Table 8.1). ToLCNDV is a bipartite begomovirus from the Geminiviridae family, characterized by two circular single-stranded DNA components, DNA-A and DNA-B, each approximately 2.7 kb in size (Srivastava et al., 1995). The virus is transmitted by the whitefly (*Bemisia tabaci*), which acquires the virus by feeding on infected plants and spreads it to healthy ones after a latent period (Moriones et al., 2017a, b; Naga et al., 2021a, b). This efficient transmission mechanism contributes to the virus's rapid proliferation across diverse agricultural regions. Infected plants exhibit severe symptoms, including leaf curling, yellowing, chlorotic spots, flower abortion, stunted growth, shortened internodes, and a bushy appearance (Kumar et al., 2021; Jeevalatha et al., 2023). The virus's ability to infect multiple crop species amplifies its destructive potential, making it a critical concern for global agriculture.

2 Detection Methods

Effective ToLCNDV management requires an integrated approach combining early detection, vector control, and resistant crop varieties. Controlling whitefly populations through insecticides or biological methods is crucial to limiting viral spread. Additionally, developing and deploying ToLCNDV-resistant crop varieties can reduce the virus's impact. Rapid diagnostics enable farmers to implement timely interventions, minimizing crop losses and sustaining agricultural productivity. One of the approaches for detecting ToLCNDV relies on visual-based diagnosis of infected plants, where ToLCNDV can be detected based on characteristic symptoms such as upward or downward leaf curling, yellowing or chlorotic patches, mosaic mottling, stunted growth with shortened internodes, and deformed or rough-skinned fruits in cucurbits. While these signs were helpful in field-level recognition and were widely used in India in the 1990s, when ToLCNDV was first reported in tomato crops (Padidam et al., 1995a, b), they lacked specificity since other begomoviruses or pests could induce similar symptoms. To overcome these limitations, molecular and serological techniques have since been developed, including conventional PCR and rolling circle amplification (RCA) for initial detection and genome

Table 8.1 ToLCNDV reported on various crops worldwide

ToLCNDV variant	Country reported from	Year identified	Affected crop	Detection technique used	Reference
ToLCNDV	India	2012	Opium poppy (*Papaver somniferum*)	PCR, RCA, NASH	Srivastava et al. (2015)
ToLCNDV	Spain	2012	Zucchini (*Cucurbita pepo*), melon (*Cucumis melo*), cucumber (*Cucumis sativus*), tomato	RCA, PCR, sequencing	Juárez et al. (2014)
ToLCNDV	Pakistan	2005	Bitter gourd	PCR	Tahir and Haider (2005)
ToLCNDV	Spain	2013	Tomato	PCR, sequencing	Moriones et al. (2017a, b)
ToLCNDV	Greece	2018	Zucchini	DAS-ELISA	Orfanidou et al. (2019)
ToLCNDV	Tunisia	2015	Melon, cucumber, zucchini	PCR, qPCR	Mnari-Hattab et al. (2017)
ToLCNDV	Thailand	2017	Tomato, chili	PCR, ELISA	Zaidi et al. (2017)
ToLCNDV	Turkey	2019	Cucumber, melon, squash (*Cucurbita pepo*)	PCR, sequencing	Fidan et al. (2023)
ToLCNDV	China	2021	Tomato	PCR, NGS	Li et al. (2023)
ToLCNDV	Morocco	2018	Zucchini, melon, cucumber	PCR, sequencing	Radouane et al. (2018)
ToLCNDV	Italy	2017	Pumpkin (*Cucurbita moschata*)	ELISA, PCR	Parrella et al. (2017)
ToLCNDV	Sri Lanka	2014	Bitter gourd	PCR, sequencing	Bandaranayake et al. (2014)
ToLCNDV	Spain	2013	Zucchini, melon	Southern blotting (DNA hybridization)	Fortes et al. (2016)
ToLCNDV	Italy	2015	Zucchini squash (*Cucurbita pepo*)	ELISA, PCR	Panno et al. (2016)
ToLCNDV	Thailand	2007	Cucumber, bottle gourd and muskmelon	ELISA, PCR	Ito et al. (2008)
ToLCNDV	Italy	2023	Zucchini, squash, tomato, pepper (*Capsicum annuum*)	LAMP, RPA	Caruso et al. (2023)
ToLCNDV	India	2017	Potato (*Solanum tuberosum*)	RCA, LAMP	Jeevalatha et al. (2014, 2018)

(continued)

Table 8.1 (continued)

ToLCNDV variant	Country reported from	Year identified	Affected crop	Detection technique used	Reference
ToLCNDV	Pakistan	2015	Cotton (*Gossypium hirsutum*)	RCA, RT-PCR	Zaidi et al. (2016)

characterization, quantitative PCR (qPCR) for sensitive quantification, and high-throughput sequencing approaches for strain identification and discovery of associated satellites. More recently, rapid field-deployable methods such as loop-mediated isothermal amplification (LAMP), recombinase polymerase amplification (RPA), and CRISPR–Cas12-based assays have been introduced, enabling quick and reliable on-site diagnosis. Although serological assays are less common due to variability in coat protein epitopes, they can complement DNA-based methods in some instances. Collectively, these advancements provide a robust toolkit for rapid, sensitive, and accurate detection of ToLCNDV across diverse crops and environments.

2.1 Visual Symptom-Based Detection

One of the earliest approaches for detecting ToLCNDV relied on visual-based diagnosis of the infected plants. Farmers and researchers noted distinct symptoms, such as upward or downward curling of leaves, yellowing or chlorotic patches on foliage, mosaic patterns or mottling, stunted growth and shortened internodes, and deformed or rough-skinned fruits in cucurbits. While these signs were indicative of ToLCNDV, they were not specific, as other begomoviruses or pests could cause similar symptoms. This method was widely used in the 1990s in India, where ToLCNDV was first reported in tomato crops (Padidam et al., 1995a, b).

2.2 Molecular Detection

ToLCNDV can be detected and identified using a variety of molecular methods, which are more reliable than the visual symptom-based diagnosis, such as polymerase chain reaction (PCR), rolling circle amplification (RCA), loop-mediated isothermal amplification (LAMP), next-generation sequencing (NGS), and recombinase polymerase amplification (RPA) (Kumar et al., 2023). Polymerase chain reaction (PCR) is a widely used nucleic acid-based technique that amplifies specific viral DNA sequences, offering exceptional sensitivity and precision. Its robustness makes it a standard for diagnosing plant pathogens in laboratory settings. Loop-mediated isothermal amplification (LAMP) provides rapid and accurate detection and is suitable for both field and lab environments. Its isothermal nature eliminates the need for sophisticated equipment, enabling on-site diagnostics. Rolling circle amplification (RCA) amplifies circular DNA without requiring sequence information, offering a simpler and more affordable alternative to PCR. It is particularly effective for detecting ToLCNDV's circular genomic components (Haible et al., 2006). Recombinase polymerase amplification (RPA) and next-generation sequencing (NGS) are advanced methods that provide detailed viral analysis and high-throughput capabilities but are less practical for routine diagnostics due to their complexity and cost.

2.2.1 Serology-Based Detection

Serology-based detection is a method used to identify and measure the presence of antibodies or antigens in a sample using immune responses as a diagnostic tool. Enzyme-linked immunosorbent assay (ELISA) is a cost-effective and reliable method for detecting viral antigens. The double antibody sandwich ELISA (DAS-ELISA) is particularly effective for plant virus detection, utilizing two antibodies to achieve high specificity. The choice of detection method depends on factors such as cost, infrastructure, and diagnostic urgency. ELISA, particularly DAS-ELISA, remains a cornerstone for its affordability and ease of use, especially in low-resource regions. However, PCR and LAMP are used for their superior sensitivity and field applicability. RCA offers a cost-effective alternative for circular DNA detection, while RPA and NGS are reserved for specialized research needs.

2.2.1.1 Enzyme-Linked Immunosorbent Assay (ELISA)

The enzyme-linked immunosorbent assay (ELISA) is a robust and widely used technique for identifying and measuring specific molecules, such as proteins, antigens, or antibodies, within a sample. ELISA operates on the principle that a target molecule (either an antigen or antibody) is anchored to a solid surface, typically a microtiter plate. A complementary antibody or antigen, linked to an enzyme, binds to this target. After removing unbound materials through washing, a substrate is added, which the enzyme converts into a measurable signal, often a color change detectable by a spectrophotometer. The intensity of this signal correlates with the concentration of the target molecule (Engvall & Perlmann, 1971). Its high accuracy and sensitivity make it invaluable in fields like agriculture, medical diagnostics, and scientific research, particularly for detecting plant pathogens (Clark & Adams, 1977). In Asia, Europe, and some regions of Africa, the ToLCNDV has become an imminent threat to crops such as tomatoes, squash, cucumbers, and muskmelon. To effectively manage tomato leaf curl disease, immediate detection and diagnosis of the virus are crucial. Enzyme-linked immunosorbent assay (ELISA) is a vital serological technique for detecting ToLCNDV in plants. It comes in several forms, tailored to specific detection needs, including Direct Antigen Coating ELISA (DAC-ELISA), Double Antibody Sandwich ELISA (DAS-ELISA), and Immuno Capture ELISA (IC-ELISA).

2.2.1.2 Direct Antigen Coating ELISA (DAC-ELISA)

Direct antigen coating enzyme-linked immunosorbent assay (DAC-ELISA) is a key serological method for identifying plant viruses, such as ToLCNDV, that severely impact crops like tomatoes, cucurbits, and solanaceous plants. DAC-ELISA begins

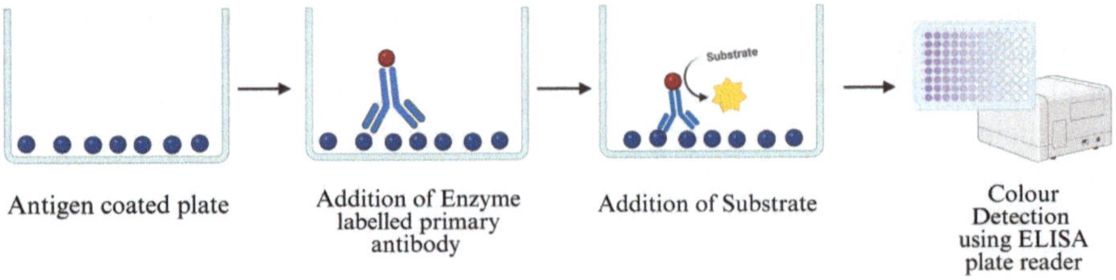

Fig. 8.1 Schematic diagram of steps involved in direct antigen coated ELISA (DAC ELISA)

with coating microtiter plates with plant extract containing viral antigens (Fig. 8.1). Specific antibodies, typically targeting the viral coat protein, are added to bind these antigens. An enzyme-conjugated secondary antibody, such as alkaline phosphatase, is then applied, followed by a substrate that induces a measurable color change, quantified as optical density (OD) using a spectrophotometer. The OD values indicate the presence and relative concentration of ToLCNDV (Clark & Adams, 1977). One of the major advancements involves utilizing recombinant versions of the virus's coat protein to produce antibodies. These recombinant proteins, when expressed in *E. coli*, enable the production of highly specific polyclonal antibodies. These antibodies are capable of detecting even low levels of viral presence in plant samples, offering a robust solution for early-stage diagnostics (Bhat et al., 2002). The integration of nanomaterials into ELISA systems has led to substantial gains in detection sensitivity. Materials like gold nanoparticles are now being used to amplify the enzyme-substrate reaction signal. This advancement allows the assay to recognize much smaller quantities of viral antigens, enhancing its usefulness for rapid and precise diagnostics (Singh & Singh, 2020). The polyclonal antibodies developed for ToLCNDV can detect other begomoviruses due to shared epitopes in their coat proteins. This broad-spectrum detection is advantageous in regions with multiple begomovirus infections, reducing the need for separate assays and enhancing diagnostic efficiency (Saeed & Samad, 2017).

2.2.1.3 Double Antibody Sandwich ELISA (DAS-ELISA)

The double antibody sandwich enzyme-linked immunosorbent assay (DAS-ELISA) is a robust and precise method used to identify viral antigens in plant samples (Fig. 8.2). Recent research has focused on the production of monoclonal antibodies (mAbs) that can recognize multiple strains or species of begomoviruses, including ToLCNDV. For example, a monoclonal antibody designated mAb D2 was developed to detect several begomoviruses such as Tomato yellow leaf curl Thailand virus (TYLCTHV), Squash leaf curl China virus (SLCCNV), and ToLCNDV. This allows for broad-spectrum detection within a single assay, improving diagnostic coverage in mixed infections or regions where several begomoviruses co-circulate (Seepiban et al., 2017). Optimizations in antibody production and assay protocols have improved DAS-ELISA's sensitivity, allowing detection of ToLCNDV antigens at concentrations as low as 0.1 μg/mL. This capability is crucial for identifying the virus in asymptomatic plants, facilitating early disease management in agricultural settings (Saeed et al., 2019).

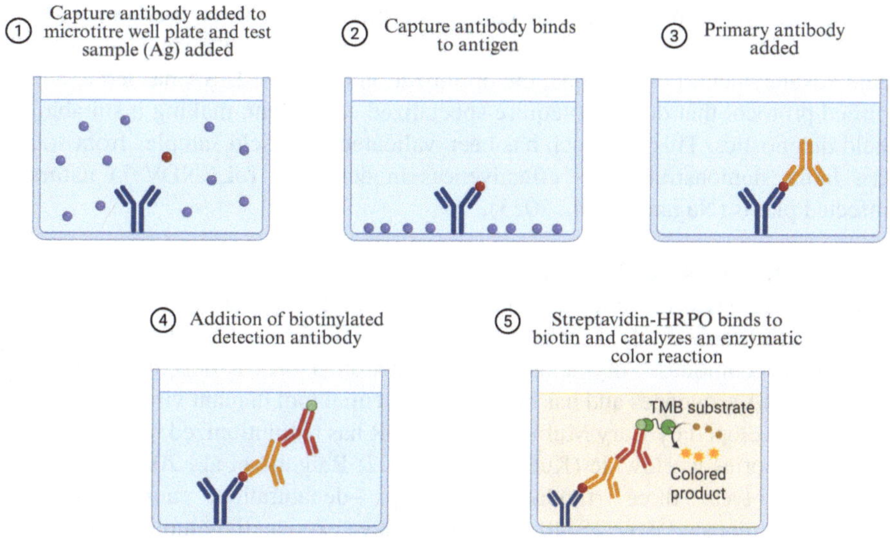

Fig. 8.2 Schematic diagram of double antibody sandwich ELISA (DAC ELISA)

2.2.1.4 Immuno Capture ELISA (IC-ELISA)

Immunocapture enzyme-linked immunosorbent assay (IC-ELISA) offers a reliable, sensitive, and relatively straightforward approach for virus detection by exploiting the specific binding between antibodies and viral antigens. Immunocapture enzyme-linked immunosorbent assay (IC-ELISA) has undergone significant advancements in recent years, enhancing its sensitivity and applicability for detecting ToLCNDV across diverse crops. One of the most notable benefits of this improved detection method is its exceptional sensitivity, particularly when integrated with immunocapture loop-mediated isothermal amplification (IC-LAMP). Recent studies have demonstrated that combining IC-ELISA with molecular techniques, such as Immunocapture Loop-Mediated Isothermal Amplification (IC-LAMP), significantly improves detection sensitivity. IC-LAMP, which captures viral particles before amplification, has shown superior sensitivity in detecting ToLCNDV, even in low-titer infections. This integration allows for the detection of multiple begomoviruses, including Tomato leaf curl Palampur virus and Squash leaf curl China virus, using a single antibody-based assay (Mandal et al., 2023). The incorporation of artificial intelligence (AI) into diagnostic platforms is an emerging trend. AI-enhanced digital nucleic acid amplification testing (dNAAT) systems offer improved image analysis and data interpretation, potentially integrating with IC-ELISA techniques to provide more accurate and efficient diagnostics (Wei et al., 2024). Advancements in antibody production have led to the development of polyclonal antibodies that can detect multiple begomoviruses. These antibodies have been utilized in various assays, such as Dot Immuno Binding Assay (DIBA), Direct Antigen Coating Enzyme Linked Immuno Sorbent Assay (DAC-ELISA), and immunocapture-based PCR and LAMP assays, facilitating the detection of these viruses in different host plants, including cucurbits and solanaceous crops (Naganur et al., 2023). The development of user-friendly diagnostic tools has been a significant advancement. For example, the optimization of DAC-ELISA has led to a simplified protocol that does not require specialized equipment, making it suitable for field diagnostics. This approach has been validated with field samples from southern India, demonstrating its effectiveness in detecting ToLCNDV in naturally infected plants (Naganur et al., 2023).

2.2.2 Nucleic Acid-Based Detection

Nucleic acid-based detection technique for ToLCNDV is the most effective, economical, and rapid method for plant virus detection. One of the most widely used molecular techniques, Polymerase Chain Reaction (PCR), is designed to amplify specific DNA sequences and has become an essential tool in plant virus diagnostics. Initially developed by Kary Mullis in 1983, PCR has revolutionized molecular biology laboratories worldwide (Kumar et al., 2022; Raigond et al., 2022). The technique involves three fundamental steps—denaturation, annealing, and extension—repeated over 25–40 cycles to achieve exponential amplification of the target DNA. During denaturation, the double-stranded DNA is separated into single strands; in the annealing step, primers specifically bind to complementary sequences in the 5′ to 3′ direction; and during extension, deoxynucleotide triphosphates (dNTPs) are incorporated by DNA polymerase to synthesize new strands based on the template sequence (Cassedy et al., 2021). For the detection of ToLCNDV, PCR assays employ primers targeting conserved regions such as the coat protein (CP) gene or the replication-associated protein (Rep) gene, allowing for precise and sensitive identification of the virus, including in plants that do not exhibit visible symptoms. Evolving from conventional PCR multiple variations of PCR, like nested PCR, multiplex PCR, real-time PCR, and nano PCR, have further enhanced detection sensitivity and specificity (Mehetre et al., 2021).

2.2.2.1 Conventional PCR

Conventional Polymerase Chain Reaction (PCR) is a widely used, reliable, and sensitive method for detecting ToLCNDV (Fig. 8.3). The technique involves designing primers that target specific genes within the virus. This is achieved by identifying highly conserved regions of the viral genome and designing primers to amplify these regions. The process begins with the isolation of total DNA from plant tissue, followed by amplification of the target regions using the designed primers. Subsequently, the amplified products are visualized using agarose gel electrophoresis. The presence of the virus is confirmed by the presence of an amplified DNA band and the amplicon size (Naidu et al., 2003). The coat protein gene is frequently targeted for ToLCNDV detection due to its conserved nature (Verma et al., 2016; Sivalingam & Varma, 2007), and this method remains a standard laboratory technique because of its simplicity, reproducibility, and cost-effectiveness (Nayaka et al., 2023). However, conventional PCR has limitations, including the necessity for high-quality DNA, the potential for false-negative results due to low viral loads, and the inability to distinguish between closely related viral strains.

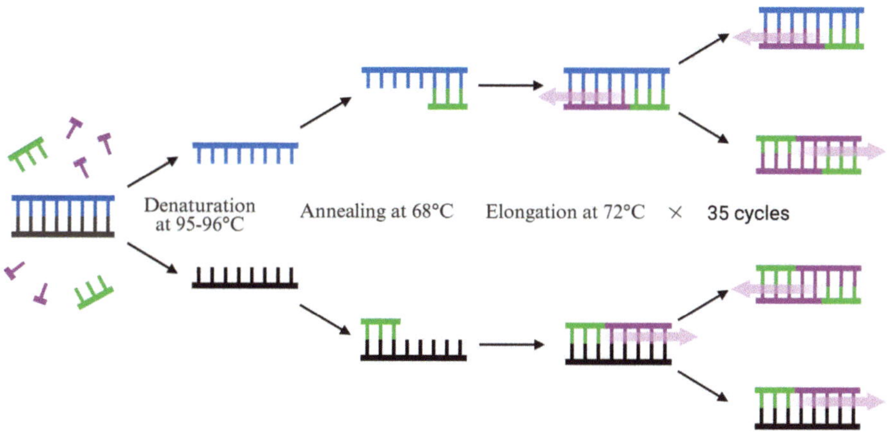

Fig. 8.3 Schematic diagram of polymerase chain reaction technique

2.2.2.2 Nested PCR

Nested PCR is an advanced version of conventional PCR that offers increased sensitivity and specificity through an additional amplification step. Initially, the first set of primers amplifies a large region of the target DNA. Then the second set of inner primers (nested primers) binds within the first amplicon for a second round of amplification. This two-step process provides a significant advantage by dramatically increasing sensitivity, allowing the detection of even very low viral concentrations (Fig. 8.4). The use of two sets of primers also enhances specificity, reducing the likelihood of false positives (Swarnalatha & Reddy, 2014). Nested PCR is particularly valuable for ToLCNDV detection, as the virus often occurs in low titers in infected plants. This method can detect viral DNA even when present in minute quantities or when obscured by plant-derived PCR inhibitors. The need for two independent primer binding events makes the technique highly specific for ToLCNDV detection, minimizing cross-reactivity with related viral strains. It has been successfully used for detecting ToLCNDV (Khan et al., 2006). However, these advantages come with increased technical demands, including the need for stringent contamination controls to prevent amplicon carryover and more complex experimental procedures compared to single-round PCR. Despite these challenges, the enhanced sensitivity of nested PCR makes it particularly useful for early infection detection, strain differentiation, and reliable diagnosis, especially in research and quarantine applications, although it requires more resources.

2.2.2.3 Multiplex PCR

Multiplex PCR is a highly effective variation of conventional PCR that allows the simultaneous amplification of multiple target sequences within a single reaction (Fig. 8.5). This technique proves particularly valuable in cases where plants are co-infected with multiple viruses, which often leads to overlapping and confusing symptoms, making it difficult to determine the specific pathogens involved. While

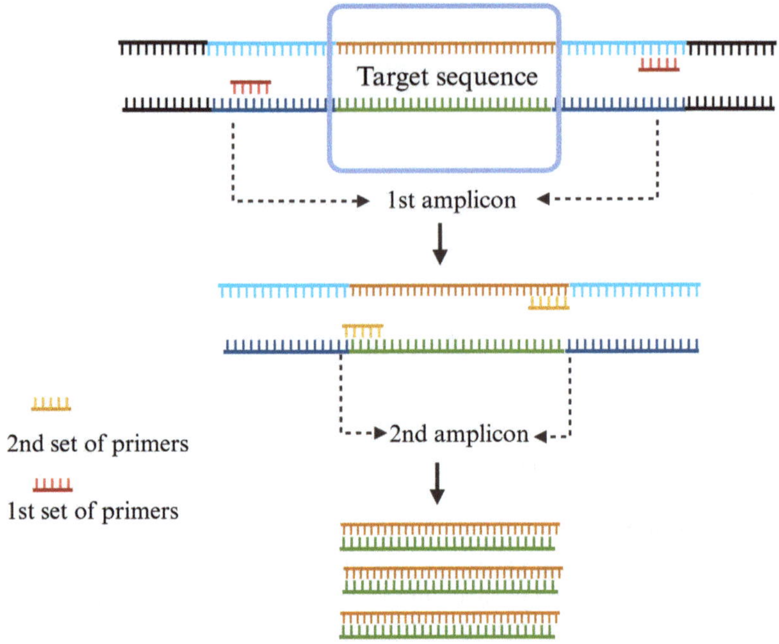

Fig. 8.4 Schematic diagram of nested PCR

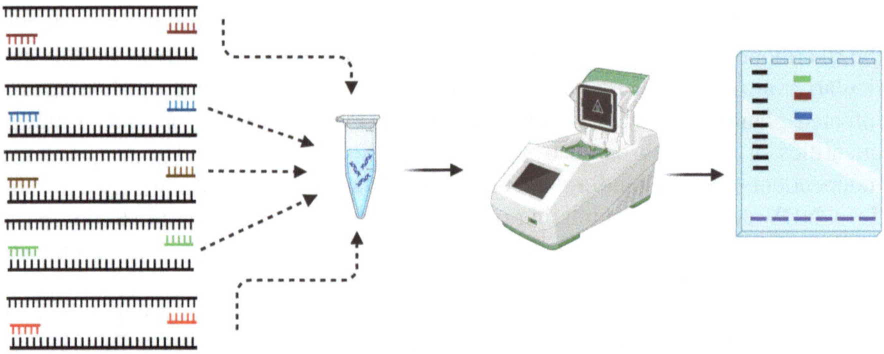

Fig. 8.5 Schematic diagram of multiplex PCR

conventional PCR can be used to detect each virus individually by employing separate primer sets, this approach is time-consuming, resource-intensive, and costly when multiple viruses must be screened. In contrast, multiplex PCR offers a more efficient and economical solution by enabling the detection of several viruses simultaneously (Roy et al., 2005). This method is especially important for diagnosing mixed infections and distinguishing between viruses with closely related genome sequences. Multiplex PCR has been successfully applied to the simultaneous detection of viruses such as tomato yellow leaf curl virus (TYLCV) and ToLCNDV (Hu et al., 2025). Additionally, it can differentiate mixed infections involving different types of plant pathogens (Quintero-Vásquez et al., 2013), highlighting its value as a critical tool for rapid and comprehensive plant virus diagnostics. Despite its advantages, multiplex PCR presents certain technical challenges. The concurrent amplification of multiple targets can lead to primer competition, which may reduce the sensitivity for detecting viruses present at low concentrations. Therefore, meticulous optimization of reaction conditions is essential to ensure the efficient and balanced amplification of all target sequences without interference.

2.2.2.4 Real-Time Quantitative PCR

Real-time quantitative PCR (qPCR) is widely recognized as a premier molecular approach for the detection and quantification of plant viruses, including ToLCNDV (Janssen & Ruiz García, 2025). Unlike conventional PCR, where amplified products are analyzed after the reaction, qPCR tracks DNA amplification in real time by measuring fluorescence signals generated during each cycle (Fig. 8.6). This real-time monitoring eliminates the need for post-amplification procedures like gel electrophoresis. Fluorescent dyes such as SYBR Green, which bind to double-stranded DNA, or sequence-specific probes like TaqMan probes, are commonly employed to achieve high sensitivity and specificity (Kokkinos & Clark, 2006). In ToLCNDV detection, primers and probes are carefully designed to target conserved regions of the viral genome, particularly the coat protein (CP) and replication-associated protein (Rep) genes. Real-time qPCR not only identifies the presence of the virus but also quantifies viral DNA by determining the cycle threshold (Ct) value—the cycle number at which the fluorescence signal surpasses a set threshold above background levels. A lower Ct value generally indicates a higher concentration of viral DNA in the sample (Mackay et al., 2002). Absolute quantification can be achieved by comparing sample Ct values against a standard curve constructed from serial dilutions of known viral DNA concentrations (Jyothsna et al., 2013). This method is especially effective for detecting early infections of ToLCNDV when viral loads are minimal and visual symptoms are absent (Luigi et al., 2020; Ali et al., 2025).

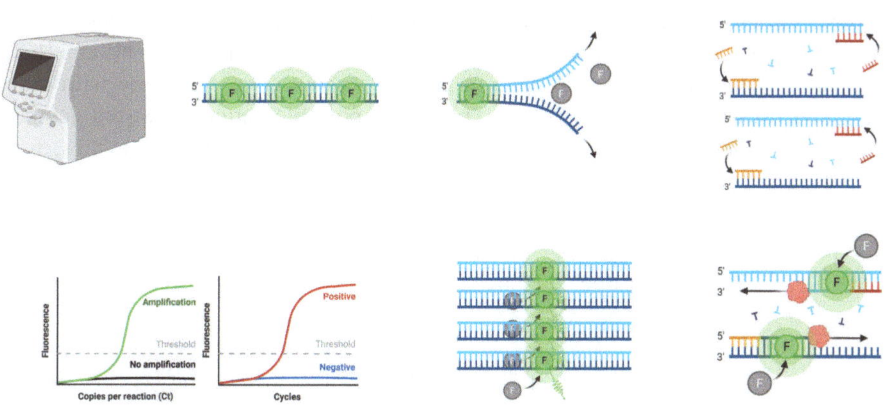

Fig. 8.6 Schematic diagram of multiplex PCR technique

Additionally, real-time qPCR is well-suited for large-scale surveys and breeding programs aimed at identifying virus-resistant tomato lines (Lata et al., 2019; Pradhan et al., 2015). The closed-tube system minimizes contamination risks, further improving the reliability of diagnostic results. However, real-time qPCR requires significant investment in specialized equipment, high-quality reagents, and trained personnel to design and interpret assays correctly. Despite these limitations, its superior accuracy, sensitivity, and ability to provide quantitative data establish real-time qPCR as the benchmark method for ToLCNDV detection in both research and applied fields.

2.2.2.5 Loop-Mediated Isothermal Amplification (LAMP)

Loop-mediated isothermal amplification (LAMP) is a nucleic acid amplification technique that operates under isothermal conditions, eliminating the need for thermal cycling equipment required in conventional polymerase chain reaction (PCR) (Fig. 8.7). LAMP uses a set of four to six specially designed primers that recognize six to eight distinct regions on the target DNA or RNA (via reverse transcription). The reaction is catalyzed by a strand-displacing DNA polymerase, such as Bst polymerase, which amplifies the target sequence at a constant temperature (typically 60–65 °C). The process generates large amounts of DNA, producing a characteristic ladder-like pattern on gel electrophoresis or detectable signals through turbidity, fluorescence, or colorimetric indicators (Notomi et al., 2000). It is a rapid, sensitive, and cost-effective molecular biology technique that has emerged as a pivotal tool for the detection of ToLCNDV, a significant pathogen affecting various crops. Unlike polymerase chain reaction (PCR), which requires thermal cycling, LAMP operates at a constant temperature, eliminating the need for specialized equipment like thermocyclers. This makes LAMP ideal for point-of-care diagnostics and field applications, particularly in resource-constrained environments (Notomi et al., 2000). Recent advancements have significantly enhanced the utility of LAMP for ToLCNDV detection, particularly in field settings and across diverse crops. A notable development is the real-time LAMP assay coupled with rapid sample preparation methods, enabling in-field detection of ToLCNDV in crops like zucchini squash, tomato, and pepper. This assay, targeting a 218-nucleotide sequence within the AV1 gene, demonstrated a sensitivity 1000 times higher than conventional PCR, detecting as few as 4.41×10^2 genome copies without cross-reactivity with other geminiviruses (Caruso et al., 2023). The use of portable battery-powered devices, such as the Genelyzer™ III fluorometer, has facilitated on-site diagnostics, with

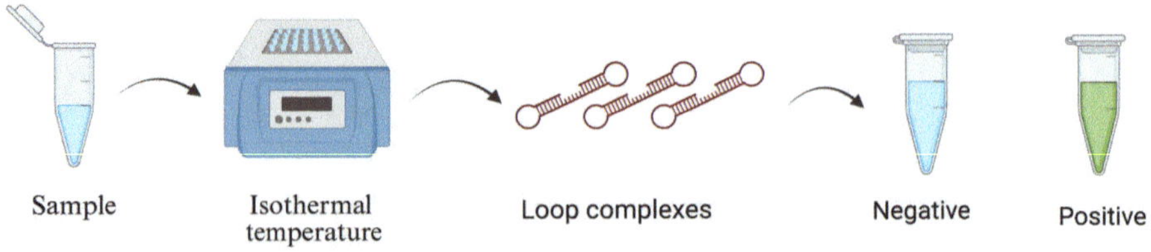

Fig. 8.7 Schematic diagram of LAMP technique

results visualized in real-time via mobile devices, enhancing practical applicability for farmers (Caruso et al., 2023). The integration of LAMP with technologies like CRISPR-Cas12a has further improved specificity and sensitivity, enabling rapid detection without the need for sophisticated laboratory infrastructure (Mahas et al., 2021). Colorimetric LAMP assays have also advanced, allowing visual detection through color changes (e.g., pink to yellow using hydroxy naphthol blue), which eliminates the need for gel electrophoresis and enhances field usability (Yang et al., 2024). Moreover, the combination of LAMP with rolling circle amplification (RCA-LAMP) has been explored to detect ultra-low virus titers, addressing challenges associated with early-stage infections where viral loads are minimal (Gawande et al., 2022). These advancements collectively make LAMP a versatile and powerful tool for ToLCNDV detection, supporting timely interventions and sustainable crop management. Portable heating devices, like battery-powered incubators, enable on-site testing in agricultural fields, making LAMP ideal for rapid ToLCNDV screening in remote regions (Saeed & Samad, 2017). LAMP's simplicity, speed, and reliability make it an invaluable tool for ToLCNDV detection, empowering farmers and researchers to manage this economically damaging virus effectively.

2.2.2.6 Recombinase Polymerase Amplification (RPA)

Recombinase Polymerase Amplification (RPA) is an innovative isothermal amplification technique that enables rapid and specific nucleic acid detection at a constant low temperature, offering a promising solution for field-based diagnostics (Piepenburg et al., 2006). RPA employs recombinase enzymes to facilitate primer binding to specific regions of the ToLCNDV genome, typically targeting conserved sequences like the coat protein or replication-associated genes. These enzymes form complexes with primers, enabling them to scan and invade double-stranded DNA without denaturation. A strand-displacing polymerase then amplifies the target sequence, producing detectable amplicons in as little as 10–20 min (Daher et al., 2016). RPA's isothermal nature allows integration with portable devices, such as battery-powered incubators or handheld fluorescence readers, enabling on-site ToLCNDV detection. For instance, RPA combined with lateral flow dipsticks can deliver results within 30 min, making it suitable for rapid screening in remote farming areas (Kapoor et al., 2017). This approach has been successfully applied to other plant viruses, such as Banana bunchy top virus, and holds similar promise for ToLCNDV. RPA offers several practical benefits for ToLCNDV detection, such as functioning in low-resource settings with minimal infrastructure and eliminating the need for thermal cyclers. It delivers results in 10–20 min, faster than most molecular diagnostic methods, enabling quick disease management (Daher et al., 2016). Therefore, recombinase polymerase amplification offers a powerful, field-deployable solution for detecting ToLCNDV, combining speed, sensitivity, and portability. Its integration with portable devices and digital platforms supports rapid and scalable diagnostics, enabling proactive disease management in agriculture.

2.2.3 Emerging Techniques

2.2.3.1 CRISPR-LAMP-Based Detection

Clustered Regularly Interspaced Short Palindromic Repeats (CRISPR)-based diagnostic platforms offer powerful tools for the sensitive and specific detection of viruses and other pathogens. These systems typically involve an initial isothermal amplification of the target sequence, followed by CRISPR–Cas-mediated recognition and collateral cleavage of a reporter molecule, signaling the presence of the target nucleic acid. Loop-mediated isothermal amplification (LAMP) is a rapid, highly specific, and cost-effective method for amplifying DNA, operating at a constant temperature—generally between 60 and 65 °C—thus eliminating the need for thermal cycling (Caruso et al., 2023). LAMP enables the amplification of target sequences within 30–60 min, making it ideal for quick diagnostics. However, despite its high sensitivity, LAMP is sometimes prone to non-specific amplification and contamination, which can limit its reliability for virus detection.

To obtain specificity coupled with sensitivity, LAMP has been combined with CRISPR-Cas12a-based detection for diagnosing ToLCNDV. This approach involves a two-step process: first, the viral genome is amplified using LAMP, taking advantage of its high sensitivity; then, CRISPR-Cas12a is employed to specifically recognize and cleave a reporter molecule in addition to the amplified region, providing a highly specific detection signal. In Cas12a-mediated detection, upon target recognition, the Cas12a enzyme induces collateral cleavage of nearby single-stranded DNA reporters alongside the amplified LAMP products. For ToLCNDV detection, primers are usually designed against conserved regions within the coat protein (CP) gene, and a complementary CRISPR RNA (crRNA) is subsequently engineered to target the LAMP amplicon (Mahas et al., 2021). This integrated system significantly enhances both sensitivity and specificity, making it a promising tool for rapid and accurate virus diagnostics.

2.2.3.2 Nanotechnology-Based Detection

The integration of nanotechnology into plant virus diagnostics has led to the development of highly sensitive, rapid, and field-adaptable detection platforms. Gold nanoparticles (AuNPs), in particular, have played a pivotal role due to their unique optical properties and their ability to be functionalized with virus-specific DNA probes. These nanoparticle-based assays provide a versatile foundation for colorimetric and optical biosensor-based detection strategies targeting ToLCNDV. One such approach involves a scanometric DNA detection system, where AuNPs are conjugated with thiol-modified oligonucleotide probes designed against the ToLCNDV coat protein (CP) gene. These probes can bind directly to viral DNA immobilized on amine-functionalized glass slides using two assay formats: Direct DNA Hybridization (DDH) and Sandwich DNA Hybridization (SDH). Signal development is enhanced through silver deposition, resulting in a visible color change that can be observed with the naked eye or captured using a flatbed scanner. The DDH assay, in particular, offers higher sensitivity, reduced assay time, and lower cost compared to SDH, as it requires only a single hybridization step.

Furthermore, bifunctional probes—capable of binding both strands of DNA—generate more intense signals than monofunctional ones, with detection limits as low as 100 zeptomolar (zM) (Dharanivasan et al., 2019).

Building on this concept, a U-bent optical fiber-based localized surface plasmon resonance (LSPR) biosensor has also been developed for label-free detection of ToLCNDV DNA. In this system, AuNPs are immobilized on a modified optical fiber surface and functionalized with ssDNA probes complementary to the viral genome. Upon hybridization with target DNA, the localized refractive index changes, producing a measurable shift in the LSPR absorption spectrum—typically monitored at 540 nm. The reverse primer-based probes demonstrated faster response times (within 24 s) and higher hybridization efficiency than forward primers. The system achieved a detection limit of 50 ± 10 ng/μL and completed the diagnostic process in less than 10 min (Das et al., 2022). Both approaches-scanometric colorimetric assays and fiber-optic LSPR biosensors—demonstrate high specificity, rapid detection, and potential for miniaturization into portable devices. Their ability to detect ToLCNDV in crude plant extracts, distinguish it from non-specific targets, and operate under isothermal or ambient conditions makes them valuable additions to the toolkit for field-based diagnostics and real-time virus surveillance in agriculture.

3 Future Prospects in ToLCNDV Detection Technologies

The persistent impact of ToLCNDV on global tomato cultivation highlights the urgent need for diagnostic systems that are not only accurate and rapid but also accessible and deployable in diverse agricultural settings. While existing diagnostic platforms such as ELISA, PCR, and LAMP have long served as the foundation for virus detection, future advancements are expected to arise from the integration of cutting-edge molecular technologies, nanoscale innovations, and digital tools tailored for field-level application. In the case of enzyme-linked immunosorbent assay (ELISA), improvements are likely to emerge through the incorporation of nanoparticle-enhanced reagents, which may significantly improve detection sensitivity. Moreover, the combination of ELISA with microfluidic systems offers potential for developing portable, user-friendly diagnostic kits with faster processing times and reduced reagent consumption, thereby making them more suitable for use in low-resource settings. Molecular techniques, particularly real-time PCR and nested PCR, are anticipated to remain integral to virus diagnostics in laboratory environments. Their high specificity and sensitivity make them valuable tools for confirmatory testing, especially when used in tandem with field-based methods. Advancements in nanotechnology, such as the use of bifunctional AuNP-ssDNA probes and U-bent fiber optic biosensors, have laid the groundwork for the development of next-generation handheld diagnostic devices. These platforms offer the possibility of label-free, real-time detection of ToLCNDV with minimal user training, enhancing their practicality in real-world agricultural settings. In parallel, the emergence of artificial intelligence (AI) and machine learning (ML) tools is expected to enhance virus detection workflows significantly. When trained on high-resolution plant imagery and environmental data, AI-based systems could serve as early screening tools to identify potential infections even before molecular confirmation, particularly valuable in remote or underserved regions. The future of ToLCNDV diagnostics lies in the synergistic integration of multiple platforms—blending molecular biology, serological methods, nanomaterial-based sensing, and digital technologies into cohesive and interoperable diagnostic systems. Such innovations hold the potential to facilitate on-site detection, improve the responsiveness of plant health management programs, and enable data-driven breeding strategies aimed at developing virus-resistant cultivars. Collectively, these advancements will play a pivotal role in sustaining tomato production amid increasing biotic pressures and changing climatic conditions.

4 Conclusion

The ToLCNDV constitutes a formidable obstacle to global agricultural productivity, adversely affecting key crops such as tomatoes, cucumbers, and squashes across diverse agroecological zones. Facilitated by whitefly vectors, the virus induces severe symptoms, including leaf curling, chlorosis, and diminished yields, thereby jeopardizing food security and economic stability. Effective mitigation necessitates prompt detection, vector control, and the deployment of resistant cultivars. A spectrum of diagnostic methodologies, encompassing cost-effective serological assays like DAC-ELISA and DAS-ELISA, alongside advanced molecular techniques such as PCR, LAMP, and emergent CRISPR-based and nanotechnology-enabled platforms, provides versatile tools for ToLCNDV identification. These approaches differ in cost, sensitivity, and applicability, with ELISA and LAMP being particularly advantageous in resource-constrained environments, while PCR variants excel in precision for laboratory-based analyses. Future innovations, including nanoparticle-augmented diagnostics, portable microfluidic systems, and artificial intelligence-driven early warning mechanisms, hold significant promise for enhancing diagnostic efficiency and accessibility. The integration of these novel technologies with comprehensive pest management strategies and resistant crop breeding programs is essential to sustainably address ToLCNDV, thereby bolstering agricultural resilience and ensuring stable crop production in the context of escalating biotic pressure.

References

Ali, M. A., Khan, M. A. U., Naz, H., Malik, M. A. M., Rashid, U., Ashraf, N. M., et al. (2025). CAS9 mediated in-planta defence strategy against tomato leaf curl New Delhi virus (ToLCNDV) in tomato. *Journal of Phytopathology, 173*(1), e70026. https://doi.org/10.1111/jph.70026

Bandaranayake, W. M. E. K., Wickramarachchi, W. A. R. T., Wickramasinghe, H. A. M., Rajapakshe, R. G. A. S., & Dissanayake, D. M. K. K. (2014). *Molecular detection and characterization of begomoviruses associated with cucurbitaceae vegetables in Sri Lanka* (Vol. 42). NSF.

Bhat, A. I., Jain, R. K., & Varma, A. (2002). Serological detection and identification of begomovirus associated with mosaic disease of eggplant (*Solanum melongena*). *Indian Phytopathology, 55*(1), 23–28. https://pubmed.ncbi.nlm.nih.gov/23454646/

Caruso, A. G., Ragona, A., Bertacca, S., Montoya, M. A. M., Panno, S., & Davino, S. (2023). Development of an in-field real-time LAMP assay for rapid detection of tomato leaf curl New Delhi virus. *Plants, 12*(7), 1487. https://doi.org/10.3390/plants12071487

Cassedy, A., Parle-McDermott, A., & O'Kennedy, R. (2021). Virus detection: A review of the current and emerging molecular and immunological methods. *Frontiers in Molecular Biosciences, 8*, 637559. https://doi.org/10.3389/fmolb.2021.637559

Clark, M. F., & Adams, A. N. (1977). Characteristics of the microplate method of enzyme-linked immunosorbent assay for the detection of plant viruses. *Journal of General Virology, 34*(3), 475–483. https://doi.org/10.1099/0022-1317-34-3-475

Daher, R. K., Stewart, G., Boissinot, M., & Bergeron, M. G. (2016). Recombinase polymerase amplification for diagnostic applications. *Clinical Chemistry, 62*(7), 947–958. https://doi.org/10.1373/clinchem.2015.245829

Das, S., Mandal, B., Rao, V. R., & Kundu, T. (2022). Detection of tomato leaf curl New Delhi virus DNA using U-bent optical fiber-based LSPR probes. *Optical Fiber Technology, 74*, 103108. https://doi.org/10.1016/j.yofte.2022.103108

Dharanivasan, G., Jesse, D. M. I., Rajamuthuramalingam, T., Rajendran, G., Shanthi, S., & Kathiravan, K. (2019). Scanometric detection of tomato leaf curl New Delhi viral DNA using mono-and bifunctional AuNP-conjugated oligonucleotide probes. *ACS Omega, 4*(6), 10094–10107. https://doi.org/10.1021/acsomega.9b00340

Engvall, E., & Perlmann, P. (1971). Enzyme-linked immunosorbent assay (ELISA) quantitative assay of immunoglobulin G. *Immunochemistry, 8*(9), 871–874. https://doi.org/10.1016/0019-2791(71)90454-X

Fidan, H. A. K. A. N., Yıldız, K., Sarıkaya, P., & Çalış, Ö. Z. E. R. (2023). First report of tomato leaf curl New Delhi virus in Türkiye. *New Disease Reports, 47*(2), e12180.

Fortes, I. M., Sánchez-Campos, S., Fiallo-Olivé, E., Díaz-Pendón, J. A., Navas-Castillo, J., & Moriones, E. (2016). A novel strain of tomato leaf curl New Delhi virus has spread to the Mediterranean basin. *Viruses, 8*(11), 307. https://doi.org/10.3390/v8110307

Gawande, S. P., Raghavendra, K. P., Monga, D., Nagrale, D. T., Prabhulinga, T., Hiremani, N., Meshram, M., Kranthi, S., Gokte-Narkhedkar, N., & Waghmare, V. N. (2022). Development of loop-mediated isothermal amplification (LAMP): A new tool for rapid diagnosis of cotton leaf curl viral disease. *Journal of Virological Methods, 306*, 114541. https://doi.org/10.1016/j.jviromet.2022.114541

Haible, D., Kober, S., & Jeske, H. (2006). Rolling circle amplification revolutionizes diagnosis and genomics of geminiviruses. *Journal of Virological Methods, 135*(1), 9–16. https://doi.org/10.1016/j.jviromet.2006.01.017

Hu, H., Zhang, J., Wu, X., Li, L., & Qian, Y. (2025). Development and application of a multiplex PCR assay for simultaneous detection of tomato yellow leaf curl virus and tomato leaf curl New Delhi virus. *Viruses, 17*(3), 322. https://doi.org/10.3390/v17030322

Ito, T., Sharma, P., Kittipakorn, K., & Ikegami, M. (2008). Complete nucleotide sequence of a new isolate of tomato leaf curl New Delhi virus infecting cucumber, bottle gourd and muskmelon in Thailand. *Archives of Virology, 153*, 611–613. https://doi.org/10.1007/s00705-007-0029-y

Janssen, D., & Ruiz García, L. (2025). Detection of tomato leaf curl New Delhi virus-ES by real-time quantitative PCR. In *Geminiviruses: Methods and protocols* (pp. 145–156). Springer. https://doi.org/10.1007/978-1-0716-4454-6_12

Jeevalatha, A., Singh, B. P., Kaundal, P., Kumar, R., & Raigond, B. (2014). RCA-PCR: A robust technique for the detection of tomato leaf curl New Delhi virus-[potato] at ultra-low virus titre. *Potato Journal, 41*, 76–80.

Jeevalatha, A., Kaundal, P., Kumar, R., et al. (2018). Optimized loop-mediated isothermal amplification assay for *tomato leaf curl New Delhi virus-[potato]* detection in potato leaves and tubers. *European Journal of Plant Pathology, 150*, 565–573. https://doi.org/10.1007/s10658-017-1300-z

Jeevalatha, A., Siddappa, S., Kumar, R., Tiwari, R. K., Lal, M. K., Sharma, S., Chakrabarty, S. K., & Singh, B. P. (2023). RNA-seq analysis reveals an early defense response to tomato leaf curl New Delhi virus in potato cultivar Kufri Bahar. *Functional & Integrative Genomics, 23*, 215. https://doi.org/10.1007/s10142-023-01138-5

Juárez, M., et al. (2014). First detection of tomato leaf curl New Delhi virus in zucchini in Spain. *Plant Disease, 98*(6), 857. https://doi.org/10.1094/PDIS-10-13-1050-PDN

Jyothsna, P., Haq, Q. M. I., Singh, P., Sumiya, K. V., Praveen, S., Rawat, R., et al. (2013). Infection of tomato leaf curl New Delhi virus (ToLCNDV), a bipartite begomovirus with betasatellites, results in enhanced level of helper virus components and antagonistic interaction between DNA B and betasatellites. *Applied Microbiology and Biotechnology, 97*, 5457–5471. https://doi.org/10.1007/s00253-012-4685-9

Kapoor, R., Srivastava, N., Kumar, S., Saritha, R. K., Sharma, S. K., Jain, R. K., & Baranwal, V. K. (2017). Development of a recombinase polymerase amplification assay for the diagnosis of banana bunchy top virus in different banana cultivars. *Archives of Virology, 162*, 2791–2796. https://doi.org/10.1007/s00705-017-3399-9

Khan, M. S., Raj, S. K., & Singh, R. (2006). First report of tomato leaf curl New Delhi virus infecting chilli in India. *Plant Pathology, 55*(2), 289.

Kokkinos, C. D., & Clark, C. A. (2006). Real-time PCR assays for detection and quantification of sweet potato viruses. *Plant Disease, 90*(6), 783–788. https://doi.org/10.1094/PD-90-0783

Kumar, R., Tiwari, R. K., Jeevalatha, A., Sundaresha, S., Shah, M. A., Sharma, S., Sagar, V., Kumar, M., & Chakrabarti, S. K. (2021). Potato apical leaf curl disease: Current status and perspectives on a disease caused by tomato leaf curl New Delhi virus. *Journal of Plant Diseases and Protection, 128*, 897–911. https://doi.org/10.1007/s41348-021-00463-w

Kumar, R., Tiwari, R. K., Sundaresha, S., Kaundal, P., & Raigond, B. (2022). Potato viruses and their management. In S. K. Chakrabarti, S. Sharma, & M. A. Shah (Eds.), *Sustainable management of potato pests and diseases*. Springer. https://doi.org/10.1007/978-981-16-7695-6_12

Kumar, R., Lal, M. K., Tiwari, R. K., Chourasia, K. N., Kumar, A., Kumar, R., Sharma, S., & Singh, B. (2023). Investigating the interplay between tomato leaf curl New Delhi virus infection, starch metabolism and antioxidant defence system in potato (*Solanum tuberosum* L.). *Antioxidants, 12*, 1447. https://doi.org/10.3390/antiox12071447

Lata, S., Hussain, Z., Mangal, M., Yadav, R. K., Jat, G. S., Gosavi, G., et al. (2019). QPCR analysis of Ty-2 and Ty-3 gene pyramided lines of tomato for resistance to tomato leaf curl New Delhi virus (ToLCNDV). *The Indian Journal of Agricultural Sciences, 89*(10), 1719–1722. https://doi.org/10.56093/ijas.v89i10.94636

Li, R., Liu, Y., Yin, C., Sun, K., & Zhang, P. (2023). Occurrence of tomato leaf curl New Delhi virus in tomato (*Lycopersicon esculentum*) in China. *Plant Disease, 107*(5), 1639.

Luigi, M., Manglli, A., Bertin, S., Donati, L., Tomassoli, L., Ferretti, L., & Faggioli, F. (2020). Development and validation of a specific real-time PCR protocol for the detection of tomato leaf curl New Delhi virus. *European Journal of Plant Pathology, 157*, 969–974. https://doi.org/10.1007/s10658-020-02038-1

Mackay, I. M., Arden, K. E., & Nitsche, A. (2002). Real-time PCR in virology. *Nucleic Acids Research, 30*(6), 1292–1305. https://doi.org/10.1093/nar/30.6.1292

Mahas, A., Hassan, N., Aman, R., Marsic, T., Wang, Q., Ali, Z., & Mahfouz, M. M. (2021). LAMP-coupled CRISPR–Cas12a module for rapid and sensitive detection of plant DNA viruses. *Viruses, 13*(3), 466. https://doi.org/10.3390/v13030466

Mehetre, G. T., Leo, V. V., Singh, G., Sorokan, A., Maksimov, I., Yadav, M. K., et al. (2021). Current developments and challenges in plant viral diagnostics: A systematic review. *Viruses, 13*(3), 412. https://doi.org/10.3390/v13030412

Mnari-Hattab, M., Salah, I. B., Bahri, R., Mezghani-Khemakhem, M., & Hajlaoui, M. R. (2017). First report of tomato leaf curl New Delhi virus infecting cucurbits in Tunisia. *Plant Disease, 101*(3), 509. https://doi.org/10.1094/PDIS-10-16-1426-PDN

Moriones, E., Praveen, S., & Chakraborty, S. (2017a). Tomato leaf curl New Delhi virus: An emerging virus complex threatening vegetable crops. *Molecular Plant Pathology, 18*(5), 601–606. https://doi.org/10.3390/v9100264

Moriones, E., Praveen, S., & Chakraborty, S. (2017b). Tomato leaf curl New Delhi virus: An emerging virus complex threatening vegetable and fiber crops. *Viruses, 9*(10), 264. https://doi.org/10.3390/v9100264

Naga, K. C., Subhash, S., Bhatnagar, A., Tiwari, R. K., Kumar, R., Verma, G., Sharma, S., & Chakrabarti, S. K. (2021a). Influence of host plants on virus acquisition and endosymbionts of whitefly, *Bemisia tabaci* (Gennadius). *Indian Journal Of Entomology, 83*, 371–375. https://doi.org/10.5958/0974-8172.2020.00206.0

Naga, K. C., Sundaresha, S., Kumar, R., Tiwari, R. K., Subhash, S., Verma, G., Buckseth, T., Bairwa, A., Sharma, S., Katare, S., & Srivastava, R. M. (2021b). A new record of Asia II 5 genetic group of *Bemisia tabaci* (Gennadius) in the major potato growing areas of India and its relationship with tomato leaf curl New Delhi virus infecting potato. *3 Biotech, 11*, 421. https://doi.org/10.1007/s13205-021-02966-7

Naganur, P., Shankarappa, K. S., Mesta, R. K., Rao, C. D., Venkataravanappa, V., Maruthi, M. N., & Reddy, L. C. N. R. (2023). Detecting tomato leaf curl New Delhi virus causing ridge gourd yellow mosaic disease, and other begomoviruses by antibody-based methods. *Plants, 12*(3), 490. https://doi.org/10.3390/plants12030490

Nayaka, S. N., Singh, O. W., Kumar, P., Roy, A., & Mandal, B. (2023). Geographical distribution of tomato-infecting begomoviruses in major cucurbits in India: A diagnostic analysis using begomovirus species specific PCR. *VirusDisease, 34*(3), 421–430. https://doi.org/10.1007/s13337-023-00837-8

Notomi, T., Okayama, H., Masubuchi, H., Yonekawa, T., Watanabe, K., Amino, N., & Hase, T. (2000). Loop-mediated isothermal amplification of DNA. *Nucleic Acids Research, 28*(12), e63. https://doi.org/10.1093/nar/28.12.e63

Orfanidou, C. G., Malandraki, I., Beris, D., et al. (2019). First report of tomato leaf curl New Delhi virus in zucchini crops in Greece. *Journal of Plant Pathology, 101*, 799. https://doi.org/10.1007/s42161-019-00265-y

Padidam, M., Beachy, R. N., & Fauquet, C. M. (1995a). Classification and identification of geminiviruses using sequence comparisons. *Journal of General Virology, 76*(9), 2491–2502. https://doi.org/10.1099/0022-1317-76-9-2491

Padidam, M., Beachy, R. N., & Fauquet, C. M. (1995b). Tomato leaf curl geminivirus from India has a bipartite genome and coat protein is not essential for infectivity. *Journal of General Virology, 76*(1), 25–35.

Panno, S., Iacono, G., Davino, M., Marchione, S., Zappardo, V., Bella, P., et al. (2016). First report of tomato leaf curl New Delhi virus affecting zucchini squash in an important horticultural area of southern Italy. *New Disease Reports, 33*(6), 2044–0588. https://doi.org/10.5197/j.2044-0588.2016.033.006

Parrella, G., Troiano, E., Formisano, G., Accotto, G. P., & Giorgini, M. (2017). First report of tomato leaf curl New Delhi virus associated with severe mosaic of pumpkin in Italy. *Plant Disease, 102*. https://doi.org/10.1094/PDIS-07-17-0940-PDN

Piepenburg, O., Williams, C. H., Stemple, D. L., & Armes, N. A. (2006). DNA detection using recombination proteins. *PLoS Biology, 4*(7), e204. https://doi.org/10.1371/journal.pbio.0040204

Pradhan, B., Naqvi, A. R., Saraf, S., Mukherjee, S. K., & Dey, N. (2015). Prediction and characterization of tomato leaf curl New Delhi virus (ToLCNDV) responsive novel microRNAs in *Solanum lycopersicum*. *Virus Research, 195*, 183–195. https://doi.org/10.1016/j.virusres.2014.09.001

Quintero-Vásquez, G. A., Bazán-Tejeda, M. L., Martínez-Peñafiel, E., Kameyama-Kawabe, L., & Bermúdez-Cruz, R. M. (2013). Multiplex PCR to detect four different tomato-infecting pathogens. *Folia Microbiologica, 58*, 269–276. https://doi.org/10.1007/s12223-012-0206-6

Radouane, N., Tahiri, A., El Ghadraoui, L., Al Figuigui, J., & Lahlali, R. (2018). First report of tomato leaf curl New Delhi virus in Morocco. *New Disease Reports, 37*(2), 2044–0588.

Raigond, B., Verma, G., Kumar, R., & Tiwari, R. K. (2022). Serological and molecular diagnosis of potato viruses: An overview. In S. K. Chakrabarti, S. Sharma, & M. A. Shah (Eds.),

Sustainable management of potato pests and diseases (pp. 337–352). Springer. https://doi.org/10.1007/978-981-16-7695-6_13

Roy, A., Fayad, A., Barthe, G., & Brlansky, R. H. (2005). A multiplex polymerase chain reaction method for reliable, sensitive and simultaneous detection of multiple viruses in citrus trees. *Journal of Virological Methods, 129*(1), 47–55. https://doi.org/10.1016/j.jviromet.2005.05.008

Saeed, S. T., & Samad, A. (2017). Emerging threats of begomoviruses to the cultivation of horticultural crops in India. *VirusDisease, 28*(1), 1–12. https://doi.org/10.1007/s13337-016-0358-0

Saeed, S. T., Kumar, S., & Khan, M. S. (2019). Monoclonal antibodies for the detection of tomato leaf curl New Delhi virus in cucurbit crops. *Phytopathology, 109*(S2), 45. https://doi.org/10.1007/s13337-016-0358-0

Seepiban, C., Chiemsombat, P., Attathom, T., & Sirithorn, P. (2017). Broad-spectrum detection of begomoviruses infecting economically important crops in Thailand using monoclonal antibodies. *Virology Journal, 14*(1), 76. https://doi.org/10.1186/s12985-017-0763-z

Singh, M. P., & Singh, R. P. (2020). Nanomaterial-based ELISA for improved detection of plant viruses. *The Analyst, 145*(12), 4125–4134. https://doi.org/10.1039/D0AN00597E

Sivalingam, P. N., & Varma, A. (2007). PCR based diagnosis of begomoviruses associated with tomato leaf curl disease in India. *Journal of Plant Biochemistry and Biotechnology, 16*, 17–22. https://doi.org/10.1007/BF03321923

Srivastava, K. M., Hallan, V., Raizada, R. K., Chandra, G., Singh, B. P., & Sane, P. V. (1995). Molecular cloning of Indian tomato leaf curl virus genome following a simple method of concentrating the supercoiled replicative form of viral DNA. *Journal of Virological Methods, 51*(2–3), 297–304. https://doi.org/10.1016/0166-0934(94)00122-W

Srivastava, A., Kumar, S., Jaidi, M., Raj, S. K., & Shukla, S. (2015). First report of tomato leaf curl New Delhi virus on opium poppy (Papaver somniferum) in India. *Plant Disease, 100*. https://doi.org/10.1094/PDIS-08-15-0883-PDN

Swarnalatha, P., & Reddy, M. K. (2014). Duplex PCR for simultaneous detection of Begomovirus and Phytoplasma from naturally infected tomato. *Pest Management in Horticultural Ecosystems, 20*(1), 59–68.

Tahir, M., & Haider, M. S. (2005). First report of *tomato leaf curl New Delhi virus* infecting bitter gourd in Pakistan. *Plant Pathology, 54*, 807–807. https://doi.org/10.1111/j.1365-3059.2005.01215.x

Verma, Y. S., Bhardwaj, S. V., & Rathore, R. (2016). Molecular diagnosis and characterization of a geminivirus causing leaf curl disease of tomato in Mid Hills of northern India. *International Journal of Agriculture, Environment and Biotechnology, 9*(6), 951–965. https://doi.org/10.5958/2230-732X.2016.00122.4

Wei, Y., et al. (2024). Artificial intelligence enhanced digital nucleic acid amplification testing for precision medicine and molecular diagnostics. *arXiv*. https://arxiv.org/abs/2407.21080

Zaidi, S. S. E. A., Shafiq, M., Amin, I., Scheffler, B. E., Scheffler, J. A., Briddon, R. W., & Mansoor, S. (2016). Frequent occurrence of tomato leaf curl New Delhi virus in cotton leaf curl disease affected cotton in Pakistan. *PLoS One, 11*(5), e0155520. https://doi.org/10.1371/journal.pone.0155520

Role of Vector Management Strategies in Mitigating Tomato Leaf Curl New Delhi Virus Infection

9

Mohd Abas Shah, Hafsa Ajaz Tramboo, and Sheikh Aafreen Rehman

Abstract

Tomato leaf curl New Delhi virus (ToLCNDV) and tomato yellow leaf curl virus (TYLCV), two begomoviruses that inflict significant financial losses in solanaceous and cucurbitaceous crops, are primarily spread by whiteflies (*Bemisia tabaci*). This chapter examines how vector control techniques, with an emphasis on integrated pest management (IPM), can help reduce begomovirus infections. The predominant approach is still chemical control, which makes extensive use of neonicotinoids (imidacloprid, thiamethoxam, acetamiprid), pyrethroids (bifenthrin, deltamethrin), and insect growth regulators (pyriproxyfen, buprofezin). But resistance development brought on by biochemical detoxification and target-site alterations calls for insecticide resistance management (IRM) techniques, such as rotating modes of action and using synergists Cultural methods including intercropping with plants that repel viruses (such Ocimum and Tagetes species), rotating crops with non-host species, and taking sanitary precautions to get rid of viral reservoirs decrease in the number of whiteflies. Physical barriers, including reflective mulches, insect-proof nets, and floating row covers, provide additional protection. Biological control, employing parasitoids and entomopathogenic fungi (*Beauveria bassiana*), offers sustainable alternatives. Advances in molecular approaches, including CRISPR/Cas9-mediated virus-resistant crops and RNA interference (RNAi)-based suppression of whitefly genes, present novel control strategies. New technologies that enable early detection and focused therapies include remote sensing and artificial intelligence (AI)-based pest surveillance. The development of next-generation biopesticides and regulatory policies restricting high-risk insecticides will shape future whitefly management strategies. A multidisciplinary approach integrating molecular biology,

M. A. Shah (✉) · H. A. Tramboo · S. A. Rehman
ICAR-Central Institute of Temperate Horticulture,
Rangreth, Srinagar, Jammu and Kashmir, India

R. Kumar et al. (eds.), *Tomato Leaf Curl New Delhi Virus (ToLCNDV)*,
https://doi.org/10.1007/978-981-95-2852-3_9

entomology, and precision agriculture is critical for effective and sustainable whitefly and begomovirus control.

Keywords

Whitefly · Begomovirus · Neonicotinoids · Pyrethroids · Biocontrol · Mulching · Trap cropping · Insecticides · Vector management

1 Introduction

Global agricultural productivity is seriously threatened by begomoviruses, especially those that are members of the tomato leaf curl virus (ToLCV) complex. These viruses cause major production losses in crops like tomato (*Solanum lycopersicum*), cotton (*Gossypium hirsutum*), and cassava (*Manihot esculenta*) (Navas-Castillo et al., 2019). Since the whitefly *Bemisia tabaci* is the only vector that can spread these viruses, vector management is the best way to contain begomovirus epidemics (Jeevalatha et al., 2021, 2023). Whiteflies acquire begomoviruses in a persistent circulative way, which allows the virus to spread to new host plants by passing through the insect's gut and hemolymph before arriving at the salivary glands. The primary strategy to lower begomovirus incidence is to limit whitefly populations using chemical, cultural, and physical management techniques (Fig. 9.1), as there are no viable antiviral treatments for plants (Perring et al., 2018a). The most effective strategy for quickly lowering whitefly numbers and, as a result, preventing the transmission of begomoviruses is still chemical control. Neonicotinoids, which include imidacloprid, thiamethoxam, and acetamiprid, are often used insecticides because of their systemic activity, which protects against whiteflies even when a plant is just starting to grow. These substances cause paralysis in whiteflies by interfering with neural signaling and acting as agonists of the nicotinic acetylcholine receptor (nAChR). To minimize early virus transmission occurrences,

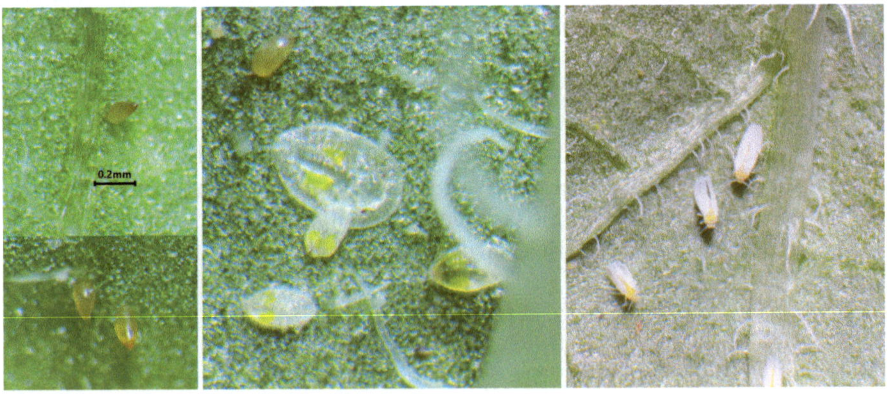

Fig. 9.1 Life stages of *Bemisia tabaci* on potato

neonicotinoids are frequently used as root drenches or seed treatments to offer prolonged protection against early-stage infections. To improve efficacy, it is necessary to rotate insecticide classes and employ synergists like piperonyl butoxide because whiteflies have become resistant to neonicotinoids in several areas (Legarrea et al., 2020). By interfering with voltage-gated sodium channels in insect neurons, pyrethroids (such as bifenthrin, cypermethrin, and deltamethrin) and neonicotinoids both produce quick knockdown effects. However, the effectiveness of these pesticides against nymphal stages is limited, and they mainly target adult whiteflies. Furthermore, in some agricultural regions, pyrethroids are less effective due to *B. tabaci*'s extensive ability to resist them. An alternate strategy is provided by insect growth regulators (IGRs), which include buprofezin and pyriproxyfen, which interfere with the growth and survival of whiteflies. A juvenile hormone analog called pyriproxen prevents metamorphosis and lowers fertility, which causes population decreases over several generations. By stopping nymphal molting, the chitin synthesis inhibitor buprofezin efficiently reduces the number of new adults that can spread begomoviruses. In IPM programs, these IGRs are especially helpful when used in conjunction with traditional pesticides to control whiteflies both immediately and over time (Wang & Liu, 2024).

For the management of whiteflies, botanical insecticides, including pyrethrins, essential oils, and azadirachtin (derived from *Azadirachta indica*), offer an eco-friendly substitute (Cuthbertson & Walters, 2015). By acting as an antifeedant, growth regulator, and oviposition deterrent, azadirachtin lowers whitefly populations while having the least adverse effect on beneficial creatures like pollinators and parasitoids. Alkaloids and flavonoids derived from plants also have insecticidal properties through metabolic disruption and neurotoxicity. Incorporating botanical pesticides into IPM systems can assist in preventing insecticide resistance development and lessen environmental contamination, even though they usually have shorter residual effects than synthetic chemicals (Liu et al., 2023). To decrease whitefly infestations and break the cycles of begomovirus transmission, cultural management techniques are crucial. By eliminating constant host availability, crop rotation with non-host plants like maize, sorghum, or legumes disrupts whitefly life cycles. It has also been demonstrated that intercropping with plant species that are undesirable or repellent inhibits the host-seeking behavior of whiteflies. Another essential element of cultural control is sanitation. A wide variety of begomoviruses may spread to both indigenous and cultivated plant species, which act as virus carriers. During non-cropping times, weeds including Parthenium hysterophorus, Ageratum conyzoides, and Euphorbia heterophylla may carry tomato leaf curl viruses and support populations of whiteflies. Removing these reservoir plants from field margins and surrounding landscapes helps to reduce viral inoculum sources and minimize the spread of begomoviruses. Additionally, prompt destruction of infected crop residues prevents whitefly-mediated transmission between planting cycles. In greenhouse environments, strict sanitation protocols, including the removal of infested plant material and regular disinfection of equipment, are necessary to prevent whitefly infestations and viral outbreaks (Stansly & Naranjo, 2010; Schuster & Smith, 2016).

Physical control methods provide additional barriers to whitefly colonization and begomovirus transmission. By interfering with their visual orientation, reflective mulches—especially those composed of aluminum or silver-colored polyethylene—have been demonstrated to prevent whiteflies from landing and feeding. Whiteflies can also be kept out of protected cropping systems with insect-proof netting and floating row covers, especially in greenhouse and nursery environments. While allowing for sufficient airflow, fine-mesh screens with pores smaller than 400 μm offer a physical barrier against whiteflies. A similar tactic that has been investigated is trap farming, which uses very appealing plant species to draw whiteflies to avoid the primary crop. In greenhouse settings, sticky traps—especially yellow-colored ones—are frequently employed for whitefly surveillance and suppression. By taking advantage of whiteflies' high sensitivity to yellow wavelengths, these traps capture them and lessen the likelihood that they will spread begomoviruses. Even if chemical, cultural, and physical control methods are efficient, their incorporation within an all-encompassing IPM framework is essential for long-term management of begomovirus and whiteflies. An over-reliance on chemical pesticides alone has harmed natural adversaries, degraded the environment, and created resistance. Therefore, combining insecticides with cultural and physical methods enhances overall control efficacy while minimizing ecological risks. Additionally, breeding programs aimed at developing tomato cultivars resistant to tomato leaf curl viruses through conventional and molecular-assisted selection have produced promising results. New developments in genome-editing technological advances, like CRISPR/Cas9, provide possibilities to impair whitefly transmission proficiency and create plants resistant to viruses. In this chapter, we discuss the various aspects of sampling whiteflies in different crops to inform management decisions. We also examine the current and previously used chemical and non-chemical options available for controlling whiteflies and the begomoviruses they transmit efficiently.

2 Monitoring Vector Populations

The use of yellow sticky traps and direct observation is an effective whitefly monitoring technique that allows for long-term population-level surveillance and early diagnosis. Comprehensive sampling protocols have been devised to monitor the widespread distribution of *B. tabaci* in the cultivation of potatoes (Shah et al., 2021a, 2022). Similarly, recommendations for *T. vaporariorum* monitoring are in the process of being produced (Jumardi et al., 2020). While the economic threshold level (ETL) for *B. tabaci* has been set at two adult whiteflies for each plant in India, the ETL for *T. vaporariorum* relies on viral frequency as opposed to the vector density (Rincon et al., 2019; Shah et al., 2020a). A periodic sampling strategy was created by Shah et al. (2022) to monitor the prevalence of whiteflies in potato crops. Exact consecutive sampling plans were developed using Green's and Kuno's approaches, and the average necessary sample size was comparable. According to both Green's and Kuno's designs, 51 samples should be collected at a predetermined precision of 0.1 for the recorded mean density of 2.10 whitefly adults per

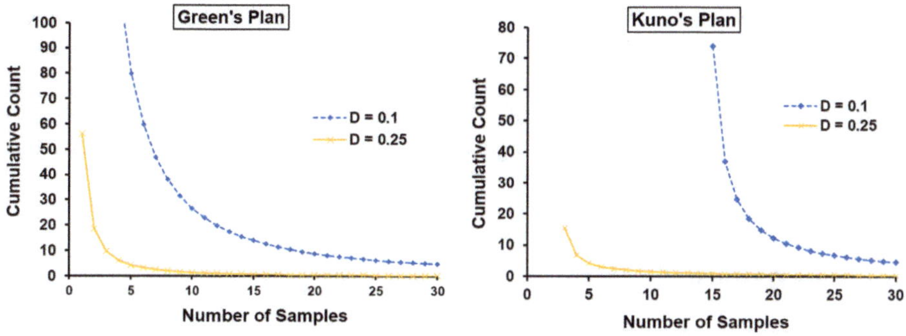

Fig. 9.2 Stop lines for sequential sampling to estimate adult *Bemisia tabaci* density with specified levels of precision, constructed using the methods of Green (1970) and Kuno (1969)

plant. In the same way, Kuno's plan recommends taking 26 samples at a predetermined precision of 0.25 for the measured mean population size of 0.5 adult whiteflies/plant, while Green's plan recommends taking 20 samples. In contrast to Green's plan, which had an average sample count of 19 with a range of 16–23 sample units and an average precision of 0.25, ranging from 0.18 to 0.28, Kuno's plan's corresponding simulation analysis produces a median sample size of 23 with a range of 14–33 sample units and an average accuracy of 0.22, ranging from 0.15 to 0.29 (Fig. 9.2).

According to Mound (1962), *B. tabaci* is attracted to two distinct light wavelength ranges: the yellow and blue/ultraviolet portions of the spectrum. While yellow light causes vegetative behavior, which may be involved in host selection, he connected the onset of migratory behavior to UV attraction. A vital method for sampling and tracking whitefly populations is their attraction to yellow light (e.g., Gerling & Horowitz, 1984). Whiteflies, along with various insect pests, make use of this visual cue, which is an effective but frequently disregarded way to control greenhouse pests (van de Veire & Vacante, 1984). Although beneficial insects like the whitefly parasitoid *Encarsia formosa* may also be caught, "yellow sticky cards" are commonly used to catch flying pests like aphids, whiteflies, thrips, leafminers, fungus gnats, and shore flies. While nymph monitoring aids in the timing of IGR treatments, tracking of adult whiteflies aids in the timing of adulticide applications. Adult whiteflies are quite mobile; thus, their migration between crops may cause viruses to spread. Therefore, it is crucial for virus prevention initiatives and for controlling whiteflies in a given area to keep an eye on adult populations, their movement, and the proportion of viruses they carry (Salati et al., 2002).

From early squaring till harvest, whiteflies on the undersides of leaves should be observed. Just adult whiteflies were monitored for treatment choices in cotton before the registration of IGRs for whitefly control. Action levels were later applied to nymphs as well (Naranjo & Flint, 1994; Ellsworth & Martinez-Carrillo, 2001; Naranjo & Ellsworth, 2009). When whiteflies are discovered in sweep samples, collection should begin using the leaf-turn approach (Naranjo & Flint, 1995; Ellsworth et al., 1995) and be maintained through harvest conclusion. Examining every region of the cotton field is crucial, but field margins should receive special attention. Field margins should be inspected for adult whiteflies up to twice a week during key times, particularly early in the season when numbers may rise when neighboring host crops are harvested or dying off. According to Naranjo et al. (1996), whitefly activity thresholds in cotton can range from three to ten individuals per leaf while monitoring on the fifth main stem leaf from the top.

3 Chemical Control

Early greenhouse whitefly research used fumigation materials such as tetrachloroethane (Parker, 1928), calcium cyanide (Miles, 1927), and hydrocyanic acid gas (Lloyd, 1922; Speyer & Owen, 1926; Miles, 1927), with suggestions to start the treatment at night and open the greenhouse at sunrise to reduce the danger to workers. Additional studies on pesticides, such as nicotine fumigation (Richardson et al., 1943), DDT, and hexaethyl tetraphosphate aerosols (Smith et al., 1947), were carried out in the 1940s and 1950s. In the 1960s and 1970s, more research was conducted on organophosphate compounds like Dibrom (Condron et al., 1962; Adamson et al., 1972). For 3 weeks, weekly applications of azodrin, guthion, and parathion were effective in controlling *T. vaporariorum* (Smith, 1970). These compounds were assessed as chemical fumigants, airborne particles, and plant sprays in greenhouses. Additionally, carbamates such as methomyl (Krueger et al., 1973) and aldicarb (Lindquist et al., 1972) were introduced as part of the management of greenhouse whiteflies. The primary method employed for conventional management of *B. tabaci* using insecticides, in agricultural and vegetable crops, is oil-based sprays that contain older active compounds. These sprays must be adequately covered and deposited (Sharaf, 1986). Although *B. tabaci* has been successfully managed using a variety of biological, physical, and other control techniques, pesticide application remains the primary means of management. Multiple treatments of different pesticides are usually used to control whiteflies. In the absence of genetic resistance, insecticides are the primary method for regulating whitefly populations and preventing the transmission of viruses (Lapidot et al., 2014). Early research (e.g., Sharaf, 1986; Dittrich et al., 1990; Horowitz & Ishaaya, 1996) provided information on over 50 common insecticides to control *B. tabaci* populations and prevent virus transmission. Conventional insecticides that were most commonly utilized were pyrethroids, carbamates, and OPs. Up until the middle of the 1990s, spray combinations of synergized pyrethroids were the most effective way to control *B. tabaci* populations (Horowitz & Ishaaya, 1996; Prabhaker et al., 1998; Palumbo et al., 2001; Castle et al., 2010). It is challenging to fully control *B. tabaci* with traditional insecticides due to the pest's under-leaf habitat for both immature and adult stages, the abundance of mature larvae on the lower canopy growth, its significant polyphagous behavior, and the frequent migration of adults (Horowitz & Ishaaya, 1996; Palumbo et al., 2001). Because of their non-persistent nature in the environment, low application rates, and specificity for target pests, new chemical insecticides are frequently chosen. The effectiveness of managing vector-virus complexes can differ, though. The most promising new chemical pesticides suggested for whitefly control are described here. Effective use of these insecticides requires careful consideration of several parameters (Horowitz et al., 2011). Field scouts who keep an eye on whitefly populations can provide the best information for selecting the right insecticide and when to apply it. The development of resistance in whitefly populations can be prevented or delayed by rotating pesticides with different mechanisms of action (Lapidot et al., 2014; Modgill et al., 2025).

3.1 Neonicotinoid Insecticides

According to Ware (2000), nicotinoids are a novel class of neurotoxins that work differently from any insecticide now used in protected farming systems, fields, or vegetable crops. Although they are collectively referred to as nitroquanidines, nitromethylenes, and chloronicotinyls, the term "neonicotinoids" is most frequently used to describe compounds in the nicotinoid class (Yamamoto et al., 1995). These were based on natural nicotine compounds and, like nicotine, they block postsynaptic nicotinergic acetylcholine receptors irreversibly in insects' central nervous systems (Bai et al., 1991; Liu & Casida, 1993). Targeting the insect nicotinic acetylcholine receptor (nAChR) within the central nervous system, neonicotinoids disrupt nerve transmission, leading to rapid paralysis and death (Tomizawa & Casida, 2005). The physicochemical characteristics of the many nicotinoid insecticides are a significant aspect of this class of pesticides. These substances work very well against *B. tabaci* and other sucking insects because of their low partition coefficients and relatively high water solubility, which also result in excellent systemic characteristics and sustained residual activity (Kagabu, 1999; Yamada et al., 1999; Maienfisch et al., 2001). Because of these qualities, they can also be applied in many ways, including paint-on plant containers, canopy sprays, soil drenches, granular and liquid formulations in-furrow and subsurface, and seed treatments. To manage *B. tabaci*, several second-generation nicotinoid chemicals are currently being progressively introduced into agricultural markets. These consist of thiacloprid, nitenpyram, acetamiprid, and thiamethoxam. When applied to foliage, they appear to have increased translaminar activity and generally have an analogous mode of action (Palumbo et al., 2001). Second-generation neonicotinoids have been shown in cotton field experiments to have superior foliar action against *B. tabaci* when compared to imidacloprid. The following is a ranking of their relative efficacy: Acetamiprid > Thiamethoxam > Thiacloprid > Imidacloprid. Particularly against sucking insects like leafhoppers, aphids, and whiteflies, neonicotinoids exhibit persistent residual activity due to their systemic and translaminar properties. To manage agricultural pests, imidacloprid was the first neurotoxin to be effectively sold.

A few review publications, such as those by Bass et al. (2015) and Nauen and Denholm (2005), have put together the global elements of neonicotinoid resistance. The Q biotype of *B. tabaci* has often been linked to tolerance to neonicotinoids (e.g., Nauen et al., 2002; Horowitz et al., 2004; Roditakis et al., 2009; Dennehy et al., 2010; Luo et al., 2010), despite a few instances of neonicotinoid resistance observed in B-biotype strains (Byrne et al., 2003; Schuster et al., 2010; Wang et al., 2010). Recently, RNAseq research has been increasingly utilized to examine *B. tabaci's* ability to resist pesticides, particularly neonicotinoids, to identify less obvious mechanisms of resistance. These studies additionally improve the gene, gene expression, and resistance-related pathway databases that are currently available. In a survey of the MED species of *B. tabaci*, RNAseq was used to characterize the expression patterns of numerous detoxification genes, including P450s, GSTs, and carboxylesterases, as likely potential genes for pesticide resistance (Ilias et al., 2015). The study found that CYP6CM1 was the most overexpressed P450 among those belonging to the CYP2, CYP3, and CYP4 groups. Comparing eleven unigenes encoding nAChR subunits to a susceptible reference strain, it was interesting to note that none of them showed polymorphisms associated with neonicotinoid resistance (Ilias et al., 2015).

3.2 Insect Growth Regulators (IGRs)

Insect growth regulators (IGRs) that are particularly important for controlling whiteflies include buprofezin (Ishaaya, 1990) and pyriproxyfen (Ishaaya & Horowitz, 1992). *B. tabaci* is also somewhat affected by novaluron, another IGR that prevents the synthesis of chitin (Ishaaya et al., 2003). Buprofezin, a thiadizine-like compound with prolonged residual activity that demonstrates both contact and vapor activity, primarily targets the nymphal stages of sucking insects, especially whiteflies (Ishaaya et al., 1988; De Cock & Degheele, 1998). One of the most effective insecticides for managing whiteflies, especially biotype B, is pyriproxen, a strong juvenile hormone mimic (Ishaaya & Horowitz, 1995; Horowitz et al., 2005; Castle et al., 2010). These non-neurotoxic insect growth regulators (IGRs) were developed in response to the need for a greater range of insecticides to control *B. tabaci* (Horowitz & Ishaaya, 1996). Their distinct modes of action and favorable biological and environmental profiles have led to their incorporation into vegetable and melon cropping systems and made them indispensable components of cotton integrated pest management (IPM) programs (Horowitz et al., 1999; Ellsworth & Martinez-Carrillo, 2001). The first targeted insect growth regulator (IGR) formulated to inhibit *B. tabaci* in cotton was buprofezin (Horowitz & Ishaaya, 1992). This novel thiadiazine chitin-synthesizing inhibitor shows selective efficacy against some homopteran pests by blocking the incorporation of N-acetyl-[D-H3] glucosamine into chitin and disrupting cuticle formation (Kanno et al., 1981). This substance causes nymphal death during ecdysis by selectively targeting immature developmental stages (Yasui et al., 1987; Ishaaya et al., 1988). Although buprofezin has no immediate impact on the longevity or oviposition of *B. tabaci* adults, it may, in certain situations, reduce the reproduction rate and egg hatch of female whiteflies that are exposed to treated foliage (Ishaaya et al., 1988; Beevi & Balasubramanian, 1991, 1995). Nevertheless, buprofezin exhibits very little translaminar transport in cotton leaves and has no soil systemic activity.

A powerful growth regulator with long-lasting effects against a variety of insects, including *B. tabaci*, is pyriproxen, an analog of a non-terpenoidal growth hormone (Schaefer et al., 1988; Ishaaya & Horowitz, 1992). By changing the normal hormonal balance, the drug inhibits embryogenesis, metamorphosis, and adult formation (Ascher & Eliyahu, 1988; Ishaaya & Horowitz, 1992, 1995) (Dhadialla et al., 1998). According to Ishaaya and Horowitz (1995), adult exposure to treated foliage can occur directly (ovicidal) or indirectly (transovarially) through contact with eggs placed on lower leaf surfaces. Both buprofezin and pyriproxyfen are foliar sprays that are used on cotton (Ellsworth & Naranjo, 1999; Henneberry et al., 1999), tomatoes (Schuster & Polston, 1997; Schuster, 2000; Stansly & Conner, 2000), cantaloupes (Palumbo et al., 1993, 1994; Palumbo, 1998), and Poinsettia (Redak et al., 1998). Two main problems have affected the development and use of decision aids for IGR-based *B. tabaci* control. First, buprofezin and pyriproxyfen are highly successful at preventing the development of the *B. tabaci* colony only when given at the appropriate life stages. Second, due to the possibility of rapid development of resistance to insecticides, including IGRs, resistance management programs have purposefully limited the use of IGRs to a single application per crop season (Horowitz et al., 1999; Denholm et al., 1998; Ellsworth et al., 1996; Dennehy et al., 1996). Also, the use of IGRs is considerably cheaper than traditional methods (Ellsworth et al., 1997; Frizvold et al., 2000). A novel inhibitor of benzoylphenylurea chitin synthesis, which has demonstrated efficacy against *B. tabaci*, is another IGR that is in the final stages of research (Ishaaya et al., 1996). Ishaaya et al. (1997) claim that the substance works by coming into contact with the eggs and nymphs of *B. tabaci*. It appears to have little effect on some parasitoids or phytoseiids, as well as other natural enemies. It has been demonstrated to be as effective as or more effective than other IGRs against *B. tabaci* in field settings (Ishaaya et al., 2001).

3.3 Diafenthiuron

Since 1998, Europe and Israel have been using diafenthiuron, a powerful whitefly control agent, in place of pyriproxyfen for the management of *B. tabaci* on cotton (Horowitz et al., 1999; Palumbo et al., 2001). When adult female whiteflies come into contact with treated plants, diafenthiuron stops the growth of their progeny (Ishaaya et al., 1993). Compared to pupae or eggs, it is very efficient against nymphs. It is among the few effective adulticides for whiteflies. The compound, a derivative of thiourea, has both insecticidal and acaricidal properties and targets a specific category of homopterans and phytophagous mites (Steinemann et al., 1990; Horowitz & Ishaaya, 1996). It suppresses *B. tabaci* in a different way than other chemicals. It directly affects insect respiration by blocking oxidative phosphorylation and compromising mitochondrial ATP synthesis (Ruder & Kayser, 1993) (Fig. 9.3). Because it is not toxic to animals and has no effect on pollinators and natural enemies, the chemical is a valuable part of integrated pest management (IPM) systems (Streibert et al., 1988; De Cock et al., 1996). Diafenthiuron's selectivity is one of its benefits. Research indicates negligible negative impacts on

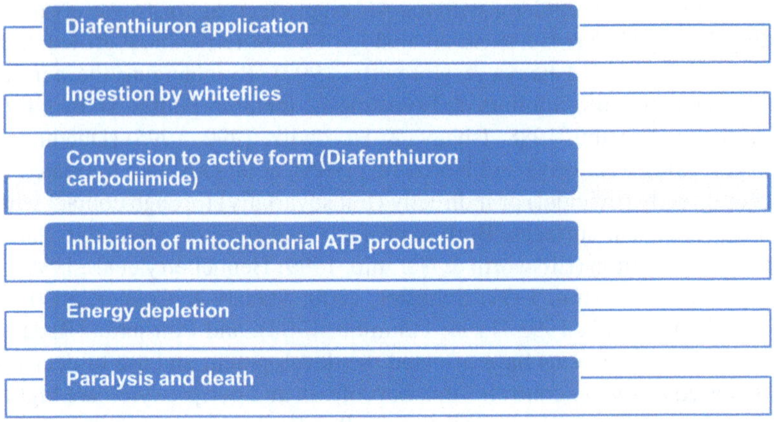

Fig. 9.3 Schematic representation illustrating the mode of action of diafenthiuron against *Bemisia tabaci*, highlighting its impact on mitochondrial function and energy metabolism

pollinators like honeybees, parasitoid wasps, and predatory beetles. Because of this, it is the best method for controlling pests (Aravind et al., 2018; Bajya et al. 2016). Diafenthiuron's environmental breakdown into non-toxic metabolites is another important characteristic. Additionally, it has an anti-feeding effect that prevents whiteflies from landing on treated plants, hence minimizing pest damage. A recent study has highlighted the compatibility of diafenthiuron with other IPM strategies, including cultural practices and biological control measures.

3.4 Pyridine Insecticides (Pymetrozine)

Sucking insect pests are the primary target of pymetrozine (Flückiger et al., 1992a, b; Fuog et al., 1998). It causes the insect to starve and perish by interfering with the neurons that control the salivary pump (Kayser et al., 1994). Pymetrozine can be given foliarly or as a drench and has both systemic and translaminar properties (Flückiger et al., 1992a, b). Pymetrozine exhibits acropetal translocation and translaminar activity in the xylem upon foliar spray treatment, although basipetal translocation in the phloem is comparatively restricted. It is effective against several whitefly species, including *B. tabaci*, and primarily targets adult whiteflies (Koenig et al., 1998; Ishaaya & Horowitz, 1998). Pymetrozine and pyriproxyfen have been shown to improve the control of *B. tabaci* in cotton, regardless of the adult and immature stages (Koenig et al., 1998). Because it is safe for many pollinators and natural enemies and has low toxicity to mammals (Flückiger et al., 1992a, b), pymetrozine appears to be a valuable addition to some IPM programs when used in conjunction or alternation with other new compounds. A recent study (Gorman et al., 2010) has surprisingly revealed the development of cross-resistance among neonicotinoids and pymetrozine in *B. tabaci*, despite their structural and functional differences. This cross-resistance is most likely caused by the chronic amplification of a

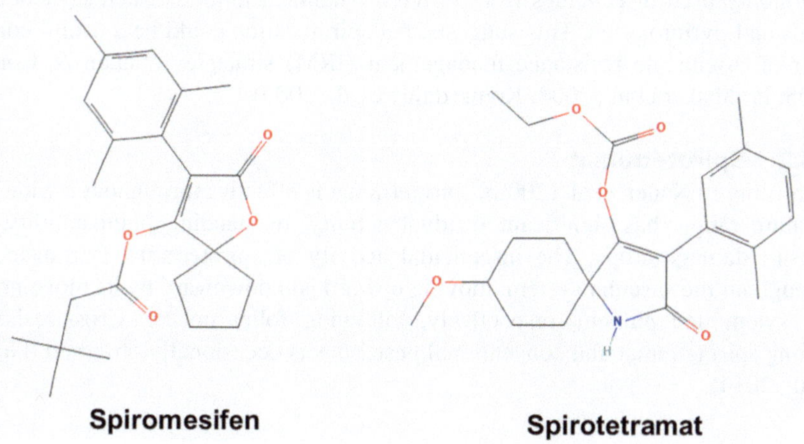

Fig. 9.4 Chemical structure of spiromesifen and spirotetramat. (Source: PubChem, 2025 update)

cytochrome-P450 monooxygenase whose activity is capable of breaking down both types of compounds. The use of neonicotinoids and pymetrozine during the same "window of opportunity" is suggested by this finding, which is critical for insecticide resistance management (IRM) methods for *B. tabaci* (Gorman et al., 2010).

3.5 Spiromesifen and Spirotetramat

Ketoenols, a new class of insecticides, are products of tetronic acids (like spiromesifen) and tetramic acids (like spirotetramat) (Fig. 9.4). Spiromesifen is a novel kind of insecticide that primarily targets mites and whiteflies. It is developed from spirocyclic tetronic acid. It is highly effective against the developing egg and the initial nymphal stages (both biotypes B and Q), though it has a minimal impact on adults and late nymphal stages of *B. tabaci* (Prabhaker et al., 2008; Kontsedalov et al., 2009). Another insecticide belonging to the ketoenol group that inhibits the synthesis of lipids is called spirotetramat, a novel spirocyclic tetramic acid derivative. Aphids, whiteflies, psyllids, and scales are among the sucking insects that can be effectively controlled with the systemic pesticide spirotetramat, which has both phloem and xylem mobility. It is particularly efficient towards the juvenile phases of sucking pests and significantly inhibits lipid biosynthesis and reduces the reproductive potential and productivity of female *B. tabaci* (Brück et al., 2009).

3.5.1 Spiromesifen

A recent study found that female *B. tabaci* treated with spiromesifen have smaller-than-normal eggs with an abnormally perforated chorion, which prevents them from completing oviposition (Kontsedalov et al., 2009). Furthermore, several studies have demonstrated that spiromesifen does not display cross-resistance with other commonly used insecticides from different chemical families, such as neonicotinoids and pyriproxyfen. This suggests that spiromesifen could be a useful component of insecticide resistance management (IRM) strategies (Nauen & Konanz, 2005; Prabhaker et al., 2008; Kontsedalov et al., 2009).

3.5.2 Spirotetramat

According to Nauen et al. (2008), spirotetramat is effective throughout a wide temperature range, has significant residual activity, outstanding photostability, and doesn't damage crops. The insecticidal activity of spirotetramat is translocated throughout the vascular system, moving upward and downward by its movement in the xylem and phloem, respectively, following foliar uptake. Cross-resistance among spirotetramat and conventional pesticides is occasionally observed (Guillén et al., 2014).

3.6 Ryanodine Receptor Insecticides

A plant-derived alkaloid, which is utilized as an organic botanical insecticide, is ryanodine. As a ligand-gated calcium channel, it aids in the mobilization of calcium through intracellular stores (Horowitz et al., 2011). Two synthetic types of medicines that target insect ryanodine receptors are currently being investigated and registered: Rynaxypyr® (Clorantraniliprole), which is highly efficient against lepidopteran pests, and CyazypyrTM (Cyantraniliprole), which targets sucking pests along with other insect pests (Sattelle et al., 2008; Lahm et al., 2009). Diamide insecticides were the latest type of chemicals to be introduced to the market, which initially became available about 13 years ago (Nauen & Steinbach, 2016). Presently available diamide insecticides include flubendiamide, a phthalic diamide, and the anthranilic diamides tetraniliprole, cyclaniliprole, cyantraniliprole, and chlorantraniliprole. These substances target the big, tetrameric calcium release channels called insect ryanodine receptors, which are present in neuromuscular tissues. In China, *B. tabaci* resistance to cyantraniliprole has spread quickly, particularly in the Q biotype (MED), where reports of low to moderate degrees have been made (Yao et al., 2017; Wang et al., 2018). However, in European countries like Spain, Italy, and Greece, no changes in cyantraniliprole sensitivity have been discovered to date (Gravalos et al., 2015).

3.7 Non-conventional Insecticides

Non-conventional pesticides are thought to be more selective than conventional ones when it comes to their effects on beneficial fauna in crops. Because they have no residual action and have little negative impact on insects and mites that are not immediately sprayed, they are compatible with biological management (Fagan et al., 2010; Curkovic, 2013). The adverse effects of pesticides can be lessened with the use of these non-conventional substances. Numerous cases of successful control of whiteflies exist with the use of non-conventional insecticides. However, large-scale commercial applications are still limited. According to Shah et al. (2019), after three and 7 days of topical potassium silicate (0.3%) spraying on potatoes, whitefly populations decreased by 53.27% and 61.42%, respectively. Potassium silicate sprays significantly increased the amount of silicon in the leaves, and the prevalence of viruses (such as mild and severe mosaic, leaf roll, and apical leaf curl) in the potassium silicate-treated plots was similar to that of the recommended insecticides but much lower than in the control plots. According to Liang and Liu (2002), there have been reports of mineral oils significantly repelling adult whiteflies from a variety of crops, including cotton, melon, squash, chrysanthemum, and tomato. According to Xue et al. (2002), the quantity of eggs laid on tomato leaves was highly impacted by the content of mineral oils. According to Schuster et al. (2009), mineral oils can hinder the spread of begomovirus by reducing the settling of *B. tabaci* adults. One possible mechanism by which field treatments lead to fewer virus-infected plants is this decrease in settling. According to Malik et al. (2020), after a series of insecticidal treatments, horticultural mineral oil spraying at a rate of 50 mL/10 L produced the most significant decrease in *B. tabaci* populations (74.5%) as compared to the untreated control, as well as a 93.0% reduction in viral infection. Shah et al. (2021b) summarize research on mineral oils' ability to repel whiteflies.

In their evaluation of the efficacy of an insecticidal soap, 15 detergents, and two horticultural oils against *B. tabaci*, Butler Jr et al. (1993) discovered that, except for three detergents, all of these substances at 1% concentration resulted in more than 85% mortality. In Florida, in infested cucumber leaves, similar results were obtained using two detergents and an insecticidal soap. These materials may be beneficial in nations where natural oils are expensive and appropriate emulsifiers are not widely available (Butler Jr et al., 1993). When treated to individual cotton leaves, Puri et al. (1994) discovered that the four common detergent powders available in India—Nirma, Rin, Surf, and Wheel—at doses of 0.25%, 0.5%, and 1.0% decreased the total population of adult *B. tabaci* by 69–91% along with nymphs by 97–99%. Similarly, soaps and detergents are effective against *Trialeurodes vaporariorum*, the greenhouse whitefly, on tomato seedlings (Sclar et al., 1999). The spiraling whitefly, *Aleurodicus dispersus*, was successfully managed on guava in Bangladesh by foliar spraying 10 and 15 g/l of Wheel powder, a widely available and reasonably priced detergent (Islam et al., 2003). Olive trees subjected to agricultural detergents (TS 2035 and Tecsafruta) at doses of 1–2% demonstrated a significant level of control against the ash whitefly, *Siphoninus phillyreae* (Haliday), as reported by Curkovic and Ballesteros (2011). These compounds do have certain drawbacks, though, such as possible toxicity to plants and their lack of residual effect. Acute toxicity is generally low for soaps and detergents (Pavela, 2007), particularly for non-ionic or anionic detergents, because they are much less toxic than traditional insecticides (Conti, 1987).

4 Physical Control

4.1 Yellow Sticky Traps

For some pests that impact different crops, including fungus gnats, leaf mining flies, shore flies, whiteflies, and thrips, yellow sticky traps (YSTs) are now an essential part of pest management strategies. YSTs are used for dispersal activity monitoring, hotspot identification, relative abundance estimation, and early detection of adult whiteflies, including those found in greenhouses such as *Trialeurodes vaporariorum* (Westwood) and *Bemisia tabaci* (Gennadius) (Gillespie & Quiring, 1992; Heinz et al., 1992; Naranjo et al., 1995). To capture adult whiteflies migrating from adjacent crops, it is recommended to place stationary yellow sticky traps surrounding the field before planting. Whitefly adult populations in potato crops can be successfully decreased by using both fixed and movable yellow sticky traps (Cisneros & Mujica, 1999; Naga et al., 2021a, b). Yellow sticky traps should be positioned 50 cm or less from the plants in tomato and cucumber crops, 15–30 cm below the top of the canopy (Shen & Ren, 2003; Qiu & Ren, 2006; Hou et al., 2006). To optimize the collection of both *T. vaporariorum* and *B. tabaci*, the traps should be placed vertically, corresponding to the plant rows (Hou et al., 2006; Gu et al., 2008; Zhang & Yu, 2009). Park et al. (2011) advise using small-sized traps whenever feasible to reduce evaluation time (e.g., insect counts), provided that there is a significant correlation across trap counts and direct counts. Additionally, there is a negative correlation between trap size and the number of adults captured per unit area (Zhang & Yu, 2009; Kim & Lim, 2011). However, in well-designed protocols, the trap size may influence the number of traps that need to be sampled; consequently, it should be considered when determining the sample size (Parrella & Jones, 1985).

4.2 Protected Cultivation

Physical methods, such as keeping insects away from vulnerable plants, can shield crops from harm caused by whiteflies and virus infection. In the worst situations, as with vegetables, the entire crop is cultivated in a screen house or greenhouse, which offers protection against whiteflies during the whole production cycle. When these structures are kept away from whiteflies and viruses, such as through the use of glass, plastic, or screening, and with double doors and vents to maintain positive pressure, both can be effectively controlled (Ausher, 1997; Berlinger et al., 2002). Shimla recommends temporary net-house structures for the production of high-quality seed potatoes, including tissue culture-based seedlings and early-generation minitubers, CPRI. Following crop harvest, these structures can be disassembled. When utilized as greenhouse walls, fifty-mesh screens successfully stop TYLCV and whiteflies from spreading among greenhouse tomatoes (Berlinger et al., 1991). Incorporating UV-absorbing shielding can also significantly improve these screens' functioning (Antignus et al., 1998; Diaz & Fereres, 2007; Ben-Yakir et al., 2008; Legarrea et al., 2010). Dense screens, however, can obstruct proper airflow, especially when summer temperatures are at their highest. These screens make use of both their optical and physical characteristics to offer a dual insect-exclusion strategy. Antignus et al. (1998) designed and published the first UV-absorbing screens (BioNet®). Compared to standard 50-mesh screens, UV-absorbing screens inhibit whitefly access and the spread of TYLCV by a ratio of four (Antignus et al., 1998). Later, several researchers verified these findings (Diaz & Fereres, 2007; Ben-Yakir et al., 2008; Legarrea et al., 2010).

4.3 Floating Row Covers

Covering young plants—whether they are transplanted or emerge from seeds—with protective netting is another popular exclusion technique. Such a net, which is made of spun-bonded polyester (marketed as Agribon or Agril), is typically placed directly over the rows of plants or transplants. 25 GSM spun-bonded row covers were assessed by Shah et al. (2020b) for their ability to keep vectors out of potato crops. Aphids and whiteflies were kept out of the covered plots by the coverings. Still, the uncovered control had noticeably more of them (Fig. 9.5). In contrast to the uncovered control, the covered plots had a lower overall tuber yield and a higher terminal intensity of late blight. When used in conjunction with insect growth regulators, row covers help minimize fruit damage and enhance the size, weight, and quality of zucchini by reducing the number of adults, eggs, and pupae/nymphs on the leaves compared to treatments without covers (Qureshi et al., 2007). These temporary pest isolation systems provide short-term damage protection by successfully preventing insect infestations during critical crop growth stages (Gilbertson et al., 2011).

Fig. 9.5 Use of polypropylene row covers as a physical barrier to prevent virus vector entry in seed potato cultivation

4.4 Vegetative Barriers and Mulch-Based Approaches for Vector Suppression

Whiteflies can be kept out of fields with crops that are susceptible by installing physical barriers. These barriers can be biological, like tall-growing plant species that are not hosts to whiteflies or the viruses they transmit, or they may consist of inanimate objects like screens or plastic (yellow plastic that has a sticky material to catch insects). Monocots such as elephant grass, sorghum, and corn are good barrier plants for stopping the spread of viruses. There is little evidence that barriers effectively prevent viruses from spreading or whiteflies from migrating. By flying or being transported by the wind, whiteflies can get past obstacles and spread viruses for extended periods because of how persistent the virus is (Hilje et al., 2001). For viruses spread by whiteflies, barriers are therefore usually not regarded as an essential component of the integrated pest management (IPM) strategy.

Open-field crops are frequently protected from aphids, thrips, and whiteflies, as well as the viruses they transmit, by mulch or plastic (polyethylene) soil covers (Weintraub & Berlinger, 2004). Plastic mulch coated in silver or aluminum, which powerfully reflects light and repels whiteflies, and colored (usually yellow) plastic, which attracts whiteflies to the mulch rather than the host plant, are the two primary techniques for plastic soil mulching. From the beginning, during the growing period, once the plant canopy completely covers the mulch, both forms of mulch work well since they make it difficult for insects to find the crop (Lapidot et al., 2014). Mulches prevent insects from identifying and landing on crops that could become infected with viruses. Mulches can be either living (plants grown with the sensitive crop) or non-living (plastic or other materials), much like barriers. It has been shown that colored or UV-reflective polymers, which reduce viral incidence and whitefly population densities, are the most effective non-living mulches (Antignus, 2000). It has been shown that UV-reflective mulch minimizes the incidence of tomato mottle virus (ToMoV) in Florida, TYLCV in Jordan (Csizinszky et al., 1995; Suwwan et al., 1988), and CuLCrV in zucchini squash in Florida (Nyoike et al., 2008). Additionally, these mulches can encourage improved crop development. They may be expensive, time-consuming, and environmentally hazardous, though. Mulches are also being advocated for high-value seed potato harvests despite these disadvantages.

4.5 Roguing

This technique is used to physically remove plants infected with a virus during the development period. Roguing is most effective when the virus prevalence is low (<5%), and it should commence as soon as the plots are established. Before roguing, pesticide treatments should be used to eradicate adult whiteflies if populations are high. The sick plants should be taken out and discarded far from the productive areas if nymphs are present. Fields should ideally be routinely inspected, and any plants that exhibit symptoms should be destroyed right away. In India, roguing of seed potato crops occurs three times: just after the crop establishes itself, once in the middle of the growing season, and once before harvest (before senescence begins), as reported by Kumar et al. (2021, 2022). It is recommended to eliminate rogue tomato plants exhibiting TYLCV signs until the second tie. Plants need to be treated for adult whiteflies before roguing. When nymphs are detected, the affected plants should be removed from the field, preferably in plastic bags, and destroyed as far away from fields as is practical (Schuster et al., 2007).

5 Conclusion and Future Outlook

Owing to the vector's strong reproductive capability, polyphagy, and capacity to acquire resistance against chemical pesticides, managing whiteflies continues to be a major concern in global agriculture. Huge losses in solanaceous and cucurbitaceous crops have resulted from the proliferation of begomoviruses, including the ToLCNDV and the TYLCV. Integrated pest management (IPM), which combines chemical, cultural, biological, and physical control techniques, is the main management approach. However, populations of *Bemisia tabaci* over the world have developed insecticide resistance as a result of a significant reliance on chemical insecticides, specifically neonicotinoids, pyrethroids, and insect growth regulators (IGRs Alternative approaches are required because, processes which include metabolic detoxification (increased cytochrome P450 monooxygenase activity), behavioral adaptations, and target-site alterations (e.g., changes in nicotinic acetylcholine receptors) induce insecticide resistance.

Cultural and physical control methods, such as crop rotation, intercropping, mulching, and the use of insect-proof netting, have demonstrated efficacy in reducing whitefly populations and virus transmission rates. The deployment of trap cropping with desirable host plants and reflective mulches has been shown to disrupt host-seeking behavior and reduce initial infestations. Additionally, the use of botanical insecticides, such as azadirachtin and essential oils, provides an eco-friendly alternative to synthetic chemicals. However, the limited residual activity of botanical insecticides and their variable efficacy under field conditions require further optimization. Biological control remains a crucial component of sustainable whitefly management. The utilization of entomopathogenic fungi (e.g., *Beauveria bassiana*, *Isaria fumosorosea*), parasitoids (e.g., *Encarsia formosa*, *Eretmocerus mundus*), and predatory insects (e.g., *Nesidiocoris tenuis*, *Orius laevigatus*) has shown promise in suppressing whitefly populations under controlled conditions. In IPM programs, combining microbial biopesticides with traditional insecticides can improve control effectiveness while reducing the possibility of resistance. Nevertheless, obstacles such as uneven field performance and the need for ideal environmental conditions for entomopathogen development call for additional applications to improve the technique.

Prioritizing insecticide resistance management (IRM) techniques is necessary to maintain the effectiveness of chemical control measures. Essential elements of IRM include the strategic use of pesticides depending on economic threshold levels

(ETLs), the rotation of insecticides with various modes of action, and the use of synergists to combat metabolic resistance. The implementation of refugia, where susceptible whitefly populations are maintained to delay resistance evolution, is another potential approach. Furthermore, real-time insecticide selection can be guided by molecular diagnostics for quick resistance diagnosis, avoiding the abuse of chemicals with diminished performance. To guarantee long-term reduction of whiteflies, biological and cultural management methods should be combined with a thorough IRM framework. Molecular strategies for controlling whiteflies should be the main focus of future studies. The development of virus-resistant crop varieties, especially those with mutations in important begomovirus receptor proteins, has been investigated using CRISPR/Cas9-mediated genome editing. A promising method for targeted pest suppression is the silencing of whitefly genes important in viral transmission, reproduction, and pesticide resistance via RNA interference (RNAi). The incorporation of precision agriculture methods, such as automated trapping systems, remote sensing, and AI-powered pest surveillance, can improve early identification and focused intervention tactics. Improving IPM efficiency will require using big data analytics to forecast whitefly outbreaks and schedule interventions optimally. Furthermore, regulatory restrictions on high-risk chemical pesticides are expected to drive the development of next-generation biopesticides and resistance-management strategies. Global collaborations between researchers, policymakers, and farmers will play a pivotal role in mitigating the impact of whiteflies and begomoviruses on agricultural productivity.

References

Adamson, R. M., Tonks, N. V., & Maas, E. F. (1972). Yields of greenhouse tomatoes treated with Naled for control of the greenhouse whitefly. *Journal of Economic Entomology, 65*, 1205.

Antignus, Y. (2000). Manipulation of wavelength dependent behavior of insects: An IPM tool to impede epidemics and restrict spread of insect-borne viruses. *Virus Research, 71*, 213–220.

Antignus, Y., Lapidot, M., Hadar, D., Messika, Y., & Cohen, S. (1998). UV absorbing screens serve as optical barriers to protect vegetable crops from virus diseases and insect pests. *Journal of Economic Entomology, 91*, 1401–1405.

Aravind, J., Samiayyan, K., & Kuttalam, S. (2018). Assessing the impact of diafenthiuron 50WP on the pest, pollinator and natural enemy diversity in cardamom ecosystem. *Jounral of Entomology and Zoology Studies, 6*(3), 398–402.

Ascher, K. R. S., & Eliyahu, M. (1988). The ovicidal properties of the juvenile hormone mimic Sumitomo S-31183 (SK-591) to insects. *Phytoparasitica, 16*, 15–21.

Ausher, R. (1997). Implementation of integrated pest management in Israel. *Phytoparasitica, 25*, 119–141.

Bai, D., Lummis, S. C. R., Leicht, W., Breer, H., & Stelle, D. B. (1991). Action of imidacloprid and a related nitromethylene on cholinergic receptors of an identified insect motor neurone. *Pesticide Science, 33*, 197–204.

Bajya, D. R., Ranjith, M., Lakharan, M. C., & Raza, S. K. (2016). Efficacy of diafenthiuron 47.8 SC against sucking pests of cotton and its safety to natural enemies. *Indian Journal of Entomology, 78*(1), 15–23.

Bass, C., Denholm, I., Williamson, M. S., & Nauen, R. (2015). The global status of insect resistance to neonicotinoid insecticides. *Pesticide Biochemistry and Physiology, 121*, 78–87.

Beevi, S. P., & Balasubramanian, M. (1991). Effect of buprofezin on adult life span, oviposition, egg hatch and progeny production of the cotton whitefly, *Bemisia tabaci*. *Phytoparasitica, 19*, 33–47.

Beevi, S. P., & Balasubramanian, M. (1995). Effect of buprofezin, a novel insect growth regulator, against cotton whitefly *Bemisia tabaci* Genn. *Entomon, 20*, 11–14.

Ben-Yakir, D., Hadar, M. D., Offir, Y., Chen, M., & Tregerman, M. (2008). Protecting crops from pests using OptiNet® and ChromatiNet® shading nets. *Acta Horticulturae, 770*, 205–212.

Berlinger, M. J., Dahan, R., Mordechi, S., Liper, A., Katz, J., & Levav, N. (1991). The use of nets to prevent the penetration of *Bemisia tabaci* into greenhouse. *Hassadeh, 71*, 1579–1583.

Berlinger, M. J., Taylor, R. A. J., Lebiush-Mordechi, J., Shalhevet, S., & Spharim, I. (2002). Efficiency of insect exclusion screens for preventing whitefly transmission of tomato yellow leaf curl virus of tomatoes in Israel. *Bulletin of Entomological Research, 92*, 367–373.

Brück, E., Elbert, A., Fischer, R., et al. (2009). Movento®, an innovative ambimobile insecticide for sucking insect pest control in agriculture: Biological profile and field performance. *Crop Protection, 28*, 838–844.

Butler, G. D., Jr., Henneberry, T. J., Stansly, P. A., & Schuster, D. J. (1993). Insecticidal effects of selected soaps, oils and detergents on the sweet potato whitefly (Homoptera: Aleyrodidae). *Florida Entomologist, 76*(1), 161–167.

Byrne, F. J., Castle, S., Prabhaker, N., & Toscano, N. (2003). Biochemical study of resistance to imidacloprid in B biotype *Bemisia tabaci* from Guatemala. *Pest Management Science, 59*, 347–352.

Castle, S. J., Palumbo, J. C., Prabhaker, N., Horowitz, A. R., & Denholm, I. (2010). Ecological determinants of *Bemisia tabaci* resistance to insecticides. In P. A. Stansly & S. E. Naranjo (Eds.), *Bemisia: Bionomics and management of a global pest*. Springer.

Cisneros, F., & Mujica, N. (1999). *Biological and selective control of the sweet potato whitefly, Bemisia tabaci (Gennadius) (Hom.: Aleyrodidae)* (pp. 255–264). International Potato Center. Program Report 1997–98.

Condron, C. H., Neiswander, R. B., & Wessel, R. D. (1962). Toxicity of Dibrom vapors to greenhouse insects. *Journal of Economic Entomology, 55*, 221–224.

Conti, E. (1987). Acute toxicity of three detergents and two insecticides in the lugworm, *Arenicola marina* (L.): A histological and a scanning electron microscopic study. *Aquatic Toxicology, 10*, 325–334.

Csizinszky, A. A., Schuster, D. J., & Kring, J. B. (1995). Color mulches influence yield and insect pest populations in tomatoes. *Journal of the American Society for Horticultural Science, 120*, 778–784.

Curkovic, T. (2013). Integrated pest management in olive orchards in Chile [in Spanish]. In T. Fichet & J. Henríquez (Eds.), *Contributions to the knowledge of the cultivation of the olive in Chile* (Vol. 23, pp. 155–204). Serie Ciencias Agronómicas Universidad de Chile.

Curkovic, T., & Ballesteros, C. (2011). Integrated pest management of key pests in olives in Copiapó and Huasco [in Spanish]. In T. Fichet, B. Razetom, & T. Curkovic (Eds.), *The olive: Agronomic study in the Atacama Región* (Vol. 16, pp. 117–144). Serie Ciencias Agronómicas Universidad de Chile.

Cuthbertson, A. G. S., & Walters, K. F. A. (2015). Efficacy of insecticides of natural origin in whitefly (*Trialeurodes vaporariorum*) control in tomato. *Journal of Applied Entomology, 139*(9), 651–659. https://doi.org/10.1111/jen.12215

De Cock, A., & Degheele, D. (1998). Buprofezin: A novel chitin synthesis inhibitor affecting specifically planthoppers, whiteflies and scale insects. In I. Ishaaya & D. Degheele (Eds.), *Insecticides with novel modes of action: Mechanism and application* (pp. 74–91). Springer.

De Cock, A., de Clercq, P., Tirry, L., & Degheele, D. (1996). Toxicity of diafenthiuron and imidacloprid to the predatory bug *Podisus maculiventris* (Heteroptera: Pentatomidae). *Environmental Entomology, 25*, 476–480.

Denholm, I., Cahill, M., Dennehy, T. J., & Horowitz, A. R. (1998). Challenges with managing insecticide resistance in agricultural pests, exemplified by the whitefly *Bemisia tabaci*. *Philosophical Transactions of the Royal Society of London B, 353*, 1757–1767.

Dennehy, T. J., Ellsworth, P. C., & Nichols, R. L. (1996). The 1996 whitefly resistance management program for Arizona cotton. *University of Arizona Cooperative Extension IPM Series, No. 8*, 16pp.

Dennehy, T. J., Degain, B. A., Harpold, V. S., Zaborac, M., Morin, S., Fabrick, J. A., Nichols, R. L., Brown, J. K., Byrne, F. J., & Li, X. C. (2010). Extraordinary resistance to insecticides reveals exotic Q biotype of *Bemisia tabaci* in the new world. *Journal of Economic Entomology, 103*, 2174–2186.

Dhadialla, T. S., Calson, G. R., & Le, D. P. (1998). New insecticides with ecdysteroidal and juvenile hormone activity. *Annual Review of Entomology, 43*, 545–569.

Diaz, B. M., & Fereres, A. (2007). Ultraviolet-blocking materials as a physical barrier to control insect pests and pathogens in protected crops. *Pest Tech, 1*, 85–95.

Dittrich, V., Ernst, G. H., Ruesh, O., & Uk, S. (1990). Resistance mechanisms in sweetpotato whitefly (Homoptera: Aleyrodidae) populations from Sudan, Turkey, Guatemala, and Nicaragua. *Journal of Economic Entomology, 83*, 1665–1670.

Ellsworth, P. C., & Martinez-Carrillo, J. L. (2001). IPM for *Bemisia tabaci*: A case study from North America. *Crop Protection, 20*, 853–869.

Ellsworth, P. C., & Naranjo, S. E. (1999). Whitefly management with insect growth regulators and the influence of Lygus controls. In J. C. Silvertooth (Ed.), *Cotton, A College of Agriculture Report, series P-116* (pp. 339–354). University of Arizona, College of Agriculture. http://ag.arizona.edu/pubs/crops/az1123/az11237e.pdf

Ellsworth, P. C., Diehl, J. W., Dennehy, T. J., & Naranjo, S. E. (1995). *Sampling sweetpotato whiteflies in cotton. IPM ser no 2. Publ no 194023*. University of Arizona, College of Agriculture and Life Sciences, Cooperative Extension. http://cals.arizona.edu/crops/cotton/insects/wf/wfsampl.html

Ellsworth, P. C., Dennehy, T. J., & Nichols, R. L. (1996). *Whitefly management in Arizona cotton 1996, IPM series no. 3. Publication no. 196004*. The University of Arizona, Cooperative Extension. http://ag.arizona.edu/crops/cotton/insects/wf/cibroch.html

Ellsworth, P. C., Diehl, J. W., Kirk, I. W., & Henneberry, T. J. (1997). *Bemisia* growth regulators: Large-scale evaluation. In P. Dugger & D. Richter (Eds.), *Proc Beltwide cotton Conf* (pp. 922–929). Natl Cotton Council.

Fagan, L. L., McLachlan, A., Till, C. M., & Walker, M. K. (2010). Synergy between chemical and biological control in the IPM of currant-lettuce aphid (*Nasonovia ribisnigri*) in Canterbury, New Zealand. *Bulletin of Entomological Research, 100*, 217–223.

Flückiger, C. R., Kristinsson, H., Senn, R., Rindlisbacher, A., Buholzer, H., & Voss, G. (1992a). A novel agent to control aphids and whiteflies. In *Brighton crop protection conference: Pests and diseases* (pp. 43–50). British Crop Protection Council.

Flückiger, C. R., Kristinsson, H., Senn, R., Rindlisbacher, A., Buholzer, H., & Voss, G. (1992b). CGA 215'944—A novel agent to control aphids and whiteflies. In *Brighton crop protection conference—Pests and diseases* (Vol. 1, pp. 43–50).

Frizvold, G., Tronstad, R., & Mortensen, J. (2000). *Economic impacts of Bt cotton adoption: A national and regional assessment*. College of Agriculture, University of Arizona, Series/Report no. AZ1170.

Fuog, D., Fergusson, S. J., & Flückiger, C. (1998). Pymetrozine: A novel insecticide affecting aphids and whiteflies. In I. Ishaaya & D. Degheele (Eds.), *Insecticides with novel modes of action: Mechanism and application* (pp. 40–49). Springer.

Gerling, D., & Horowitz, A. R. (1984). Yellow traps for evaluating the population levels and dispersal patterns of *Bemisia tabaci* (Gennadius) (Homoptera: Aleyrodidae). *Annals of the Entomological Society of America, 77*, 753–759.

Gilbertson, R. L., Rojas, M., & Natwick, E. (2011). Development of integrated pest management (IPM) strategies for whitefly (*Bemisia tabaci*)-transmissible geminiviruses. In W. M. O. Thompson (Ed.), *The whitefly, Bemisia tabaci (Homoptera: Aleyrodidae) interaction with geminivirus-infected host plants: Bemisia tabaci, host plants and geminiviruses* (pp. 323–356). Springer.

Gillespie, D. R., & Quiring, D. J. M. (1992). Flight behavior of the greenhouse whitefly, *Trialeurodes vaporariorum* (Westwood) (Homoptera: Aleyrodidae), in relation to yellow sticky traps. *Canadian Entomologist, 124*, 907–916.

Gorman, K., Slater, R., Blande, J., Clarke, A., Wren, J., McCaffery, A., & Denholm, I. (2010). Cross-resistance relationships between neonicotinoids and pymetrozine in *Bemisia tabaci* (Hemiptera: Aleyrodidae). *Pest Management Science, 66*, 1186–1190.

Gravalos, C., Fernandez, E., Belando, A., Moreno, I., Ros, C., & Bielza, P. (2015). Cross-resistance and baseline susceptibility of Mediterranean strains of *Bemisia tabaci* to cyantraniliprole. *Pest Management Science, 71*, 1030–1036.

Green, R. H. (1970). On fixed precision sequential sampling. *Researches on Population Ecology, 12*(2), 249–225.

Gu, X., Bu, W., Xu, W., Bai, Y., Liu, B., & Liu, T. (2008). Population suppression of *Bemisia tabaci* (Hemiptera: Aleyrodidae) using yellow sticky traps and *Eretmocerus* nr. *Rajasthanicus* (Hymenoptera: Aphelinidae) on tomato plants in greenhouses. *Insect Science, 15*, 263–270.

Guillén, J., Navarro, M., & Bielza, P. (2014). Cross-resistance and baseline susceptibility of spirotetramat in *Frankliniella occidentalis* (Thysanoptera: Thripidae). *Journal of Economic Entomology, 107*, 1239–1244.

Heinz, K. M., Parrella, M. P., & Newman, J. P. (1992). Time-efficient use of yellow sticky traps in monitoring insect populations. *Journal of Economic Entomology, 85*, 2263–2269.

Henneberry, T. J., Jech, L., Hendrix, D. L., & Steele, T. (1999). *Bemisia argentifolii* (Homoptera: Aleyrodidae): Factors affecting adult and nymph honeydew production. *South West Entomology, 24*, 207–231.

Hilje, L., Costa, H. C., & Stansly, P. A. (2001). Cultural practices for managing *Bemisia tabaci* and associated viral diseases. *Crop Protection, 20*, 801–812.

Horowitz, A. R., & Ishaaya, I. (1992). Susceptibility of the sweetpotato whitefly (Homoptera: Aleyrodidae) to buprofezin during the cotton season. *Journal of Economic Entomology, 85*, 318–324.

Horowitz, A. R., & Ishaaya, I. (1996). Chemical control of Bemisia: Management and application. In D. Gerling & R. T. Mayer (Eds.), *Bemisia: 1995 taxonomy, biology, damage, control and management* (pp. 537–556). Intercept Ltd.

Horowitz, A. R., Denholm, I., Groman, K., & Ishaaya, I. (1999). Insecticide resistance in whiteflies: Current status and implications for management. In I. Denholm & P. Ioannidis (Eds.), *Proceedings of an ENMARIA symposium: Combating insecticide resistance, Thessalonikis, Greece.*, May (pp. 8996–8998).

Horowitz, A. R., Kontsedalov, S., & Ishaaya, I. (2004). Dynamics of resistance to the neonicotinoids acetamiprid and thiamethoxam in *Bemisia tabaci* (Homoptera: Aleyrodidae). *Journal of Economic Entomology, 97*, 2051–2056.

Horowitz, A. R., Kontsedalov, S., Khasdan, V., & Ishaaya, I. (2005). Biotypes B and Q of *Bemisia tabaci* and their relevance to neonicotinoid and pyriproxyfen resistance. *Archives of Insect Biochemistry and Physiology, 58*, 216–225.

Horowitz, A. R., Antignus, Y., & Gerling, D. (2011). Management of *Bemisia tabaci* whiteflies. In *The whitefly, Bemisia tabaci (Homoptera: Aleyrodidae) interaction with geminivirus-infected host plants: Bemisia tabaci, host plants and geminiviruses* (pp. 293–322). Springer.

Hou, M. L., Lu, W., & Wen, J. (2006). Trap catches and control efficiency of *Bemisia tabaci* (Homoptera: Aleyrodidae) adults in greenhouse by yellow sticky traps. *Scientia Agricultura Sinica, 39*, 1934–1939.

Ilias, A., Lagnel, J., Kapantaidaki, D. E., Roditakis, E., Tsigenopoulos, C. S., Vontas, J., & Tsagkarakou, A. (2015). Transcription analysis of neonicotinoid resistance in Mediterranean (MED) populations of *B. tabaci* reveals novel cytochrome P450s, but no nAChR mutations associated with the phenotype. *BMC Genomics, 16*, 939.

Ishaaya, I. (1990). Buprofezin and other IGRs for controlling cotton pests. *Pesticide Outlook, 1*(2), 30–33.

Ishaaya, I., & Horowitz, A. R. (1992). Novel phenoxy juvenile hormone analog (pyriproxyfen) suppresses embryogenesis and adult emergence of sweetpotato whitefly (Homoptera, Aleyrodidae). *Journal of Economic Entomology, 85*, 2113–2117.

Ishaaya, I., & Horowitz, A. R. (1995). Pyriproxyfen, a novel insect growth regulator for controlling whiteflies: Mechanisms and resistance management. *Pesticide Science, 43*, 227–232.

Ishaaya, I., & Horowitz, A. R. (1998). Insecticides with novel modes of action: An overview. In I. Ishaaya & D. Degheele (Eds.), *Insecticides with novel modes of action, mechanisms and application* (pp. 1–24). Springer.

Ishaaya, I., Mendelson, Z., & Melamed-Madjar, V. (1988). Effect of buprofezin on embryogenesis and progeny formation of sweetpotato whitefly (Homoptera: Aleyrodidae). *Journal of Economic Entomology, 81*, 781–784.

Ishaaya, I., Mendelson, Z., & Horowitz, A. R. (1993). Toxicity and growth-suppression exerted by diafenthiuron in the sweetpotato whitefly, *Bemisia tabaci. Phytoparasitica, 21*, 199–204.

Ishaaya, I., Yablonski, S., Mendelson, Z., Mansour, Y., & Horowitz, A. R. (1996). Novaluron (MCW-275), a novel benzoylphenyl urea, suppressing developing stages of lepidopteran, whitefly, and leafminer pests. In *Brighton crop protection conference: pests and diseases*, British Crop Protection Council, Farnham (pp. 1013–1020).

Ishaaya, I., Yablonski, S., Mendeleson, Z., Mansour, Y., Kontsedalov, S., Pleuschkell, U., & Horowitz, A. R. (1997). Novel control agents: Mechanisms and application. *Phytoparasitica, 25*, 144.

Ishaaya, I., Kontsedalov, S., Mazirov, D., & Horowitz, A. R. (2001). Biorational agents: Mechanisms and importance in IPM and IRM programs for controlling agricultural pests. In *Proceedings of the 53rd international symposium on crop protection* (Vol. 66). Meded Fac Landbouww, Univ Gent. (in press).

Ishaaya, I., Kontsedalov, S., & Horowitz, A. R. (2003). Novaluron (Rimon), a novel IGR: Potency and cross-resistance. *Archives of Insect Biochemistry and Physiology, 54*, 157–164.

Islam, M. N., Rahim, M. A., Alam, M. S., & Naher, N. (2003). Control of spiraling whitefly (*Aleurodicus dispersus*) of guava by spraying detergent. *Asian Journal of Plant Sciences, 2*(3), 270–272.

Jeevalatha, A., Vanishree, G., Siddappa, S., Kumar, R., Kaundal, P., Kumar, A., & Chakrabarti, S. K. (2021). Molecular characterization and infectivity analysis of tomato leaf curl New Delhi virus isolates infecting potato. *3 Biotech, 11*, 203. https://doi.org/10.1007/s13205-021-02752-5

Jeevalatha, A., Siddappa, S., Kumar, R., Tiwari, R. K., Lal, M. K., Sharma, S., Chakrabarty, S. K., & Singh, B. P. (2023). RNA-seq analysis reveals an early defense response to tomato leaf curl New Delhi virus in potato cultivar Kufri Bahar. *Functional & Integrative Genomics, 23*, 215. https://doi.org/10.1007/s10142-023-01138-5

Jumardi, J., Melina, M., & Nasruddin, A. (2020). Vertical distribution of the greenhouse whitefly, *Trialeurodes vaporariorum* Westwood, within potato plant canopy. *IOP Conference Series: Earth and Environmental Science, 486*(1), 012144.

Kagabu, S. (1999). Studies on the synthesis and insecticidal activity of neonicotinoid compounds. *Pesticide Science, 21*, 231–239.

Kanno, H. K., Asi, T., & Maekawa, S. (1981). 2-tert-butylimino-3-isopropyl-perhydro-1,3,5-thiadiazin-4-one (NNI 750), a new insecticide. In *Proceedings, Brighton crop protection conference, pests and diseases* (pp. 56–69).

Kayser, H., Kaufmann, L., & Schürmann, F. (1994). Pymetrozine (CGA 215'944): A novel compound for aphid and whitefly control. An overview of its mode of action. In *Proceedings of 1994 Brighton crop protection conference—Pests and diseases* (Vol. 2, pp. 737–742).

Kim, S., & Lim, U. T. (2011). Evaluation of a modified sticky card to attract *Bemisia tabaci* (Hemiptera: Aleyrodidae) and a behavioral study on their visual response. *Crop Protection, 30*, 508–511.

Koenig, J. P., Lawson, D. S., & White, S. M. (1998). Utility of fulfill 50WG for aphid and whitefly management in cotton. In P. Dugger & D. Richter (Eds.), *Proceedings Beltwide cotton conferences* (pp. 997–999). National Cotton Council.

Kontsedalov, S., Gottlieb, Y., Ishaaya, I., Nauen, R., Horowitz, A. R., & Ghanim, M. (2009). Toxicity of spiromesifen to the developmental stages of *Bemisia tabaci* biotype B. *Pest Management Science, 65*, 5–13.

Krueger, H. R., Lindquist, R. K., Mason, J. F., & Spadafora, R. R. (1973). Application of methomyl to greenhouse tomatoes: Greenhouse whitefly control and residues in foliage and fruits. *Journal of Economic Entomology, 66*, 1223–1224.

Kumar, R., Tiwari, R. K., Jeevalatha, A., Sundaresha, S., Shah, M. A., Sharma, S., Sagar, V., Kumar, M., & Chakrabarti, S. K. (2021). Potato apical leaf curl disease: Current status and perspectives on a disease caused by tomato leaf curl New Delhi virus. *Journal of Plant Diseases and Protection, 128*, 897–911. https://doi.org/10.1007/s41348-021-00463-w

Kumar, R., Tiwari, R. K., Sundaresha, S., Kaundal, P., & Raigond, B. (2022). Potato viruses and their management. In S. K. Chakrabarti, S. Sharma, & M. A. Shah (Eds.), *Sustainable management of potato pests and diseases*. Springer. https://doi.org/10.1007/978-981-16-7695-6_12

Kuno, E. (1969). A new method of sequential sampling to obtain the population estimates with a fixed level of precision. *Researches on Population Ecology, 11*(2), 127–136.

Lahm, G. P., Cordova, D., & Barry, J. D. (2009). New and selective ryanodine receptor activators for insect control. *Bioorganic & Medicinal Chemistry, 17*, 4127–4133.

Lapidot, M., Legg, J. P., Wintermantel, W. M., & Polston, J. E. (2014). Management of whitefly-transmitted viruses in open-field production systems. In *Advances in virus research* (Vol. 90, pp. 147–206). Academic Press.

Legarrea, S., Karnieli, A., Fereras, A., & Weintraub, P. G. (2010). Comparison of UV-absorbing nets in pepper crops, spectral properties, effects on plants and pest control. *Photochemistry and Photobiology, 86*, 324–330.

Legarrea, S., Barman, A., Marchant, W., Diffie, S., Srinivasan, R., & Culbreath, A. K. (2020). Physical and chemical tactics for management of whiteflies and whitefly-transmitted viruses in tomato production. *Crop Protection, 134*, 105178. https://doi.org/10.1016/j.cropro.2020.105178

Liang, G., & Liu, T. X. (2002). Repellency of a kaolin particle film, surround, and a mineral oil, sunspray oil, to silverleaf whitefly (Homoptera: Aleyrodidae) on melon in the laboratory. *Journal of Economic Entomology, 95*(2), 317–324.

Lindquist, R. K., Krueger, H. R., Spadafora, R. R., & Mason, J. F. (1972). Application of aldicarb to greenhouse tomatoes: Plant growth, fruit yields, greenhouse whitefly control and residues in fruits. *Journal of Economic Entomology, 65*, 862–864.

Liu, M. Y., & Casida, J. E. (1993). High affinity binding of [3H]imidacloprid in the insect acetylcholine receptor. *Pesticide Biochemistry and Physiology, 46*, 40–46.

Liu, S., Zhang, P., Wang, Y., & Tang, L. (2023). Repellent effects of essential oils against *Bemisia tabaci*: Patchouli oil as a promising botanical insecticide. *Journal of Economic Entomology, 116*(4), 1324–1333. https://doi.org/10.1111/jen.13128

Lloyd, L. (1922). The control of the greenhouse whitefly (*Asterochiton vaporariorum*) with notes on its biology. *The Annals of Applied Biology, 9*, 1–32.

Luo, C., Jones, C. M., Zhang, F., Denholm, I., & Gorman, K. (2010). Insecticide resistance in *Bemisia tabaci* biotype Q (Hemiptera: Aleyrodidae) from China. *Crop Protection, 29*, 429–434.

Maienfisch, P., Huerlimann, H., Rindlisbacher, A., Gsell, L., Dettwiler, H., Haettenschwiler, J., Sieger, E., & Walti, M. (2001). The discovery of thiamethoxam: A second-generation neonicotinoid. *Pest Management Science: formerly Pesticide Science, 57*(2), 165–176.

Malik, K., Raghavendra, K. V., & Kumar, M. (2020). Vector management strategies against *Bemisia tabaci* (Gennadius) transmitting potato apical leaf curl virus in seed potatoes. *Potato Research*. https://doi.org/10.1007/s11540-020-09470-0

Miles, H. W. (1927). On the control of glasshouse insects with calcium cyanide. *The Annals of Applied Biology, 14*, 240–246.

Modgill, N., Naga, K. C., Firake, D. M., Tiwari, R. K., Shivaramu, S., Kumar, R., Bhatnagar, A., Buckseth, T., Bairwa, A., Maharana, C., Sharma, S., & Singh, B. (2025). Flupyradifurone application: A promising approach to curtail tomato leaf curl New Delhi virus (ToLCNDV) transmission vectored by *Bemisia tabaci* in potato farming. *Crop Protection, 187*, 106951.

Mound, L. (1962). Studies on the olfaction colour sensitivity of *Bemisia tabaci* Genn. Aleyrodidae. *Entomología Experimentalis et Applicata, 5*, 99–104.

Naga, K. C., Subhash, S., Bhatnagar, A., Tiwari, R. K., Kumar, R., Verma, G., Sharma, S., & Chakrabarti, S. K. (2021a). Influence of host plants on virus acquisition and endosymbionts of whitefly, *Bemisia tabaci* (Gennadius). *Indian Journal Of Entomology, 83*, 371–375. https://doi.org/10.5958/0974-8172.2020.00206.0

Naga, K. C., Sundaresha, S., Kumar, R., Tiwari, R. K., Subhash, S., Verma, G., Buckseth, T., Bairwa, A., Sharma, S., Katare, S., & Srivastava, R. M. (2021b). A new record of Asia II 5 genetic group of *Bemisia tabaci* (Gennadius) in the major potato growing areas of India and its relationship with tomato leaf curl New Delhi virus infecting potato. *3 Biotech, 11*, 421. https://doi.org/10.1007/s13205-021-02966-7

Naranjo, S. E., & Ellsworth, P. C. (2009). Fifty years of the integrated control concept: Moving the model and implementation forward in Arizona. *Pest Management Science, 65*, 1267–1286.

Naranjo, S. E., & Flint, H. M. (1994). Spatial distribution of premaginal *Bemisia tabaci* (Homoptera: Aleyrodidae) in cotton and development of fixed-precision, sequential sampling plans. *Environmental Entomology, 24*, 254–266.

Naranjo, S. E., & Flint, H. M. (1995). Spatial distribution of adult *Bemisia tabaci* (Homoptera: Aleyrodidae) in cotton, and development and validation of fixed-precision sampling plans for estimating population density. *Environmental Entomology, 24*, 261–270.

Naranjo, S. E., Flint, H. M., & Henneberry, T. J. (1995). Comparative analysis of selected sampling methods for adult *Bemisia tabaci* (Homoptera, Aleyrodidae) in cotton. *Journal of Economic Entomology, 88*, 1666–1678.

Naranjo, S. E., Chu, C. C., & Henneberry, T. J. (1996). Economic injury levels for *Bemisia tabaci* (Homoptera: Aleyrodidae) in cotton: Impact of crop price, control costs, and efficacy of control. *Crop Protection, 15*, 779–788.

Nauen, R., & Denholm, I. (2005). Resistance of insect pests to neonicotinoid insecticides: Current status and future prospects. *Archives of Insect Biochemistry and Physiology, 58*, 200–215.

Nauen, R., & Konanz, S. (2005). Spiromesifen as a new chemical option for resistance management in whiteflies and spider mites. *Pflanzenschutz-Nachrichten Bayer, 58*, 485–502.

Nauen, R., & Steinbach, D. (2016). Resistance to diamide insecticides in lepidopteran pests. In A. R. Horowitz & I. Ishaaya (Eds.), *Advances in insect control and resistance management* (pp. 219–240). Springer.

Nauen, R., Stumpf, N., & Elbert, A. (2002). Toxicological and mechanistic studies on neonicotinoid cross resistance in Q-type *Bemisia tabaci* (Hemiptera: Aleyrodidae). *Pest Management Science, 58*, 868–875.

Nauen, R., Reckmann, U., Thomzik, J., & Thielert, W. (2008). Biological profile of spirotetramat (Movento®)—A new two-way systemic (ambimobile) insecticide against sucking pest species. *Bayer Crop Science Journal, 61*, 245–278.

Navas-Castillo, J., Fiallo-Olivé, E., & Sánchez-Campos, S. (2019). Emerging virus diseases transmitted by whiteflies. *Annual Review of Phytopathology, 57*, 181–203. https://doi.org/10.1146/annurev-phyto-080417-050036

Nyoike, T. W., Liburd, O. E., & Webb, S. E. (2008). Suppression of whiteflies, *Bemisia tabaci* (Hemiptera: Aleyrodidae) and incidence of cucurbit leaf crumple virus, a whitefly-transmitted virus of zucchini squash new to Florida, with mulches and imidacloprid. *Florida Entomologist, 91*, 460–465.

Palumbo, J. C. (1998). Evaluation of insect growth regulators for management of whiteflies in melons. In N. Oebker (Ed.), *1998 vegetable report, series P-115, AZ 1101* (pp. 35–40). University of Arizona, College of Agriculture. http://ag.arizona.edu/pubs/crops/az1101/az11015.html

Palumbo, J. C., Mullis, C. H., & Reyes, F. J. (1993). Control of sweetpotato whitefly in cantaloupe with various pesticides, 1992. *Insect Acaricide Tests, 18*, 117.

Palumbo, J. C., Mullis, C. H., & Reyes, F. J. (1994). Control of sweetpotato whitefly in cantaloupe with various pesticides, 1992. *Arthropod Management Tests, 19*, 80–81.

Palumbo, J. C., Horowitz, A. R., & Prabhaker, N. (2001). Insecticidal control and resistance management for *Bemisia tabaci*. *Crop Protection, 20*(9), 739–765.

Park, J., Lee, J., Shin, K., Lee, S. E., & Cho, K. (2011). Geostatistical analysis of the attractive distance of two different sizes of yellow sticky traps for greenhouse whitefly, *Trialeurodes vaporariorum* (Westwood) (Homoptera: Aleyrodidae), in cherry tomato greenhouses. *Australian Journal of Entomology, 50*, 144–151.

Parker, T. (1928). The use of tetrachloroethane for commercial glasshouse fumigation. *The Annals of Applied Biology, 15*, 251–257.

Parrella, M. P., & Jones, V. P. (1985). Yellow traps as monitoring tools for *Liriomyza trifolii* (Diptera: Agromyzidae) in Chrysanthemum greenhouses. *Journal of Economic Entomology, 78*, 53–56.

Pavela, R. (2007). Possibilities of botanical insecticide exploitation in plant protection. *Pest Tech, 1*, 47–52.

Perring, T. M., Stansly, P. A., & Liu, T. X. (2018a). A new era of *Bemisia tabaci* in North America: Impacts, prospects, and management. *Annual Review of Entomology, 63*, 367–386. https://doi.org/10.1146/annurev-ento-020117-043229

Perring, T. M., Stansly, P. A., Liu, T. X., Smith, H. A., & Andreason, S. A. (2018b). Whiteflies: Biology, ecology, and management. In W. Wakil, G. E. Brust, & T. M. Perring (Eds.), *Sustainable management of arthropod pests of tomato* (pp. 73–110). Academic Press.

Prabhaker, N., Toscano, N. C., & Henneberry, T. J. (1998). Evaluation of insecticide rotations and mixtures as resistance management strategies for *Bemisia argentifolii* (Homoptera: Aleyrodidae). *Journal of Economic Entomology, 91*, 820–826.

Prabhaker, N., Castle, S. J., Buckelew, L., & Toscano, N. C. (2008). Baseline susceptibility of *Bemisia tabaci* B biotype (Hemiptera: Aleyrodidae) populations from California and Arizona to spiromesifen. *Journal of Economic Entomology, 101*, 174–181.

Puri, S. N., Bhosle, B. B., Ilyas, M., Butler, G. D., Jr., & Henneberry, T. J. (1994). Detergents and plant-derived oils for control of the sweet potato whitefly on cotton. *Crop Protection, 13*(1), 45–48.

Qiu, B. L., & Ren, S. X. (2006). Using yellow sticky traps to inspect population dynamics of *Bemisia tabaci* and its parasitoids. *Chinese Bulletin of Entomology, 43*, 53–56.

Qureshi, M. S., Midmore, D. J., Syeda, S. S., & Playford, C. L. (2007). Floating row covers and pyriproxyfen help control silverleaf whitefly *Bemisia tabaci* (Gennadius) biotype B (Homoptera: Aleyrodidae) in zucchini. *Australian Journal of Entomology, 46*, 313–319.

Redak, R. A., Bethke, J. A., & Costa, H. S. (1998). Control of silverleaf whitefly on poinsettia under greenhouse conditions, summer 1997. *Arthropod Management Tests, 23*, 350.

Richardson, H., Bulger, J. W., Busbey, R. L., Nelson, R. H., & Weigel, C. A. (1943). Studies on nicotine fumigation in greenhouses. *Circular: United States Department of Agriculture, 684*, 1–15.

Rincon, D. F., Vasquez, D. F., Rivera-Trujillo, H. F., Beltrán, C., & Borrero-Echeverry, F. (2019). Economic injury levels for the potato yellow vein disease and its vector, *Trialeurodes vaporariorum* (Hemiptera: Aleyrodidae), affecting potato crops in the Andes. *Crop Protection, 119*, 52–58. https://doi.org/10.1016/j.cropro.2019.01.002

Roditakis, E., Grispou, M., Morou, E., Kristoffersen, J. B., Roditakis, N., Nauen, R., Vontas, J., & Tsagkarakou, A. (2009). Current status of insecticide resistance in Q biotype *Bemisia tabaci* populations from Crete. *Pest Management Science, 65*, 313–322.

Ruder, F. J., & Kayser, H. (1993). The carbodiimide product of diafenthiuron inhibits mitochondria in vivo. *Pesticide Biochemistry and Physiology, 46*, 96–106.

Salati, R., Nahkla, M. K., Rojas, M. R., Guzman, P., Jaquez, J., Maxwell, D. P., & Gilbertson, R. L. (2002). Tomato yellow leaf curl virus in The Dominican Republic: Characterization of an infectious clone, virus monitoring in whiteflies, and identification of reservoir hosts. *Phytopathology, 92*(5), 487–496.

Sattelle, D. B., Cordova, D., & Cheek, T. R. (2008). Insect ryanodine receptors: Molecular targets for novel pest control chemicals. *Invertebrate Neuroscience, 8*, 107–119.

Schaefer, C. H., Miura, T., Dupras, E. F., Jr., Mulligan, F. S., III, & Wilder, W. H. (1988). Efficacy, non-target effects and chemical persistence of S-31183, a promising mosquito (Dipter: Culicidae) control agent. *Journal of Economic Entomology, 81*, 1648–1655.

Schuster, D. J. (2000). Insecticides applied on demand for managing the silverleaf whitefly on fresh market tomatoes in spring 1998. *Arthropod Management Tests, 25*, 100.

Schuster, D. J., & Polston, J. E. (1997). Management of whiteflies and armyworm on fresh market tomatoes fall 1995. *Arthropod Management Tests, 22*, 182–183.

Schuster, D. J., & Smith, H. A. (2016). *Management of whiteflies, whitefly-vectored plant virus, and insecticide resistance for vegetable production in southern Florida*. University of Florida IFAS Extension, ENY-735.

Schuster, D. J., Stansly, P. A., Polston, J. E., Gilreath, P. R., & McAvoy, E. (2007). *Management of whiteflies, whitefly-vectored plant virus, and insecticide resistance for vegetable production in southern Florida*. ENY-735 (IN695), IFAS extension. University of Florida.

Schuster, D. J., Thompson, S., Ortega, L. D., & Polston, J. E. (2009). Laboratory evaluation of products to reduce settling of sweetpotato whitefly adults. *Journal of Economic Entomology, 102*(4), 1482–1489.

Schuster, D. J., Mann, R. S., Toapanta, M., Cordero, R., Thompson, S., Cyman, S., Shurtleff, A., & Morris, R. F., II. (2010). Monitoring neonicotinoid resistance in biotype B of *Bemisia tabaci* in Florida. *Pest Management Science, 66*, 186–195.

Sclar, D. C., Gerace, D., Tupy, A., Wilson, K., Spriggs, S. A., Bishop, R. J., & Cranshaw, W. S. (1999). Effects of application of various reduced-risk pesticides to tomato, with notes on control of greenhouse whitefly. *HortTechnology, 9*, 185–189.

Shah, M. A., Sharma, S. & Kumar, R. (2022). Efficacy of new-chemistry insecticides against whitefly and aphids in potato. *Potato Journal, 49*(1), 41–47.

Shah, M. A., Sharma, S., & Sharma, J. (2019). Bio-efficacy of potassium silicate against aphids and whitefly in potato. *Potato Journal, 46*(2), 132–137.

Shah, M. A., Aarti, B., Naga, K. C., Subhash, S., et al. (2020a). Important potato pests and their management. In A. K. Singh, S. K. Chakrabarti, B. Singh, J. Sharma, & V. K. Dua (Eds.), *Potato science and technology for sub-tropics* (pp. 295–326). NIPA.

Shah, M. A., Sharma, S., Kumar, R., & Singh, R. K. (2020b). Evaluation of polypropylene row covers for excluding virus vectors and their effect on the incidence of diseases and yield in potato. *Indian Phytopathology, 73*, 751–757.

Shah, M. A., Sharma, S., & Kumar, R. (2021a). Seasonal population dynamics and flight activity of sweetpotato whitefly, *Bemisia tabaci* on potato in the sub-tropical plains of India. *Arthropod-Plant Interactions, 15*, 605–613. https://doi.org/10.1007/s11829-021-09835-3

Shah, M. A., Naga, K. C., Subhash, S., et al. (2021b). Use of petroleum-derived spray oils for the management of vector-virus complex in potato. *Potato Research*. https://doi.org/10.1007/s11540-021-09505-0

Shah, M. A., Sridhar, J., & Sharma, S. (2022). Studies on the within-plant distribution of white-fly, *Bemisia tabaci* (Gennadius) in potato and development of a sequential sampling plan. *International Journal of Pest Management*. https://doi.org/10.1080/09670874.2020.1823520

Sharaf, N. (1986). Chemical control of *Bemisia tabaci*. *Agriculture, Ecosystems and Environment, 17*, 111–127.

Shen, B. B., & Ren, S. X. (2003). Yellow card traps and its effects on population of *Bemisia tabaci*. *Journal of South China Agricultural University, 24*, 40–43.

Smith, F. F. (1970). Identifying and controlling the greenhouse whitefly. *American Vegetable Grower, 18*, 41–42.

Smith, F., Fulton, R. A., & Brierly, P. (1947). Use of DDT and H.E.T.P. as aerosols in greenhouses. *Agricultural Chemistry, 2*, 28–31.

Speyer, E. R., & Owen, O. (1926). The fumigation of tomato houses with hydrocyanic acid gas. *The Annals of Applied Biology, 13*, 144–147.

Stansly, P. A., & Conner, J. M. (2000). Impact of insecticides on silverleaf whitefly and tomato yellow leaf curl virus on staked tomato, 1999. *Arthropod Management Tests, 25*, 104.

Stansly, P. A., & Naranjo, S. E. (2010). Cultural practices for managing *Bemisia tabaci* and associated viral diseases. *Crop Protection, 29*(10), 1157–1166.

Steinemann, H. P., Stamm, E., & Frei, B. (1990). Chemodynamics in research and development of new plant protection agents. *Pesticide Outlook, 1*, 3–7.

Streibert, H. P., Drabek, J., & Rindlisbacher, A. (1988). CGA 106630 – A new type of acaricide/insecticide for the control of the sucking pest complex in cotton and other crops. In *Proceedings of the Brighton Crop Protection Conference – Pests and Diseases*, pp. 25–33.

Suwwan, M. A., Akkawi, M., Al-Musa, A. M., & Mansour, A. (1988). Tomato performance and incidence of tomato yellow leaf curl (TYLC) virus as affected by type of mulch. *Scientia Horticulturae, 37*, 39–45.

Tomizawa, M., & Casida, J. E. (2005). Neonicotinoid insecticide toxicity: Mechanisms of selective action. *Annual Review of Pharmacology and Toxicology, 45*, 247–268.

van de Veire, M., & Vacante, V. (1984). Greenhouse whitefly control through the combined use of the colour attraction system with the parasite wasp *Encarsia formosa* (Hym.: Aphelinidae). *Entomophaga, 29*, 303–310.

Wang, Y., & Liu, S. (2024). A review of interactions between plants and whitefly-transmitted begomoviruses: From virus transmission to disease management. *Frontiers in Plant Science, 15*, 10648457.

Wang, Z., Yan, H., Yang, Y., & Wu, Y. (2010). Biotype and insecticide resistance status of the whitefly *Bemisia tabaci* from China. *Pest Management Science, 66*, 1360–1366.

Wang, R., Wang, J., Che, W., & Luo, C. (2018). First report of field resistance to cyantraniliprole, a new anthranilic diamide insecticide, on *Bemisia tabaci* MED in China. *Journal of Integrative Agriculture, 16*, 60345–60347.

Ware, G. W. (2000). *The pesticide book* (5th ed.). Thompson Publications.

Weintraub, P. G., & Berlinger, M. J. (2004). Physical control in greenhouses and field crops. In *Insect pest management: Field and protected crops* (pp. 301–318). Springer.

Xue, Y. G., Watson, D. M., Nicetic, O., & Beattie, G. A. C. (2002). Impact of nC24 horticultural mineral oil deposits on oviposition by greenhouse whitefly *Trialeurodes vaporariorum* (Westwood) (Hemiptera: Aleyrodidae). *General and Applied Entomology: The Journal of the Entomological Society of New South Wales, 31*, 59–64.

Yamada, T., Takahashi, H., & Hatano, R. (1999). A novel insecticide, acetamiprid. In I. Yamamoto & J. E. Casida (Eds.), *Nicotinoid insecticides and the nicotinic acetylcholine receptor* (pp. 149–176). Springer.

Yamamoto, I., Yabuta, G., Tomizawa, M., Saito, T., Miyamotot, T., & Kagabu, S. (1995). Molecular mechanisms for selective toxicity of nicotinoids and neonicotinoids. *Journal of Pesticide Science, 20*, 33–40.

Yao, F. L., Zheng, Y., Huang, X. Y., Ding, X. L., et al. (2017). Dynamics of *Bemisia tabaci* biotypes and insecticide resistance in Fujian province in China during 2005–2014. *Scientific Reports, 7*, 40803. https://doi.org/10.1038/srep40803

Yasui, M., Fakuda, M., & Maekawa, S. (1987). Effects of buprofezin on reproduction of the greenhouse whitefly, *Trialeurodes vaporarium* (Westwood) (Homoptera: Aleyrodidae). *Applied Entomology and Zoology, 22*, 266–271.

Zhang, N., & Yu, L. (2009). Impact and control efficiency of yellow trap to *Bemisia tabaci* adults on tomato in greenhouse. *Hubei Agricultural Sciences, 48*, 1884–1886.

10 Nanoparticle-Based Delivery Systems for Antiviral Compounds and Nanotechnological Interventions in Tomato Leaf Curl New Delhi Virus Management

Ayushi Malik, Ramswaroop Saini, Vijay Vardhan Pandey, Haripriya Kavidayal, Krishnaveer Singh Jhala, and Avinash Marwal

Abstract

The *Tomato leaf curl New Delhi virus* (ToLCNDV), classified under the genus *Begomovirus* and the family *Geminiviridae*, constitutes a significant impediment to tomato cultivation, being the foremost and economically vital disease affecting tomatoes in the Indian subcontinent. In recent years, however, ToLCNDV has been rapidly broadening its host range and disseminating into new geographical territories, including the Middle East and the western Mediterranean Basin. In the realm of ToLCNDV management, nanoparticle-based delivery systems exhibit substantial potential. Encapsulation of viral-inhibiting agents within affected tomato plants can efficiently mitigate viral components, thereby reducing potential crop loss. Nanotechnology-integrated methods can be combined with various management approaches, utilizing resistant cultivars to enhance pest management and produce a broad spectrum of control parameters for ToLCNDV. This comprehensive study provides a detailed historical examination

A. Malik · K. S. Jhala · A. Marwal (✉)
Nanoparticle Synthesis and Bioinformatics Laboratory, Department of Biotechnology, Vigyan Bhawan, Block – B, New Campus, Mohanlal Sukhadia University, Udaipur, Rajasthan, India

R. Saini
Department of Biotechnology, School of Life and Allied Health Sciences, Joy University, Tirunelveli, Tamil Nadu, India

V. V. Pandey
Jawaharlal Nehru Custom House Laboratory, Raigarh, Maharashtra, India

H. Kavidayal
Genetics and Tree Improvement Division, ICFRE-Forest Research Institute, Dehradun, Uttarakhand, India

R. Kumar et al. (eds.), *Tomato Leaf Curl New Delhi Virus (ToLCNDV)*,
https://doi.org/10.1007/978-981-95-2852-3_10

of the strategies required to manage viral diseases and limit their growth. Furthermore, it elucidates the superficial role of nano-formulations that encompass viral-inhibiting compounds and their noteworthy role in ToLCNDV management.

Keywords

Tomato cultivation · Crop management · Nano-encapsulation · Viral diseases · *Begomovirus*

1 Introduction

1.1 Overview of Geminiviruses Infections in Plants and the Emergence of ToLCNDV

Tomato leaf curl New Delhi virus (ToLCNDV) is characterized within the *Geminiviridae* family category of plant viruses. It is classified based on its quasi-icosahedral conjugated particles, with an approximate size of 18 by 30 nm^2, that compress spherical-shaped, single-stranded DNA of length 2.5 to 3.1 kb (Stanley, 1985). Classification of Gemini virus family into 14 different genera, *Begomovirus* (Navas-Castillo et al., 2011), *Curtovirus* (Varsani et al., 2014), *Topocuvirus* (Zerbini et al., 2017), *Mastrevirus* (Muhire et al., 2013), *Becurtovirus* (Varsani et al., 2014), *Turncurtovirus* (Kamali et al., 2016), *Eragrovirus* (Varsani et al., 2014), *Topilevirus*, *Opunvirus*, *Mulcrilevirus*, *Maldovirus, Grablovirus, Citlodavirus,* and *Capulavirus* (Roumagnac et al., 2022)-dependent on the vector-host relationship and their relative genome organizations. Notably, most economically significant geminiviruses are adherents of the genus *Begomovirus*. Currently, 288 recognized species under begomovirus (www.ictvonline.org) are categorized into two distinct groups based on their genomic composition: viruses of the monopartite category possess only a genome constituent, while viruses of the bipartite category consist of two genome components (Prajapat et al., 2013; Marwal et al., 2014a). Transmission of begomoviruses occurs predominantly through plant species such as tobacco and sweet potato. At the same time, it can also be transmitted by silver leaf whitefly, for instance, *Bemisia tabaci* known as Gennadius in everyday language, in a calculative tenacious way, with a primary localization in the vasculatory system of infested crops (Gilbertson et al., 2015; Navas-Castillo et al., 2011). These viral species pose a magnificent risk to numerous significant dicotyledonous crops belonging to various families comprising *Cucurbitaceae* (e.g., gourds), *Euphorbiaceae* (e.g., cassava), *Solanaceae* (e.g., chilli), *Malvaceae* (e.g., cotton), and *Fabaceae* (e.g., lima bean) across diverse geographical regions (Srivastava et al., 2022; Marwal et al., 2021; Nehra et al., 2019a; Chauhan et al., 2023).

Tomato leaf curl disease (ToLCD) appeared as a highly destructive ailment affecting tomato crops on the Indian subcontinent, resulting in potential complete yield loss in plants infected at an early stage (Chakraborty, 2008; Prasanna et al., 2015). The foremost report of this disease in India dates back to 1948 (Vasudeva &

Raj, 1948); it is now recognized in various regions of Asia as well as the Pacific Rim, further in North Africa and Europe. Furthermore, ToLCD is considered a significant ailment affecting tomatoes in sub-Saharan Africa; it is observed in association with several complicated monopartite begomoviruses (Abhary et al., 2007; Kon & Gilbertson, 2012; Leke et al., 2011; Osei et al., 2008; Zhou et al., 2008). Within the subcontinents of India, a minimum of 13 discrete species of virus were linked to ToLCD (Brown et al., 2015), among which ToLCNDV is of particular importance. With the accumulation of contaminating exclusive tomato cultivars (Sahu et al., 2010), ToLCNDV affects at least 43 supplementary dicotyledonous crops, including weeds and economically relevant vegetable and ornamental species (Nehra et al., 2014; Marwal et al., 2013; Gaur et al., 2011). Another significant leaf curl disease affecting tomatoes is tomato yellow leaf curl disease (TYLCD) (Abhary et al., 2007), generally associated with the group of begomoviruses commonly considered as *Tomato yellow leaf curl viruses* (TYLCVs) (Brown et al., 2015). The dynamics inside the cells and evolutionary aspects of the TYLCV association are comprehensively appraised in the literature, including works by Gafni (2003) and Diaz-Pendon et al. (2010).

1.2 Conventional Disease Management Strategies for Plant Viruses

Viruses are obligate cytoplasmic bedbugs whose replication as well as pathogenicity are entirely dependent upon the cellular machinery of their host organisms (Kumar et al., 2021a, b; Jeevalatha et al., 2023). These viral entities can cause considerable harm to crops, livestock, and human health, presenting significant threats to living organisms (Sahu et al., 2015; Prajapat et al., 2012). Additionally, many wild plant species are frequently found to be associated with viral infections (Roossinck, 2012). The *Tobacco mosaic virus* (TMV) is the earliest plant virus to be subjected to scientific study, and it is known to be responsible for causing mosaic disease affecting tobacco plants (Beijerinck, 1898). The International Committee on the Taxonomy of Viruses has identified approximately 900 distinct species of plant viruses (King et al., 2012), and research suggests that numerous additional plant viruses need to be studied (Roossinck et al., 2010). Most viruses that inflict impairment in refined plants exhibit acute characteristics, resulting in severe infections over a short duration; conversely, a significant number of viruses in wild plants possess a persistent lifecycle, allowing them to maintain a relationship with their host plants' machinery (Roossinck, 2012). Viral infections in plants can lead to considerable reductions in both crop yield and quality (Prajapat et al., 2011; Nehra et al., 2019b). For example, plants infected with TMV, *Papaya ring spot virus*, and *Potato virus Y* exhibit symptoms such as leaf distortion, yellowing, stunted growth, and abnormalities in flowers and fruits (Kumar et al., 2022, 2023). Within the agronomic sector, the appraised annual yield losses attributed to plants affected with viruses are more than $30 billion on a global scale (Sastry & Zitter, 2014). Some of the conventional techniques to manage the spread of viral diseases in plants are given below (Fig. 10.1).

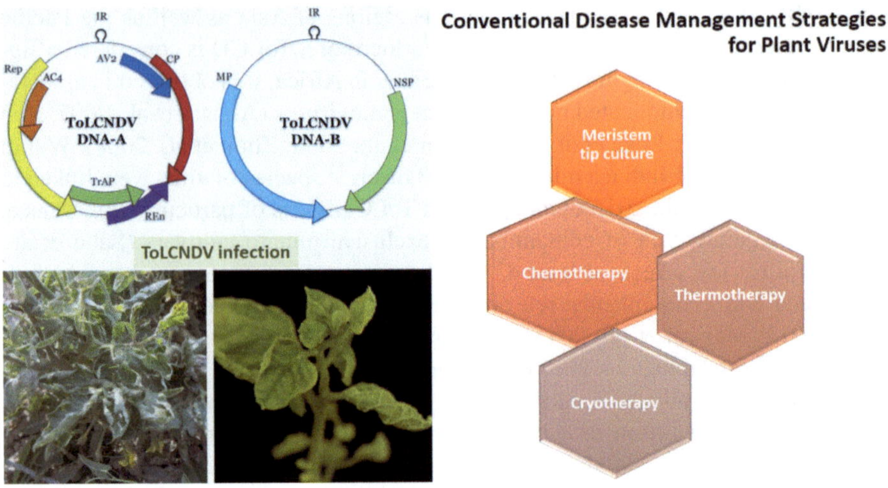

Fig. 10.1 ToLCNDV virulence and its historic management strategies

1.2.1 Meristem Tip Culture

Practice of meristem tip culture entails the excision of a systematized shoot apex taken from the parent plant for ensuing in vitro cultivation. This methodology has been confirmed as the most effective technique for the elimination of phloem-associated viruses (Lassois et al., 2013). The process originated with the meristem tips of Nasturtium (*Tropaeolum majus), culminating in the successful development of a* whole plant (Ball, 1946). In numerous circumstances, meristem tip culture has proven operative in the eradication of plant affecting viruses, including the complete removal of *Sugarcane yellow leaf virus* from sugarcane (Fitch et al., 2001), *Peanut stripe virus* (PStV) from patchouli plants (Singh et al., 2009), and *Piper yellow mottle virus* (PYMoV) shows an 84% success rate in black pepper (Sasi & Bhat, 2018). The benefits of the stated methodology comprise the utilization of disease-free explants excised from the mother plant for lab-based culture. Moreover, it preferably exhibits the inheritance of stable genes from the parent plant (Ancora et al., 1981; Murashige, 1974). As it also poses certain limitations, such as experimental cost, associated challenges to grow and acclimatise, somaclonal variation, concerns with production forecasts, and the risk of contamination.

1.2.2 Chemotherapy

Utilization of chemical-based viral inhibitory agents can be beneficial for the administration of viral ailments in plants. Chemical entities, for instance, ribavirin (RBV) (virazole), azidothymidine, and 2-thiouracil (Mathews, 1953), moreover, a few viral inhibiting drugs comprising inosine monophosphate dehydrogenase (IMPDH) inhibitors, S-adenosyl homocysteine hydrolase inhibitors, and neuraminidase (NA) inhibitors (Panattoni et al., 2013), can be employed on a routine basis for chemotherapy procedures. Furthermore, these compounds are also known for their ease of permeability inside plants through the soaking method, which can lead to the prevention of viral replication (Lal et al., 2015). Successful eradication of *Grapevine leaf roll-associated virus-1* and *-3* (GLRaV-1 and 3) was observed while applying discerning chemotherapy. Increased efficiency of IMPDH inhibitors was highlighted, in contrast to GLRaV-1. Conversely, nucleic acid or majorly purine biosynthesis-inhibiting factors reveal increased efficacy against GLRaV-3 (Panattoni et al., 2011). Moreover, *Grape-wine rupestris stem pitting-associated virus* (GRSPav) was eliminated after the utilization of IMPDH inhibitors as antiviral compounds, together with tiazofurin, RBV, and mycophenolic acid, within laboratory chemotherapy circumstances (Skiada et al., 2013). Before the implementation of meristem tip culture, chemotherapy was utilized, resulting in the complete eradication of Lily symptomless virus (Chinestra et al., 2015). Despite their effectiveness, these treatments exhibit certain limitations, given that they possess diverse modes of action (De Clercq, 2005; Gubareva, 2004; Kumar et al., 2022) and exhibit reduced efficacy ex vitro, even at elevated concentrations (Chinestra et al., 2015).

1.2.3 Cryotherapy

Researchers employ saline combined with crumpled ice at temperatures between −18° and − 24 °C in the treatment of various human tumours, including those of the breast, cervical region, and skin, which correspondingly serve to alleviate discomfort (Arnott, 1850). Similarly, in the field of crop cryotherapy, intruders like viruses, phytoplasma, and bacteria can be subjected to extremely low temperatures (−196 °C) for extended periods, effectively eliminating virus complexes and resulting in a higher frequency of obtaining virus-free plants compared to traditional meristem tip culture (Vieira et al., 2015). This technique inhibits the occurrence of thermally directed metabolic reactions. Notably, three *Closteroviridae* viruses responsible for leaf roll disease in grapevines have been successfully eradicated through vitrification-based cryotherapy that utilizes dehydrating materials on buds from contaminated clones (Pathirana et al., 2015). The advantages of cryotherapy include the capability to treat large quantities of plantlets and the applicability of the method regardless of shoot tip size; however, a significant disadvantage is the substantial consumption of certain gases, including argon and nitrogen (Littrup et al., 2006).

1.2.4 Thermotherapy

Thermotherapy involves the application of heat treatment for a specific duration, which effectively exterminates the targeted pathogens while exerting minimal negative impact on the host. Heat is primarily administered through mediums such as water, air, or vapor (Grondeau et al., 1994). The optimum temperature series for the successful conduction of this technique is reported to be 52 °C to 55 °C, with acquaintance times varying between 10 and 30 min. Research has indicated that an increase in temperature significantly reduces the incidence of virus-related diseases by disrupting the synthesis of viral ssRNA and dsRNA. Specifically, exposing tubers at 37 °C for 4 days, followed by 34 °C for an additional 3 days over 2 weeks, yields an endurance proportion of approximately [insert percentage]. 50% among diseased plants (Astarini et al., 2018). The effectiveness of thermotherapy is markedly enhanced when integrated with additional conventional therapies; for instance, the *Apple chlorotic leaf spot virus*, *Apple stem grooving virus*, and *Apple stem pitting virus*, which affect apple plants, are exterminated using thermotherapy at various temperatures combined along with chemotherapy using differing concentrations of RBV (Hu et al., 2015). Moreover, the joint application of chemotherapy and thermotherapy has proven successful in eliminating *Arabis mosaic virus* (ArMV), *Prunus necrotic ringspot virus* (PNRSV), and the combination of ArMV and PNRSV via rose-infected plants, achieving attainment rates of 63.33%, 90.09%, and 85.18%, respectively (Modarresi Chahardehi et al., 2016).

Advancement in cultivation nanotechnology has significant potential for designing or developing diverse types of nanoparticles that can efficiently deliver and target specific, organized applications of fertilisers, pesticides, herbicides, and insecticides for plant growth and the management of microbial plant diseases. This chapter signifies that we will appraise the advantages of the above-mentioned nanoparticles in controlling the infection and administration of virus-causing plant diseases.

2 Nanoparticle-Based Delivery Systems

To attain global food safety and overcome the challenges associated with climate change, it is essential to develop advanced strategies that enhance the productivity of key crop species. Pathogenic entities like bacteria, fungi, and viruses are the common microbes causing several plant diseases and causing a substantial loss in food fabrication across the world. The overall total crop losses are caused by microbial pathogens globally, with plant viruses contributing around 47% of crop loss by causing certain plant diseases (Frank & Schanz, 2022). Conventional practices to control viral infection and plant viral diseases involve using certain chemical pesticides. Conversely, the inadequacy of pesticide application and distribution is a giant concern; for instance, above 50% of pesticides applied aerially on the plant leaves and stems do not extend to the preferred target sites in the yields (Pimentel & Burgess, 2012), causing severe threats to the ecosystem and non-targeted creatures (Schäfer et al., 2019). Consequently, the development of pioneering and advanced technology, which has the potential to enhance the efficiency of supply and efficacy of pesticides, is required. The application of nanotechnology-based systems to control plant viral diseases is an auspicious tool to address the mentioned inadequacies (Tailor et al., 2022, 2025a).

2.1 Applications of Nanotechnology

Nanoparticles are tiny entities, devices, and systems that come in a variety of sizes ranging from 1 nm (nano-meter) to 100 nm. Owing to their minimal size and increased apparent area, nanoparticles exhibit important physical, chemical, and biological assets like amended thermal conductivity, enhanced firmness, catalytic reactivity corresponding with biological activity, as matched to their massive counterparts (Taniguchi, 1974; Goyal, 2017). Nanophytovirology (Marwal & Gaur, 2019) involves the application of nanotechnology in plant virology, encompassing biological stimulation, medication delivery, nucleic acid modification, therapy-based agents, and induction of defence mechanisms (Fig. 10.2). Because of their unique physicochemical properties, nanoscale particles can interact more easily and have the potential to eliminate virus particles. It has been demonstrated that nanoparticles reduce viral infectivity and exhibit virucidal activity against a range of viruses. Among the NPs that can stop plant viral infections are nonmetals, metalloids, metallic oxides, and carbon nanomaterials (Elmer & White, 2018; Rajini & Ferdinand, 2022). For years, nanotechnology has shown great potential to design or engineer different types of nanoparticles, and their application in the field of medicine and pharmacology has achieved great success in drug delivery and diagnosis of various diseases, but in the agriculture domain, nanotechnology research is not reaching its potential due to the scarcity of commercial applications (Worrall et al., 2018). In the past few years, agricultural nanotechnology has shown to be growing, especially for disease management and crop protection (Sastry et al., 2011; Gogos et al., 2012).

Various types of nanoparticles can be engineered to meet specific requirements, enabling them to have desired physiognomies, such as physiological assets, which

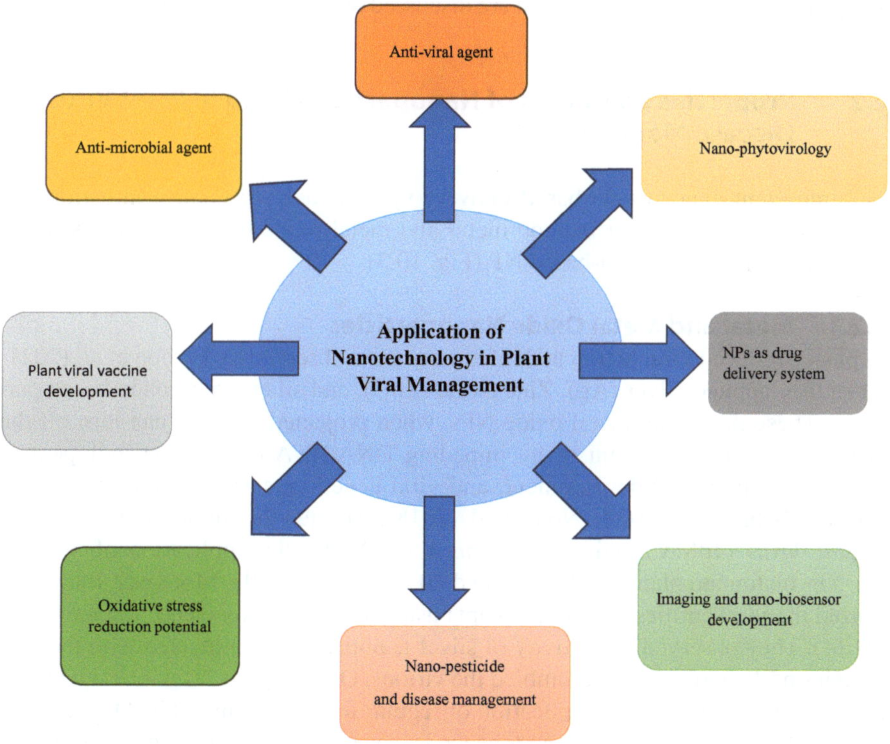

Fig. 10.2 Application of nanotechnology in viral management

can then be utilized as protectants or for targeted transport through adsorption, encapsulation, and conjugation of pesticides and herbicides (Khandelwal et al., 2016; Tailor et al., 2025b). All type of nanoparticles performs their function in two different ways: (A) Nanoparticles alone as protectants by showing antibacterial, antifungal, and antiviral activity and by inducing plant immune response to protect plants from pests and pathogens. Several metal nanoparticles have been extensively studied for their antimicrobial activity. Silver, copper, zinc oxide, and titanium oxide have been found to have significant antiviral activity (Gogos et al., 2012; Kah & Hofmann, 2014; Kim et al., 2018) (B). Nanoparticles act as carriers to entrap, encapsulate, or attach active molecules to develop different agricultural formulations. Silica, chitosan, SLN (solid lipid nanoparticle), and LDH (layered double hydroxide) are commonly used nanoparticles as carriers for disease management in plants (Barik et al., 2008; Ekambaram et al., 2012; Kashyap et al., 2015). The above nanoparticles are designed to act as carriers for insecticides, fungicides, herbicides, and dsRNA-induced RNA interference by controlled release and targeted delivery of the attached active molecules to plants. Several benefits are provided by the nanoparticles used as carrier molecules, such as enhanced durability, enhanced solubility of less water-soluble pesticides, minimized toxicity, and heightening site-specific intake of pesticides inside the targeted pest (Hayles et al., 2017). Important properties and types of nanoparticle-based systems used to control viral infection and plant viral disease management are discussed in the section below in more detail.

2.2 Properties and Types of Nanoparticles Used in Plant Virus Disease Management

The important nanoparticle-based delivery systems used to control infection and viral disease management include metal and metal oxide NP, polymer-based NP, lipid-based NP, and carbon-based NP (Fig. 10.3).

2.2.1 Metal and Metal Oxide Nanoparticles

In plants, reduced viral action and infectivity were observed by Farooq et al. (2021) when they applied silver (Ag), Zinc oxide (ZnO), and silicon monoxide-based particles. These metal and metal oxide NPs, when projected directly, can turn off the virus proliferating mechanism by impeding DNA/RNA duplication, polypeptide amalgamation, capsid development, and viral genetic material packing inside the capsid (Mishra et al., 2022). Noha et al. (2018), reported the suppression of *Tomato mosaic Virus* (ToMV) and *Potato virus Y* (PVY) by the application of 50 ppm Ag-NPs on tomato plants by 17.50% and 20.25% respectively. Moreover, they scrutinized the viral entities within the plants using a transmission electron microscope (TEM). They observed the activity of silver nanoparticles conjugated with the coat proteins of the virus, which crumbled the virions. Other reports, as witnessed by Cai et al. (2019), state that the application of regular foliar spraying of ZnO nanoparticles and SiO2 nanoparticles upon tobacco leaves for a 12-day duration caused inhibition of replication in TMV as a result. The plant antioxidant mechanism activates

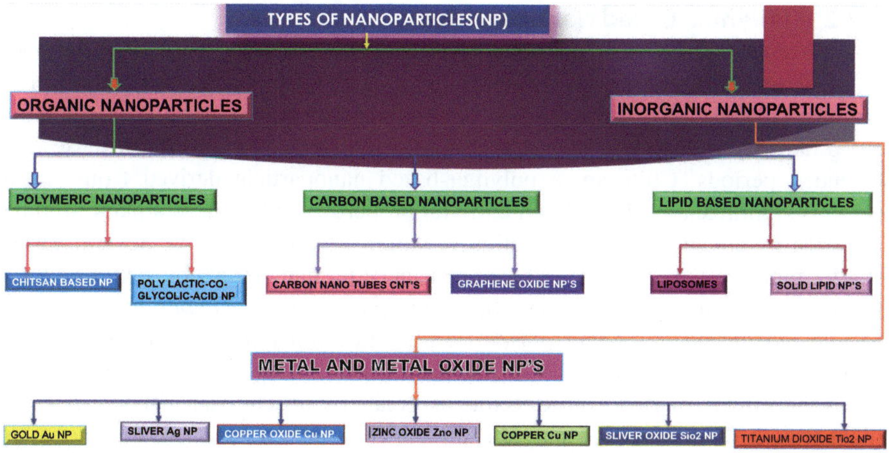

Fig. 10.3 Types of nanoparticle-based delivery systems in plants

to scavenge ROS, which are released when metal NPs are applied to the foliage, facilitating their absorption and dispersion within the plant cells. Jain and Kothari (2014) experimented and revealed the effect of spraying 50 ppm silver nano-formulations on cluster beans, which depicted a sensitized vira- inhibiting effect with absolute suppression of Sun Hemp Rosette Virus (SHRV) disease. Applied concepts of Si- NP studied within the tomato species resulted in gradual higher level of ROS dealing antioxidant enzyme peroxidase (POD), polyphenol oxidase (PPO), with simultaneous deduction within the extent of brutality of the disease affected by *Tomato yellow leaf curl virus* (TYLCV) through adjourning the uprising symptoms of viral infection in treated crops (El-Sawy et al., 2017a). Within the greenhouse environment, ZnO-NPs exhibit augmented applications in enhancing resistance to the *Cucumber mosaic virus* (CMV) in brinjal plants (El-Sawy et al., 2017b). Likewise, previous reports have highlighted the consequences of gold nanoformulations, which have demonstrated excellent efficiency in combating various infections caused by plant viral entities. For example, beans are affected by the bean mild mosaic virus, tobacco necrosis virus in tobacco, and barley yellow mosaic virus in barley crops. The therapeutically effective concentration of poly-dispersed gold nanoparticles introduced in Barley induced resistance in plants, contrary to the disease caused by the *Barley yellow mosaic virus*. The gold nanoparticles (Au-NP) were familiarized via a power-driven coarse and observed that Au-NPs melted and dissolved the virus particle, conferring resistance to the plants (Alkubaisi et al., 2015; Bhardwaj et al., 2019). Abdelkhalek and Al-Askar (2020), tested the antiviral activity of biologically synthesized ZnO-NPs against TMV. They observed a significant reduction in the viral accumulation level and disease severity of around 90% by foliar application of ZnO-NPs twice, 24 h before and after TMV inoculation in tobacco plants.

2.2.2 Polymer-Based Nanoparticles

Polymeric nanoparticles are made from biodegradable polymers and can be designed to carry encapsulated antiviral agents or active molecules that modulate plant immune response. Properties such as controlled release of antiviral agents and targeted delivery make these nanoparticles effective in managing viral disease over extended periods. Chitosan, a polymer-based nanoparticle derived from chitin, exhibits significant antifungal and antiviral properties and induces a plant immune response by enhancing the production of defence-related proteins. Several earlier studies reported that chitosan induced resistance in many plants by shielding them from infection triggered by the *Alfalfa mosaic virus, Cucumber mosaic virus, tomato mosaic virus, Peanut mosaic virus,* and *Snuff mosaic virus* (Pospieszny et al., 1991; Kochkina et al., 1995; Chirkov, 2002). Murphy et al. (2003) investigated the effects of two PGPR strains associated with chitosan nanoparticles on plant growth. They observed a significant reduction in CMV-induced disease severity, along with increased plant height, root and shoot weight, fruit, and flower production, in experimental plants compared to control plants.

2.2.3 Lipid-Based Nanoparticles

Liposomes and solid lipid nanoparticles (SLNs) are two important lipid-based nanoparticles, which are made of lipid bilayers that can encapsulate the antiviral agents, and deliver them to plant tissue with reduced toxicity and improved stability, providing a more efficient means of viral disease control (Worrall et al., 2018; Das et al., 2015a). Regulated discharge matrices of nano-formulated pesticides have become a promising approach for a sustainable future. Proliferation in productivity, decreased volatile properties, and further decline in toxicity, along with eco-contamination, are illustrated characteristics depicted via entrapped pesticides on the crops (González-Melendi et al., 2008; Sopeña et al., 2009). Lipid-based solid nano-encapsulations are ideal for the directed delivery of dsRNA entities, which initiate RNA interference in contrast to various viral infections by inhibiting the synthesis of a few mRNAs by the virus's pathogenic genome, thereby safeguarding crops against infections (Hannon, 2002). Polymeric nanoparticles (PNPs) have been reported to exhibit both hydrophobic and hydrophilic properties, which have been altered to function as a superficial transporter of inorganic-based nanoparticles (Sharma, 2019).

2.2.4 Carbon-Based Nanoparticles

Graphene oxide (GO) as well as carbon nanotubes (CNTs) are known for their utilization as carbon-centered nanoparticles. CNTs can be bio-structured as a carrier entity to transport viral inhibitory agents directly into the plant tissues, thereby inhibiting viral multiplication. Oxides of graphene unveil the inhibition of viruses through direct interaction with plant viruses, thereby preventing the connectivity of viral entities with plant cell membranes. Additionally, they can serve as carriers for the target-based delivery of compounds known for their antiviral capabilities, with an increase in their strength and bioavailability. Historical reports highlight that CNTs and various organic and inorganic nano-encapsulation unveil antiviral activity and are collected with essential elements required for plant growth. Thus, the destruction of certain diseases in plants and promotion of growth in plants could be simultaneously achieved while using these nutrients in the form of nanoparticles (Wang et al., 2014; Al-Zubaidi et al., 2019; Adeel et al., 2021). Adeel et al. (2021) examined the effect of different NPs (TiO_2, Ag, C60 fullerenes, and CNTs) application against TMV infection on tobacco plants by spraying 100, 200, and 500 mg/L concentrations. They observed that C60 and CNT-treated plants exhibit normal phenotypes, and viral symptoms were not evident, while TiO_2- and Ag-NP-treated plants showed no sign of viral infection suppression. The viral inhibitory activity of graphene-based silver nano-composites, applied both through soil and foliar methods, was demonstrated against the tomato bushy stunt virus (TBSV) by reducing the viral concentration and disease severity in lettuce plants (Elazzazy et al., 2017).

2.3 Efficacy of Different Nanoparticle Delivery Systems in Crops

Nanoparticle (NP)-based delivery systems in agriculture have gained significant attention for their potential to improve crop productivity, enhance nutrient delivery, and contribute to the control of various plant diseases (Tailor et al., 2021). Distinct characteristics mark the system, which features an improved surface area with superficial reactivity and the capability to hold numerous active biomolecules. This enhances the efficiency of forecasting and monitoring the discharge of nutrients, plant growth regulators, insecticides, pesticides, and various valuable compounds to the plants. Nanoparticles can be engineered for specific uses, including biosensors for early disease detection, protectants as bio-stimulants to induce plant immune responses, and delivery vehicles for genetic materials, nano-fertilizers, and nano-pesticides (Elmer & White, 2018; Tarafder et al., 2020; Ahmadian et al., 2021). Most of the metal and metal oxide-based nanoparticles, chitosan-NP, and CNT-NPs are used alone (protectants) to prevent viral infections and viral disease management, as they show a significant antibacterial, antifungal, and antiviral activity by inducing plant immune response or by activation of the antioxidant system (Chirkov, 2002; Alkubaisi et al., 2015; Cai et al., 2019). Nanoparticles are also utilized to improve effective agricultural formulations and frequently act as carriers to encapsulate, capture, or attach active molecules. The often utilized NPs act as carriers for fungicides, herbicides, insecticides, and dsRNA, including Si-NP, SLNs, LDHs, and chitosan-NPs (Worrall et al., 2018).

To protect plants from viral diseases, various types of NPs exhibit distinct modes of action against viruses. Earlier revisions have established that NPs primarily hinder premature virus replication and promote SAR in plants. Conversely, future research should focus on understanding the precise mechanisms by which each NP affects virus replication, mobility, encapsidation, and transmission. There is moreover, a need to conduct future studies to find the properties of nanocomposites and an amalgamation of polymer-based NPs (organic) with metal NPs on complex viral diseases for effective disease management under field conditions. A study by Dutta et al. (2021), found an inhibitory mechanism of nanoparticles by releasing metal ions, shadowed by oxidative and non-oxidative stress. Further, this stress leads to the denaturation of capsid protein, nucleic acid (DNA or RNA), and other protein constituents. Nanoparticles exhibit virucidal activity by preventing the binding and fusion of virion particles with plant cells, as well as weakening their effectiveness and reproduction ability through ROS production and subsequent protein oxidation. NPs can also prevent the host cells from recognizing the virus particles by attaching to the virus through surface glycoproteins (Cai et al., 2019, 2020). Hao et al. (2018) stated the consequences of distinctly manufactured nano-formulations with core metal elements Fe_2O_3 and TiO_2, as well as CNTs or C60 (carbon nano-forms), by regular treatment on tobacco plants for defeating the interference of *Turnip mosaic virus* (TuMV), resulting in a substantial inhibition of TuMV multiplication with 15–60% decline in the coat protein of TuMV. A similar effect was reported concerning the silver nanoparticles that can attach to the coat proteins of PVY and ToMV (Noha et al., 2018).

Interfering RNA is considered an alternative way to induce plant immune responses. Encapsulated nano-forms serve as a carrier for dsRNA within plant immune cells, inducing immunity and enhancing resistance to viral pathogenicity through RNA interference. Furthermore, the resulting dsRNA undergoes a process to synthesise siRNAs, which bind to similar mRNA entities transcribed from the virus's genome, thereby inhibiting the processes necessary for viral multiplication, including protein synthesis, capsid formation, and viral replication (Mitter et al., 2017). Shielding against *Pepper mild mottle virus* (PMMoV) and CMV was achieved using layered double hydroxide (LDH) clay nano-sheets for the transport of dsRNA. Moreover, they stated the mechanism of entrapment of dsRNA molecules by the cells of plants and the destruction of similar mRNA molecules in the cells (Mitter et al., 2017). This implies that NPs have a heightened capacity to counteract the viral pathogenic mechanism, leading to the effective management of viral diseases.

3 Antiviral Compounds and Their Delivery

The potential applications of nanoparticles as antiviral agents are quite promising. Nanoparticles can be classified into different types, which are currently preferred to be used as antiviral drugs or may be utilized in the future. Because of their ability to prevent viral reproduction, metallic nanoparticles, including ZnO NPs, SiO2 NPs, and Ag NPs, can be employed as antiviral agents (Vargas-Hernandez et al., 2020). NPs work against viral infection in plants (Table 10.1) in essentially two ways.

Direct approach: This approach states that the nanoparticles directly interact efficiently with the viruses. They may hinder the production of viral proteins, particularly the viral coat protein, or interfere with the viral genome's ability to replicate. In addition, NPs may interfere with the wrapping of viral genomes and the breakdown of proteins present in the capsid, which is crucial for virus particle packing. In addition to disrupting the electron transport system of virus-infected plant cells,

Table 10.1 Antiviral properties of potential nanoparticles

Sr. No.	NP	Diameter of NP	Target virus	Host plant	Effect	References
1	Silver nanoparticles (AgNPs)	12.6 ± 5 nm	(TSWV) *Tomato spotted wilt virus*	*Solanum tuberosum*	Inhibition of local spots and decreases in TSWV infection	Shafie et al. (2018)
2	AgNPs	12 nm	(PVY) *Potato virus Y*	*S. tuberosum*	Virus resistance induction in the host plant.	El-Shazly et al. (2017)
3	Schiff base nanosilver NPs	6–45 nm	TMV	*Nicotiana tabacum*	Decrease in TMV-induced infection	Wang et al. (2016)
4	Zinc oxide NPs (ZnONPs)	18 nm	*Tobacco mosaic virus* (TMV)	*Nicotiana benthamiana*	Inhibition of the TMV invasion's pace	Cai et al. (2019)
5	Iron oxide NPs (Fe3O4)	0.19 nm	TMV	*N. benthamiana*	Upregulating SA genes causes plants to become more resistant to TMV and activates antioxidants	Cai et al. (2019)
6	Graphene oxide-silver NPs (GO-AgNPs)	3050 nm	(TBSV) *Tomato bushy stunt virus*	*Lactuca sativa*	Reduction in the severity of the disease, infection rate, and virus concentration.	Elazzazy et al. (2017)
7	Gold NPs (AuNPs)	1–400 nm	(BaYMV) *Barley yellow mosaic virus*	*Hordeum vulgare*	In vitro viral particle dissociation	Aref et al. (2012)
8	Silicon dioxide NPs (SiO2NPs)	100 nm	(TYLCV) *Tomato yellow leaf curl virus*	*Solanum lycopersicum*	Decrease in TYLCV concentration and disease severity	El-Shazly et al. (2017)
9	Carbon Nanotubes (CNTs)	0.4–2 nm	TMV	*N. benthamiana*	Resistance is induced and TMV mobility and replication are inhibited	Adeel et al. (2021)

(continued)

Table 10.1 (continued)

Sr. No.	NP	Diameter of NP	Target virus	Host plant	Effect	References
10	Titanium dioxide NPs (TiO2NPs)	~3–5 µm	(BBSV) *Broad bean stain virus*	*Vicia faba L. Fabaceae*	Disease severity reduction	Derbalah and Elsharkawy (2019)
11	Nickel oxide NPs (NiONPs)	15–20 nm	(CMV) *Cucumber mosaic virus*	*Cucumis sativus*	Disease severity reduction and CMV concentration	Derbalah and Elsharkawy (2019)
12	Cerium oxide NPs (CeO2NPs)	2–20 nm	TMV	*Datura stramonium* and *Nicotiana tabacum*	Viral symptoms suppression	Eugene and Zholobak (2016)

NPs may also be the cause of an increase in barriers like cell walls or lignin deposition. Additionally, they can prevent viral DNA replication and entrance into host cells (Yasamineh et al., 2022).

Indirect Method: This approach states that the viruses are indirectly rendered inactive by the NPs. But with this approach, the nanoparticles trigger the plant defence mechanisms that combat the viral infection. According to Vargas-Hernandez et al. (2020), the presence of NPs in plants, along with virus pathogenic activity, leads to an increase in the activity of antioxidant enzymes such as peroxidase, polyphenol oxidase, phenylalanine ammonia-lyase, and superoxide dismutase.

3.1 Common Antiviral Compounds for Plant Viruses

The tomatoes (*Solanum lycopersicum* L.) are the second most economically important vegetable crop worldwide. They are considered a major agricultural cultivation activity in South Asia, including India and Pakistan (Sahu et al., 2012). The ToLCNDV does not spread infection alone; it sometimes interacts with other viruses and infects other agricultural species (Table 10.2).

The infection of viruses in tomato crops results in a variety of yield reductions, causing 10–90% crop loss. Saikia and Muniyappa (1989) reported that crop suffers about 6.4–52% yield loss in the summer and 52–100% in the winter. The devastating pathogen ToLCNDV not only reduces the yield but also degrades the quality of the tomatoes, affecting their market. People want tomatoes with less contamination at a lower price. The current approaches for managing viral pathogens are based on chemicals (Modgill et al., 2025). To control viral pathogens like ToLCNDV, farmers often use intensive pesticide spraying. However, these prevention methods increased the cost of the crop, which in turn fluctuates the market. These market dynamics lead to higher prices, which impact both farmers and consumers. The plant viruses,

Table 10.2 The interaction of *Tomato leaf curl New Delhi virus* (ToLCNDV) with other viruses and betasatellites

ToLCNDV genome	Interacting virus	Interacting betasatellite	Field host	Experimental host	References
DNA-A and DNA-B	–	Cotton Leaf Curl Multan Betasatellite (CLCuMuB)	*Solanum lycopersicum*	*N. benthamiana and S. lycopersicum*	Jyothsna et al. (2013)
DNA-A and DNA-B	–	Tomato Yellow Leaf Curl Thailand Betasatellite (TYLCTHB)	*S. lycopersicum*	–	
DNA-A and DNA-B	–	Papaya Leaf Curl β-Satellite (PaLCuB)	*S. lycopersicum and S. tuberosum*	–	
DNA-A and DNA-B	–	Bhendi Yellow Vein Mosaic Betasatellite (BYVB)	*S. tuberosum*	–	
DNA-A and DNA-B	–	Ludwigia Leaf Distortion Betasatellite (LuLDB)	*Luffa cylindrical, M. charantia, Cucumis sativus and Lagenaria siceraria*	*N. benthamiana*	
DNA-A	*Tomato leaf curl palampur virus (ToLCPaIV) and Squash leaf curl China virus (SLCCNV)*	–	*Cucurbita moschata*	–	Jaiswal et al. (2012)

(continued)

Table 10.2 (continued)

ToLCNDV genome	Interacting virus	Interacting betasatellite	Field host	Experimental host	References
DNA-A and DNA-B	*Tomato leaf curl Palampur virus-A (ToLCPalV-A) and Tomato leaf curl Palampur virus-B (ToLCPalV-B)*	–	*S. lycopersicum*	*S. lycopersicum*	Kanakala et al. (2013)
DNA-B	*Tomato leaf curl Palampur virus-A (ToLCPalV-A)*	–	–	*S. lycopersicum and Cucumis sativus*	
DNA-A	*Tomato leaf curl Palampur virus-B (ToLCPalV-B)*	–	–	*S. lycopersicum and C. sativus*	
DNA-B	*Tomato leaf curl disease Ranchi virus (ToLCRnV)*	–	–	*N. benthamiana and S. lycopersicum*	Kumari et al. (2011)
DNA-B	*Tomato leaf curl Palampur virus-A (ToLCPalV-A)*	–	*Cucurbita melo*	*N. benthamiana*	Malik et al. (2011)
DNA-B	*Pepper leaf curl Lahore virus (PepLCLV)*	Chilli Leaf Curl Betasatellite (ChLCB)	*C. annum*	*Nicotiana benthamiana, Nicotiana tabacum, and C. annum*	Shafiq et al. (2010)
DNA-B	*Pepper leaf curl Lahore virus (PepLCLV)*	–	–	*N. benthamiana, N. tabacum, and C. annum*	
DNA-B	*Zucchini yellow mosaic virus (ZYMV)*	–	*Momordica charantia*	–	Sharma et al. (2015)

DNA-A	*Croton yellow vein mosaic virus (CYVMV)*	Croton Yellow Vein Mosaic Betasatellite (CYVMB)	–	*N. benthamiana and S. lycopersicum*	Singh et al. (2012)
DNA-A and DNA-B	*Croton yellow vein mosaic virus (CYVMV)*	Croton Yellow Vein Mosaic Betasatellite (CYVMB)	–	*N. benthamiana and S. lycopersicum*	
DNA-A	*Chilli leaf curl virus-A (ChLCV-A) and Tomato leaf curl Guam virus-B (ToLCGuV-B)*	Tobacco Leaf Curl Betasatellite (TbLCB)	*Capsicum annum* and *C. chinense*	*Nicotiana benthamiana and C. annum*	Singh et al. (2016)
DNA-B	*Bhendi yellow vein mosaic virus (BYVMV)*	–	*Abelmoschus esculentus*	–	Venkataravanappa et al. (2015)
DNA-B	*Tomato leaf curl Gujarat virus-A (ToLCGuV-A)*	–	*G. hirsutum*	–	Zaidi et al. (2015)
DNA-A and DNA-B	*Cotton leaf curl Kokhran virus –Burewala (CLCuKoV-Bur)*	Cotton Leaf Curl Multan Betasatellite (CLCuMuB)	*Gossypium hirsutum*	–	Zaidi et al. (2016)

such as ToLCNDV, pose a significant threat to agriculture due to their rapid spread and the lack of effective chemical antiviral treatments. Conversely, conventional chemical approaches, such as spraying pesticides, are quite challenging to control plant viruses, as they are not susceptible to the same treatments as bacteria and fungi. According to the Sustainable Development Goals, outlined in the 2030 Agenda for Sustainable Development, the use of pesticides should be reduced by 2030. Thus, a lot of approaches such as biological compound approach, antiviral approach, and molecular approaches (in-vitro and transgenic approach, RNAi technology, CRISPR technology, etc.) (Das et al., 2011; Sahu et al., 2014; Verma et al., 2024) are effectively implemented to control and manage the plant viral pathogen, ToLCNDV.

3.1.1 Antiviral Proteins

Plants have the potential to fight against pathogens. They produce a variety of antiviral proteins against pathogenic infections (Marwal et al., 2014b). The natural products, viz., flavonoids, alkaloids, saponins, tannins, and essential oils, derived from plants have antipathogenic properties. Furthermore, when the viruses interact with hosts, plants activate their defence mechanisms like posttranscriptional silencing of genes, onset of hypersensitive responses (HR) and onset of systemic acquired resistance (SAR) to protect themselves from the viral infections (Sahu et al., 2010; Ding & Voinnet, 2007, and Seo et al., 2004). Additionally, these responses are accompanied by changes in gene-expression patterns of pathogenesis-related (PR) proteins and of several other proteins involved in cell signalling (Culver & Padmanabhan, 2007).

3.1.2 Biological Agents

In addition to chemical compounds, biological controls, or biocontrols, are used to control and manage plant viruses. Many of the plant growth-promoting rhizobacteria (PGPR) genera including *Bacillus, Enterobacter,* and *Pseudomonas* genera are implemented against the viruses CMV, TMV, *Tomato chlorotic spot virus* (TCSV), *Tomato mottle virus* (ToMoV), *Tomato spotted wilt virus* (TSWV), *Tomato yellow leaf curl virus* (TYLCV), ToLCNDV, *Potato virus Y* (PVY), and *Groundnut bud necrosis virus* (GBNV)) in economically significant crops, like potato, tomato, cucumber, tobacco, etc. The delivery of PGPRs is done in many ways as a biocontrol: (1) by modifying the root morphology to make a long root system to enhance the water and nutrient uptake, (2) unswervingly promote growth of the plants by refining the nutrients within the environment (nitrogen fixation) or construction of phyto-hormones (Indole-3-acetic acid). However, the most widely used mechanism of biocontrol is the conquest of pathogens by the synthesis of anti-pathogen compounds like antibiotics. A similar analysis was reported, stating that *Agrobacterium*-mediated plant transformation is an optimistic approach, where *Agrobacterium* tumefaciens is used as a control to manage ToLCNDV virus in crops. Natural phenomena of gene transfer capability can be enhanced by disrupting the inherent cycle in viruses or through modification within the immune pathway of plants. Gene delivery through the *Agrobacterium*-mediated delivery method assists in building up transgenic plants that possess viral inhibitory genes or proteins.

3.1.3 RNA Interference (RNAi) Approach

RNA interference (RNAi) represents an eco-friendly biotechnological approach for plant defence mechanisms, particularly in response to viral pathogenicity. This is achieved through the use of double-stranded RNAs (dsRNAs) or small interfering RNA (siRNA) molecules, which target and silence the virus's infectious genes, thereby preventing viral replication (Marwal et al., 2012). The molecules of siRNA show maximum similarity with the targeted genome of ToLCNDV and inhibit the infective mechanism to a broad extent within the plants. Reports configure it as an auspicious tool that decreases the infectious power of the pathogen and maintains the crop quality. The RNAi technology develops viral resistance in crops, such as by suppressing the AC4 gene in ToLCNDV-infected tomatoes, and controls leaf curl disease expression by deregulation of miR159/319 and miR172. This is a novel, effective, and environmentally friendly method that can lead to sustainable crop protection and food security, as well as effective management of ToLCNDV and improved crop resilience.

3.1.4 CRISPR Approach

Genome editing is an emerging technology in crop science, utilizing endonucleases to cleave or silence specific sites in the genome sequence (DNA or RNA). The clusters regularly interspaced short palindromic repeats (CRISPRs) system is one of the most advanced gene editing systems, which enables the targeted double-stranded breaks, creating a site-specific gene modification (Sattelmacher, 2001). It could be used in plants to detect and confer molecular immunity against DNA viruses, such as the *Tomato yellow leaf curl virus* (TYLCV) and ToLCNDV. It targets the Origin of replication or intergenic region, which shows delayed or reduced accumulation of viral DNA, or reduces the symptoms of infection (Vinodkumar et al., 2018).

3.2 Encapsulation of Antiviral Compounds in Nanoparticles and Their Release Kinetics for Effective ToLCNDV Management

Encapsulation is a practice of enfolding a substance (like an antiviral compound) inside a protective material and immersing it in heterogeneous or homogeneous matrices (Sagalowicz & Leser, 2010) (Fig. 10.4).

Similarly, nanoencapsulation is a technology of packaging nanoparticles of solid, liquid, or gas within a secondary material to form nanocapsules. The antivirals unencumbered from nanoparticles will combine to receptors of the virus present on the surface of the host cells, or they can be unconstrained within the cell, thereby disturbing the series of viral replication (Fig. 10.5) (Cojocaru et al., 2020; Maginnis, 2018; Jain et al., 2022).

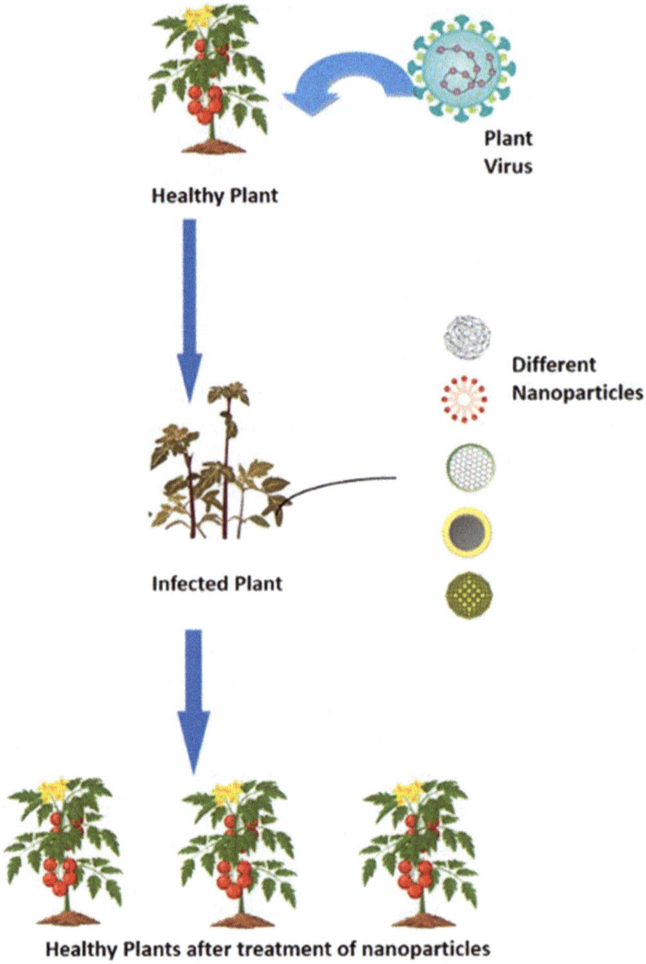

Fig. 10.4 Delivery system of encapsulated antiviral compounds of nanoparticles for the protection and management of ToLCNDV

Due to their small dimensions and tunable surface characteristics, nanoparticles can overcome biological barriers and deliver antivirals to infected sites in plants (Li et al., 2019). The drug delivery to plants primarily relies on xylem capillary action, particularly in roots, and diffusion processes. The physicochemical properties of nanoparticles, including size, charge, low water dispersibility, and controlled synthesis, pose challenges for their impending application in drug delivery. Thus, nanoparticle dimension and surface charge can play a superficial role in passage within the plant cells and tissues (Sangwan et al., 2023). The encapsulation helps protect the active compounds from the environmental factors such as pH, enzymes, heat, and light. Furthermore, it is used to design and develop novel delivery systems,

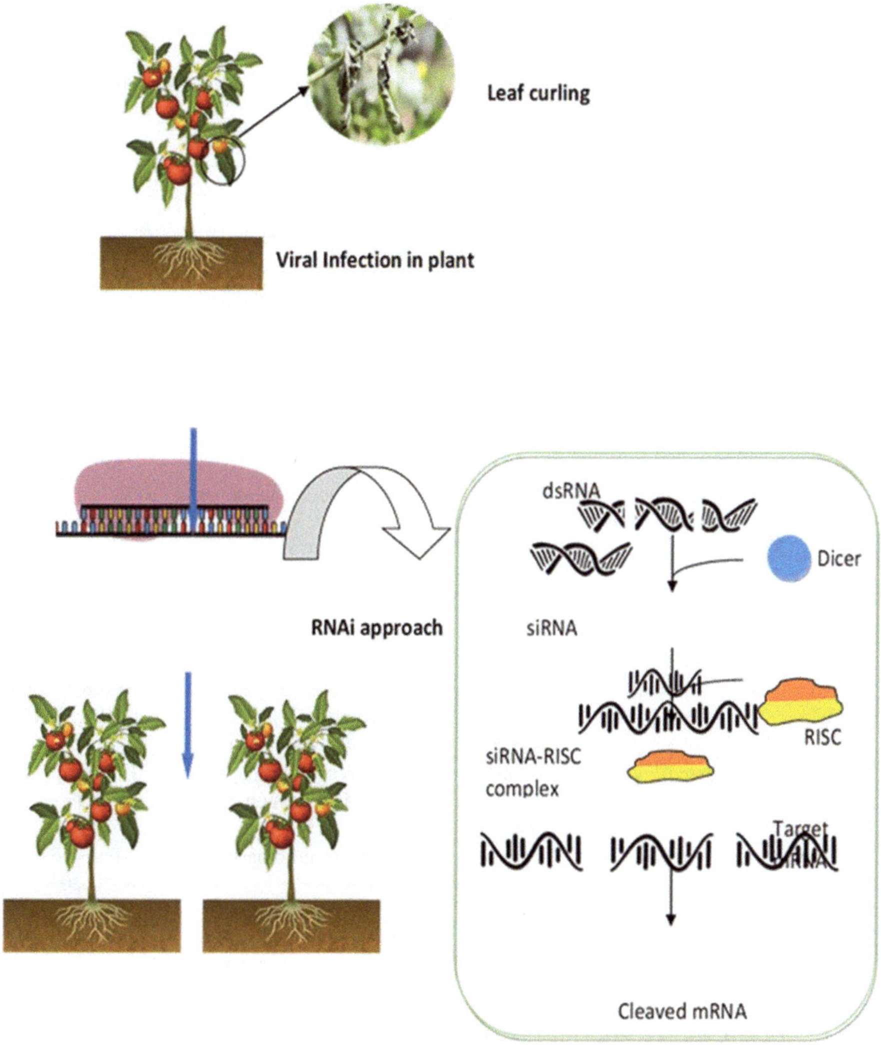

Fig. 10.5 RNAi approach in delivering the antiviral compounds into the infected plants from ToLCNDV

thereby improving the stability, solubility, and target delivery of the antiviral compound over a long period (Table 10.3).

In the past, several studies have been done on the utilization of engineered nanoparticles in agronomy against pest or pathogenic microbes like chitosan nanoparticles, gold nanoparticles (AuNPs), carbon nanotubes, layered double hydroxides (LDH) nanosheets (Mitter et al., 2017; Worrall et al., 2018), and mesoporous silica nanoparticles (MSNPs).

Table 10.3 The effect of nanoparticle-based approaches in reducing viral infection

Nanoparticle	Host Plant	Virus	Antiviral Activity	References
Silver nanoparticles (AgNPs)	*Cymopsis tetragonaloba*	Sunhemp rosette virus (SHRV)	Complete suppression of the disease	Jain and Kothari (2014)
	Solanum lycopersicum	Tomato mosaic virus (TMV)	Reduction in disease severity and virus infection	El-Dougdoug et al. (2018)
	Vicis faba	Bean yellow mosaic virus (BYMV)	Decrease in virus concentration, percentage of infection, and disease severity reduction in lesions on infected leaves	Elbeshehy et al. (2015)
Silicon dioxide NPs (SiO2NPs)	*Cucumis sativus*	Papaya ringspot virus (PRSV)	Reduction in disease severity and virus accumulation	Elbeshehy et al. (2015)
	Solanum lycopersicum	Tomato yellow leaf curl virus (TYLCV)	Reduced disease severity and TYLCV concentration	El-Sawy et al., (2017a, b)
	Nicotiana tabacum	*Tobacco mosaic virus* (TMV)	Suppression of the speed of TMV invasion	Cai et al. (2019)
ZnONPs	*Solanum melongena*	Cucumber mosaic virus (CMV)	Disease severity and incidence was significantly reduced	El-Sawy et al., (2017a, b)
	Nicotiana benthamiana	*Tobacco mosaic virus* (TMV)	Suppression of the speed of TMV invasion	Hao et al. (2018)
AuNPs	*Hordeum vulgare*	Barley yellow dwarf virus-PAV (BYDV-PAV)	Destroys gold barley yellow Dwarf virus and eliminates Virus infectivity hazards	Alkubaisi and Aref (2017)

3.2.1 Lipid-Based Nanoparticles

The lipid-based system is one of the most promising delivery systems. A variety of lipid nanoparticles, such as solid lipid nanoparticles (SLN), liposomes, nanoemulsions, and nanostructured lipid carriers, are the lipid carriers for antiviral compounds. These nanoparticles are biodegradable, biocompatible, non-toxic, and cheaper, which are employed in delivering antiviral agents to plants to encapsulate both hydrophilic and hydrophobic antiviral compounds (Puri et al., 2009; del Pozo-Rodriguez et al., 2009, 2013; Das et al., 2015b; Müller et al., 2002). For example, polymeric micelles assist in dispersing poorly water-soluble antiviral compounds in water by encapsulating them within their hydrophobic cores while interacting with water through their hydrophilic shells (Chakravarty & Vora, 2021; Batrakova et al., 2006; Yu et al., 1998). The lipid-based nanoparticles facilitate the targeted delivery of antiviral agents to specific cells/tissues, enhancing the effectiveness of treatments. Furthermore, their ability to penetrate plant cell membranes enables efficient absorption of antiviral compounds, improving virus control and reducing crop damage (Ekambaram et al., 2012; Borel & Sabliov, 2014; Bruna et al., 2014).

3.2.1.1 Polymeric Nanoparticles

The polymeric nanoparticles, made from biocompatible and biodegradable materials like poly (lactic-co-glycolic acid) (PLGA), are used to encapsulate antiviral compounds, providing stable and controlled release (Kashyap et al., 2015; Li et al., 2011). These nanoparticles can deliver antiviral agents directly to infected plant cells, enhancing the efficacy of the treatment. Polymeric nanoparticles protect the encapsulated compounds from degradation, ensuring prolonged activity against plant viruses. By reducing the frequency and number of antiviral applications needed, polymeric nanoparticles contribute to sustainable plant virus management practices and help maintain crop health.

3.2.1.2 Metallic Nanoparticles

Metallic nano-formulations, synthesized using gold (Au), silver (Ag), aluminium (Al), cadmium (Cd), cobalt (Co), copper (Cu), and zinc oxide (ZnO), illustrate the effective properties against the viral entities which were extracted for plant disease management against virus (Cai et al., 2019). Quantum effects, as well as a massive surface-to-volume ratio, are markers to detect nanoparticles through UV-sensitization, electric, catalytic, thermal, and anti-bacterial properties (Wang, 2000). As a result, it leads to the disruption of the virus, with continual inhibition of replication and a substantial reduction in the extent of pathogenicity in the plants. Ag-NPs are studied for the disruption of the viral entities due to the successful interaction with their proteins and genetic material (Luceri et al., 2023). Alkubaisi and Aref (2017) demonstrated the effect of gold nanoparticles (AuNPs), leading to the possible deformity of the virus-like particles (VLPs) of the barley yellow dwarf virus-PAV. Au-NPs play a crucial role in the deterioration of VLPs, validated by TEM.

Furthermore, Sangwan et al. (2023) observed that the relative size of mesoporous silica nanoparticles (MSNPs) ominously distresses the efficiency to uptake and subsequently deliver the AC2-consequent dsRNA inside the cells of the plant, leading to the maximal deduction of the viral adverse effects. Conventional studies prove the integration of metallic nanoparticles for antiviral conduct to shield plants effectively.

3.2.1.3 Carbon-Based Nanoparticles

Graphene oxide (GO) and carbon nanotubes (CNTs) are analyzed for their capabilities in dealing with plant-affecting viruses. They are directed toward the target expression and suppression of the infection-causing site due to an increase in the immune response of the plants, which eventually boosts resistance towards the virus (Adeel et al., 2021). This suggests that carbon encapsulation is a leading approach for managing plant viral diseases, thereby enhancing crop health without relying on conventional chemical-based applications.

3.2.1.4 Release Kinetics of Encapsulated Antiviral Compounds

The kinetics of releasing nano-compounds inside plants would imply a distinct pattern of discharge from encapsulation within the soil or plant tissues. This vital practice ensures the optimum collection of substances within the plant at a particular time, thereby enhancing growth, health, and defence against pests and microbial diseases. For instance, fertilizers with optimized release of water-soluble nutrients are progressively released within the soil gradually with time (Lawrencia et al., 2021). This supports minimizing nutrient loss due to mineral leaching or volatilization, thereby expanding nutrient utilization efficiency. Therefore, the captured antiviral components are engineered to secrete their active antiviral entities steadily, confirming the long-term protection from viruses. Understanding the release kinetics of antiviral compounds for ToLCNDV management is crucial for designing effective and sustainable agriculture practices by enhancing the delivery of necessary compounds to plants.

Modern researchers investigated the release kinetics of trapped antiviral compounds inside the plants. Urbaniak et al. (2024) contributed their efforts in analyzing the optimized release of lamivudine utilizing a double-layer coating of macromolecular prodrug entities. Observations revealed that the release of the compounds idealized zero-order kinetics, presenting a persistent release of the compounds over time, which is considered useful in attaining proficient therapeutic effects with lower spurt release. This technique demonstrates its significance in maintaining a constant level of antiviral activity in crops, thereby providing a persistent shield against ToLCNDV. Yin et al. (2023a, b) presented experimental results on pH-dependent release kinetics of drugs in cancer, which precisely target the disease and enhance treatment effectiveness while minimizing off-target effects.

However, several factors, such as environmental factors (temperature, humidity, and soil pH), encapsulated materials (such as chitosan, rosin-paraffin wax, and PLGA), and the physicochemical properties of the antiviral compounds (solubility, stability, and molecular weight), may influence the release kinetics of encapsulated antiviral compounds into the plants. These factors can affect the stability and release rate of antiviral compounds, leading to accelerated degradation of the encapsulating material. The rapid release of antiviral compounds is detrimental to both plant health and the environment. Therefore, understanding these factors is necessary to design encapsulation systems that provide a sustained release of antiviral compounds, thereby enhancing their efficacy in managing ToLCNDV and reducing the frequency of applications.

4 Challenges and Limitations

4.1 Stability and Shelf Life of Nanoparticle Systems

Stability of nanoparticles concerning shape is stated as maintaining the integrity of the innovative native construction and radius of curvature at the atomic as well as nanoscales, as these magnitudes impact the free energy of the surface directly. The period it takes for a medicine to lose 90% of its initial concentration is known as its shelf life. Once kept in the prescribed storage conditions, the resultant will stay steady. The stability of metal/metal oxide nanoparticles can be examined before, during, and/or after utilization to monitor modifications to their elemental or molecular structure. One effective method for determining the primary and interfering chemical composition of these materials is energy-dispersive X-ray spectroscopy (EDS) (Goldstein et al., 2017; Scimeca et al., 2018).

4.1.1 Factors Affecting the Stability and Shelf Life of NPs

The deterioration of nanoparticles can be accelerated by temperature, humidity, light exposure, and the presence of pollutants. Their surface chemistry greatly influences the stability of nanoparticles. Instability may result from modifications to the surface functional groups that impact how they interact with the environment. Storage conditions can affect the stability of nanoparticle systems. For instance, solid lipid nanoparticles (SLNs) can become unstable due to shear pressures, temperature, and light. mRNA lipid nanoparticles (LNPs) need to be stored in a frozen state to increase their shelf life. Over weeks or months, mRNA LNP activity may decrease. For them to continue operating at their best, they must be stored frozen. The activity of mRNA LNPs may decrease over brief intervals (weeks to months).

4.1.1.1 Effect of Size on NPs' Shelf Life

Their size may impact the stability of the nanoparticles. For instance, after roughly 2 months, tiny silver nanoparticles stabilized with PVP may start to re-nucleate. The amount of time a nanoparticle system stays stable under the suggested parameters from the manufacturer is known as its shelf life. For instance, under normal laboratory settings, gold nanoparticles (GNPs) normally have a shelf life of 6 months to a year.

4.1.1.2 Effect of pH on NPs' Shelf Life

The majority of degraded products remain entangled in the polymer matrix; however, degradation in water occurs about three times faster at 25 °C. At 37 °C, the production of total breakdown products is comparable in PBS and water, but they leave the nanoparticle twice as quickly in PBS. These findings raise the possibility that ions may have an impact on the hydrophobic core's stability. The stability of nanoparticles may also be impacted by stowage in a buffered solution that is somewhat acidic and contains additional ions (Ca 2+, K +), according to dilapidation and discharge rates in RLS. Since there is evidence that PLGA hydrolysis in hydrophobic matrices can result in pH-dependent autocatalysis (Siepmann et al., 2005), pH buffering may also affect the processes that take place inside nanoparticles.

4.1.1.3 Radiolabelling: An Alternative Approach for Increasing the Shelf Life of NPs

SERP (size exclusion of radioactive polymers), which syndicates size exclusion chromatography and radiolabelling. Size exclusion chromatography highlights chemical breakdown by tracking changes in the molecular weight, while radiolabelling offers a precise and quantity-based indication of the material concentration (Dikpati et al., 2022). High-resolution pictures of the structure and morphology of nanoparticles can be obtained using transmission electron microscopy (TEM). Measurements of the zeta potential reveal the surface charge of nanoparticles, which affects their stability. Thermo-gravimetric analysis (TGA) assesses the nanoparticles' thermal stability. Since both radioactive and non-radioactive polymers make up nanoparticles, radioactive breakdown produces only stand-ins for complete hydrolysis and matrix erosion. One benefit of SERP, however, is that it offers a numerical measure to depict these occurrences; these standards may be incorporated into mathematical models to investigate the integrity of a polymer matrix at the nanoscale and under various circumstances. To forecast the stability of small molecules (MacFaul et al., 2011), as well as the shelf life and storage conditions of pharmacological products (Waterman, 2011), Arrhenius' equation, for instance, experimentally links the reaction rate with temperature.

4.1.1.4 Lyophilization of NPs for Increasing their Shelf Life and Stability

Eliminating water from nanoparticle suspensions, lyophilization (freeze-drying) can stop oxidation and aggregation while the particles are being stored. Lyophilization would make sense as a means of enhancing the long-term stability of mRNA-LNP formulations because the presence of water starts breakdown events in these molecules. For live, attenuated viral vaccines, freeze drying is frequently utilized (Hansen et al., 2015). Additionally, it has already been studied for naked mRNA formulations, proving its usefulness and positive impact on mRNA stabilization. According to earlier studies, trehalose-formulated freeze-dried mRNA can remain stable at 4 °C for up to 10 months (Jones et al., 2007). It is also possible to successfully freeze-dry lipid nanoparticles. The structures are subjected to tension throughout the freeze-drying procedure. The design should comprise lyoprotectants that alleviate the colloidal particles. For that, trehalose as well as sucrose are used (Abdelwahed et al., 2006). Stability of the nanoparticles can be achieved by utilizing an inert environment and a lower storage temperature. Formulating the nanoparticles with supplementary preservative techniques can help in extending the shelf life of freshly produced particles (Liu et al., 2020).

4.1.1.5 Effect of Nanocomposites on Stability and Shelf Life of Fruits and Vegetables

Nanoformulations are recently preferred for maximal preservation of fruits and vegetables due to their durability of freshness, degradability, and potential antimicrobial qualities (Zhang et al., 2023; Priyanka et al., 2023). Nanocomposites of silver are favorable for the preservation of food because of their unbeatable antimicrobial properties, less toxicity and stability (Tailor et al., 2025; Kumar et al., 2021a, b). Silver nanoparticles impact the oxidation and aggregation to achieve the nanocomposite stability and antimicrobial qualities. They are superficially combined with other biomolecules, such as proteins, chitosan (CS), polyvinyl alcohol (PVA), chitosan (CS), TiO2, and magnesium oxide (MgO), to synthesize Ag composites with varied utilization, which can assist in determining underlying issues (Xiao et al., 2023). It is promising to precisely control the silver composites in terms of size, shape, surface characteristics, and various factors by adjusting key elements, such as morphology and composite components. Features of single as well as multicomponent composites contain a broad range of characteristics, like water-insulating, antioxidant, and antibacterial qualities. Silver nanoparticles were earlier produced by El Gohary et al. (2023) through biosynthesis, followed by the sol-gel preparation of TiO_2 nanoparticles and their subsequent addition to a PVA and sodium alginate (SA) blend for a reaction. They successfully generated PVA/SA/Ag-TiO_2 NPs, which significantly enhanced PVA/SA's antibacterial activity and thermal stability.

4.1.1.6 MOF (Metal-Organic Framework) NPs for Increasing Stability

Metal-organic frameworks (MOFs) offer several benefits, including enhanced biocompatibility, chemical stability, thermal stability, and optical stability, through the combination of organic and inorganic components (Zhang et al., 2022; Yin et al. 2023a, b). MOFs are crystalized compounds comprising metal clusters as well as organic ligands. They possess various functional chemicals, a wide specific surface area, a highly tunable pore structure, and good stability (Khalil et al., 2023). Silver MOFs are ideally preferred for fruits and vegetables preservation as well as packaging (Khan et al., 2024). Using 2-methylimidazole as a ligand, Chen et al., 2024) synthesized Ag-MOF in situ on carboxymethyl cellulose filter paper (CMFP), improving CMFP's resilience to UV light. The UV transmittance of Ag-MOF/CMFP was only 0.09–1.4% in the 250–400 nm wavelength range, whereas that of CMFP was 2.6–20.3%. Additionally, peaches and tomatoes were used in preservation experiments. The findings demonstrated that after 6 days, mould and rot had appeared on both the control and CMFP-packed tomatoes and peaches. Nonetheless, there was little to no deterioration in the attractiveness of the fruits enclosed in Ag-MOF/CMFP composites.

5 Regulatory and Safety Concerns

Nanoparticles (NPs) can cause cellular toxicity, which can result in harmful side effects, including increased production of reactive oxygen species (ROS), disturbance of redox equilibrium, lipid peroxidation, compromised mitochondrial function, and membrane damage. Nanopesticides have the potential to pollute soil and water systems because of their increased reactivity and prolonged persistence (Jazayeri et al., 2016). Some nanoparticles can be transferred and accumulate in plants, causing phytotoxicity. The processes of plant toxicity caused by nanoparticles are summarized as follows: (1) ROS production increases during the interaction between NPs and plants, resulting in oxidative damage (Marslin et al., 2017); (2) NP-induced transcriptional response (Xun et al., 2017); and (3) NP-DNA or organelle interaction, resulting in genotoxicity (Ghosh et al. 2019a, b).

5.1 Oxidative Stress Damage of NPs

Plants, like other aerobic species, use ROS as a signal molecule to drive cellular responses to environmental change (Mullineaux et al., 2018). When NPs reach plant cells, they can alter the quantity of ROS, disrupt cell metabolism, deplete the anti-oxidant system, and impair plant growth. ROS is naturally and continually created by cell organelles during metabolism, and it contains hydroxyl radical, singlet oxygen, superoxide anion, and hydrogen peroxide (Postiglione & Muday, 2020). ROS has two important roles: activating defence signals at low concentrations and worsening damage at high concentrations, resulting in oxidative damage to amino acids, lipids, and nucleic acids (Kapoor et al., 2019). Plant damage or even cell death can result from oxidative stress, cell structural damage, and disruption caused by elevated ROS levels (Rastogi et al., 2017; Nair & Chung, 2014).

5.2 Transcriptomic Response of NPs

Researchers highlighted that the nanoparticles have an inhibitory effect on plants due to their tenacious behavior. Interaction through stabilized link within the gene expression as well as cellular metabolism enlightens the concept of transcriptomics (Singh et al., 2017). For instance, Hossain et al. elaborated on the study of exposure to Ag NPs that can cause a reduction in the expression of subtilizing family proteins in soybean leaves, causing a significant decline in the stem length of the seedling (Hossain et al., 2016).

5.3 Genotoxic Effect of NPs

The substantial release of nanoparticles leads to the accretion at varied nutritional stages. Sensitivity of plants is observed against the toxic behavior of absorbed and reactive nanoparticles (Ghosh et al. 2019a, b). Effects on plants can be recognized with the superficial genetic simulations required for analyzing the surrounding toxic compounds (Iqbal, 2016). Nanoparticles exhibit distinct interactions with plant biomolecules, leading to cytotoxic and genotoxic effects, characterized by a declining mitotic index and an enhanced micronucleus and chromosome aberration index (Youssef & Elamawi, 2020).

5.4 Phytotoxicity from Nanoparticles

The plant's mineral transportation mechanism is typically obstructed by higher nanoparticle concentrations, which can hamper the flow of materials throughout the plant (Hong et al., 2021). Enhanced concentrations of NPs can lead to shock or stress within the plants, leading to the isolation of resources distant from the growth and development, with subsequent decline in the ideal plant health (Mittal et al.,

2020; Rani et al., 2022). Phytotoxic behavior was induced through the NPs saturation within the plants at a significant concentration. Furthermore, silver nanoparticles created adverse effects on *Arabidopsis thaliana*, as targeted silver ions were unable to monitor the phytotoxic silver nanoparticles passively (Geisler-Lee et al., 2013). Historical evidence has proven the adverse effects of ZnO NPs in plants, leading to the toxic impacts on cortical and root epidermal cells, which in turn lower their biomass (Lin & Xing, 2007). NPs are also reported to create massive oxidative stress in the plants (Rastogi et al., 2017). Natural as well as synthetic nanoparticles may cause impairment in plants, with the variance of their amount, part of exposure, and type of species (Ali et al., 2021). Phytotoxic NPs can be toxic to cells or tissues of the plant and deteriorate it (Yan & Chen, 2019). This comprises alterations in the length of the root, surface area, and root hair construction, resulting in the absorption of water and minerals (Wang et al., 2023).

5.5 Phytodamage Due to the Delivery System Using NPs

The probability of destruction in plant cells is a shortcoming of supplemented nanoparticles distribution methods (Cunningham et al., 2018). Stress is caused within plants as a result of associated delivery methods, mainly which encompass the electrical pulses or forces (Rivera et al., 2012). It can ultimately lead to cell death and can be time-consuming with multiple trial requirements to achieve ideal parameters with sustainable applications.

5.6 Bioaccumulation of NPs and their Effects on Plants

Nanoparticles can be introduced for a prolonged period in conjunction with plants. During this period, bioaccumulation is evident in plant tissues, which can have a severe impact on subsequent food chains (Abbas et al., 2020). Determining the optimal nano-dosage requires testing with both sensors and biomarkers (Singh et al., 2021; Halamoda et al., 2022). Nanoformulations can remain stable and persistent for an extended period, causing subsequent harm to all living entities in the environment (Oberdörster, 2010; Hemmrich & McNeil, 2023). Higher concentrations of nanoparticles can cause disintegration of the plant cell membrane (Wang et al., 2013). NPs can eventually disturb the symplastic pathway of plants by causing a change in the integrity of the root cell membrane or ion transport activity (Rossi et al., 2018).

5.7 Chemically Synthesized Nanoparticles and Their Phytotoxicity

Nanoparticles synthesized through physical and chemical methods have been used as catalysts in biomedicine since prehistoric times (Song et al., 2019). The gradual increase in their construction, evaluated at approximately 260,000 tonnes per year globally, has raised concerns about their safe consumption (Tolaymat et al., 2017). The NPs accumulated on the surface, producing a curl of root hairs and changes in the cell structure of root apical meristems, indicating that CuO NPs could be carried by wheat roots and had toxic effects (Adams et al., 2017).

6 Innovations and Prospects of Nanotechnology in Viral Management

Nanophytovirology offers an intriguing option for the ecological protection of crops, contrary to viruses that affect plants (Marwal & Gaur, 2019). The development of nanoparticles by the field's pioneers is especially important and advantageous for people and the agricultural sector, which sustains the lives of the expanding population (Marwal et al., 2018; Marwal & Gaur, 2019). However, several in vitro studies have suggested that NPs may help manage plant viruses. To obtain thorough results for future use, more in vivo experiments are necessary. Once more, the bulk of research has focused on RNA viruses, although DNA viruses continue to be a constant danger to crop productivity. Furthermore, the interaction of the illness cycle's components informs the development of disease management techniques. Plant-virus, vector, and host environment interactions lead to plant viral illness. The process is therefore highly complex and challenging to analyze due to the quadripartite interaction. It is still unclear how NPs interact with the plant-virus-vector-host system. Thus, more research is required to have a better understanding of the host-virus relationship. Once more, the bulk of research is concentrated on individual NPs; yet, nanocomposites have the potential to revolutionize the treatment of complicated viral illnesses. Targeting several plant viruses at once may be possible with nanocomposites, which would increase the advantages. Once more, NPs are effective against viruses when administered as a preventative measure to host plants. In other circumstances, however, they are effective when applied after the virus has been inoculated. One of another intriguing perspective of nanoparticles is that at very low concentrations, nanoparticles have an effect on the pathogens that is comparable to that of chemical pesticides (Arts et al., 2014) which provides it with the potential disease detection tool for the future pathogen detection and its management without damaging or giving stress to the plant as seen in conventional chemical pesticides dependent practices. To target viral infections, nanomaterials have been employed as carriers of active components in insecticides, compounds that induce host defence, etc. (Ghosh et al., 2017; Tailor et al., 2024). The precise method by which NPs interact with virus particles and the impact of NPs on vector acquisition, transmission, and mobility are among the questions surrounding this intricate relationship. While other metal and metal oxide NPs specifically bind with phosphorous and sulfur in proteins and nucleic acids, resulting in impaired membrane permeability (Benelli et al., 2018), AuNPs, for example, have an impact on insect development and reproduction by inhibiting trypsin (Patil et al., 2016).

7 Conclusion

Nano-encapsulation dependent delivery mechanism involving the advancement in the target-based outreach of the ToLCNDV management. Amalgamation of nano-formulations with antiviral compounds can lead to a successive approach for enhancing the stability and solubility of on-site delivered NPs, leading to subsequent mitigation of adverse side effects of the historical methods. Current studies suggest that NPs can be referred to as the best option for targeted, optimized release phenomena that can further ensure persistent antiviral activity with sustained enhancement in plant health and resilience. Moreover, the pioneering advantages of nanotechnology define the underlying potential strategies to combat diseases. After augmenting the applications of nanocarriers carrying antiviral compounds, the consequences of battling ToLCNDV are increased, leading to reduced crop damage and improved crop yield quality. This approach is creating a superficial path to step up for upcoming research targeting a sustainable future by aiming to utilize nanoformulations in agriculture. The ultimate aim of disease management and enhanced crop yield can be achieved by adopting nano-formulation-based approaches that help address challenges associated with viral disease management. Future research can define the dose-dependent behaviour of NPs for the management of ToLCNDV, so that a proficient dose of antiviral compounds in combination with nano-formulations can lead to healthier plants in the upcoming era.

References

Abbas, Q., Yousaf, B., Ullah, H., Ali, M. U., Ok, Y. S., & Rinklebe, J. (2020). Environmental transformation and nano-toxicity of engineered nanoparticles (ENPs) in aquatic and terrestrial organisms. *Critical Reviews in Environmental Science and Technology, 50*(23), 2523–2581. https://doi.org/10.1080/10643389.2020.1738455

Abdelkhalek, A., & Al-Askar, A. A. (2020). Green synthesized ZnO nanoparticles mediated by Mentha spicata extract induce plant systemic resistance against Tobacco mosaic virus. *Applied Sciences, 10*(15), 5054. https://doi.org/10.3390/app10155054

Abdelwahed, W., Degobert, G., Stainmesse, S., & Fessi, H. (2006). Freeze-drying of nanoparticles: Formulation, process and storage considerations. *Advanced Drug Delivery Reviews, 58*(15), 1688–1713. https://doi.org/10.1016/j.addr.2006.09.003

Abhary, M., Patil, B. L., & Fauquet, C. M. (2007). Molecular biodiversity, taxonomy, and nomenclature of tomato yellow leaf curl-like viruses. In H. Czosnek (Ed.), *Tomato yellow leaf curl virus disease* (pp. 85–118). Springer. https://doi.org/10.1007/978-1-4020-5822-6_5

Adams, J., Wright, M., Wagner, H., Valiente, J., Britt, D., & Anderson, A. (2017). Cu from dissolution of CuO nanoparticles signals changes in root morphology. *Plant Physiology and Biochemistry, 110*, 108–117. https://doi.org/10.1016/j.plaphy.2016.10.014

Adeel, M., Farooq, T., White, J. C., Hao, Y., He, Z., & Rui, Y. (2021). Carbon-based nanomaterials suppress tobacco mosaic virus (TMV) infection and induce resistance in *Nicotiana benthamiana. Journal of Hazardous Materials, 404*, 124167. https://doi.org/10.1016/j.jhazmat.2020.124167

Ahmadian, K., Jalilian, J., & Pirzad, A. (2021). Nano-fertilizers improved drought tolerance in wheat under deficit irrigation. *Agricultural Water Management, 244*, 106544. https://doi.org/10.1016/j.agwat.2020.106544

Ali, S., Mehmood, A., & Khan, N. (2021). Uptake, translocation, and consequences of nanomaterials on plant growth and stress adaptation. *Journal of Nanomaterials, 2021*, 6677616. https://doi.org/10.1155/2021/6677616

Alkubaisi, N. A., & Aref, N. M. A. (2017). Dispersed gold nanoparticles potentially ruin gold barley yellow dwarf virus and eliminate virus infectivity hazards. *Applied Nanoscience, 7*, 31–40. https://doi.org/10.1007/s13204-016-0582-5

Alkubaisi, N. A. O., Aref, N. M. M. A., & Hendi, A. A. (2015). Method of inhibiting plant virus using gold nanoparticles. *Google Patents*. Retrieved from https://patents.google.com/patent/US20150012385A1/en

Al-Zubaidi, S., Al-Ayafi, A., & Abdelkader, H. (2019). Biosynthesis, characterization and antifungal activity of silver nanoparticles by *Aspergillus niger* isolate. *Journal of Nano Research, 1*(1), 23–36.

Ancora, G., Belli-Donini, M. L., & Cuozzo, L. (1981). Globe artichoke plants obtained from shoot apices through rapid in vitro micropropagation. *Scientia Horticulturae, 14*(3), 207–213. Available from http://linkinghub.elsevier.com/retrieve/pii/0304423881900145

Aref, N., Alkubaisi, N., Marraiki, N., & Hindi, A. (2012, April). Multi-functional effects of gold nanoparticles inducing plant virus resistance crops. In *Proceedings of the 5th annual world congress of industrial biotechnology - 2012* (Vol. 25). Xi'an.

Arnott, J. (1850). Practical illustrations of the remedial efficacy of a very low or anæsthetic temperature.?I. In cancer. *The Lancet, 56*(1409), 257–259. Available from http://linkinghub.elsevier.com/retrieve/pii/S0140673602898749

Arts, J. H., Hadi, M., Keene, A. M., Kreiling, R., Lyon, D., Maier, M., et al. (2014). A critical appraisal of existing concepts for the grouping of nanomaterials. *Regulatory Toxicology and Pharmacology, 70*(2), 492–506. https://doi.org/10.1016/j.yrtph.2014.09.001

Astarini, I. A., Margareth, D., & Maya Temaja, I. G. R. (2018). In vivo thermotherapy: Attempt to eliminate virus in potato tuber. *IOP Conference Series: Earth and Environmental Science, 130*, 012021. Available from http://stacks.iop.org/1755-1315/130/i=1/a=012021?key=crossref.24fb7cf2f0f90971d9fd7f353d84b1a3

Ball, E. (1946). Development in sterile culture of stem tips and subjacent regions of *Tropaeolum majus* L. and of *Lupinus albus* L. *American Journal of Botany, 33*(5), 301–318. Available from http://doi.wiley.com/10.1002/j.1537-2197.1946.tb10379.x

Barik, T. K., Sahu, B., & Swain, V. (2008). Nanosilica–from medicine to pest control. *Parasitology Research, 103*, 253–258. https://doi.org/10.1007/s00436-008-0942-2

Batrakova, E. V., Bronich, T. K., Vetro, J. A., & Kabanov, A. V. (2006). Polymer micelles as drug carriers. In V. P. Torchilin (Ed.), *Nanoparticulates as drug carriers* (pp. 57–93). Imperial College Press.

Beijerinck, M. W. (1898). On a contagium vivum fluidum causing the spot disease of the tobacco leaves. *Phytopathological Classics, 7*(1), 33–52.

Benelli, G., Maggi, F., Pavela, R., Murugan, K., Govindarajan, M., Vaseeharan, B., et al. (2018). Mosquito control with green nanopesticides: Towards the one health approach? A review of non-target effects. *Environmental Science and Pollution Research, 25*(11), 10184–10206. https://doi.org/10.1007/s11356-018-1650-4

Bhardwaj, V., Sood, S., Kumar, A., Vanishree, G., Sharma, S., Sundaresha, S., Raigond, B., Kumar, R., Bairwa, A., Lal, M., & Chakrabarti, S. K. (2019). Efficiency and reliability of marker-assisted selection for resistance to major biotic stresses in potato. *Potato Journal, 46*, 56–66.

Borel, T., & Sabliov, C. (2014). Nanodelivery of bioactive components for food applications: Types of delivery systems, properties, and their effect on ADME profiles and toxicity of nanoparticles. *Annual Review of Food Science and Technology, 5*, 197–213. https://doi.org/10.1146/annurev-foodmoni-030713-100141

Brown, J. K., Zerbini, F. M., Navas-Castillo, J., Moriones, E., Ramos-Sobrinho, R., Silva, J. C., Fiallo-Olive, E., Briddon, R. W., Hernandez-Zepeda, C., Idris, A., Malathi, V. G., Martin, D. P., Rivera-Bustamante, R., Ueda, S., & Varsani, A. (2015). Revision of Begomovirus taxonomy based on pairwise sequence comparisons. *Archives of Virology, 160*, 1593–1619. https://doi.org/10.1007/s00705-015-2489-2

Bruna, V. B., Patricia, B. S., Matheus, A. S. R., Kamila, M. S. N., Taís, M. B., & Marlus, C. (2014). Nanotechnology-based drug delivery systems and herbal medicines: A review. *International Journal of Nanomedicine*, 1–15. https://doi.org/10.2147/IJN.S62922

Cai, L., Liu, C., Fan, G., Liu, C., & Sun, X. (2019). Preventing viral disease by ZnONPs through directly deactivating TMV and activating plant immunity in *Nicotiana benthamiana*. *Environmental Science: Nano, 6*(12), 3653–3669. https://doi.org/10.1039/C9EN00681F

Cai, L., et al. (2020). Foliar exposure of Fe3O4 nanoparticles on *Nicotiana benthamiana*: Evidence for nanoparticles uptake, plant growth promoter and defense response elicitor against plant virus. *Journal of Hazardous Materials, 393*, 122415. https://doi.org/10.1016/j.jhazmat.2020.122415

Chakraborty, S. (2008). Tomato leaf curl viruses from India. In B. W. J. Mahy & M. H. V. van Regenmortel (Eds.), *Encyclopedia of virology* (pp. 124–133). Academic Press.

Chakravarty, M., & Vora, A. (2021). Nanotechnology-based antiviral therapeutics. *Drug Delivery and Translational Research, 11*, 748–787. https://doi.org/10.1007/s13346-021-00945-5

Chauhan, P. K., Mehta, N., Chauhan, R. S., Kumar, A., Singh, H., Lal, M. K., Tiwari, R. K., & Kumar, R. (2023). Utilization of primary and secondary biochemical compounds in cotton as diagnostic markers for measuring resistance to cotton leaf curl virus (CLCuV). *Frontiers in Plant Science, 14*, 1185337. https://doi.org/10.3389/fpls.2023.1185337

Chen, N., Wang, C., Kong, F., & Wang, S. (2024). In situ facile synthesis and antibacterial activity of ag-MOFs/cellulose filter paper composites for fruit fresh-keeping. *International Journal of Biological Macromolecules, 256*, 128424. https://doi.org/10.1016/j.ijbiomac.2023.128424

Chinestra, S. C., Curvetto, N. R., & Marinangeli, P. A. (2015). Production of virus-free plants of *Lilium* spp. from bulbs obtained *in vitro* and *ex vitro*. *Scientia Horticulturae, 194*, 304–312. https://doi.org/10.1016/j.scienta.2015.08.017

Chirkov, S. N. (2002). The antiviral activity of chitosan. *Applied Biochemistry and Microbiology, 38*, 1–8. https://doi.org/10.1023/A:1015034915486

Cojocaru, F. D., Botezat, D., Gardikiotis, I., Uritu, C. M., Dodi, G., Trandafir, L., Rezus, C., Rezus, E., Tamba, B. I., & Mihai, C. T. (2020). Nanomaterials designed for antiviral drug delivery transport across biological barriers. *Pharmaceutics, 12*(2), 171. https://doi.org/10.3390/pharmaceutics12020171

Culver, J. N., & Padmanabhan, M. S. (2007). Virus-induced disease: Altering host physiology one interaction at a time. *Annual Review of Phytopathology, 45*, 221–243. https://doi.org/10.1146/annurev.phyto.45.062806.094444

Cunningham, F. J., Goh, N. S., Demirer, G. S., Matos, J. L., & Landry, M. P. (2018). Nanoparticle-mediated delivery towards advancing plant genetic engineering. *Trends in Biotechnology, 36*(9), 882–897. https://doi.org/10.1016/j.tibtech.2018.03.009

Das, S., Marwal, A., Choudhary, D. K., Gupta, V. K., & Gaur, R. K. (2011). Mechanism of RNA interference (RNAi): Current concept. *International Proceedings of Chemical, Biological and Environmental Engineering, 9*, 240–245.

Das, A. K., Marwal, A., & Pareek, V. (2015a). Nanoparticles-protein hybrid based magnetic liposome. *International Journal of Chemical, Molecular, Nuclear, Materials and Metallurgical Engineering, 9*(2), 230–233.

Das, A. K., Marwal, A., & Verma, R. (2015b). Preparation and characterization of nano-bio hybrid based magneto liposome. *International Journal of Pharmaceutical Sciences and Research, 6*(1), 367.

De Clercq, E. (2005). Antiviral drug discovery and development: Where chemistry meets with biomedicine. *Antiviral Research, 67*(2), 56–75. https://doi.org/10.1016/j.antiviral.2005.03.001

del Pozo-Rodriguez, A., Solinis, M. A., Gascón, A. R., & Pedraz, J. L. (2009). Short- and long-term stability study of lyophilized solid lipid nanoparticles for gene therapy. *European Journal of Pharmaceutics and Biopharmaceutics, 71*(2), 181–189. https://doi.org/10.1016/j.ejpb.2008.08.007

del Pozo-Rodriguez, A., Delgado, D., Gascón, A. R., & Solinis, M. A. (2013). Lipid nanoparticles as drug/gene delivery systems to the retina. *Journal of Ocular Pharmacology and Therapeutics, 29*(2), 173–188. https://doi.org/10.1089/jop.2012.0156

Derbalah, A., & Elsharkawy, M. (2019). A new strategy to control Cucumber mosaic virus using fabricated NiO-nanostructures. *Journal of Biotechnology, 306*, 134–141. https://doi.org/10.1016/j.jbiotec.2019.07.016

Diaz-Pendon, J. A., Canizares, M. C., Moriones, E., Bejarano, E. R., Czosnek, H., & Navas-Castillo, J. (2010). Tomato yellow leaf curl viruses: Menage a trois between the virus complex, the plant and the whitefly vector. *Molecular Plant Pathology, 11*, 441–450. https://doi.org/10.1111/j.1364-3703.2010.00626.x

Dikpati, A., Gaudreault, N., Chénard, V., Grenier, P., Boisselier, É., & Bertrand, N. (2022). Size Exclusion of Radioactive Polymers (SERP) informs on the biodegradation of trimethyl chitosan and biodegradable polymer nanoparticles in vitro and in vivo. *Journal of Controlled Release, 346*, 20–31. https://doi.org/10.1016/j.jconrel.2022.01.021

Ding, S. W., & Voinnet, O. (2007). Antiviral immunity directed by small RNAs. *Cell, 130*, 413–426. https://doi.org/10.1016/j.cell.2007.07.039

Dutta, P., Das, G., Boruah, S., Kumari, A., Mahanta, M., Yasin, A., et al. (2021). Nanoparticles as nano-priming agent for antifungal and antibacterial activity against plant pathogens. *Biological Forum – An International Journal, 13*, 476–482.

Ekambaram, P., Sathali, A. A. H., & Priyanka, K. (2012). Solid lipid nanoparticles: A review. *Scientific Reviews and Chemical Communications, 2*, 80–102.

El Gohary, H. G., Alhagri, I. A., Qahtan, T. F., Al-Hakimi, A. N., Saeed, A., Abolaban, F., et al. (2023). Reinforcement of structural, thermal and electrical properties and antibacterial activity of PVA/SA blend filled with hybrid nanoparticles (Ag and TiO_2 NPs): Nanodielectric for energy storage and food packaging industries. *Ceramics International, 49*(12), 20174–20184. https://doi.org/10.1016/j.ceramint.2023.04.164

Elazzazy, A., Elbeshehy, E., & Betiha, M. (2017). *In vitro* assessment of activity of graphene silver composite sheets against multidrug-resistant bacteria and Tomato Bushy Stunt Virus. *Tropical Journal of Pharmaceutical Research, 16*(11), 2705–2711. https://doi.org/10.4314/tjpr.v16i11.24

Elbeshehy, E. K. F., Elazzazy, A. M., & Aggelis, G. (2015). Silver nanoparticles synthesis mediated by new isolates of *Bacillus* spp.: Nanoparticle characterization and their activity against Bean Yellow Mosaic Virus and human pathogens. *Frontiers in Microbiology, 6*, 453. https://doi.org/10.3389/fmicb.2015.00453

El-Dougdoug, N. K., Bondok, A. M., & El-Dougdoug, K. A. (2018). Evaluation of silver nanoparticles as antiviral agents against ToMV and PVY in tomato plants. *Middle East Journal of Applied Sciences, 8*, 100–111.

Elmer, W., & White, J. C. (2018). The future of nanotechnology in plant pathology. *Annual Review of Phytopathology, 56*(1), 111–133. https://doi.org/10.1146/annurev-phyto-080516-035217

El-Sawy, M., Elsharkawy, M., Abass, J., & Hagag, E. (2017a). Inhibition of Tomato yellow leaf curl virus by *Zingiber officinale* and *Mentha longifolia* extracts and silica nanoparticles and its reflection on tomato growth and yield. *International Journal of Antiviral & Antiretroviral Research, 1*, 1–6.

El-Sawy, M., Mahmoud, M., et al. (2017b). Antiviral activity of 2-nitromethyl phenol, zinc nanoparticles, and seaweed extract against Cucumber mosaic virus (CMV) in Eggplant. *Journal of Virology & Antiviral Research, 6*(2).

El-Shazly, M. A., Attia, Y. A., Kabil, F. F., Anis, E., & Hazman, M. (2017). Inhibitory effects of salicylic acid and silver nanoparticles on potato virus Y-infected potato plants in Egypt. *Middle East Journal of Agriculture Research, 6*(3), 835–848.

Eugene, K., & Zholobak, N. (2016). Antiviral activity of cerium dioxide nanoparticles on tobacco mosaic virus model. In *Proceedings of the topical issues of new drugs development, Kharkiv, Ukraine* (p. 21).

Farooq, T., et al. (2021). Nanotechnology and plant viruses: An emerging disease management approach for resistant pathogens. *ACS Nano, 15*(4), 6030–6037. https://doi.org/10.1021/acsnano.0c09116

Fitch, M. M. M., Lehrer, A. T., Komor, E., & Moore, P. H. (2001). Elimination of sugarcane yellow leaf virus from infected sugarcane plants by meristem tip culture visualized by tissue blot immunoassay. *Plant Pathology, 50*(6), 676–680. https://doi.org/10.1046/j.1365-3059.2001.00639.x

Frank, L., & Schanz, H. (2022). Three perspectives on regime destabilization governance: A metatheoretical analysis of German pesticide policy. *Environmental Innovation and Societal Transitions, 44*, 245–264. https://doi.org/10.1016/j.eist.2021.10.008

Gafni, Y. (2003). Tomato yellow leaf curl virus: The intracellular dynamics of a plant DNA virus. *Molecular Plant Pathology, 4*, 9–15. https://doi.org/10.1111/j.1464-6722.2003.00121.x

Gaur, R. K., Prajapat, R., Marwal, A., Sahu, A., & Rathore, M. S. (2011). First report of a begomovirus infecting *Mimosa pudica* in India. *Journal of Plant Pathology., 93*(Suppl.4), S4.80.

Geisler-Lee, J., Wang, Q., Yao, Y., Zhang, W., Geisler, M., Li, K., et al. (2013). Phytotoxicity, accumulation and transport of silver nanoparticles by *Arabidopsis thaliana*. *Nanotoxicology, 7*(3), 323–337. https://doi.org/10.3109/17435390.2012.658925

Ghosh, S. K. B., Hunter, W. B., Park, A. L., & Gundersen-Rindal, D. E. (2017). Double strand RNA delivery system for plant-sap-feeding insects. *PLoS One, 12*(2), e0171861. https://doi.org/10.1371/journal.pone.0171861

Ghosh, I., Sadhu, A., Moriyasu, Y., Bandyopadhyay, M., & Mukherjee, A. (2019a). Manganese oxide nanoparticles induce genotoxicity and DNA hypomethylation in the moss *Physcomitrella patens*. *Mutation Research/Genetic Toxicology and Environmental Mutagenesis, 842*, 146–157. https://doi.org/10.1016/j.mrgentox.2018.08.002

Ghosh, M., Ghosh, I., Godderis, L., Hoet, P., & Mukherjee, A. (2019b). Genotoxicity of engineered nanoparticles in higher plants. *Mutation Research/Genetic Toxicology and Environmental Mutagenesis, 842*, 132–145. https://doi.org/10.1016/j.mrgentox.2018.08.001

Gilbertson, R. L., Batuman, O., Webster, C. G., & Adkins, S. (2015). Role of the insect supervectors *Bemisia tabaci* and *Frankliniella occidentalis* in the emergence and global spread of plant viruses. *Annual Review of Virology, 2*, 67–93. https://doi.org/10.1146/annurev-virology-100114-055102

Gogos, A., Knauer, K., & Bucheli, T. D. (2012). Nanomaterials in plant protection and fertilization: Current state, foreseen applications, and research priorities. *Journal of Agricultural and Food Chemistry, 60*(39), 9781–9792. https://doi.org/10.1021/jf3027557

Goldstein, J. I., Newbury, D. E., Michael, J. R., Ritchie, N. W., Scott, J. H. J., & Joy, D. C. (2017). *Scanning electron microscopy and X-ray microanalysis*. Springer.

González-Melendi, P., et al. (2008). Nanoparticles as smart treatment-delivery systems in plants: Assessment of different techniques of microscopy for their visualization in plant tissues. *Annals of Botany, 101*(1), 187–195. https://doi.org/10.1093/aob/mcm221

Goyal, R. K. (2017). *Nanomaterials and nanocomposites: Synthesis, properties, characterization techniques, and applications*. CRC Press.

Grondeau, C., Samson, R., & Sands, D. C. (1994). A review of thermotherapy to free plant materials from pathogens, especially seeds from bacteria. *Critical Reviews in Plant Sciences, 13*(1), 57–75. https://doi.org/10.1080/07352689409701908

Gubareva, L. V. (2004). Molecular mechanisms of influenza virus resistance to neuraminidase inhibitors. *Virus Research, 103*(1–2), 199–203. https://doi.org/10.1016/j.virusres.2004.05.003

Halamoda, K. B., Rolland, E., Piovesan, J., Puertas, G. A., & Bremer, S. (2022). Toxic effects of nanomaterials for health applications: How automation can support a systematic review of the literature? *Journal of Applied Toxicology, 42*, 41–51.

Hannon, G. J. (2002). RNA interference. *Nature, 418*(6894), 244–251. https://doi.org/10.1038/nature01052

Hansen, L. J. J., Daoussi, R., Vervaet, C., Remon, J. P., & De Beer, T. R. M. (2015). Freeze-drying of live virus vaccines: A review. *Vaccine, 33*(42), 5507–5519.

Hao, Y., Yuan, W., Ma, C., White, J., Zhang, Z., Adeel, M., Zhou, T., Yukui, R., & Xing, B. (2018). Engineered nanomaterials suppress Turnip mosaic virus infection in tobacco (Nicotiana benthamiana). *Environmental Science: Nano, 5*(7), 1685–1693.

Hayles, J., et al. (2017). Nanopesticides: A review of current research and perspectives. In *New pesticides and soil sensors* (pp. 193–225).

Hemmrich, E., & McNeil, S. (2023). Active ingredient vs excipient debate for nanomedicines. *Nature Nanotechnology, 18*(7), 692–695.

Hoai, P. T. T., Dai Lam, T., Huong, N. T. M., & Van Anh, M. T. (2023). Removal of ethylene by synthesized Ag/TiO_2 photocatalyst under visible light irradiation. *Chemosphere, 329*, 138607.

Hong, J., Wang, C., Wagner, D. C., Gardea-Torresdey, J. L., He, F., & Rico, C. M. (2021). Foliar application of nanoparticles: Mechanisms of absorption, transfer, and multiple impacts. *Environmental Science: Nano, 8*(5), 1196–1210.

Hossain, Z., Mustafa, G., Sakata, K., & Komatsu, S. (2016). Insights into the proteomic response of soybean towards Al_2O_3, ZnO, and Ag nanoparticles stress. *Journal of Hazardous Materials, 304*, 291–305.

Hu, G., Dong, Y., Zhang, Z., Fan, X., Ren, F., & Zhou, J. (2015). Virus elimination from in vitro apple by thermotherapy combined with chemotherapy. *Plant Cell, Tissue and Organ Culture, 121*(2), 435–443. https://doi.org/10.1007/s11240-015-0714-6

Iqbal, M. (2016). Vicia faba bioassay for environmental toxicity monitoring: A review. *Chemosphere, 144*, 785–802.

Jain, D., & Kothari, S. L. (2014). Green synthesis of silver nanoparticles and their application in plant virus inhibition. *Journal of Mycology and Plant Pathology, 44*(1), 21–24.

Jain, E., Tripathi, A. D., Agarwal, A., Mauraya, K. K., Rai, D. C., Mishra, R., & Singh, R. B. (2022). Anticancerous compounds in fruits, their extraction, and relevance to food. In *Functional foods and nutraceuticals in metabolic and non-communicable diseases* (pp. 517–532). Academic Press.

Jaiswal, N., Saritha, R. K., Datta, D., Singh, M., Dubey, R. S., Rai, A. B., & Rai, M. (2012). Mixed infections of begomoviruses in pumpkins with yellow vein mosaic disease in North India. *Archives of Phytopathology and Plant Protection, 45*(8), 938–941.

Jazayeri, M. H., Amani, H., Pourfatollah, A. A., Pazoki-Toroudi, H., & Sedighimoghaddam, B. (2016). Various methods of gold nanoparticles (GNPs) conjugation to antibodies. *Sensing and Bio-Sensing Research, 9*, 17–22.

Jeevalatha, A., Siddappa, S., Kumar, R., Tiwari, R. K., Lal, M. K., Sharma, S., Chakrabarty, S. K., & Singh, B. P. (2023). RNA-seq analysis reveals an early defense response to tomato leaf curl New Delhi virus in potato cultivar Kufri Bahar. *Functional & Integrative Genomics, 23*, 215. https://doi.org/10.1007/s10142-023-01138-5

Jones, K. L., Drane, D., & Gowans, E. J. (2007). Long-term storage of DNA-free RNA for use in vaccine studies. *BioTechniques, 43*(5), 675–681.

Jyothsna, P., Haq, Q. M. I., Singh, P., Sumiya, K. V., Praveen, S., Rawat, R., Briddon, R. W., & Malathi, V. G. (2013). Infection of Tomato leaf curl New Delhi virus (ToLCNDV), a bipartite begomovirus with betasatellites, results in enhanced level of helper virus components and antagonistic interaction between DNA B and betasatellites. *Applied Microbiology and Biotechnology, 97*, 5457–5471.

Kah, M., & Hofmann, T. (2014). Nanopesticide research: Current trends and future priorities. *Environment International, 63*, 224–235.

Kamali, M., Heydarnejad, J., Massumi, H., Kvarnheden, A., Kraberger, S., & Varsani, A. (2016). Molecular diversity of turncurtoviruses in Iran. *Archives of Virology, 161*, 551–561.

Kanakala, S., Jyothsna, P., Shukla, R., Tiwari, N., Veer, B. S., Swarnalatha, P., Krishnareddy, M., & Malathi, V. G. (2013). Asymmetric synergism and heteroencapsidation between two bipartite begomoviruses, Tomato leaf curl New Delhi virus and tomato leaf curl Palampur virus. *Virus Research, 174*, 126–136.

Kapoor, D., Singh, S., Kumar, V., Romero, R., Prasad, R., & Singh, J. (2019). Antioxidant enzymes regulation in plants in reference to reactive oxygen species (ROS) and reactive nitrogen species (RNS). *Plant Gene, 19*, 100182.

Kashyap, P. L., Xiang, X., & Heiden, P. (2015). Chitosan nanoparticle based delivery systems for sustainable agriculture. *International Journal of Biological Macromolecules, 77*, 36–51.

Khalil, I. E., Fonseca, J., Reithofer, M. R., Eder, T., & Chin, J. M. (2023). Tackling orientation of metal-organic frameworks (MOFs): The quest to enhance MOF performance. *Coordination Chemistry Reviews, 481*, 215043.

Khan, S., Hashim, S. B., Arslan, M., Zhang, K., Siman, L., Mukhtar, A., et al. (2024). Development of an active biogenic silver nanoparticles composite film based on berry wax and chitosan for rabbit meat preservation. *International Journal of Biological Macromolecules, 275*, 133128.

Khandelwal, N., et al. (2016). Budding trends in integrated pest management using advanced micro- and nano-materials: Challenges and perspectives. *Journal of Environmental Management, 184*, 157–169.

Kim, D., et al. (2018). Recent developments in nanotechnology transforming the agricultural sector: A transition replete with opportunities. *Journal of the Science of Food and Agriculture, 98*(3), 849–864.

King, A. M., Lefkowitz, E., Adams, M. J., & Carstens, E. B. (2012). Virus taxonomy. In *Ninth report of the International Committee on Taxonomy of Viruses* (pp. 193–210). Elsevier.

Kochkina, Z. M., Pospieszny, H., & Chirkov, S. N. (1995). Inhibition by chitosan of productive infection of the T-series bacteriophages in an *Escherichia coli* culture. *Microbiology, 64*(2), 173–176.

Kon, T., & Gilbertson, R. (2012). Two genetically related begomoviruses causing tomato leaf curl disease in Togo and Nigeria differ in virulence and host range but do not require a betasatellite for induction of disease symptoms. *Archives of Virology, 157*, 107–120.

Kumar, R., Tiwari, R. K., Jeevalatha, A., Sundaresha, S., Shah, M. A., Sharma, S., Sagar, V., Kumar, M., & Chakrabarti, S. K. (2021a). Potato apical leaf curl disease: Current status and perspectives on a disease caused by tomato leaf curl New Delhi virus. *Journal of Plant Diseases and Protection, 128*, 897–911. https://doi.org/10.1007/s41348-021-00463-w

Kumar, S., Basumatary, I. B., Sudhani, H. P., Bajpai, V. K., Chen, L., Shukla, S., & Mukherjee, A. (2021b). Plant extract mediated silver nanoparticles and their applications as antimicrobials and in sustainable food packaging: A state-of-the-art review. *Trends in Food Science & Technology, 112*, 651–666.

Kumar, R., Tiwari, R. K., Sundaresha, S., Kaundal, P., & Raigond, B. (2022). Potato viruses and their management. In S. K. Chakrabarti, S. Sharma, & M. A. Shah (Eds.), *Sustainable management of potato pests and diseases*. Springer. https://doi.org/10.1007/978-981-16-7695-6_12

Kumar, R., Lal, M. K., Tiwari, R. K., Kumar, A., Behera, B., Nayak, L., Dash, G. K., Sahoo, S. K., Jena, J., Lal, P., & Altaf, M.A. (2023). Traditional and emerging climate-resilient agricultural practices for enhancing food production and nutritional quality.In T. Aftab (Ed.), *New frontiers in plant-environment interactions. Environmental Science and Engineering*. Springer.https://doi.org/10.1007/978-3-031-43729-8_21

Kumari, P., Singh, A. K., Sharma, V. K., Chattopadhyay, B., & Chakraborty, S. (2011). A novel recombinant tomato-infecting begomovirus capable of transcomplementing heterologous DNA-B components. *Archives of Virology, 156*, 769–783.

Kumari, S., Kumari, A., & Sharma, R. (2024). Safe and sustainable food packaging: Argemone albiflora mediated green synthesized silver-carrageenan nanocomposite films. *International Journal of Biological Macromolecules, 264*, 130626.

Lal, A., Pant, M., & Rani, A. (2015). The who's who of plant viruses: A cognitive approach. *Asian Journal of Pharmaceutical and Clinical Research, 8*(1), 60–68. https://doi.org/10.22159/ajpcr.2015.v8i1.3721

Lassois, L., Lepoivre, P., Swennen, R., van den Houwe, I., & Panis, B. (2013). Thermotherapy, chemotherapy, and meristem culture in banana. In *In methods in molecular biology, 11013* (pp. 419–433). Humana Press.

Lawrencia, D., et al. (2021). Controlled release fertilizers: A review on coating materials and mechanism of release. *Plants, 10*(2), 1–26.

Leke, W. N., Kvarnheden, A., Ngane, E. B., Titanji, V. P. K., & Brown, J. K. (2011). Molecular characterization of a new begomovirus and divergent alphasatellite from tomato in Cameroon. *Archives of Virology, 156*, 925–928.

Li, M., Huang, Q., & Wu, Y. (2011). A novel chitosan-poly (lactide) copolymer and its submicron particles as imidacloprid carriers. *Pest Management Science, 67*(7), 831–836.

Li, A., Wang, J., Wang, Y., Gao, H., Wei, G., Huang, Y., Yu, H., Gan, Y., & Mei, L. (2019). Recent progress in drug delivery. *Acta Pharmaceutica Sinica B, 9*(6), 1145–1162.

Lin, D., & Xing, B. (2007). Phytotoxicity of nanoparticles: Inhibition of seed germination and root growth. *Environmental Pollution, 150*(2), 243–250.

Littrup, P. J., Babkin, A. V., Duncan, R., & Boldarov, S. (2006). *Cryotherapy system*. U.S. Patent No. US 7,083,612. https://patentimages.storage.googleapis.com/ad/e7/f3/b641c611fdf9c1/US7083612.pdf

Liu, W., Zhang, M., & Bhandari, B. (2020). Nanotechnology: A shelf life extension strategy for fruits and vegetables. *Critical Reviews in Food Science and Nutrition, 60*(10), 1706–1721.

Luceri, A., Francese, R., Lembo, D., Ferraris, M., & Balagna, C. (2023). Silver nanoparticles: Review of antiviral properties, mechanism of action and applications. *Microorganisms, 11*(3), 629.

MacFaul, P. A., Ruston, L., & Wood, J. M. (2011). Activation energies for the decomposition of pharmaceuticals and their application to predicting hydrolytic stability in drug discovery. *MedChemComm, 2*(2), 140–142.

Maginnis, M. S. (2018). Virus–receptor interactions: The key to cellular invasion. *Journal of Molecular Biology, 430*(17), 2590–2611.

Malik, A. H., Briddon, R. W., & Mansoor, S. (2011). Infectious clones of Tomato leaf curl Palampur virus with a defective DNA B and their pseudo-recombination with tomato leaf curl New Delhi virus. *Virology Journal, 8*, 173.

Marslin, G., Sheeba, C. J., & Franklin, G. (2017). Nanoparticles alter secondary metabolism in plants via ROS burst. *Frontiers in Plant Science, 8*, 832. https://doi.org/10.3389/fpls.2017.00832

Marwal, A., & Gaur, R. K. (2019). Nanophytovirology: An emerging field for disease management. In *Plant Diseases-Current Threats and Management Trends*. https://doi.org/10.5772/intechopen.86653

Marwal, A., Sahu, A. K., Prajapat, R., Choudhary, D. K., & Gaur, R. K. (2012). RNA silencing suppressor encoded by Betasatellite DNA associated with Croton yellow vein mosaic virus. *Open Access Scientific Reports, 1*(2), 1–2.

Marwal, A., Sahu, A. K., Prajapat, R., Choudhary, D. K., & Gaur, R. K. (2013). Molecular and recombinational characterization of Begomovirus infecting an ornamental plant *Alternanthera sessilis*: A new host of tomato leaf curl Kerala virus reported in India. *Science International, 1*(3), 51–56.

Marwal, A., Sahu, A. K., & Gaur, R. K. (2014a). Association of begomovirus and an alphasatellite with leaf curl disease of ornamental plant, *Vinca alba* in Punjab, India. *Journal of Agricultural Research, 52*(3), 339–356.

Marwal, A., Sahu, A. K., & Gaur, R. K. (2014b). Transmission and host interaction of Geminivirus in weeds. In *Plant virus–host interaction* (pp. 143–161). Academic.

Marwal, A., Gaur, R. K., & Khurana, S. M. P. (2018). Possible approaches for developing different strategies to prevent transmission of Geminiviruses to important crops. In *Plant Viruses* (pp. 301–320). CRC Press.

Marwal, A., Nehra, C., Verma, R. K., Mishra, M., Srivasatava, D., Choudhary, P., & Gaur, R. K. (2021). First report of papaya leaf curl virus and its associated papaya leaf curl betasatellite infecting *Catharanthus roseus* plants in India. *The Journal of Horticultural Science and Biotechnology, 96*(6), 808–813.

Mathews, R. E. (1953). Chemotherapy and plant viruses. *Journal of General Microbiology, 8*, 277–288. https://doi.org/10.1099/00221287-8-2-277

Mishra, P., et al. (2022). Role of nanotechnology in management of plant viral diseases. *Materials Today: Proceedings, 69*, 1–10. https://doi.org/10.1016/j.matpr.2022.01.001

Mittal, D., Kaur, G., Singh, P., Yadav, K., & Ali, S. A. (2020). Nanoparticle-based sustainable agriculture and food science: Recent advances and future outlook. *Frontiers in Nanotechnology, 2*, 579954. https://doi.org/10.3389/fnano.2020.579954

Mitter, N., et al. (2017). Clay nanosheets for topical delivery of RNAi for sustained protection against plant viruses. *Nature Plants, 3*(2), 1–10. https://doi.org/10.1038/nplants.2016.207

Modarresi Chahardehi, A., Rakhshandehroo, F., Mozafari, J., & Mousavi, L. (2016). Efficiency of a chemo-thermotherapy technique for eliminating arabis mosaic virus (ArMV) and prunus necrotic ringspot virus (PNRSV) from in vitro rose plantlets. *Journal of Crop Protection, 5*(4), 497–506.

Modgill, N., Naga, K. C., Firake, D. M., Tiwari, R. K., Shivaramu, S., Kumar, R., Bhatnagar, A., Buckseth, T., Bairwa, A., Maharana, C., Sharma, S., & Singh, B. (2025). Flupyradifurone application: A promising approach to curtail tomato leaf curl New Delhi virus (ToLCNDV) transmission vectored by Bemisia tabaci in potato farming. Crop protection. 187.

Muhire, B., Martin, D. P., Brown, J. K., Navas-Castillo, J., Moriones, E., Zerbini, F. M., et al. (2013). A genome-wide pairwise-identity-based proposal for the classification of viruses in the genus Mastrevirus (family Geminiviridae). *Archives of Virology, 158*, 1411–1424.

Müller, R. H., Radtke, M., & Wissing, S. A. (2002). Solid lipid nanoparticles (SLN) and nanostructured lipid carriers (NLC) in cosmetic and dermatological preparations. *Advanced Drug Delivery Reviews, 54*(1), 131–155. https://doi.org/10.1016/S0169-409X(01)00218-7

Mullineaux, P. M., Exposito-Rodriguez, M., Laissue, P. P., & Smirnoff, N. (2018). ROS-dependent signalling pathways in plants and algae exposed to high light: Comparisons with other eukaryotes. *Free Radical Biology and Medicine, 122*, 52–64. https://doi.org/10.1016/j.freeradbiomed.2018.01.012

Murashige, T. (1974). Plant propagation through tissue cultures. *Annual Review of Plant Physiology, 25*(1), 135–166. https://doi.org/10.1146/annurev.pp.25.060174.001031

Murphy, J. F., et al. (2003). Rhizobacteria-mediated growth promotion of tomato leads to protection against cucumber mosaic virus. *Phytopathology, 93*(10), 1301–1307. https://doi.org/10.1094/PHYTO.2003.93.10.1301

Nair, P. M. G., & Chung, I. M. (2014). Physiological and molecular level effects of silver nanoparticles exposure in rice (*Oryza sativa* L.) seedlings. *Chemosphere, 112*, 105–113. https://doi.org/10.1016/j.chemosphere.2014.03.056

Navas-Castillo, J., Fiallo-Olive, E., & Sanchez-Campos, S. (2011). Emerging virus diseases transmitted by whiteflies. *Annual Review of Phytopathology, 49*, 219–248. https://doi.org/10.1146/annurev-phyto-072910-095314

Nehra, C., Sahu, A. K., Marwal, A., & Gaur, R. K. (2014). Natural occurrence of Clerodendron yellow mosaic virus on Bougainvillea in India. *New Disease Reports, 30*(19), 2044–0588.

Nehra, C., Marwal, A., Verma, R. K., Mishra, M., Sharma, P., & Gaur, R. K. (2019a). Papaya yellow leaf curl virus: A newly identified begomovirus infecting *Carica papaya* L. from the Indian subcontinent. *The Journal of Horticultural Science and Biotechnology, 94*(4), 475–480.

Nehra, C., Marwal, A., Verma, R. K., & Gaur, R. K. (2019b). Molecular characterization of begomoviruses DNA-A and associated beta satellites with new host *Ocimum sanctum* in India. *Proceedings of the National Academy of Sciences, India Section B: Biological Sciences, 89*, 903–910.

Noha, K., Bondok, A. M., & El-Dougdoug, K. A. (2018). Evaluation of silver nanoparticles as antiviral agent against ToMV and PVY in tomato plants. *Sciences, 8*(01), 100–111.

Oberdörster, G. (2010). Safety assessment for nanotechnology and nanomedicine: Concepts of nanotoxicology. *Journal of Internal Medicine, 267*(1), 89–105. https://doi.org/10.1111/j.1365-2796.2009.02187.x

Osei, M. K., Akromah, R., Shih, S. L., Lee, L. M., & Green, S. K. (2008). First report and molecular characterization of DNA A of three distinct begomoviruses associated with tomato leaf curl disease in Ghana. *Plant Disease, 92*, 1585. https://doi.org/10.1094/PDIS-92-10-1585C

Panattoni, A., Luvisi, A., & Triolo, E. (2011). Selective chemotherapy on Grapevine leafroll-associated virus-1 and -3. *Phytoparasitica, 39*(5), 503–508. https://doi.org/10.1007/s12600-011-0185-1

Panattoni, A., Luvisi, A., & Triolo, E. (2013). Review: Elimination of viruses in plants: Twenty years of progress. *Spanish Journal of Agricultural Research, 11*(1), 173. https://doi.org/10.5424/sjar/2013111-3201

Pathirana, R., McLachlan, A., Hedderley, D., Carra, A., Carimi, F., & Panis, B. (2015). Removal of leafroll viruses from infected grapevine plants by droplet vitrification. *Acta Horticulturae, 1083*, 491–498. https://doi.org/10.17660/ActaHortic.2015.1083.64

Patil, C. D., Borase, H. P., Suryawanshi, R. K., & Patil, S. V. (2016). Trypsin inactivation by latex fabricated gold nanoparticles: A new strategy towards insect control. *Enzyme and Microbial Technology, 92*, 18–25.

Pimentel, D., & Burgess, M. (2012). Small amounts of pesticides reaching target insects. In *Environment, Development and Sustainability* (pp. 1–2). Springer. https://doi.org/10.1007/s10668-011-9296-9

Pospieszny, H., Chirkov, S., & Atabekov, J. (1991). Induction of antiviral resistance in plants by chitosan. *Plant Science, 79*(1), 63–68. https://doi.org/10.1016/0168-9452(91)90004-3

Postiglione, A. E., & Muday, G. K. (2020). The role of ROS homeostasis in ABA-induced guard cell signaling. *Frontiers in Plant Science, 11*, 968. https://doi.org/10.3389/fpls.2020.00968

Prajapat, R., Gaur, R. K., & Marwal, A. (2011). Homology modeling and docking studies between AC1 rep protein of Begomovirus and Whey á-lactalbumin. *Asian Journal of Biological Sciences, 4*(4), 352–361.

Prajapat, R., Marwal, A., Shaikh, Z., & Gaur, R. K. (2012). Geminivirus database (GVDB): First database of family Geminiviridae and its genera Begomovirus. *Pakistan Journal of Biological Sciences: PJBS, 15*(14), 702–706.

Prajapat, R., Marwal, A., & Gaur, R. K. (2013). Evidence of the Association of Solanum leaf curl lakshmangarh virus with a Weed Plant *Solanum nigrum* in Rajasthan, India. *Science International, 1*(11), 379–383.

Prasanna, H., Sinha, D., Rai, G., Krishna, R., Kashyap, S. P., Singh, N., Singh, N. K., Singh, M., & Malathi, V. G. (2015). Pyramiding Ty-2 and Ty-3 genes for resistance to monopartite and bipartite tomato leaf curl viruses of India. *Plant Pathology, 64*, 256–264. https://doi.org/10.1111/ppa.12224

Priyanka, S., Arvind Bharani, R. S., & John, A. (2023). Biocompatible green technology principles for the fabrication of food packaging material with noteworthy mechanical and antimicrobial properties: A sustainable developmental goal towards the effective, safe food preservation strategy. *Chemosphere, 336*, 139240. https://doi.org/10.1016/j.chemosphere.2023.139240

Puri, A., Loomis, K., Smith, B., Lee, J. H., Yavlovich, A., Heldman, E., et al. (2009). Lipid-based nanoparticles as pharmaceutical drug carriers: From concepts to clinic. *Critical Reviews in Therapeutic Drug Carrier Systems, 26*, 523–580. https://doi.org/10.1615/CritRevTherDrugCarrierSyst.v26.i6.10

Rajini, R., & Ferdinand, A. C. (2022). Structural, morphological and magnetic properties of (c-ZnFe2O4 and t-CuFe2O4) ferrite nanoparticle synthesized by reactive ball milling. *Chemical Data Collections, 38*, 100825. https://doi.org/10.1016/j.cdatc.2022.100825

Rani, N., Kumari, K., Sangwan, P., Barala, P., Yadav, J., Vijeta, et al. (2022). Nano-iron and nano-zinc induced growth and metabolic changes in *Vigna radiata*. *Sustainability, 14*(14), 8251. https://doi.org/10.3390/su14148251

Rastogi, A., Zivcak, M., Sytar, O., Kalaji, H. M., He, X., Mbarki, S., & Brestic, M. (2017). Impact of metal and metal oxide nanoparticles on plant: A critical review. *Frontiers in Chemistry, 5*, 78. https://doi.org/10.3389/fchem.2017.00078

Rivera, A. L., Gómez-Lim, M., Fernández, F., & Loske, A. M. (2012). Physical methods for genetic plant transformation. *Physics of Life Reviews, 9*(3), 308–345. https://doi.org/10.1016/j.plrev.2012.07.002

Roossinck, M. J. (2012). Plant virus metagenomics: Biodiversity and ecology. *Annual Review of Genetics, 46*, 359–369. https://doi.org/10.1146/annurev-genet-110711-156305

Roossinck, M. J., Saha, P., Wiley, G. B., Quan, J., White, J. D., Lai, H., et al. (2010). Ecogenomics: Using massively parallel pyrosequencing to understand virus ecology. *Molecular Ecology, 19*(Suppl 1), 81–88. https://doi.org/10.1111/j.1365-294X.2010.04577.x

Rossi, L., Sharifan, H., Zhang, W., Schwab, A. P., & Ma, X. (2018). Mutual effects and in planta accumulation of co-existing cerium oxide nanoparticles and cadmium in hydroponically grown soybean (*Glycine max* (L.) Merr.). *Environmental Science: Nano, 5*(1), 150–157. https://doi.org/10.1039/C7EN00281J

Roumagnac, P., Lett, J. M., Fiallo-Olivé, E., Navas-Castillo, J., Zerbini, F. M., Martin, D. P., & Varsani, A. (2022). Establishment of five new genera in the family Geminiviridae: Citlodavirus, Maldovirus, Mulcrilevirus, Opunvirus, and Topilevirus. *Archives of Virology, 167*, 695–710.

Sagalowicz, L., & Leser, M. E. (2010). Delivery systems for liquid food products. *Current Opinion in Colloid & Interface Science, 15*, 61–72. https://doi.org/10.1016/j.cocis.2010.04.005

Sahu, P. P., Rai, N. K., Chakraborty, S., Singh, M., Chandrappa, P. H., Ramesh, B., Chattopadhyay, D., & Prasad, M. (2010). Tomato cultivar tolerant to Tomato Leaf Curl New Delhi Virus infection induces virus-specific short interfering RNA accumulation and defence-associated host gene expression. *Molecular Plant Pathology, 11*, 531–544. https://doi.org/10.1111/j.1464-6722.2010.00615.x

Sahu, P. P., Rai, N. K., Puranik, S., Roy, A., Khan, M., & Prasad, M. (2012). Dynamics of defense-related components in two contrasting genotypes of tomato upon infection with Tomato Leaf Curl New Delhi Virus. *Molecular Biotechnology, 52*, 140–150. https://doi.org/10.1007/s12033-011-9425-7

Sahu, A. K., Marwal, A., Nehra, C., Choudhary, D. K., Sharma, P., & Gaur, R. K. (2014). RNAi mediated gene silencing against betasatellite associated with croton yellow vein mosaic begomovirus. *Molecular Biology Reports, 41*, 7631–7638.

Sahu, A. K., Nehra, C., Marwal, A., & Gaur, R. K. (2015). First report of a begomovirus associated with betasatellites infecting spinach (*Spinacia oleracea*) in India. *Journal of General Plant Pathology, 81*, 146–150.

Saikia, A. K., & Muniyappa, V. (1989). Epidemiology and control of tomato leaf curl virus in southern India. *Indian Journal of Virology, 5*(1), 55–60.

Sangwan, A., Gupta, D., Singh, O. W., Roy, A., Mukherjee, S. K., Mandal, B., & Singh, N. (2023). Size variations of mesoporous silica nanoparticle control uptake efficiency and delivery of AC2-derived ds RNA for protection against Tomato Leaf Curl New Delhi Virus. *Plant Cell Reports, 42*(10), 1571–1587. https://doi.org/10.1007/s00299-023-03178-5

Sasi, S., & Bhat, A. I. (2018). In vitro elimination of Piper yellow mottle virus from infected black pepper through somatic embryogenesis and meristem-tip culture. *Crop Protection, 103*, 39–45. https://doi.org/10.1016/j.cropro.2017.10.014

Sastry, K. S., & Zitter, T. A. (2014). Management of virus and viroid diseases of crops in the tropics. In *Plant virus and viroid diseases in the tropics* (pp. 149–480). Springer. https://doi.org/10.1007/978-94-017-8767-0_9

Sastry, R. K., Rashmi, H. B., & Rao, N. H. (2011). Nanotechnology for enhancing food security in India. *Food Policy, 36*(3), 391–400. https://doi.org/10.1016/j.foodpol.2011.01.001

Sattelmacher, B. (2001). The apoplast and its significance for plant mineral nutrition. *New Phytologist, 149*, 167–192. https://doi.org/10.1046/j.1469-8137.2001.00021.x

Schäfer, R. B., et al. (2019). Future pesticide risk assessment: Narrowing the gap between intention and reality. *Environmental Sciences Europe, 31*, 1–5. https://doi.org/10.1186/s12302-019-0194-0

Scimeca, M., Bischetti, S., Lamsira, H. K., Bonfiglio, R., & Bonanno, E. (2018). Energy dispersive X-ray (EDX) microanalysis: A powerful tool in biomedical research and diagnosis. *European Journal of Histochemistry, 62*(1). https://doi.org/10.4081/ejh.2018.2855

Seo, Y. S., Gepts, P., & Gilbertson, R. L. (2004). Genetics of resistance to the geminivirus, Bean dwarf mosaic virus, and the role of the hypersensitive response in common bean. *Theoretical and Applied Genetics, 108*, 786–793. https://doi.org/10.1007/s00122-003-1486-7

Shafie, R. M., Salama, A. M., & Farroh, K. Y. (2018). Silver nanoparticles activity against tomato spotted wilt virus. *Middle East Journal of Agricultural Research, 7*, 1251–1267. https://www.researchgate.net/publication/326644885

Shafiq, M., Asad, S., Zafar, Y., Briddon, R. W., & Mansoor, S. (2010). Pepper leaf curl Lahore virus requires the DNA B component of Tomato leaf curl New Delhi virus to cause leaf curl symptoms. *Virology Journal, 7*, 367. https://doi.org/10.1186/1743-422X-7-367

Sharma, M. (2019). Transdermal and intravenous nano drug delivery systems: Present and future. In *Applications of targeted Nano drugs and delivery systems* (pp. 499–550). Elsevier. https://doi.org/10.1016/B978-0-12-813710-9.00023-0

Sharma, S., Kang, S., & Sharma, A. (2015). First report of mixed infection of Zucchini yellow mosaic virus and tomato leaf curl New Delhi virus in bittergourd in India. *Journal of Plant Pathology, 97*, 397. https://doi.org/10.4454/JPP.V97I3.084

Siepmann, J., Elkharraz, K., Siepmann, F., & Klose, D. (2005). How autocatalysis accelerates drug release from PLGA-based microparticles: A quantitative treatment. *Biomacromolecules, 6*(4), 2312–2319. https://doi.org/10.1021/bm050177c

Singh, M. K., Chandel, V., Hallan, V., Ram, R., & Zaidi, A. A. (2009). Occurrence of peanut stripe virus on patchouli and raising of virus-free patchouli plants by meristem tip culture. *Journal of Plant Diseases and Protection, 116*(1), 2–6. https://doi.org/10.1007/BF03356278

Singh, A., Chattopadhyay, B., & Chakraborty, S. (2012). Biology and interactions of two distinct monopartite begomoviruses and betasatellites associated with radish leaf curl disease in India. *Virology Journal, 9*, 43. https://doi.org/10.1186/1743-422X-9-43

Singh, A. K., Kushwaha, N., & Chakraborty, S. (2016). Synergistic interaction among begomoviruses leads to the suppression of host defense-related gene expression and breakdown of resistance in chilli. *Applied Microbiology and Biotechnology, 100*, 4035–4049. https://doi.org/10.1007/s00253-016-7543-0

Singh, S., Vishwakarma, K., Singh, S., Sharma, S., Dubey, N. K., Singh, V. K., et al. (2017). Understanding the plant and nanoparticle interface at transcriptomic and proteomic level: A concentric overview. *Plant Gene, 11*, 265–272. https://doi.org/10.1016/j.plgene.2017.07.006

Singh, N. A., Rai, N., & Marwal, A. (2021). Nanosensors for the detection of chemical food adulterants. In *Nanotoxicology and Nanoecotoxicology* (Vol. 2, pp. 25–53). Springer International Publishing.

Skiada, F. G., Maliogka, V. I., Katis, N. I., & Eleftheriou, E. P. (2013). Elimination of Grapevine rupestris stem pitting-associated virus (GRSPaV) from two *Vitis vinifera* cultivars by in vitro chemotherapy. *European Journal of Plant Pathology, 135*(2), 407–414. https://doi.org/10.1007/s10658-012-0097-z

Song, W., Zhao, B., Wang, C., Ozaki, Y., & Lu, X. (2019). Functional nanomaterials with unique enzyme-like characteristics for sensing applications. *Journal of Materials Chemistry B, 7*(6), 850–875. https://doi.org/10.1039/C8TB01024H

Sopeña, F., Maqueda, C., & Morillo, E. (2009). Controlled release formulations of herbicides based on micro-encapsulation. *Ciencia e Investigación Agraria, 36*(1), 27–42.

Srivastava, A., Pandey, V., Verma, R. K., Marwal, A., Mishra, R., Briddon, R. W., et al. (2022). First complete genome sequence of Tomato leaf curl virus (ToLCV) from Salvia splendens in India. *Journal of Phytopathology, 170*(7–8), 479–491.

Stanley, J. (1985). The molecular biology of geminiviruses. *Advances in Virus Research, 30*, 139–177. https://doi.org/10.1016/S0065-3527(08)60005-7

Tailor, S., Marwal, A., & Meena, M. (2021). Application of nanotechnology in management of various plant diseases. In *Innovative approaches in diagnosis and management of crop diseases* (pp. 1–17). Apple Academic Press.

Tailor, S., Jain, K., Marwal, A., Meena, M., Anbarasu, K., & Gaur, R. K. (2022). Outlooks of nanotechnology in organic farming management. *Defence Life Science Journal, 7*(1), 52–60.

Tailor, S., Jain, K., Malik, A., Suthar, M., Mishra, A., Gaur, R., et al. (2024). New insight of nanotechnology in combating plant stresses: Scope and potential applications. In *Molecular dynamics of plant stress and its management* (pp. 475–490). Springer.

Tailor, S., Malik, A., Gaur, R. K., & Marwal, A. (2025a). Characterisation, synthesis, and antimicrobial activities of silver nanoparticles formed via green synthesis approach using aqueous extract derived from the bark of *Mangifera indica*: A sustainable resource. *Defence Life Science Journal, 10*(2), 111–120.

Tailor, S., Malik, A., Gaur, R. K., & Marwal, A. (2025b). Report on characterization and antimicrobial activity of biogenic synthesis of silver nanoparticles from Amchur (Indian spice) obtained from *Mangifera indica*: A sustainable resource. *Research Journal of Biotechnology, 20*(6), 126–136.

Tailor, S., Malik, A., Suthar, M., Jain, K., Gaur, R. K., Meena, M., & Marwal, A. (2025). The role of nanotechnology in mitigating food insecurity among defence personnel to provide quality nutrition. *Defence S & T Technical Bulletin, 18*(1), 65–77.

Taniguchi, N. (1974). On the basic concept of 'nano-technology'. In *Proceedings of the International Conference on Production Engineering. Tokyo, Part II*. Japan Society of Precision Engineering.

Tarafder, C., et al. (2020). Formulation of a hybrid nanofertilizer for slow and sustainable release of micronutrients. *ACS Omega, 5*(37), 23960–23966. https://doi.org/10.1021/acsomega.0c03403

Tolaymat, T., El Badawy, A., Genaidy, A., Abdelraheem, W., & Sequeira, R. (2017). Analysis of metallic and metal oxide nanomaterial environmental emissions. *Journal of Cleaner Production, 143*, 401–412. https://doi.org/10.1016/j.jclepro.2016.12.143

Urbaniak, T., Milasheuski, Y., & Musiał, W. (2024). Zero-order kinetics release of lamivudine from layer-by-layer coated macromolecular prodrug particles. *International Journal of Molecular Sciences, 25*(23), 12921.

Vargas-Hernandez, M., Macias-Bobadilla, I., Guevara-Gonzalez, R. G., Rico-Garcia, E., Ocampo-Velazquez, R. V., Avila-Juarez, L., & Torres-Pacheco, I. (2020). Nanoparticles as potential antivirals in agriculture. *Agriculture, 10*(10), 444.

Varsani, A., Navas-Castillo, J., Moriones, E., Hernandez-Zepeda, C., Idris, A., Brown, J. K., Murilo Zerbini, F., & Martin, D. P. (2014). Establishment of three new genera in the family Geminiviridae: Becurtovirus, Eragrovirus, and Turncurtovirus. *Archives of Virology, 159*, 2193–2203.

Vasudeva, R., & Raj, J. S. (1948). A leaf-curl disease of tomato. *Phytopathology, 38*, 364–369.

Venkataravanappa, V., Lakshminarayana Reddy, C. N., Jalali, S., & Krishna Reddy, M. (2015). Association of Tomato leaf curl New Delhi virus DNA-B with bhendi yellow vein mosaic virus in okra showing yellow vein mosaic disease symptoms. *Acta Virologica, 59*, 125–139.

Verma, R. K., Mishra, M., Marwal, A., Akhtar, A., Manchev, N., & Petrov, R. K. (2024). CRISPR/Cas9 protection system for Chilli Leaf Curl Virus. *Acta Microbiologica Bulgarica, 40*(1), 122–115.

Vieira, R. L., da Silva, A. L., Zaffari, G. R., Steinmacher, D. A., Fraga, H. P. H., & Guerra, M. P. (2015). Efficient elimination of virus complex from garlic (*Allium sativum L.*) by cryotherapy of shoot tips. *Acta Physiologiae Plantarum, 37*, 1733. Available from http://link.springer.com/10.1007/s11738-014-1733-3

Vinodkumar, S., Senthilraja, C., Vanthana, M., & Nakkeeran, S. (2018). Novel biocontrol strategies with multifaceted approach for the management of plant diseases.

Wang, Z. L. (2000). Characterizing the structure and properties of individual wire-like nanoentities. *Advanced Materials, 12*, 1295–1298.

Wang, Q., Ebbs, S. D., Chen, Y., & Ma, X. (2013). Trans-generational impact of cerium oxide nanoparticles on tomato plants. *Metallomics, 5*(6), 753–759.

Wang, Q., et al. (2014). Fluorescent carbon dots as an efficient siRNA nanocarrier for its interference therapy in gastric cancer cells. *Journal of Nanobiotechnology, 12*, 1–12.

Wang, Y., Sun, C., Xu, C., Wang, Z., Zhao, M., Wang, C., et al. (2016). Preliminary experiments on nano-silver against tobacco mosaic virus and its mechanism. *Tobacco Science and Technology, 49*(1), 22–30.

Wang, X., Xie, H., Wang, P., & Yin, H. (2023). Nanoparticles in plants: Uptake, transport, and physiological activity in leaf and root. *Materials, 16*(8), 3097.

Waterman, K. C. (2011). The application of the accelerated stability assessment program (ASAP) to quality by design (QbD) for drug product stability. *AAPS PharmSciTech, 12*, 932–937.

Worrall, E. A., et al. (2018). Nanotechnology for plant disease management. *Agronomy, 8*(12), 285.

Xiao, Y., Ahmad, T., Belwal, T., Aadil, R. M., Siddique, M., Pang, L., & Xu, Y. (2023). A review on protein-based nanocarriers for polyphenols: Interaction and stabilization mechanisms. *Food Innovation and Advances, 2*(3), 193–202.

Xun, H., Ma, X., Chen, J., Yang, Z., Liu, B., Gao, X., et al. (2017). Zinc oxide nanoparticle exposure triggers different gene expression patterns in maize shoots and roots. *Environmental Pollution, 229*, 479–488.

Yan, A., & Chen, Z. (2019). Impacts of silver nanoparticles on plants: A focus on the phytotoxicity and underlying mechanism. *International Journal of Molecular Sciences, 20*(5), 1003.

Yasamineh, S., Kalajahi, H. G., Yasamineh, P., Gholizadeh, O., Youshanlouei, H. R., Matloub, S. K., et al. (2022). Spotlight on therapeutic efficiency of mesenchymal stem cells in viral infections with a focus on COVID-19. *Stem Cell Research & Therapy, 13*(1), 257.

Yin, A. Y., Xu, J., Yang, C. S., & Hsu, S. H. (2023a). Antiviral and antibacterial sulfated polysaccharide-chitosan nanocomposite particles as a drug carrier. *Molecules, 28*(5), 2105.

Yin, Y., Zhang, J., Fan, R., Zhu, K., Jiang, X., Ji, C., et al. (2023b). Terbium-functionalized silver-based metal-organic frameworks for efficient antibacterial and simultaneous monitoring of bacterial spores. *Journal of Hazardous Materials, 446*, 130753.

Youssef, M. S., & Elamawi, R. M. (2020). Evaluation of phytotoxicity, cytotoxicity, and genotoxicity of ZnO nanoparticles in *Vicia faba*. *Environmental Science and Pollution Research, 27*(16), 18972–18984.

Yu, B. G., Okano, T., Kataoka, K., & Kwon, G. (1998). Polymeric micelles for drug delivery: Solubilization and haemolytic activity of amphotericin B. *Journal of Controlled Release, 53*, 131–136.

Zaidi, S. S. A., Iqbal, Z., Amin, I., & Mansoor, S. (2015). First report of Tomato leaf curl Gujarat virus, a bipartite begomovirus on cotton showing leaf curl symptoms in Pakistan. *Plant Disease, 99*, 1655–1656.

Zaidi, S. S. A., Shafiq, M., Amin, I., Scheffler, B. E., Scheffler, J. A., Briddon, R. W., & Mansoor, S. (2016). Frequent occurrence of tomato leaf curl New Delhi virus in cotton leaf curl disease affected cotton in Pakistan. *PLoS One, 11*, e0155520.

Zerbini, F. M., Briddon, R. W., Idris, A., Martin, D. P., Moriones, E., Navas-Castillo, J., et al. (2017). ICTV virus taxonomy profile: Geminiviridae. *Journal of General Virology, 98*(2), 131–133.

Zhang, W., Ye, G., Liao, D., Chen, X., Lu, C., Nezamzadeh-Ejhieh, A., et al. (2022). Recent advances of silver-based coordination polymers on antibacterial applications. *Molecules, 27*(21), 7166.

Zhang, W., Ahari, H., Zhang, Z., & Jafari, S. M. (2023). Role of silica (SiO2) nano/micro-particles in the functionality of degradable packaging films/coatings and their application in food preservation. *Trends in Food Science & Technology, 133*, 75–86.

Zhou, Y. C., Noussourou, M., Kon, T., Rojas, M., Jiang, H., Chen, L. F., Gamby, K., Foster, R., & Gilbertson, R. L. (2008). Evidence of local evolution of tomato infecting begomovirus species in West Africa: Characterization of tomato leaf curl Mali virus and tomato yellow leaf crumple virus from Mali. *Archives of Virology, 153*, 693–706.

Employing CRISPR-Cas9 Genome Editing Tool Against Tomato Leaf Curl New Delhi Virus: Prospects and Retrospect

11

Sanskriti Shiwani, Mobassareen, Auroshikha Mohanty, Ravi Ranjan Kumar, Mohammad Ansar, Kumari Shashi Bala, Chandan Kishore, and Tushar Ranjan

Abstract

Tomato leaf curl New Delhi virus (ToLCNDV) is one of the most devastating viruses affecting our tomato crops. Besides the tomato crops, this virus has also been seen to infect other crops of the *Solanaceae* family and other similar families like *Fabaceae* and *Malvaceae*. Many traditional methods, along with the use of heavy pesticides, seem to be ineffective in combating this particular virus. In this book chapter, CRISPR/Cas9 is being used as a powerful technology against this deadly virus. CRISPR/Cas9 has evolved as a robust genome editing tool, following zinc finger nucleases and transcription activator-like effector nucleases. The mechanism of CRISPR/Cas9 follows three basic steps, which include target search and recognition, target cleavage, and the repair of double-strand breaks (homology-directed repair and nonhomologous end joining). This book chapter highlights the effective use of CRISPR/Cas9 against the *Tomato leaf curl*

S. Shiwani · Mobassareen · K. S. Bala · T. Ranjan (✉)
Department of Molecular Biology and Genetic Engineering, Bihar Agricultural University, Sabour, Bhagalpur, Bihar, India

A. Mohanty
Department of Molecular Biology and Biotechnology, Acharya Narendra Dev University of Agriculture and Technology, Ayodhya, Uttar Pradesh, India

R. R. Kumar
Department of Molecular Biology and Genetic Engineering, Bhola Paswan Shastri Agricultural College, Purnea, Bihar Agricultural University, Sabour, Bhagalpur, Bihar, India

M. Ansar
Department of Plant Pathology, Bihar Agricultural University, Sabour, Bhagalpur, Bihar, India

C. Kishore
Department of Plant Breeding and Genetics, Bihar Agricultural University, Sabour, Bhagalpur, Bihar, India

R. Kumar et al. (eds.), *Tomato Leaf Curl New Delhi Virus (ToLCNDV)*,
https://doi.org/10.1007/978-981-95-2852-3_11

New Delhi virus. The last section of this chapter also includes the limitations of the system along with the strategies used in overcoming them. Besides this, the use of prime editing and Cas9 nuclease as an advanced alternative to CRISPR/Cas9 is also being discussed.

Keywords

Genome editing CRISPR-Cas9 · ToLCNDV · Prime editing

1 Introduction

In 1995, the *Tomato leaf curl New Delhi virus* (ToLCNDV) emerged as a deadly pathogen affecting an extensive range of solanaceous crops in India. This pathogen belongs to the genus *Begomovirus* and family *Geminiviridae*. Its genome consists of a bipartite single-stranded DNA (Janssen et al., 2022). ToLCNDV is one of the most significant among 13 different virus species that are prevalent in India. ToLCNDV is also known to infect plants of various cultivars, including those from the *Malvaceae*, *Solanaceae*, and *Euphorbiaceae* families, as well as almost 43 other plant species besides the tomato, which is considered the main cultivated host of ToLCNDV. Thickening of the vein, curling of the leaf, darkening at the margins of the leaf, shortening of the leaf area and internodes, and stunting are the symptoms of ToLCNDV infection (Zaidi et al., 2017). ToLCNDV may exhibit varying degrees of pathogenicity based on the type of host plant. Consequently, a relative study of ToLCNDV pathogenesis conducted on a variety of solanaceous family members, such as tomato, *Nicotiana benthamiana*, *Nicotiana tabacum*, and pepper (*C. annuum*), revealed that the ToLCNDV strain under investigation exhibited differential host adaptation (Jeevalatha et al., 2021, 2023). Among these, *N. benthamiana* was the most susceptible, whereas the isolate was incapable of infecting pepper. In *N. tabacum* and tomato, the virus caused noticeable symptoms and accumulated at significantly high levels. These findings suggest that the degree of ToLCNDV accumulation and the intensity of its symptoms are influenced by the varying degree of expression caused by the host plants in their defense, which differs across different host plants. In tomatoes, symptoms of ToLCNDV infection include leaf curling in both upward and downward directions, veins becoming translucent, yellow spots on the leaves, and leaf crinkling. When infection progresses, plants become stunted with reduced internode length and exhibit poor fruit setting, often resulting in total crop loss. In cucurbits, ToLCNDV is responsible for leaf distortion and mosaic diseases. In natural conditions, ToLCNDV is vectored by *Bemisia tabaci*, a whitefly, and can also be transmitted through different cryptic species within the *B. tabaci* complex, contributing to its widespread spread. Early studies explored the transmission characteristics of ToLCNDV through its whitefly vector, while subsequent research demonstrated that the virus is transmitted through a persistent, circulative mechanism. One whitefly may spread the virus, and females are better at carrying it than males. The virus can be acquired within 30 min, and longer feeding durations further increase transmission efficiency. After 24 h of feeding, both male and female whiteflies can retain the virus and remain infectious for extended periods. Research suggests close cellular and molecular interactions between genes and proteins of *Begomovirus* with those of whitefly. Key protein receptors in the gut of *Bemisia tabaci*, a whitefly, have some protein receptors present in its stomach, which play a crucial role in facilitating the virus's entry into the insect's hemolymph (Moriones et al., 2017; Naga et al., 2021; Modgill et al., 2025).

As ToLCNDV is a type of *Begomovirus*, which has a bipartite genome having two different genomic fragments that are DNA-B and DNA-A, each encoding multiple genes responsible for producing essential proteins with diverse functions (detailed and summarized in Table 11.1). DNA-A of the ToLCNDV genome encodes several key genes that are responsible for viral replication and symptom development. The *AV1* and *AV2* genes, crucial for forming the coat or capsid protein of the virus and its proper functioning, are encoded in the sense orientation (Kumar et al., 2021; Sarkar et al., 2024). In contrast, DNA-A encodes genes such as AC2, AC1, AC4, and AC3 in a complementary orientation. The genes encoded by the DNA-A fragment of the genome contribute to and aid in various molecular functions, including DNA replication, protein formation, and alterations of host defensive mechanisms, all to enhance virus propagation and infection in the host plant (Jeevalatha et al., 2021, 2023).

On the other hand, the DNA-B fragment of the genome encodes genes that are solely responsible for the virus's movement throughout the host plant. DNA-A fragment encodes two genes, BV1 and BC1. These genes have been found in the sense orientation and complementary orientation of the virion, respectively. The BV1

Table 11.1 Exhibiting the genomic composition of the ToLCNDV

Genome component	Name of gene	Encoding protein	Function of the protein
DNA-A	AC1	Replication initiator protein	It attaches to the origin of replication of viral species and recruits host factors to initiate DNA replication of the virus
	AC2	Transcriptional activator protein	Activates transcription of the complementary-sense genes and helps in enhancing the virulence of the viruses. (Moriones et al., 2017)
	AC3	Replication enhancer protein	Cooperates with Rep to improve the effectiveness of replication (Moriones et al., 2017)
	AC4	AC4 protein	Enhances virulence by countering host defenses
	AV1	Capsid protein or coat protein (CP)	Essential for virus particle assembly
	AV2	Pre-coat protein	Assist coat protein in viral encapsulation
DNA-B	BC1	Movement protein (MP)	Interacts with host factors to allow the propagation of viral genome across the plasmodesmata
	BV1	Nuclear shuttle protein (NSP)	Aids in the viral genome's (DNA) transit between the cytoplasm and nucleus. (Moriones et al., 2017)

gene plays a vital role in aiding the virus to move across the different cells of the host plant, whereas the BC1 gene is responsible for viral movement within the host plant only. These genes work in coordination to help the virus produce infection and propagation throughout the plant. There is a common intervening location with the sequence (TAATATT/AC) that is identical on both DNA-A and DNA-B, and it separates the genes in the sense and complementary sense orientations of the virion, respectively (Zaidi et al., 2017). The genomic organization of ToLCNDV is highly regulated and efficient because of the overlapping and embedded ORFs. Because of that kind of organization, the ToLCNDV can manage the production and activity of its eight viral proteins while making the best use of its small genome's ability to code. The multifunctionality of these proteins is responsible for the ability of the virus to persist for a longer period as well as its adaptability in the host plant (Moriones et al., 2017).

To investigate the pathogenicity of ToLCNDV in various tomato varieties, scientists conducted an experiment involving the exchange of genome fragments from two distinct strains isolated from Italy and Pakistan. After that, the formed hybrid varieties were introduced into various tomato varieties. Since the genomic component DNA-A has been demonstrated to be the primary factor driving ToLCNDV infectivity in tomatoes, a significant number of mutant clones are produced by switching the common intergenic region along with the 6 ORFs between the two ToLCNDV strains. Using an experiment, it was discovered that the capsid protein-containing clone was the main source of virulency of ToLCNDV in tomatoes. In the coat protein sequence, the amino acid residue at position 143 was shown to have a considerable impact on the infectivity of tomato ToLCNDV-ES when site-directed mutagenesis of CP was employed to evaluate the functions of amino acid residues in ToLCNDV infection. Using yeast two-hybrid screening, the relationship between the Coat protein of each ToLCNDV isolate and the proteins of the tomato host plant was examined, revealing that the CPs of two ToLCNDV isolates interacted with the host plant protein Ring-finger protein 44-like to varying degrees. This interaction shows that different strains pf ToLCNDV can show varying levels of pathogenicity in tomatoes. This preliminary study examines the pathogenic features of two ToLCNDV strains in a single host plant to identify the primary factor responsible for viral infectivity (Vo et al., 2023). When the scientists investigated the interaction between the plants and the virus in various tomato host plants, they found that several significant host genes play a crucial role in the plant's defensive system against ToLCNDV. The three most important resistant Ty genes are Ty-5, Ty-1, and Ty-3. The Ty-3 and Ty-1 genes are allelic and encode an RNA-dependent RNA polymerase (RDRP), which initiates the RNA interference response in the host plant against viral infections. Besides these genes, other resistant genes, such as Ty4 and Ty-2 from *S. chilense* and S. *habrochaites*, respectively, have also been identified in various resistant sources of host plants. While the majority of known resistance genes prevent viral accumulation to varying degrees, they do not prevent viral replication. Using some of the resistance genes together to build broad-spectrum, long-lasting resistance, a technique known as pyramiding would make sense because these various resistance genes give diverse types of resistance. In tomatoes, it has been demonstrated that pyramiding the resistant genes Ty-3 and Ty-2 increases resistance against ToLCNDV. Recently, a few selected specific resistant genes, also known as R genes, were examined thoroughly to investigate the underlying mechanisms of the biological defense reactions against ToLCNDV in many host plants, such as *N. benthamiana, C. annuum, S. lycopersicum*, and *N. tabacum*. ToLCNDV significantly alters the expression and function of genes critical for gene silencing, like SGS3, RDR6, and AGO1, as well as key elements of the defense system of plants, including lipid transfer protein (LTP), leucine-rich repeat protein, and NSP-interacting kinase (NIK). These disruptions compromise the plant's antiviral defenses, potentially facilitating viral replication, systemic movement, and pathogenicity, thereby exacerbating the impact of infection (Zaidi et al., 2017). Traditional ToLCNDV management strategies, such as physical barriers, use of pesticides, and natural predators, are limited by things like whitefly movement and virus evolution. Although some progress has been achieved using RNA interference (RNAi) and resistance genes from wild plants, these techniques do not offer total protection (Zaidi et al., 2017). Comparing the miRNA composition of healthy and diseased tomato plants has revealed new miRNAs that may contribute to resistance, potentially aiding in the development of virus-resistant tactics, albeit to a certain extent (Pradhan et al., 2015; Kumar & Naqvi, 2016). To overcome the limitations of these techniques, a new technique, CRISPR/Cas9, has been introduced. This technique targets the essential viral genes, such as CP and Rep genes, to provide complete and precise protection of host plants against ToLCNDV.

2 Use of the CRISPR/Cas9 System as a Tool for Genome Editing

CRISPR/Cas is a natural adaptive immune system used by archaea and bacteria as a defense mechanism against invading plasmids and viruses. These defensive mechanisms use short RNAs to identify and silence invading foreign nucleic acids based on their specific sequence (Jinek et al., 2012). A group of CRISPR-associated genes (Cas) that produce proteins linked to the immune response is necessary for CRISPR to work. The Cas genes are always present near CRISPR. There are three phases through which CRISPR-Cas technology works. The first one is adaptation, the second is biogenesis, and the third is interference. In the first phase, the Cas proteins record a molecular memory of previous invaders by capturing brief segments of foreign nucleic acids (known as protospacers) and encoding them as "spacers" into the CRISPR locus. In Biogenesis, pre-CRISPR RNA (pre-crRNA) is produced through the transcription of the CRISPR locus and then transformed into the mature form of crRNA. Every mature crRNA contains a transcribed sequence known as a spacer that is joined to a partial repeat sequence. During the interference phase, a complex forms between crRNA and the protospacer of the foreign DNA through base pairing. Finally, the invading DNA from the crRNA spacer sequence is cleaved by Cas proteins at a position corresponding to the protospacer (Rath et al., 2015).

The CRISPR/Cas system is highly specific, enabling precise cutting at a precise location on the genome through the use of endonucleases guided by RNA. This process generates double-strand breaks in the DNA, facilitated by Cas proteins that are directed to specific DNA sequences under the guidance of guide RNA molecules. This highly specific nature of this system allows it to be used to target almost any genomic sequences with a high degree of precision, specificity, and efficiency. This straightforward design has driven the evolution of the CRISPR/Cas system into a powerful genetic tool, enabling researchers to rapidly target specific genes or regions by altering the RNA sequence. This results in the formation of an entirely new genome, allowing for precise targeting of distinct genes or areas. In 2012, it was declared a genome editing tool, and its popularity for precise genome and gene editing has increased steadily since then. Its extremely high ability to produce highly specific genetic changes leads to its vast applications in the field of transformed molecular biology, crop enhancement, gene therapy, and various other medicinal treatments. Nowadays, its potential application in the research area of gene therapy and crop enhancement has become a great matter of study and extensive research (Li et al., 2021). A recent study classifies the CRISPR/Cas system into two classes, comprising six types and 33 subtypes. This categorization is based on various factors, like phylogenetic analysis, homology of the sequences and specific structural and functional features of the components, which help to differentiate the multiple systems within the broader CRISPR/Cas framework (Zhang et al., 2021).

Class I systems can be categorized into three types and 16 subtypes. Class I systems often contain several Cas enzymes working together to target DNA. This system is rarely employed as a genome editing tool because of the difficulty of designing and delivering many Cas restriction endonuclease enzymes into the cell. Comparatively, Class II systems are easier to modify because they usually only need one big, multipurpose Cas enzyme. Consequently, a great deal of study has been done to develop contemporary Class II systems and gain further insights. Like Class I, there are presently three types and seventeen subtypes described in Class II. Type II systems have been the most thoroughly researched. Since the discovery of Cas9, the endonuclease is currently most frequently being utilized in the CRISPR/Cas system (Li et al., 2021). Before CRISPR/Cas9, scientists primarily used two techniques for gene editing that involved restriction enzymes: zinc finger nucleases (ZFN) and transcription activator-like effector nucleases (TALENs). ZFNs have a zinc finger DNA-binding domain that attaches to a specific DNA sequence, paired with a domain of restriction endonuclease that induces a cut in the DNA at its target site. Like ZFNs, TALENs also contain a restriction domain and a DNA-binding domain. Still, the DNA-binding domain of TALENs allows them to target a widespread variety of sequences compared to ZFNs. Despite their utility, both methods were challenging to engineer, costly, and time-consuming. For years, researchers have aimed to evolve a more versatile and reliable gene editing tool for living cells. After uncovering the mechanism of CRISPR in prokaryotes, scientists discovered that it may be used in humans, plants, and other living things (Asmamaw & Zawdie, 2021).

2.1 CRISPR/Cas9: Components and Working Mechanism

The two most essential components of the CRISPR-Cas9 system are CRISPR-associated proteins (Cas9) and a single-stranded guide RNA (sgRNA). Cas9 protein was the first protein to be used in genome editing, and it was isolated from a bacterial species named Streptococcus pyogenes (SpCas-9). This protein is a type of DNA endonuclease having multiple domains and is made up of 1368 amino acid residues. Because of its ability to create a double-strand break in target DNA, it is also known as a genetic scissor. The Cas9 protein is made up of two lobes. The first lobe is called the nuclease lobe (NUC), and the other lobe is called the recognition lobe (REC). The first lobe, i.e., the NUC lobe, is made up of the RuvC domain, the PAM interaction domain, and the HNH domain. The primary function of the PAM (Protospacer Adjacent Motif) interaction domain is to provide specificity to the PAM. It initiates the binding process on the targeted DNA, while the other two domains are involved in cutting the complementary and noncomplementary strands of DNA. The NUC lobe, another lobe, comprises two domains, REC1 and REC2, which attach to the DNA to be targeted.

The guide RNA consists of two parts: tracrRNA and crRNA. The guide RNA consists of two parts: trans-activating CRISPR RNA & CRISPR RNA. The crRNA (CRISPR RNA) contains approximately 18 to 20 bp, utilizing complementary base pairing to guide the entire CRISPR-Cas9 complex to the targeted DNA. TracrRNA helps Cas9 bind by forming a structural scaffold. In genome editing, the crRNA and tracrRNA are fused to form sgRNA, which can cut any genomic sequences. In the case of bacteria, the gRNA targets explicitly the virus's DNA (Asmamaw & Zawdie, 2021; Jinek et al., 2012). By making some changes in the sequence of amino acid residues, located in HNH and RuvC domains, which have the capacity of undergoing catalytic reactions, it can be concluded that the targeting and cleavage activity of the CRISPR Cas9 system can be separated from each other. This offers a flexible framework to use this system for different uses beyond conventional genome editing, including RNA-targeting treatments, transcriptional control, and epigenetic alteration (Jiang & Doudna, 2017). Nowadays, the CRISPR-Cas9 System is primarily used to develop virus-resistant plants (Table 11.2). CRISPR/Cas9 genome editing involves three primary processes: target sequence recognition, target cleavage, and cleaved DNA repair (Fig. 11.2). The single guide RNA facilitates the Cas9 protein's access to the targeted DNA sequence, where the crRNA utilizes base pairing to identify it. Cas9 becomes active only after its association with the sgRNA. Double-strand breaks are specifically induced through the Cas9 endonuclease at the location 3 bp ahead of the PAM sequence. The Protospacer Adjacent Motif contains a conserved gene sequence having a length of around 3 to 5 bp, and is located downstream of the cut location. The common 5′-NGG-3′ sequence is recognized by Cas9 (Zhang et al., 2021).

Table 11.2 Use of CRISPR/Cas9 as a genome editing tool to produce virus resistant host plants

S. No.	Host plant species	Symptoms observed	Targeted viral species	Targeted gene region	A study done by different scientists
1	Cucumber (*Cucumis sativus*)	Yellowing of the vein	*Potyviruses*	elF4E	Chandrasekaran et al. (2016)
2	*Nicotiana benthamiana*	Curling of the leaf	TYLCV	Gene encoding the capsid protein	Ali et al. (2015)
3	Arabidopsis (*Arabidopsis thaliana*)	Mosaic pattern of leaves	*Turnip mosaic virus*	elF(iso)4E	Pyott et al. (2016)
4	Banana (*Musa* spp.)	Yellow streaks that gradually turn brown on the leaves	*Banana streak virus*	ORF1, 2, 3	Tripathi et al. (2019)
5	Mexican cotton (*Gossypium hirsutum*)	Leaf curling	*Begomovirus*	CLCuD rep and IR regions	Iqbal et al. (2016)
6	Tomato (*Solanum lycopersicum*)	Mild mosaic symptoms on leaves	Potato virus X	SlDCL2a and SlDCL2b	Wan et al. (2020)
7	Rice (*Oryza sativa*)	Discoloration, stunting, and a reduced no. of tillers as well as grains	*Rice tungro spherical virus*	eIF4G	Macovei et al. (2018)
8	*Solanum lycopersicum* (tomato)	Chlorosis, curling, distortion, and dwarfing of the leaves, fruit, and even the whole plant itself	*Pepper mottle virus*	eIF4E1	Yoon et al. (2020)
9	Wheat (*Triticum aestivum*)	Powdery mildew disease	*Blumeria graminis f.* sp. *tritici*	TaMLO-A1	Wang et al. (2014)
10	Cassava	Chlorotic mosaic of leaf (S-shaped)	*African cassava mosaic virus*	AC2 and AC3 (TrAp/REn)	Mehta et al. (2019)
11	*Arabidopsis thaliana*	Leaf crinkles and chlorotic spots (yellow spots)	*Clover yellow vein virus(potyvirus)*	eIF4E gene	Bastet et al. (2019)

(continued)

Table 11.2 (continued)

S. No.	Host plant species	Symptoms observed	Targeted viral species	Targeted gene region	A study done by different scientists
12	*Morchella esculenta*	Brown steak lesions, infected leaf shows feathery chlorosis associated with secondary veins	*Cassava brown streak virus*	Isoforms of eIF4E-nCBP-2 & nCBP-1	Gomez et al. (2019)
13	*N. benthamiana*	Leaf curling, veinal swelling, and distorted growth	*Beet severe curly top virus*	Coat protein and replication initiator protein	Ji et al. (2018)
14	*S. tuberosum*	Mild yellow mottling on leaves to tuber necrosis	*Potyviruses* PVY	Coilin gene	Makhotenko et al. (2019)
15	Barley (*Hordeum vulgare* L.)	Leaves turn yellow, stunted growth, and reduced spike number	*Wheat dwarf virus*	MP, CP, Rep, RepA, and LIR	Kis et al. (2019)
16	*N. benthamiana*	Leaf-like structures (enations) form along leaf veins	*Cotton leaf curl virus* and *betasatellite*	Rep	Khan et al. (2020)
17	*N. benthamiana*	Leaves turn a pale yellow color and curl downward	*Bean yellow dwarf virus*	IR gene	Makarova et al. (2018)

2.2 Target Search and Recognition

The Cas9 proteins combine with guide RNA to form a complex that attaches to a site complementary to the targeted DNA. Besides this, the complementary base pairing. PAM sequence is another set of conserved sequences present near the target sites, which aids in the easier identification of the sequence to be targeted and its precise cutting. The PAM sequence is necessary for differentiating the self-sequences from the nonself ones. A single mutation in the PAM can lower the host's total adaptive immunity and impair the cleavage activity of the Cas9 protein. TheSpyCas9 is the most frequently used Cas9 protein, and its PAM sequence consists of 5′-NGG-3′, in which the letter "N" denotes any random nucleotide. Cas9 starts to search the target by looking for the right PAM sequence, followed by a check for guide RNA complementarity in the surrounding DNA, according to single-molecule tests. Three-dimensional collisions are used for target identification, indicating that the Cas9 protein binds only to DNA with a proper PAM sequence. There should be a high degree of complementarity between the target DNA and the guide RNA, as it increases the binding affinity between them and gives this system enough time to work efficiently. The Cas 9 protein initiates strand separation of the target DNA at the location near the PAM sequence, to form a hybrid of RNA and DNA, along with single-stranded DNA forming a "R" loop-like structure. After this process, the process of strand invasion initiates. The method of cleavage depends on the high level of complementarity between the seed region of the single guide RNA and the DNA to be targeted. At the same time, some sort of non-complementarity is required in the non-seed region of guide RNA for ensuring a high level of target specificity (Jiang & Doudna, 2017).

2.3 Target Cleavage

The Cas9 protein binds to DNA at the location where the hybrid structure of guide RNA and target DNA is present. Cas9 becomes activated by this contact and starts its nuclease activity. Two separate nuclease domains found in Cas9 unite to break the DNA. There are three split RuvC motifs in the first domain, the RuvC domain. This domain is responsible for creating a nick in the single strand of DNA. In contrast, the HNH domain, located in the middle region of the Cas9 protein, is responsible for cutting the complementary strand of DNA. Both domains together create DSBs in the target DNA, which have blunt ends. The repair machinery of the cell is now activated, which fixes the double-strand break and results in genome editing and forced mutation (Jiang & Doudna, 2017).

2.4 Double-Strand Break Repair Mechanism

The two most important methods of DSB are the nonhomologous end joining (NHEJ) method and the homology-directed repair (HDR) method. The NHEJ method of DNA repair involves the use of enzymes for joining the DNA fragments without the need for template DNA. This mechanism is dynamic in almost all stages of the cell cycle. NHEJ is the most popular and efficient method of DNA repair. Still, its biggest drawback is that it is error-prone, causing small insertions or deletions (indels) at the damaged location, which may cause frameshift mutations, resulting in nonfunctional proteins. HDR ensures precise, site-specific alterations by directing repair using a homologous DNA template, such as single-stranded oligodeoxynucleotides. Although HDR is incredibly precise, it can only be applied in a few stages of the cell cycle, such as the late S and G2 stages. As the name itself suggests, the HDR method of DSB repairing needs a significant proportion of exogenous donor DNA templates comprising the desired sequence. This donor DNA facilitates targeted gene insertion or replacement by aligning with the DSB site (Asmamaw & Zawdie, 2021; Talakayala et al., 2022). NHEJ is the preferred repair machinery due to its high proficiency and ability to function throughout the cell cycle, though it is error-prone and often introduces indels. HDR, while precise and

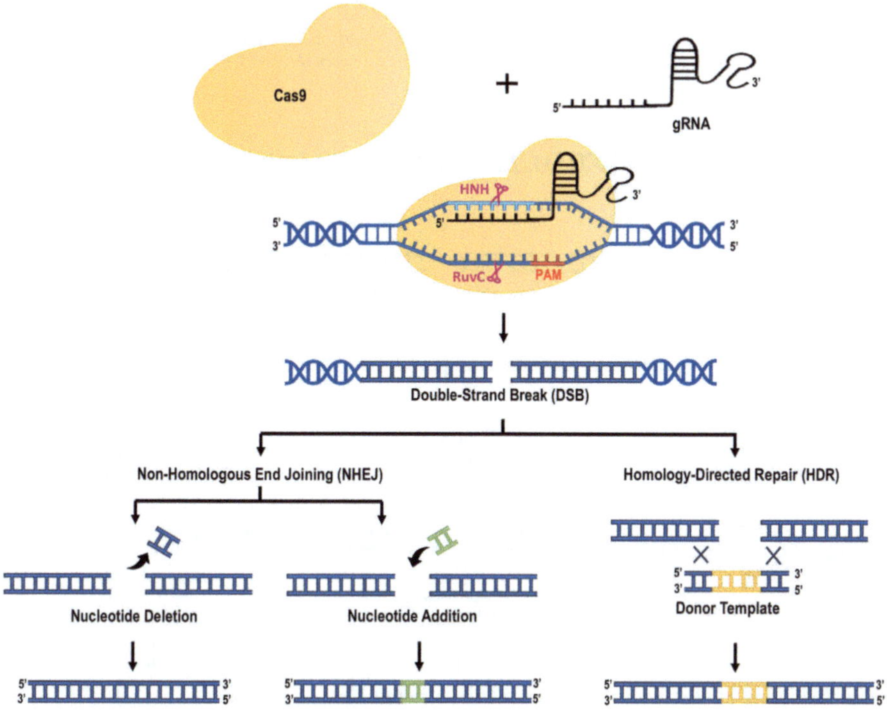

Fig. 11.1 The overall mechanism of CRISPR/Cas 9 system

template-dependent, has limited efficiency and is active only in specific cell cycle phases. Both mechanisms are used with CRISPR for improvement in crops, but HDR's low proficiency in plants poses challenges for its application in agricultural biotechnology (Ahmad et al., 2021) (Fig. 11.1).

3 SDN Systems: Precise Genome Editing Techniques

Site-directed nuclease (SDN) techniques involve creating DSBs at specific locations on the genome using customized SDNs. These nucleases recognize a certain set of conserved sequences in DNA and cleave the DNA either within the target sequence or in its vicinity. DNA target recognition is facilitated either by the protein component of the nuclease (in protein-directed SDNs) or by aRNA molecule that guides the specific nuclease to the target site (in RNA-directed SDNs) (Podevin et al., 2013; Naegeli et al., 2020). Transcription activator-like effector nucleases (TALENs), zinc finger nucleases (ZFNs), and CRISPR/Cas technology are examples of SDN approaches. SDNs lessen the possibility of random mutagenesis by precisely altering certain regions of a genome. Compared to plant breeding and genetic engineering, SDNs offer previously unheard-of control over precise genome alterations at a lower cost and with higher efficiency. SDNs are categorized as SDN1, SDN2, and SDN3 (given in Fig. 11.2) based on the results of DSB repair and

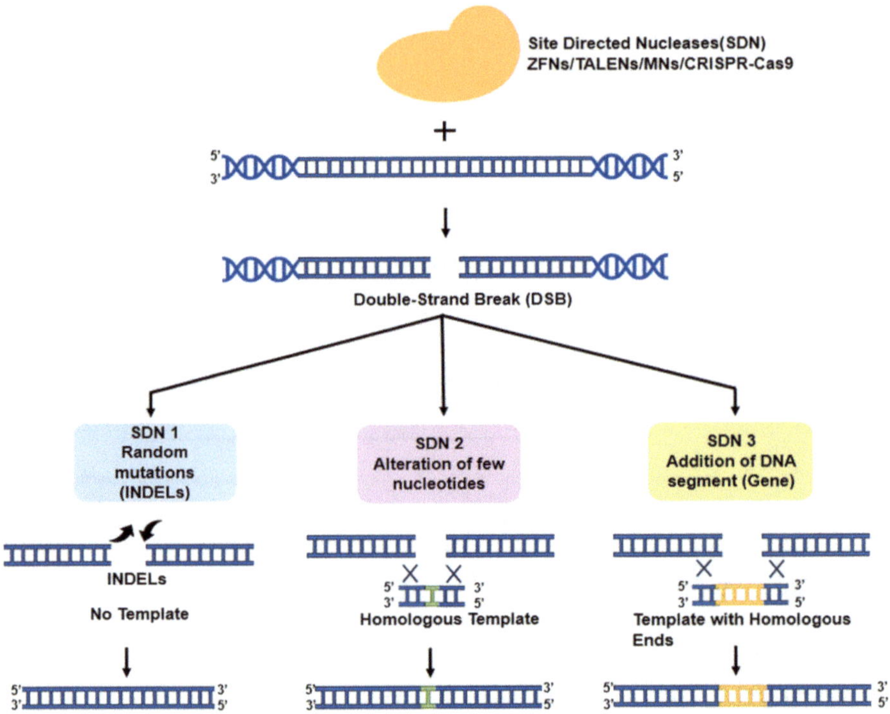

Fig. 11.2 Mechanisms of SDN1, SDN2, and SDN3: SDN1 induces small indels via NHEJ, SDN2 achieves precise edits with HDR and a repair template, while SDN3 enables gene replacement or insertion, using HDR and a donor template

genomic alteration (Li et al., 2021). In the SDN1 technique, a single DSB is typically used to create site-specific random mutations, which plants primarily repair by nonhomologous end joining. In the SDN-2 technique, an externally provided DNA template, commonly referred to as donor DNA, is used for the repair process, which is carried out by homologous recombination (HR). In addition to having the specific desirable mutations (a few nucleotide substitutions or brief insertions/deletions) in donor DNA, the endogenous gene must have high sequence identity for HR. The goal of SDN-3 approaches is to use donor DNA, which contains the integrated sequence along with overlapping sequences that exhibit similarity to the target locus, in conjunction with the site-directed nucleases to insert lengthy DNA fragments (such as transgenes) at a predetermined locus (Podevin et al., 2013).

4 CRISPR-Cas9 Is a Potent Weapon Against the ToLCNDV

The CRISPR-Cas9 system was successfully engineered in tomato plants to target the ToLCV New Delhi strains. This system primarily targeted the viral capsid protein (CP), which is essential for infection, and demonstrated effective interference in tomato plants across multiple generations. Although CRISPR-Cas9 also affected the replication-associated protein (REP), its impact was less significant compared to its effect on the coat protein. *Agrobacterium* strains with ToLCV New Delhi variations were injected into thirty-day-old tomato seedlings, while a control set of plants received no treatment. After a few days, the genomic RNA of the infected plants was extracted using an RNA extraction kit, and then cDNA was formed from this RNA. In that cDNA, an antiviral protein named rgsCaM (regulator of gene silencing calmodulin) was assessed by the researchers. Later, it was found that this particular protein helps in preventing viral infections in plants.

To improve the accuracy of genome editing, the scientists created an inducible CRISPR/Cas9 system. In this inducible system, the single-guide RNAs are driven by the AtU6 promoter, which further increases their activity as an effective genome editing tool. The main aim of this inducible system was to minimize any harmful consequences on the plant while efficiently targeting and stopping the viral reproduction. For further investigation, the single guide RNAs were amplified, digested, and then ligated. The ligated product was then added to *E. coli.* When the RT-PCR data were examined, it was shown that the plants treated with the CRISPR-Cas9 system had a decreased level of viral presence compared to the normal ones. Besides this, various sgRNA samples that were infiltrated also showed lower virus levels. The most innovative aspect of this system was the use of inducible promoters, which led to the activation of Cas9 during infection, resulting in efficient and transient genome editing activity. This method can be used to reduce the risk of off-target effects and Cas9 activation for a more extended period (Shankar & Kumar, 2021). CRISPR-Cas9 can also be used to target the AC4 protein, in addition to the capsid protein and replication-associated protein. According to molecular research, AC4 interacts with the host plant's Argonaute 4 (AGO4) protein to influence its cytosine methylation patterns (Loriato et al., 2020). This relationship implies that using CRISPR-Cas9 to target AC4 could further impair the virus's ability to interfere with the plant's gene-silencing processes, thereby adding another line of defense against viral invasion.

There are some specific phases involved in the CRISPR/Cas9-based genome editing method for ToLCNDV (shown in Fig. 11.3). The first stage consists of creating a vector construct with the Cas9 protein and the appropriate gRNA sequences that target the critical genes of the virus that encode the CP Rep and AC4 proteins. This vector construct is then converted into a strain of *Agrobacterium.* In the second step, explants are taken from the virus-infected host plant and then used for callus induction. The induced callus and the transformed *Agrobacterium* strain are then co-cultured to introduce the *Agrobacterium* strain containing the vector construct into the callus. In this step, the gRNA recognizes the CP and Rep protein of the virus and degrades it through the CRISPR/Cas9 mechanism, which we have already discussed in the previous section of this paper. The next step involves selecting the antibiotic-resistant co-cultured plant sample to isolate the transformed cells that have successfully incorporated the CRISPR/Cas9 vector. The samples are then regenerated through tissue culture technique into a plant by transferring the samples to different media with varying compositions at regular intervals. The plant grown from this process is completely resistant to ToLCNDV.

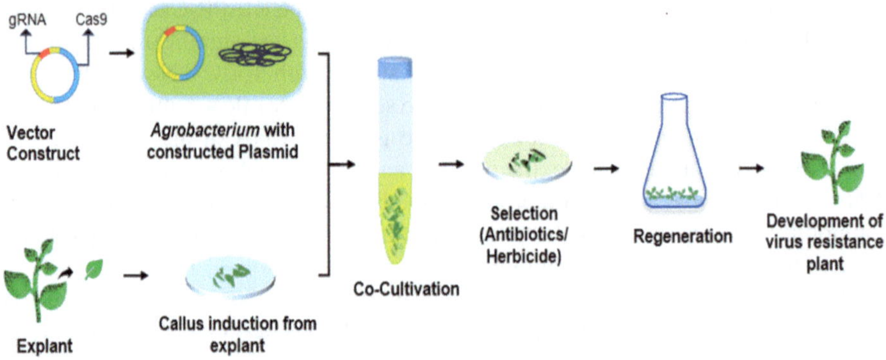

Fig. 11.3 Overall mechanism of the detailed workflow of CRISPR-Cas9-based development of virus-resistant plants: From vector construction and *Agrobacterium* transformation to callus induction, coculturing, antibiotic selection, and the regeneration of plants resistant to ToLCNDV

5 Conclusion and Prospects

In this study, a detailed explanation of the genomic structure of the ToLCNDV, along with the vital functions of the proteins that are linked to it during infection, has been discussed. This chapter also sheds light on how this virus is reducing the country's agricultural yield by expanding its host range, initially starting with the tomato plant. The traditional and conventional methods for stopping ToLCNDV infection, such as using pesticides to control whitefly *Bemisia tabaci* and the RNA interference technique, have been discussed briefly, along with their limitations, including ineffectiveness, low long-term efficacy, and difficulties in achieving long-term virus epidemic control. Considering these drawbacks, we present CRISPR-Cas9 technology as a highly novel, inventive, and promising substitute. It is a valuable tool for fighting ToLCNDV because of its accuracy, versatility, and effectiveness. The mechanisms, components, and functions of this system, focusing on its capability to target and disrupt the viral DNA specifically, are thoroughly discussed in this paper. The last section of this paper emphasizes the in-depth analysis of how the system's RNA-guided methodology permits the highly accurate targeted cleavage of the virus's genome. Further highlighting CRISPR-Cas9's ability to transform plant virology, the study's results also show the step-by-step, precise course of action of CRISPR/Cas9 in combating ToLCNDV. As CRISPR-Cas9 research continues to progress and is optimized for plant systems, this technique has great potential to lessen the harmful effects of ToLCNDV and related viral diseases in agricultural fields. Although CRISPR-Cas9 has given new hope to genome editing, it has some disadvantages, such as random mutations and off-target effects that can harm host plants. To overcome these limitations, researchers have developed various Cas9 variants to enhance its precision and broaden its applications. One crucial advancement involves base editors, exemplified by cytosine and adenine targeting base. For example, CBE leads to the conversion of cytosine into thymine.

In contrast, ABE leads to the conversion of adenine into guanine directly within the DNA sequence, avoiding the requirement for double-strand breaks (DSBs) and offering higher precision in certain scenarios. Another significant development is Cas9 nickases, which produce one strand break (nicks) in DNA, reducing off-target effects compared to traditional Cas9. In this method, two Cas9 nuclease enzymes are used in tandem to target opposite DNA strands (complementary and noncomplementary strands), which leads to the creation of a DSB. Additionally, the chances of off-target mutations can be diminished by utilizing two separate guide RNAs. The other breakthrough technique, the use of prime editors, is a significant advancement since they do not require homology-directed repair (HDR) and enable accurate deletions, insertions, and all 12 potential base-to-base changes. Besides these variants, variants like CRISPR/Cas12a and CRISPR/Cas13 show significant promise in antiviral defense. Cas12a can process multiple guide RNAs for simultaneous targeting of viral genes, while Cas13 targets RNA, making it ideal for RNA-based viruses like ToLCNDV. These variants collectively enhance the flexibility and precision of this system, establishing it as a more powerful tool for genetic research and therapeutic purposes.

References

Ahmad, A., Munawar, N., Khan, Z., Qusmani, A. T., Khan, S. H., Jamil, A., et al. (2021). An outlook on global regulatory landscape for genome-edited crops. *International Journal of Molecular Sciences, 22*(21), 11753.

Ali, Z., Abulfaraj, A., Idris, A., Ali, S., Tashkandi, M., & Mahfouz, M. M. (2015). CRISPR/Cas9-mediated viral interference in plants. *Genome Biology, 16*, 238.

Asmamaw, M., & Zawdie, B. (2021). Mechanism and applications of CRISPR/Cas-9-mediated genome editing. *Biologics: Targets and Therapy*, 353–361.

Bastet, A., Zafirov, D., Giovinazzo, N., Guyon-Debast, A., Nogu'e, F., Robaglia, C., & Gallois, J. L. (2019). Mimicking natural polymorphism in eIF 4E by CRISPR-Cas9 base editing is associated with resistance to potyviruses. *Plant Biotechnology Journal, 17*, 1736–1750.

Chandrasekaran, J., Brumin, M., Wolf, D., Leibman, D., Klap, C., Pearlsman, M., et al. (2016). Development of broad virus resistance in non-transgenic cucumber using CRISPR/Cas9 technology. *Molecular plant pathology., 17*(7), 1140–1153.

EFSA Panel on Genetically Modified Organisms (EFSA GMO Panel), Naegeli, H., Bresson, J. L., Dalmay, T., Dewhurst, I. C., Epstein, M. M., et al. (2020). Applicability of the EFSA opinion on site-directed nucleases type 3 for the safety assessment of plants developed using site-directed nucleases type 1 and 2 and oligonucleotide-directed mutagenesis. *EFSA Journal, 18*(11), e06299.

Gomez, M. A., Lin, Z. D., Moll, T., Chauhan, R. D., Hayden, L., Renninger, K., Beyene, G., Taylor, N. J., Carrington, J. C., Staskawicz, B. J., & Bart, R. S. (2019). Simultaneous CRISPR/Cas9mediated editing of cassava eIF 4E isoforms nCBP-1 and nCBP-2 reduces cassava brown streak disease symptom severity and incidence. *Plant Biotechnology Journal, 17*, 421–434.

Iqbal, Z., Sattar, M. N., & Shafiq, M. (2016). CRISPR/Cas9: A tool to circumscribe cotton leaf curl disease. *Frontiers in Plant Science., 7*, 475.

Janssen, D., Simón, A., Boulares, M., & Ruiz, L. (2022). Host species-dependent transmission of tomato leaf curl New Delhi virus-ES by *Bemisia tabaci*. *Plants., 11*(3), 390.

Jeevalatha, A., Vanishree, G., Siddappa, S., Kumar, R., Kaundal, P., Kumar, A., & Chakrabarti, S. K. (2021). Molecular characterization and infectivity analysis of tomato leaf curl New Delhi virus isolates infecting potato. *3 Biotech, 11*, 203. https://doi.org/10.1007/s13205-021-02752-5

Jeevalatha, A., Siddappa, S., Kumar, R., Tiwari, R. K., Lal, M. K., Sharma, S., Chakrabarty, S. K., & Singh, B. P. (2023). RNA-seq analysis reveals an early defense response to tomato leaf curl New Delhi virus in potato cultivar Kufri Bahar. *Functional & Integrative Genomics, 23*, 215. https://doi.org/10.1007/s10142-023-01138-5

Ji, X., Si, X., Zhang, Y., Zhang, H., Zhang, F., & Gao, C. (2018). Conferring DNA virus resistance with high specificity in plants using virus-inducible genome-editing system. *Genome Biology, 19*, 197.

Jiang, F., & Doudna, J. A. (2017). CRISPR–Cas9 structures and mechanisms. *Annual Review of Biophysics, 46*(1), 505–529.

Jinek, M., Chylinski, K., Fonfara, I., Hauer, M., Doudna, J. A., & Charpentier, E. (2012). A programmable dual-RNA–guided DNA endonuclease in adaptive bacterial immunity. *Science, 337*(6096), 816–821.

Khan, S., Mahmood, M. S., Rahman, S. U., Rizvi, F., & Ahmad, A. (2020). Evaluation of the CRISPR/Cas9 system for the development of resistance against cotton leaf curl virus in model plants. *Plant Protection Science, 56*, 154–162.

Kis, A., Hamar, É., Tholt, G., Bán, R., & Havelda, Z. (2019). Creating highly efficient resistance against wheat dwarf virus in barley by employing CRISPR/Cas9 system. *Plant Biotechnology Journal, 17*, 1004–1006.

Kumar, V., & Naqvi, A. R. (2016). Tomato leaf curl New Delhi virus (ToLCNDV) encoded AC2 associates with host miRNAs by directly interacting with AGO1. *Journal of RNAi & Gene Silencing., 12*.

Kumar, R., Tiwari, R. K., Jeevalatha, A., Sundaresha, S., Shah, M. A., Sharma, S., Sagar, V., Kumar, M., & Chakrabarti, S. K. (2021). Potato apical leaf curl disease: Current status and perspectives on a disease caused by tomato leaf curl New Delhi virus. *Journal of Plant Diseases and Protection, 128*, 897–911. https://doi.org/10.1007/s41348-021-00463-w

Li, C., Brant, E., Budak, H., & Zhang, B. (2021). CRISPR/Cas: A Nobel prize award-winning precise genome editing technology for gene therapy and crop improvement. *Journal of Zhejiang University-SCIENCE B, 22*(4), 253–284.

Loriato, V. A., Martins, L. G., Euclydes, N. C., Reis, P. A., Duarte, C. E., & Fontes, E. P. (2020). Engineering resistance against geminiviruses: A review of suppressed natural defenses and the use of RNAi and the CRISPR/Cas system. *Plant Science, 292*, 110410.

Macovei, A., Sevilla, N. R., Cantos, C., Jonson, G. B., Slamet-Loedin, I., Čermák, T., Voytas, D. F., Choi, I. R., & Chadha-Mohanty, P. (2018). Novel alleles of rice eIF4G generated by CRISPR/Cas9-targeted mutagenesis confer resistance to Rice tungro spherical virus. *Plant Biotechnology Journal, 16*, 1918–1927.

Makarova, S. S., Khromov, A. V., Spechenkova, N. A., Taliansky, M. E., & Kalinina, N. O. (2018). Application of the CRISPR/Cas system for generation of pathogen-resistant plants. *Biochemistry (Moscow), 83*(12), 1552–1562.

Makhotenko, A. V., Khromov, A. V., Snigir, E. A., Makarova, S. S., Makarov, V. V., Suprunova, T. P., Kalinina, N. O., & Taliansky, M. E. (2019). Functional analysis of coilin in virus resistance and stress tolerance of potato Solanum tuberosum using CRISPRCas9 editing. *Doklady Biochemistry and Biophysics, 484*, 88–91.

Mehta, D., Stürchler, A., Anjanappa, R. B., Zaidi, S. S. E. A., Hirsch-Hoffmann, M., Gruissem, W., & Vanderschuren, H. (2019). Linking CRISPR-Cas9 interference in cassava to the evolution of editing-resistant geminiviruses. *Genome Biology., 20*, 1–10.

Modgill, N., Naga, K. C., Firake, D. M., Tiwari, R. K., Shivaramu, S., Kumar, R., Bhatnagar, A., Buckseth, T., Bairwa, A., Maharana, C., Sharma, S., & Singh, B. (2025). Flupyradifurone application: A promising approach to curtail tomato leaf curl New Delhi virus (ToLCNDV) transmission vectored by Bemisia tabaci in potato farming. *Crop Protection*, 187.

Moriones, E., Praveen, S., & Chakraborty, S. (2017). Tomato leaf curl New Delhi virus: An emerging virus complex threatening vegetable and fiber crops. *Viruses, 9*(10), 264.

Naga, K. C., Sundaresha, S., Kumar, R., Tiwari, R. K., Subhash, S., Verma, G., Buckseth, T., Bairwa, A., Sharma, S., Katare, S., & Srivastava, R. M. (2021). A new record of Asia II 5 genetic group of Bemisia tabaci (Gennadius) in the major potato growing areas of India and its relationship with tomato leaf curl New Delhi virus infecting potato. *3 Biotech, 11*, 421. https://doi.org/10.1007/s13205-021-02966-7

Podevin, N., Davies, H. V., Hartung, F., Nogué, F., & Casacuberta, J. M. (2013). Site-directed nucleases: A paradigm shift in predictable, knowledge-based plant breeding. *Trends in Biotechnology, 31*(6), 375–383.

Pradhan, B., Naqvi, A. R., Saraf, S., Mukherjee, S. K., & Dey, N. (2015). Prediction and characterization of tomato leaf curl New Delhi virus (ToLCNDV) responsive novel microRNAs in *Solanum lycopersicum. Virus Research., 195*, 183–195.

Pyott, D. E., Sheehan, E., & Molnar, A. (2016). Engineering of CRISPR/Cas9-mediated potyvirus resistance in transgene-free Arabidopsis plants. *Molecular Plant Pathology., 17*(8), 1276–1288.

Rath, D., Amlinger, L., Rath, A., & Lundgren, M. (2015). The CRISPR-Cas immune system: Biology, mechanisms and applications. *Biochimie, 117*, 119–128.

Sarkar, M., Gupta, D., Singh, O. W., Paul, S., Kumar, R., Mandal, B., & Roy, A. (2024). Identification of pathogenicity determinants in ToLCNDV and their RNAi-based knockdown for disease management in Nicotiana benthamiana and tomato plants. *Frontiers in Microbiology, 15*, 1481523. https://doi.org/10.3389/fmicb.2024.1481523

Shankar, A., & Kumar, A. (2021). In the fight against the tomato leaf curl virus in tomato: Contenders and scenarios.

Talakayala, A., Ankanagari, S., & Garladinne, M. (2022). CRISPR-Cas genome editing system: A versatile tool for developing disease resistant crops. *Plant Stress, 3*, 100056.

Tripathi, J. N., Ntui, V. O., Ron, M., Muiruri, S. K., Britt, A., & Tripathi, L. (2019). CRISPR/Cas9 editing of endogenous banana streak virus in the B genome of Musa spp. overcomes a major challenge in banana breeding. *Communications Biology, 2*, 460.

Vo, T. T., Lal, A., Nattanong, B., Tabassum, M., Qureshi, M. A., Troiano, E., et al. (2023). Coat protein is responsible for tomato leaf curl New Delhi virus pathogenicity in tomato. *Frontiers in Plant Science, 14*, 1206255.

Wan, D. Y., Guo, Y., Cheng, Y., Hu, Y., Xiao, S., Wang, Y., & Wen, Y. Q. (2020). CRISPR/Cas9-mediated mutagenesis of VvMLO3 results in enhanced resistance to powdery mildew in grapevine (Vitis vinifera). *Horticulture Research, 7*, 116.

Wang, Y., Cheng, X., Shan, Q., Zhang, Y., Liu, J., Gao, C., & Qiu, J. L. (2014). Simultaneous editing of three homoeoalleles in hexaploid bread wheat confers heritable resistance to powdery mildew. *Nature Biotechnology., 32*, 94–951. https://doi.org/10.1038/nbt.2969

Yoon, Y. J., Venkatesh, J., Lee, J. H., Kim, J., Lee, H. E., Kim, D. S., & Kang, B. C. (2020). Genome editing of eIF4E1 in tomato confers resistance to pepper mottle virus. *Frontiers in plant science., 11*, 1098.

Zaidi, S. S. E. A., Martin, D. P., Amin, I., Farooq, M., & Mansoor, S. (2017). Tomato leaf curl New Delhi virus: A widespread bipartite begomovirus in the territory of monopartite begomoviruses. *Molecular Plant Pathology., 18*(7), 901–911.

Zhang, D., Zhang, Z., Unver, T., & Zhang, B. (2021). CRISPR/Cas: A powerful tool for gene function study and crop improvement. *Journal of Advanced Research, 29*, 207–221.

Plant Viruses in a Changing Climate: Tomato Leaf Curl New Delhi Virus Dynamics, Evolution, and the Myth and Reality of Beneficial Interaction

12

Bibhu Prasad Behera, Krishnamayee Mallick, Akanksha Behera, Ankita Das, Rajit Kumar Pal, Gulab Chand Arya, Aditya Mallick, and Namisha Sharma

Abstract

Climate change is radically remodeling plant-pathogen relationships, particularly in the context of virus-plant interactions. Long viewed as devastating, new evidence has revealed a complex role for plant viruses, from pathogenic villains to possible mutualists that provide stress tolerance. This chapter discusses the dynamic interaction between viruses and plants under abiotic stresses, such as heat, drought, and salinity. It presents the burgeoning positive impacts of some viruses on plant productivity and health. Through the discussion of some illustrative case studies, mechanisms such as hormone regulation, osmolyte accumulation, and genome editing technologies are presented. The chapter also elaborates on cross-protection, CRISPR-based viral genome editing, and the eco-evolutionary dynamics that foster plant adaptation to the changing climate. Comprehension of these interactions provides the gateway for sustainable agriculture and the formation of crops that can sustain a harsh climate. The identification of viruses as mutualism, biocontrol, and biotechnology agents represents a paradigm shift in how we perceive plant virology in the context of global food security.

B. P. Behera · K. Mallick · A. Das · R. K. Pal · N. Sharma (✉)
Institute of Life Sciences, Bhubaneswar, Odisha, India

A. Behera · A. Mallick
Department of Life Sciences, Central University of Karnataka, Kalaburagi, Karnataka, India

G. C. Arya
School of Biotechnology, Kalinga Institute of Industrial Technology, Deemed to be University, Bhubaneswar, Odisha, India

R. Kumar et al. (eds.), *Tomato Leaf Curl New Delhi Virus (ToLCNDV)*,
https://doi.org/10.1007/978-981-95-2852-3_12

Keywords

Climate · Dynamics · Plant virus · ToLCNDV · Management · Plant virus interaction

1 Introduction

The major problem of the world now is climate change because of how it impacts agriculture, farming, and food security. Changes in Earth's temperature, abnormal rainfall, and an increase in extreme weather events have shifted the dynamics between plants and pathogens, changed the physiology of crops, and worsened the stress conditions (Lahlali et al., 2024). These transformations have a profound impact on various crops and plant cultivations globally, and, at the same time, change the intensity, distribution, and propagation of several plant infections, thus threatening food security (Moullec et al., 2019). Climate change poses a substantial threat to humankind, resulting in an annual revenue loss of approximately 1.2 trillion US dollars and leading to approximately 0.4 million losses in agricultural resources (EPA, 2023). However, some of the pathogens might have advantageous effects on plants by providing tolerance in response to climate change. Recognizing the interconnection between climate change and the study of plant diseases and their phenomena is crucial for maintaining agricultural and ecological stability, which in turn ensures food security and minimizes the impact of diseases on key food crops. This mitigation should involve the creation of tolerant cultivars, new disease control strategies, and adaptive methods of reducing possible damage.

Plant viruses are obligate parasites, comprising nucleic acid encapsulated by a protein coat, and are dependent on the host cell for multiplication (Trebicki, 2020). Numerous viruses cause severe pathological disorders by initiating infections that produce significant yield losses and reduced quality of crops nationwide. At the global level, these viral infections inflict a $30 billion loss of agriculturally important crops, damaging the profitability and ecological viability of agriculture (Jones, 2021). However, plant viruses are also reported to have less adverse effects on plant hosts and be beneficial, improving host fitness and survival under severe conditions (Roossinck, 2011). In addition, virus epidemiology greatly depends on varied climatic conditions. The significance of climate change lies in understanding how the viruses behave when plants are under stress and how a few of these viruses might have practical functions. As a result, plant viruses, previously known mostly as causative agents of diseases, are increasingly being reassessed for their possible ability to provide adaptive benefits to plants, such as enhanced tolerance to abiotic stressors like drought, salinity, and temperature variations. Such "beneficiary viruses" shake the traditional paradigm and propose a multifaceted role for viral symbiosis in the ecology of plants, especially under climate-stressed environments. This chapter summarizes the evolving impacts of plant viruses in agroecosystems under climate change, emphasizing instances where viral infections positively influence plant growth. By examining molecular and ecological mechanisms, we highlight the potential role of beneficial plant viruses in enhancing crop resilience, promoting sustainable agriculture, and ensuring a world with food security for everyone.

2 Climate Change and Its Impact on Plant-Virus Interactions

Climate change, characterized by increasing global temperatures, shifted precipitation regimes, and intensified incidences of extreme weather events, is drastically remaking plant-pathogen relationships within ecosystems (Kumar et al., 2023a, b). Of these, plant-virus interactions are going through intricate evolution. Temperature increases, drought stress, and CO_2 enrichment can influence host susceptibility, vector performance, and virus replication, frequently in complicated ways, as illustrated in Fig. 12.1. In a study, genetically diverse genotypes of tomato were screened under field conditions to determine the influence of a changing climate on tomato leaf curl virus (ToLCV), a disease closely related to ToLCNDVs. It was also reported that ToLCV prevalence in tomato is more frequent and severe in the winter season (November–March) when compared to the rainy season (June–October). However, among the six genotypes screened, only three genotypes, LA 3969, EC 519730, and LA 4040-1, were found to be stable and relatively resistant in both winter and rainy seasons, could be utilized freely in breeding programs, and can be cultivated in any ToLCV-infested area (Singh et al., 2015).

Researchers also explored the responses of tomato plants to higher temperatures and stress induced by viral infection. High temperature exposure to *Tomato leaf curl New Delhi virus* (ToLCNDV) infestation alone had modest effects on photosynthesis and transpiration rates. However, the combined application resulted in a substantial reduction in rates and increased water uptake. The findings show potential for

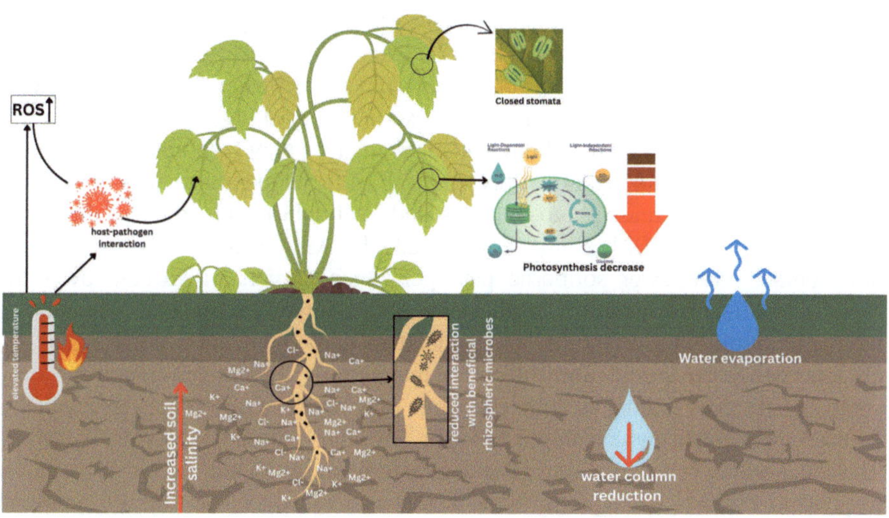

Fig. 12.1 Schematic representation of the effect of abiotic stresses such as heat, salinity, drought, and low soil moisture on plant systems. These stresses impact plant physiological processes, cell homeostasis, and structural integrity and interfere with the plant-microbiome interactions, thus influencing the nutrient acquisition, signaling pathways, and stress resilience mechanisms

producing stress-tolerant tomato cultivars by understanding the consequences of individual or combined biotic and abiotic stressors and targeting the genes that contribute to susceptibility, to increase the yield and plant growth (Ali et al., 2024). Whitefly-transmitted geminivirus diseases have also been reported to increase drastically over the years. Additionally, the geographic distribution of specific geminiviruses is also reported to be expanding, including in different parts of Asia, Europe, and America (Zerbini et al., 2017; Sharma & Prasad, 2017). Researchers believe that the fundamental reason for this increase is the appearance of the vector *Bemisia tabaci* biotype B and the changing climate, which may have favored the spread of these vectors over the years.

Modern scientists found a link between begomovirus incidence and weather conditions in soybean and bean fields in northwestern Argentina (Reyna et al., 2023). The lower temperatures, reduced levels of precipitation, and humidity before crop planting were associated with a huge prevalence of *Bean golden mosaic virus* (BGMV). In soybeans, elevated temperatures in the pre-sowing seasons and late winter, as well as little spring precipitation, were determined to be good indicators of high Soyabean blistering mosaic virus (SbBMV) occurrence. These results can be applied in early warning systems to enhance crop management and reduce the impact of viruses on these plants. It also indicates that there is a considerable correlation between climate conditions and virus infection. Studies also discovered that the infectivity of *Bud necrosis begomoviruses*, ToLCNDV, *Tomato leaf curl Karnataka virus* (ToLCKV), and *Tomato leaf curl Palampur virus* (ToLCPalV) could be enhanced by dry conditions and hot temperatures in India's Indo-Gangetic eastern plains (Nagendran et al., 2019; Sarkar et al., 2024).

Elad and Pertot (2014) stated that plant-pathogen interactions are greatly affected by climate change, which can also influence plant physiology and the surrounding microenvironment (Elad & Pertot, 2014). Plants adjust their gene expression profiles to adapt to a fluctuating environment, resulting in changes in essential properties such as leaf size, stomatal density, root exudation, total sugar, and starch content of the plant. These physical and physiological changes may favorably or negatively influence the plant; for example, drought and high temperature stress can cause stomatal closure, decrease photosynthesis, and inhibit the development of the leaves and roots (Lal et al., 2021; Kumar et al., 2023a, b). As a result, the roots may come into contact with fewer soilborne pathogens than earlier, which can lower the chances of infection. However, similar pressures can weaken plants too, increasing their susceptibility to some other pathogen attack and colonization. Apart from this, climate change can significantly affect the interactions of pathogens that infect only some specific host plants, as climate change may result in differences in the host's defense mechanism, which might favor or not favor the pathogen's proliferation. The balance between temperature and moisture plays a vital role in how plants respond to disease. Changes in these factors can influence the time of symptom arrival and the severity of the disease. For instance, rising carbon dioxide (CO_2) levels often improve plant growth by enhancing the photosynthesis rate. This leads to bigger leaves, taller plants, and an increase in overall biomass and yield of the plant. As a result, the structure of the plant can change, and larger organs may create new opportunities or barriers for pathogen invasion. On the other side, ozone in the atmosphere tends to have damaging effects on plant health, affecting everything from leaf tissue to photosynthetic efficiency. Higher levels of CO_2 can sometimes help in counteracting the negative impact of ozone. However, the overall outcome of whether these changes make a plant vulnerable to disease depends heavily on the specific plant, the type of pathogen, and the surrounding environmental conditions.

Climate change is leading to significant alterations in the worldwide distribution of viral and bacterial plant diseases. This can lead to an alteration in host morphology, physiology, resistance to vectors, and life cycles (Jones & Barbetti, 2012). Additionally, climate change will reduce the efficacy of control measures and render epidemics more unstable. To make the global food system resilient to projected climate conditions, research needs to address the unique challenges of climate variability. As indicated by varied studies on various crops and locations, climatic conditions like increased temperature, changed precipitation patterns, and rising atmospheric CO_2 can redefine virus virulence, vector dynamics, and host susceptibility in complex ways. These stresses occur in combination, which tests the resilience of current cultivars and conventional disease management practices (Hernández et al., 2021). Hence, a comprehensive understanding of the interactions in the changing climate conditions is critical for the production of crops that can withstand harsh climate and tolerate viruses.

3 ToLCNDV: A Climate-Sensitive Emerging Threat

ToLCNDV is a whitefly (*B. tabaci*) transmitted bipartite begomovirus, which is considered one of the most virulent begomoviruses, infecting plants including tomato and other solanaceous crops in the Mediterranean region and South Asia. In the current climate change scenarios, the epidemiology, distribution, and severity of ToLCNDV have all undergone significant alterations, making it an ideal model virus for investigating the climate-mediated evolution of viruses and host-plant interactions (Jeevalatha et al., 2021, 2023). Higher temperatures were found to facilitate a larger whitefly population, the primary vector of ToLCNDV, which in turn increases the prevalence of tomato leaf curl disease in large tomato-producing states such as Karnataka (Naga et al., 2021a, b). The seasonal fluctuation of whiteflies is strongly related to intensive cultivation and climate change. Additionally, their endosymbiotic microbe partners, such as *Candidatus Portiera*, *Hamiltonella*, and *Rickettsia*, contribute significantly to heat tolerance and vector competence, allowing the insects to survive and transmit the virus more effectively. This indicates that climate change is not merely accelerating whitefly epizootics but also enhancing their efficiency as viral vectors, so there is a need for more climate-resilient pest and disease control strategies (Sujatha et al., 2025). In addition, this further indicates that plant stress due to climate has been revealed to compromise plant immunity and weaken the plant and increase susceptibility to infection by ToLCNDV (Ali et al., 2024). ToLCNDV is also evolving quickly, utilizing genetic recombination and host jumps. Its capacity to infect non-solanaceous crops like cucurbits still highlights its expanding host range, with the trend being fanned by changing agroclimatic regions. The virus's DNA-A and DNA-B segments are also known to reassort their segments, enabling them to adapt quickly to new hosts and climatic stresses (Vo et al., 2025; Moriones et al., 2017).

In the current climate change scenario, there is a risk that traditional resistance mechanisms such as Ty-gene-mediated resistance may no longer be significant to protect the plant against ToLCNDV, either due to temperature-mediated breakdown or through the evolution of the virus. This calls for climate-resilient cultivars and vector control strategies adapted to warming climates. ToLCNDV thus demonstrates the potential of begomoviruses to adapt swiftly to changes in the environment, and how this can create substantial challenges for disease control. The example underscores the need to integrate plant pathology with climate modeling to predict and mitigate viral disease risks to tomatoes and other economically significant crops.

4 Emerging Understanding of Beneficial Plant Viruses

Although viruses are commonly recognized for their capacity to infect and cause disease in plants, some recent studies discovered that viruses can help plants resist environmental stresses like heat and drought, as summarized in Table 12.1. In some cases, viruses establish symbiotic or even mutualistic relationships with their hosts. Further, virus-plant-fungus interactions can enhance crop resistance and enable biological control processes. For example, the *Curvularia thermal tolerance virus* (CThTV) infecting the fungus *Curvularia protuberata* enables it to help protect its host plant, *Dichanthelium lanuginosum*, from the extreme heat of geothermal soils (Poudel et al., 2021). The fungus lacks this protective ability in the absence of the virus. Similarly, bacteriophages, which kill bacterial hosts, offer a sustainable idea for the management of bacterial plant diseases.

The study by Roossinck (2015a, b) indicates that infection with *Tomato yellow leaf curl virus* (TYLCV) can enhance drought tolerance in tomato plants (Roossinck, 2015a, b). When exposed to drought stress, infected plants were found to survive better, remain physiologically stable, and redistribute sugars and amino acids from shoots to roots in a strategic manner. This redistribution supports root metabolism and osmoprotection, allowing the plant to survive prolonged drought. TYLCV also controls stress response mechanisms, including heat shock proteins, transcription factors, and autophagy-related genes, which interact to buffer the effect of drought and control virus accumulation. These findings reveal a shift from a parasitic to a mutualistic role of TYLCV under abiotic stress, and it can be speculated that certain plant-virus interactions might become beneficial under stressful conditions and can be exploited as tools for enhancing crop resilience. Xu et al., in 2008, stated that viruses can redirect their interactions with plants toward mutualism under certain environmental conditions through a process called conditional mutualism. This is not a viral-specific process and has also been reported in other symbiotic systems. For instance, some fungal pathogens, such as *Colletotrichum* spp., can act as

Table 12.1 Climate-driven changes in plant virus dynamics along with host interactions

Virus name	Function	Reference
Tomato leaf curl virus	More frequent and severe during the winter season, and also it induces heat	Bhandari et al. (2021)
Pepper mild mottle virus (PMMV)	It can be shed in human feces and remains infectious, showing environmental stability. It is also used as a viral indicator of fecal pollution in water	Roossinck et al. (2015)
ToLCV	Increased incidence in cooler months	Singh et al. (2015)
ToLCNDV	Exposure to high temperatures or ToLCNDV infestations had significant effects on photosynthesis and transpiration rates	Ali et al. (2024)
Chilli leaf curl virus	Reduction in growth and yield, variance in prevalence b/w northern and southern India, indicating virus adaptation to climatic conditions	Jayanthi et al. (2024)
Bean golden mosaic virus	Climatic conditions induce the spreading of viruses, such as high winter temperatures and low spring humidity	Reyna et al. (2023)
Soybean blistering mosaic virus	High temperatures in late winter and pre-sowing seasons, as well as little spring precipitation, were indicators of virus	Reyna et al. (2023)
Tomato yellow leaf curl virus (TYLCV)	It thrives under warmer conditions and increases its outbreaks. Additionally, warmer temperatures increase whitefly populations and their movement	Elad and Pertot (2014)
Potato virus Y (PVY)	Climate-induced stress (heat, drought) can increase susceptibility and symptom expression Elevated temperatures can enhance virus replication or reduce host resistance	Elad and Pertot (2014); Bhardwaj et al. (2019)
Geminivirus	Increased transmission due to geographic distribution expansion and a new vector for transmission	Polston and Anderson (1997)
Barley yellow dwarf virus (BYDV)	High atmospheric CO2, temperature, and water status influence plant growth, quality, and plant-pathogen interactions. Abiotic factors also play a role in plant virus epidemiology, providing an approach to studying virus epidemiology	Trebicki (2020)
Tomato yellow leaf curl virus (TYLCV)	Abiotic stress, such as soil salinity, is a subject of study as it is influential to the phytomicrobes in the roots of the plants, which counteract chlorosis	Rojas et al. (2018)
Curvularia thermal tolerance virus (CThTV)	Enables the fungus to protect its host plant from the extreme geothermal soils	Poudel et al. (2021)
Tobacco mosaic virus	The host resistance is altered under increased temperature due to the interaction; the HR type resistance is overcome when plants are kept in high temperatures	Sharma et al. (2024)
Tomato spotted wilt virus (TSWV)	In capsicum plants, at elevated temperature, viruses overcame the Tsw-mediated resistance	Tsai et al. (2022)

(continued)

Table 12.1 (continued)

Virus name	Function	Reference
Curvularia thermal tolerance virus	A holobiont can thrive at temperatures up to 65 °C in the presence of a virus and fungus as a mutual interaction	Stobbe and Roossinck (2014)
Turnip mosaic virus (TuMV)	Increase drought tolerance and survival in plants, with effects varying between host genotypes, and accumulate mutations in viral proteins, allowing them to evolve in stress environments	González et al. (2021)

mutualists depending on host genotype, while endophyte-infected grasses, such as tall fescue, show improved recovery from drought (Xu et al., 2008). Virus-induced differences in plant physiology and metabolism, such as reduction in plant size, increase in tissue water content, reduction in water loss, and accumulation of protective metabolites, can promote survival of the plant under drought and cold stress. Elucidation of the mechanisms through which virus infections regulate plant stress responses can have the potential to benefit agriculture, especially under the scenario of climate change, where drought is an increasing risk to crop yield.

5 Mechanistic Insights on the Role of Viruses in Providing Abiotic Stress Tolerance

Plant viruses can provide resistance to abiotic stresses like drought, temperature variations, and salinity by various mechanisms by modulating the phenotypic, morphological and genotypic features of the host plant and other physiological processes like modulation of ROS, tuning of phytohormones, regulating the formation of stress-responsive proteins such as HSPs, and RNAi mechanisms (Fig. 12.2). In the following section, detailed insights are provided on these mechanisms.

5.1 Reactive Oxygen Species (ROS): Double-Edged Signals in Stress Tolerance

Plant viruses regulate reactive oxygen species (ROS) metabolism in complex ways, which are not trivial by-products of infection but are an integral part of a reprogrammed host response, especially under abiotic stress. ROS, including hydrogen peroxide (H_2O_2) and superoxide (O_2^-), function as essential signaling molecules in plants that induce systemic acquired resistance and adaptation to environmental extremes (Sharma et al., 2021). Viral infection can modify the intensity and spatiotemporal distribution of ROS generation under drought or salinity stress conditions to modulate stress responses (Hasanuzzaman et al., 2020). For instance, in *Sugarcane mosaic virus* (SCMV)-infected sugarcane plants, various genotypes respond differentially to ROS. The susceptible genotype "Badila" indicated high overexpression of ROS-inducing genes, including NADH oxidase and malate dehydrogenase,

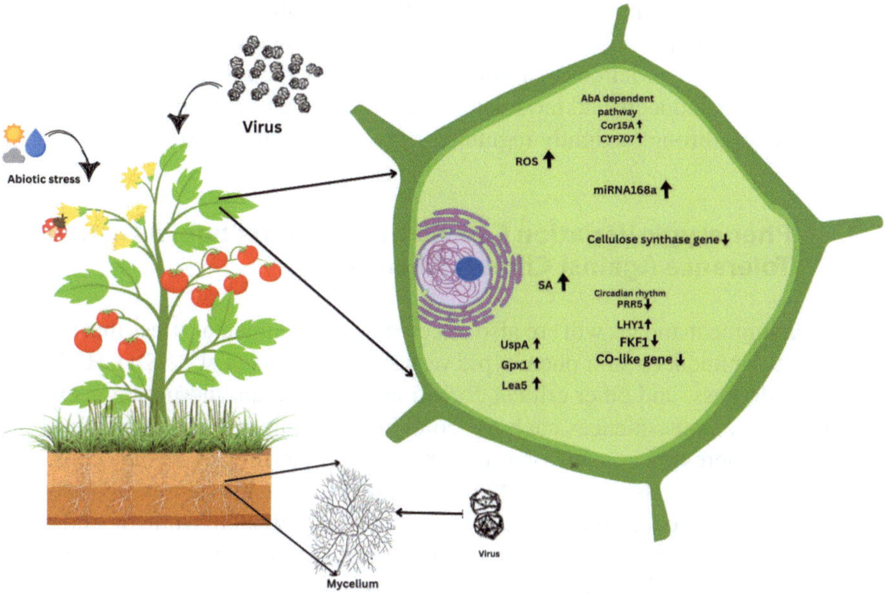

Fig. 12.2 Schematic representation of virus-induced plant abiotic stress tolerance, including heat, drought, and salinity. Through mechanisms such as the interference of the viral 2b protein in ABA signaling, the plant regulates stress reactions. A tripartite symbiosis between virus-plant-fungi, enhancing plant resistance against drought, and the virus-induced modulations of ROS, miRNA, various stress-induced proteins, and the expression levels of circadian rhythm and phytohormone-related genes

along with higher levels of glutathione S-transferase (GST), a protective antioxidant enzyme commonly associated with pathogen stress. Conversely, the transgenic line B-48 showed lower GST activity. Major transcription factors like WRKY, AP2, NAC, bZIP, and bHLH were also upregulated in B-48, indicating a more proficient ROS management system that probably helps it to resist. This implies that viruses assist plants in transitioning from a hyperreactive to a better-regulated redox balance under stress conditions (Akbar et al., 2020). In virus-infected *Nicotiana benthamiana*, notably when the plants are co-inoculated with fungal endophytes, antioxidant enzyme activity in response to drought significantly increased. These are catalase (CAT), peroxidase (POD), and polyphenol oxidase (PPO), which individually or in concert decrease the amount of ROS accumulation, restrict lipid peroxidation, and preserve membrane integrity. Such coordinated regulation of ROS helps in osmotic balance and enhanced drought tolerance (Dastogeer et al., 2018).

Furthermore, a stress-reactive network is produced by the interaction of ROS with signaling elements such as calcium ions, protein kinases, and NADPH oxidases (e.g., *Arabidopsis* AtrbohD and AtrbohF). These NADPH oxidases are triggered under salt stress, resulting in improved stress survival and ionic equilibrium. To enable plants to respond adaptively, ROS also causes the downstream activation of genes for transcription factors, antioxidant systems, and heat shock protein synthesis (Sachdev et al., 2021). Curiously, ROS regulation via virus interaction is highly context dependent. Whereas some viruses stimulate ROS buildup for signaling, others suppress its actions to minimize cytotoxicity. This duality enables viruses to be both stress inducers and modulators, promoting host survival and tolerance to abiotic stresses through a tightly regulated redox balance (Lee, 2018).

5.2 Phenotypic Variation Caused by Virus Infection, Assisting Tolerance Against Climate Change

Plant development and growth are also affected by environmental stress that directly or indirectly influences their phenotypic variations (Sharma et al., 2024). High temperatures, dryness, and other critical conditions disturb plant morphology, exhibiting phenotypic consequences such as scorching of stems and leaves, reduced life processes, suppressed growth of roots and shoots, sterility of sexual cells, and depleted soil quality, as well as fruit quality. At elevated temperatures, cell physiology is also influenced, primarily marked by a decrease in photosynthetic and respiratory rates, as well as alterations in membrane structure that cause enhanced permeability and fluidity (Hu et al., 2020). Viruses are identified to perform a critical function in causing stress tolerance in plants, providing solutions to overcome their effects on crop yield. The molecular mechanisms and physiological alterations caused by viruses lead to drought tolerance, such as changes in hormone signaling, antioxidant defense, RNA silencing and miRNA pathway, and alterations in the expression of various genes like heat shock proteins and cellulose synthase (Saharan et al., 2022). In a study, the C4 protein of *Tomato yellow leaf curl Sardinia virus* (TYLCSV) induces tolerance against drought in plants via morphological adaptations, suggesting the possibility of viral proteins in increasing the plant resistance to water deficiency (Pagliarani et al., 2022). Further, the C4 protein of *Begomovirus Tomato yellow leaf curl virus* (TYLCV) also proved to confer drought tolerance in model systems of plants (Corrales-Gutierrez et al., 2020). Research by González et al. (2021) demonstrated that the occurrence of severe drought conditions affects Turnip Mosaic Virus (TuMV) evolution and, surprisingly, alters the host-virus interaction at various physiological levels. Phenotypic and genotypic shifts resulting from TuMV evolution under well-watered or drought conditions, and differences in the transcriptome and hormonal profiles of the host induced by the infection were evaluated (González et al., 2021). Consequently, a drought environment has been shown to select TuMV lineages that provide increased drought tolerance. In another study, it was reported that different plant species, when inoculated with four viruses—namely, *Brome mosaic virus (BMV)*, *Tobacco mosaic virus*, *Cucumber mosaic virus (CMV)*, and *Tobacco rattle virus*—showed stressed symptoms similar to those of withholding water (Xu et al., 2008). In each case, the viral infection postponed the onset of drought symptoms. CMV-infected beet plants also showed considerably enhanced freezing tolerance. CMV infection elevated the water content in aerial tissues. The water loss was observed to be less in the case of leaves detached from virus-infected plants, in contrast to the leaves detached from mock-inoculated plants, a sign of improved water holding capacity in infected plants. Water retention may be attributed to decreased transpiration rate due to stomatal closure in virus-infected plants (Beattie, 2011). Metabolite profiling analysis indicated high levels of many osmoprotectants and antioxidants in BMV- and CMV-infected plants under pre- and post-drought stress conditions. These findings confirm the improved plant abiotic stress resistance upon viral infection.

5.3 Hormonal Crosstalk: Virus-Modulated Phytohormones in Abiotic Stress Alleviation

The interaction of phytohormones and viral infection is surfacing as a key regulator of abiotic stress tolerance. Viruses tend to hijack the plant hormone environment to facilitate their life cycle or enhance host survival in stressful environmental conditions. Of these, abscisic acid (ABA), salicylic acid (SA), cytokinins (CKs), jasmonic acid (JA), and auxins are key players in orchestrating stress responses (Alazem & Lin, 2015). ABA is most likely the best-documented phytohormone that responds to abiotic stress, such as viruses and drought. Notably, many plant viruses appear to possess mechanisms to modulate ABA pathways that simulate or augment drought tolerance. Infections with *Tobacco mosaic virus (TMV)*, *Cucumber mosaic virus (CMV)*, and *Bamboo mosaic virus* (BaMV) have been found to increase ABA levels in infected plants. This increase encourages stomatal closure, facilitates water retention, and results in the increased concentration of osmoprotectants like proline, which are typical characteristics of the plant's drought stress adaptations. These virus-induced alterations eventually enhance the plant's survivability under water-deficient conditions (Muhammad et al., 2022). However, this positive effect is not invariable, and the response tends to be virus-specific. For instance, *Rice black-streaked dwarf virus* (RBSDV) represses the ABA-responsive genes, leading to enhanced susceptibility as opposed to tolerance to drought stress. This emphasizes the complex and sometimes contradictory manner in which viruses interact with host hormonal systems (Zhang et al., 2019). One particularly interesting example exists from CMV 2b protein, which is a multifunctional viral RNA silencing suppressor that also has a crucial role to play in the regulation of ABA signaling. Research has indicated that the 2b protein interferes with the ABA signaling components directly by interacting with them. Specifically, 2b protein inhibits the ABA-responsive transcription factor ABA INSENSITIVE 5 (ABI5), affecting its ability to make RNA and take part in transcription. In this way, CMV regulates the ABA signaling pathway, not to prevent it from happening at all but to prevent it from being overstimulated under drought conditions. This regulated modulation assists the plant in balancing development and stress adaptation and thereby improving survival during drought without completely halting development (Xu et al., 2013). SA has a biphasic role with its traditional antiviral defense mechanism; it again aids abiotic stress responses through the induction of ROS formation, hypersensitive response (HR), and systemic acquired resistance (SAR). Induction of RNA silencing-associated genes (DCLs, RDRs) by SA can decelerate viral replication while at the same time increasing stress tolerance. Interestingly, methyl salicylate, one of the volatile derivatives of SA, is an airborne signal that preconditions neighboring plants for defense, spreading the protective effect beyond the initially infected host (Gao et al., 2015). JA signaling also has a paradoxical function in virus-infected plants subjected to stress. In some situations, JA suppresses SA pathways and antiviral defense, but in others, its biosynthesis helps bring about stress alleviation. Viral proteins like the C2 protein of geminiviruses can interfere with JA signaling by inhibiting the SCF ubiquitin ligase complex, suggesting that viruses would downregulate JA responses whenever they are at odds with each other over replication strategies (Wu & Ye, 2020). Overall, viruses appear to rewire hormonal networks in plants to create a state of physiological balance that supports both survival under abiotic stress and optimal conditions for limited viral replication. This strategy may tilt the interaction from parasitism to mutualism under environmental pressure.

Apart from these defense-related hormones, growth-related hormones, such as cytokinin and auxin, have also been reported to respond to virus infection, leading to tolerance against climate change. Cytokinins are modulated in response to viral infections, and viral proteins are found to inhibit cytokinin biosynthesis (e.g., by adenosine kinase inhibition by begomovirus proteins). In contrast, others indirectly facilitate cytokinin biosynthesis to induce SA-dependent defense genes. The outcome, resistance or susceptibility, is host and viral-strain-dependent. For instance, cytokinin accumulation in tobacco plants infected with CMV increases resistance against subsequent TMV infection (Baliji et al., 2010). Similarly, auxins, which are essential regulators of shoot and root growth, are viral protein targets, including the C4 protein of *Tomato yellow leaf curl Sardinia virus* (TYLCSV). In drought, this protein reduces auxin response factor (ARF) activity, lowering auxin signaling and root morphology to enhance water uptake. Drought tolerance, water-use efficiency, and rehydration recovery were improved in transgenic tomato plants expressing C4 (Pagliarani et al., 2022). These findings highlight the role of virus-mediated regulation of hormones, which assists in activating tolerance mechanisms against different climate changes in the host plants.

5.4 Activation of RNA Interference Upon Virus Infection as a Shield Against Climate Change

In a coevolutionary arms race, the pathogen's virulence and the host's defense mechanism are constantly changing during the host-pathogen interaction. In plants, the antiviral innate immune system uses RNA interference (RNAi) to silence virus genes. To counter this antiviral plant defense mechanism, viruses produce viral suppressor of RNA silencing factors, which either interact with the small RNAs or with the biosynthesis machinery, thus regulating the plant defense mechanism.

Climate change makes plants more vulnerable to viruses. According to studies, plants that have antiviral defense mechanisms are better able to withstand abiotic stress. Cucumber mosaic virus produces 2b protein, which suppresses the RNA silencing mechanism by interacting with dsRNA and AGONAUTE1 (AGO1), thus limiting its availability to the silencing machinery. In wild-type plants, during drought conditions, ABA treatment stimulates stress-responsive genes such as RD29A (responsive to desiccation 29A) and NCED3 (9-cis-epoxycarotenoid dioxygenase3). However, RD29A and NCED3 induction is lessened in transgenic 2b and AGO1-mutated plants. Thus, the study demonstrates that whereas the 2b protein does not affect ABA hormone accumulation, it does interact with ABA hormone signaling. 2b protein-mutated plants exhibit drought tolerance by having improved water absorption or retention capacity, despite suppressing the primary abiotic stress hormone ABA (Westwood et al., 2013). Grapevine rupestris stem pitting-associated virus (GRSPaV) infects *Vitis vinifera*. Reports show that during infection, GRSPaV alters the miRNA regulations in the host. One of the significant findings is that in the infected plant, miR164, miR409712_2, and miRC102 are unregulated, which leads to the induction of three members of the stress-responsive transcription factors NAC family (VvNAC05, VvNAC11, VvNAC17) under drought conditions (Pantaleo et al., 2016). *Cowpea mild mottle virus* has a triple gene block 1 (TGB1), which facilitates the virus's infection in common bean. The TGB1 gene derived a small interfering RNA called vsiRNA6163, which suppresses PvTCP2, which is a repressor of the autophagy gene PvATG8c. PvATG8c promotes autophagy of the positive regulator of stomatal opening gene PvERD15. During drought stress, activation of this pathway benefits the plant to conserve water (Wu et al., 2024). Investigations into the relationship between RNAi, virus, and abiotic stresses can help us develop stress-resistant plants in the future.

5.5 Signaling Pathways and Proteins: Heat Shock Proteins (HSPs) and Viral Modulators

Viruses hijack and reprogram host signaling pathways and molecular chaperones to regulate both their life cycles and abiotic stress responses. Of these, heat shock proteins (HSPs) are multifunctional molecular scaffolds, protein-stabilizing under stress conditions and helping in viral replication. HSP70 and HSP90 are well-characterized for their role in plant development and response against different environmental stresses, while other members, such as HSP23.6 and HSP101, are also involved in defense and stress tolerance (Zhang & Yu, 2022). Heat stress, viral infection, or a combination of the two usually triggers the expression of HSPs. For instance, HSP70 is induced in *Nicotiana benthamiana* upon infection with RNA viruses such as TMV, CMV, and PVX. Its functions range from protein folding and stress alleviation to viral protein translocation and replication. Interestingly, this induction is greater when caused by viral protein buildup rather than heat stress alone, as indicated by pHSP70::GUS constructs (Chen et al., 2008). *Tomato yellow leaf curl virus* (TYLCV) also shows how viruses hijack heat shock pathways. TYLCV also downregulates drought-induced heat shock factor (HSF) transcription of genes such as HSFA1a, HSFA2, and HSFB1, keeping protein homeostasis intact and preventing proteotoxic stress (Sarkar et al., 2024). This viral preconditioning increases drought tolerance by preserving chaperone function, particularly in root tissues. At the same time, TYLCV infection avoids extreme downregulation of HSP70 and HSP90 under long-term drought, maintaining cell integrity (Shteinberg et al., 2021). Virus-encoded proteins such as CMV 2b, PVX P25, and TYLCV C4 also engage core stress responses. CMV 2b stimulates lateral root emergence and regulates ABA-independent response modification, while PVX P25 impacts host metabolic activities. These proteins also stimulate proline and trehalose synthesis, contributing to cellular dehydration tolerance. In the meantime, the C4 protein not just suppresses auxin responses but also impacts allocation of metabolites toward the roots, conditioning plants for potential water deficiency (Zhang et al., 2006).

Apart from HSPs, other TFs and proteins are involved. WRKY, NAC, AP2, and bZIP families are upregulated uniformly in virus-infected, stress-acclimated plants. Heat-killed viruses like TMV (HKTMV) are capable of replicating these protective effects without causing viral symptoms. HKTMV-treated plants exhibit enhanced levels of stress-related transcription factors like DREB2, WRKY1, and ERF1B, and decreased ADF gene levels, which increases drought tolerance. These plants also demonstrate enhanced membrane stability and water retention capacity, characteristics lacking in untreated controls (Augustine et al., 2022). These TFs orchestrate downstream mechanisms like antioxidant activation, hormone crosstalk, and RNA silencing. In addition, viruses can fine-tune host epigenetic landscapes; for example, 24-nt siRNAs are associated with transgenerational stress tolerance as seen in PVX-infected *N. benthamiana* progeny (Llorca et al., 2014). The research suggests that viruses are subtle regulators of cellular signaling, utilizing the host molecular machinery to move and overcome abiotic stresses. Through stabilizing significant proteins, modulating transcriptional networks, and regulating homeostasis, viruses are not just pathogens but also prospective mutualists in a demanding environment.

6 Implications for Crop Resilience and Agriculture

Considering the various abiotic and biotic stressors affecting plants, their resilience is crucial for feeding the exponentially increasing population. Soil salinity is a significant parameter that has a profound impact on food production. Given that 1/5th of the world's agricultural land is covered with saline soil, it is estimated that in the next decades, this will cause a significant loss in agricultural production (Rojas et al., 2018) and harm food security for the increasingly growing population. Apart from salinity, temperature variation and drought can also be a potent risk for crop productivity. Although the term "microbes" has had numerous negative connotations in the past, they are the only solution to the puzzle of improving food production resilience. Stress responses of plants, especially with viral infections, are necessary to know to engineer stress-tolerant crops. Plant virus evolution is also influenced by environmental factors like drought, which can shift virus-host relations from parasitism to mutualism. Viruses like CMV, TMV, BMV, TYLCV, and GFLV can help plants withstand stress; on the contrary, viruses like CaMV and TuMV can augment the drought effect (Aguilar & Lozano-Duran, 2022). Even the same virus may differ in its impact based on how it has evolved and the host plant's condition. For example, TuMV can increase drought tolerance and survival in plants, with effects varying between host genotypes (González et al., 2021). These viruses accumulate mutations in viral proteins, allowing them to evolve toward stressful environments.

The most crucial area where viral tools have arisen with high promise is CRISPR/Cas-based genome editing. This tool, originally taken from the bacterial defense system, has been designed to achieve efficient and specific genome editing in plants. The CRISPR/Cas9 system, based on *Streptococcus pyogenes*, employs a single-guide RNA (sgRNA) that assists the Cas9 nuclease to a specific DNA target site to generate double-stranded breaks (DSBs). Specificity of such an interaction is defined by the existence of a protospacer adjacent motif (PAM), typically the NGG sequence. Plant viruses are effective delivery vectors for CRISPR/Cas components to deliver targeted gene editing that can convey desirable agronomic features, such as disease resistance, tolerance against heat and drought, and improved nutritional content. Several delivery methods have been developed to introduce these viral- or virus-derived editing machinery into plant cells, including *Agrobacterium*-mediated transformation, protoplast transfection, and particle bombardment. Viral vectors facilitate transient and systemic editing machinery for gene delivery, typically avoiding stable transformation and significantly reducing regulatory hurdles. Viral-mediated genome editing offers new opportunities for producing more durable crops with increased tolerance to climate change-mediated stressors, such as drought, salinity, and temperature fluctuations. Virus-mediated gene editing also provides a fast functional analysis of plant immunity genes, stress response genes, and growth regulation genes, thus speeding up crop improvement programs. The viral-induced genome editing (VIGE) platform exploits the intrinsic ability of plant viruses to transfer genetic material into plant cells. Since most viral vectors have minimal cargo capacity, VIGE is primarily used to deliver single-guide RNAs (sgRNAs) into transgenic plants pre-expressing Cas9. This allows for high systemic levels of sgRNA expression, leading to successful gene editing. Plant viruses such as *Potato virus X* (PVX), *Tobacco rattle virus* (TRV), *Cotton leaf crumple virus* (CLCrV), and *Barley stripe mosaic virus* (BSMV) have been used as effective delivery vectors. TRV, because of its ease of use and broad host range, has been used for single and multiplex gene editing in model plants like *Nicotiana benthamiana* and *Arabidopsis thaliana* (Kumar et al., 2021). Interestingly, heritable gene transformations were observed when sgRNAs were combined with mobile FLOWERING LOCUS T (FT) mRNA, exhibiting transmission efficiencies of up to 65–100%, effectively arriving in the reproductive organs. In *Solanaceae* crops, PVX-based systems have been up-and-coming. Engineered PVX has been used to induce specific mutations in *N. benthamiana*, where the mutation was directly detected in virus-infected tissues (Lei et al., 2021).

The viral vector-based precise editing allows for faster development of traits for drought tolerance, disease resistance, and yield improvement. Another important innovation is the creation of virus-based biocontrol agents (VBAs). This approach entails that mild or weakened strains of viruses are used to defend plants against their more virulent relatives, a phenomenon referred to as cross-protection. Working like live attenuated vaccines in medicine, this technique primes plant defenses without inducing disease. First used with Tobacco mosaic virus (TMV), cross-protection is today used against viruses such as citrus tristeza virus, zucchini yellow mosaic virus, and papaya ringspot virus (Pianka and Izhevsky, 1979). The protection mechanism includes a combination of RNA-mediated silencing, protein interference, and superinfection exclusion. Significantly, this protection remains even in RNA-silencing-deficient plants, showing multiple layers of antiviral defense reprogramming (Charudattan, 2024). The discovery of site-directed mutagenesis enables researchers to design these benign strains intentionally, eliminating the necessity of looking for naturally occurring variants. This technique accelerates the creation of customized VBAs, making them more predictable and responsive to emerging threats. Viruses, aside from their role in plants, are important in the biological control of pests and bacterial diseases in agriculture. *Baculoviruses*, including the *Nucleopolyhedroviruses* (NPVs) and the *Granuloviruses* (GVs), are double-stranded DNA viruses infecting lepidopteran larvae only, prominent agricultural pests such as *Spodoptera frugiperda* (fall armyworm) in maize and codling moth in apple production. These viruses are highly insect host-specific and do not affect humans, animals, or beneficial insects, positioning them as ideal candidates for environmentally friendly biopesticides. Wettable powders or liquid suspensions are usually employed to apply baculoviruses onto the plants, which have been found to cut crop loss due to pests by as much as 50% (Martinez et al., 2025). At the same time, bacteriophages, or phages, are being more widely applied in phage therapy to combat antibiotic-resistant bacterial diseases in plants. These phages represent a targeted, biodegradable, and sustainable option over chemical antibiotics, thus promoting plant health with minimal environmental footprint. Altogether, such virus-based solutions represent vital tools for contemporary, sustainable agriculture.

7 Future Directions and Technological Innovations

Modern-day research provides us with links among viruses of plants, fungi, humans, and arthropods, which could share a common ancestor. If we examine viral genomes in different ecological niches, we will learn more about the ancient origins of major groups of viruses. This research also provides insights into how viruses function in ecosystems where human activities are more prevalent. When scientists gather samples from insects, plants, and soil, they can deeply investigate the way viruses disperse and how they evolve. Improved techniques can help significantly in uncovering the connections between viruses among various organisms and ecosystems, so that we can better utilize the potential of the beneficial viruses (Lefeuvre et al., 2019). Scientists have discussed the influence of heat stress and *Tomato spotted wilt virus* (TSWV) infection on tomato plants, leading to extensive damage to photosynthetic capacity and chlorophyll content (Lefeuvre et al., 2019). This study recommends the need for transcriptomic and physiological analysis to explain stress responses, and molecular breeding, genome editing, and transgenics to create tomato cultivars with improved resistance to combined abiotic and viral stresses to provide yield stability under future climate conditions. Although some advances in research have been made over the last few years, there are still many unknown things regarding the impacts of climate change on plant virus diseases, particularly when compared with similar studies for fungal pathogens and insect pests. While viruses (such as BYDV, CMV, TYLCV, and TuMV) and their insect vectors (such as whiteflies and aphids) have been researched under increased CO_2, temperature, and rainfall, much of this evidence remains unexplored. Promising developments are field-based experiments and research on concurrent climate stresses, mixed infections, and virus-vector interactions. Plant-virus interactions remain an area with many unanswered questions; metagenomics can be a powerful approach to explore these relationships. By analyzing individual plant samples rather than pooled environmental samples, a method known as "ecogenomics," researchers can gain valuable insights into the diversity, distribution, and ecological roles of plant viruses, including potential host- or environment-specific interactions (Stobbe & Roossinck, 2014). Climate change is also found to shift host plant characteristics and disease transmission patterns, making virus outbreaks more volatile and more difficult to control, particularly in tropical and subtropical areas of high food insecurity. Therefore, there is a need for technological advancements and deeper insights into plant-virus-vector interactions under the stress of climate, to provide promises for improved control of diseases in the future. A few studies also reported that viral infections augment plant stress and inhibit growth in some climatic conditions (Sharma et al., 2024). These conflicting and perplexing results imply that the direct connection between abiotic stress, viral pathogenesis, and associated microorganisms remains to be unveiled. To gain clarification on these issues, additional scientific research in this area is required. Apart from this, recent research has placed CRISPR/Cas systems at the center stage as an innovative platform for designing resistance to both DNA and RNA plant viruses (Zhan et al., 2024). Cas13 and CRISPR/Cas9 have dominated the research focus, indicating significant promise toward new antiviral technologies. However, there is a need for further research into different CRISPR/Cas variants that have increased selectivity and efficacy in developing virus-based delivery systems to generate climate-resilient plants. Thus, the incorporation of advanced technologies with ecological and molecular research will be critical to fully comprehend the intricate dynamics of plant-virus interactions and harnessing useful viruses. This integrated approach will ultimately guarantee food security worldwide in a time of rising climate change.

8 Conclusion

Plant viruses have long been regarded as agricultural enemies. Nevertheless, amazingly, some of them are proving to be unlikely assistants, particularly when plants are stressed by factors such as drought, heat, or salty soil. These viruses possess a kind of dual personality; they can be harmful to plants, but in some circumstances, they can help plants survive adverse abiotic conditions. Studies have found that certain viruses can enhance a plant's resilience by modifying its growth, energy use, and cellular response. This newfound knowledge is revolutionizing the way we perceive plant disease and providing imaginative opportunities for agriculture under a new climate. Researchers are working on promising technologies such as virus-based genome editing, employing benign viruses to guard against more aggressive ones (cross-protection) and even making viruses natural pesticides. These strategies could become valuable options for more resilient and sustainable agriculture. Yet several unanswered questions remain. Viruses don't act identically in every plant, and viruses can rapidly adapt to changes through evolution. Therefore, we must be wary of how all these changes impinge on the environment. If we are going to proceed, scientists will have to use novel tools such as metagenomics, systems biology, and ecological modeling to analyze these intricate relations. If we begin to envision plant viruses less as dangers, but as partners, perhaps we will enable the development of new avenues in crop protection, and in a changing world, food security as well.

Acknowledgments The authors' work in this area is supported by research grants from the Ministry of Science and Technology, Government of India (SRG/2023/001019) and the Core Grant of the Institute of Life Sciences, Bhubaneswar, India. GCA acknowledges the Department of Biotechnology, Government of India, for Ramalingaswami Re-entry Fellowship (Grant Number-BT/RLF/Re-entry/16/2022). The authors are grateful to the DBT eLibrary Consortium (DeLCON) for providing access to e-resources.

References

Aguilar, E., & Lozano-Duran, R. (2022). Plant viruses as probes to engineer tolerance to abiotic stress in crops.*Stress biology, 2*(1), 20. https://doi.org/10.1007/s44154-022-00043-4

Akbar, S., Wei, Y., Yuan, Y., Khan, M. T., Qin, L., Powell, C. A., Chen, B., & Zhang, M. (2020). Gene expression profiling of reactive oxygen species (ROS) and antioxidant defense system following Sugarcane mosaic virus (SCMV) infection. *BMC Plant Biology, 20*(1), 532. https://doi.org/10.1186/s12870-020-02737-1

Alazem, M., & Lin, N. S. (2015). Roles of plant hormones in the regulation of host-virus interactions. *Molecular Plant Pathology, 16*(5), 529–540. https://doi.org/10.1111/mpp.12204

Ali, M., Khan, M., & Rao, A. (2024). Response of tomato plant to abiotic (heat stress) and biotic (tomato leaf curl New Delhi virus, ToLCNDV) factors. *International Journal of Phytopathology, 13*, 99–107. https://doi.org/10.33687/phytopath.013.01.5127

Augustine, S. M., Tzigos, S., & Snowdon, R. (2022). Heat-killed tobacco mosaic virus mitigates plant abiotic stress symptoms. *Microorganisms, 11*(1), 87. https://doi.org/10.3390/microorganisms11010087

Baliji, S., Lacatus, G., & Sunter, G. (2010). The interaction between geminivirus pathogenicity proteins and adenosine kinase leads to increased expression of primary cytokinin-responsive genes. *Virology, 402*(2), 238–247. https://doi.org/10.1016/j.virol.2010.03.023

Beattie, G. A. (2011). Water relations in the interaction of foliar bacterial pathogens with plants. *Annual Review of Phytopathology, 49*, 533–555. https://doi.org/10.1146/annurev-phyto-073009-114436

Bhandari, R., Neupane, N., & Adhikari, D. P. (2021). Climatic change and its impact on tomato (Lycopersicum esculentum L.) production in plain area of Nepal. *Environmental Challenges, 4*, 100129. https://doi.org/10.1016/j.envc.2021.100129

Bhardwaj, V., Sood, S., Kumar, A., Vanishree, G., Sharma, S., Sundaresha, S., Raigond, B., Kumar, R., Bairwa, A., Lal, M., & Chakrabarti, S. K. (2019). Efficiency and reliability of marker-assisted selection for resistance to major biotic stresses in potato. *Potato Journal, 46*, 56–66.

Charudattan, R. (2024). Use of plant viruses as bioherbicides: The first virus-based bioherbicide and future opportunities. *Pest Management Science, 80*(1), 103–114. https://doi.org/10.1002/ps.7760

Chen, Z., Zhou, T., Wu, X., Hong, Y., Fan, Z., & Li, H. (2008). Influence of cytoplasmic heat shock protein 70 on viral infection of Nicotiana benthamiana. *Molecular Plant Pathology, 9*(6), 809–817. https://doi.org/10.1111/j.1364-3703.2008.00505.x

Corrales-Gutierrez, M., Medina-Puche, L., Yu, Y., Wang, L., Ding, X., Luna, A. P., Bejarano, E. R., Castillo, A. G., & Lozano-Duran, R. (2020). The C4 protein from the geminivirus tomato yellow leaf curl virus confers drought tolerance in Arabidopsis through an ABA-independent mechanism. *Plant Biotechnology Journal, 18*(5), 1121–1123. https://doi.org/10.1111/pbi.13280

Dastogeer, K. M. G., Li, H., Sivasithamparam, K., Jones, M. G. K., & Wylie, S. J. (2018). Fungal endophytes and a virus confer drought tolerance to Nicotiana benthamiana plants through modulating osmolytes, antioxidant enzymes and expression of host drought responsive genes. *Environmental and Experimental Botany, 149*, 95–108. https://doi.org/10.1016/j.envexpbot.2018.02.009

Elad, Y., & Pertot, I. (2014). Climate change impacts on plant pathogens and plant diseases. *Journal of Crop Improvement, 28*(1), 99–139. https://doi.org/10.1080/15427528.2014.865412

EPA. (2023). *Climate change impacts on agriculture and food supply*. U.S. Environmental Protection Agency. https://www.epa.gov/climateimpacts/climate-change-impacts-agriculture-and-food-supply

Gao, Q. M., Zhu, S., Kachroo, P., & Kachroo, A. (2015). Signal regulators of systemic acquired resistance. *Frontiers in Plant Science, 6*, 228. https://doi.org/10.3389/fpls.2015.00228

González, R., Butković, A., Escaray, F. J., Martínez-Latorre, J., Melero, Í., Pérez-Parets, E., Gómez-Cadenas, A., Carrasco, P., & Elena, S. F. (2021). Plant virus evolution under strong drought conditions results in a transition from parasitism to mutualism. *Proceedings of the National Academy of Sciences, 118*(6), e2020990118. https://doi.org/10.1073/pnas.2020990118

Hasanuzzaman, M., Bhuyan, M. H. M. B., Zulfiqar, F., Raza, A., Mohsin, S. M., Mahmud, J. A., Fujita, M., &Fotopoulos, V. (2020). Reactive Oxygen Species and Antioxidant Defense in Plants under Abiotic Stress:Revisiting the Crucial Role of a Universal Defense Regulator. *Antioxidants (Basel, Switzerland), 9*(8), 681.https://doi.org/10.3390/antiox9080681

Hernández, J. A., Diaz-Vivancos, P., Acosta-Motos, J. R., & Barba-Espín, G. (2021). Where biotic and abiotic stress responses converge: Common patterns in response to salinity and Plum pox virus infection in pea and peach plants. *Annals of Applied Biology, 178*(2), 281–292. https://doi.org/10.1111/aab.12667

Hu, S., Ding, Y., & Zhu, C. (2020). Sensitivity and responses of chloroplasts to heat stress in plants. *Frontiers in Plant Science, 11*, 375. https://doi.org/10.3389/fpls.2020.00375

Jayanthi, P., Kumar, P., Roy, A., Mandal, B., & Geetanjali, A. S. (2024). Detection of multiple Begomoviruses in chilli crop in the five agro-climatic zones of Tamil Nadu state of India. *Tropical Plant Pathology, 49*(5), 649–661. https://doi.org/10.1007/s40858-024-00659-z

Jeevalatha, A., Vanishree, G., Siddappa, S., Kumar, R., Kaundal, P., Kumar, A., & Chakrabarti, S. K. (2021). Molecular characterization and infectivity analysis of tomato leaf curl New Delhi virus isolates infecting potato. *3 Biotech, 11*, 203. https://doi.org/10.1007/s13205-021-02752-5

Jeevalatha, A., Siddappa, S., Kumar, R., Tiwari, R. K., Lal, M. K., Sharma, S., Chakrabarty, S. K., & Singh, B. P. (2023). RNA-seq analysis reveals an early defense response to tomato leaf curl New Delhi virus in potato cultivar Kufri Bahar. *Functional & Integrative Genomics, 23*, 215. https://doi.org/10.1007/s10142-023-01138-5

Jones, R. A. C. (2021). Global plant virus disease pandemics and epidemics. *Plants, 10*(2), 233. https://doi.org/10.3390/plants10020233

Jones, R., & Barbetti, M. (2012). Influence of climate change on plant disease infections and epidemics caused by viruses and bacteria. *CAB Reviews: Perspectives in Agriculture,*

Veterinary Science, Nutrition and Natural Resources, 7, 1–31. https://doi.org/10.1079/PAVSNNR20127022

Kumar, R., Tiwari, R. K., Jeevalatha, A., Sundaresha, S., Shah, M. A., Sharma, S., Sagar, V., Kumar, M., & Chakrabarti, S. K. (2021). Potato apical leaf curl disease: Current status and perspectives on a disease caused by tomato leaf curl New Delhi virus. *Journal of Plant Diseases and Protection, 128*, 897–911. https://doi.org/10.1007/s41348-021-00463-w

Kumar, R., Lal, M. K., Tiwari, R. K., Chourasia, K. N., Kumar, A., Kumar, R., Sharma, S., & Singh, B. (2023a). Investigating the interplay between tomato leaf curl New Delhi virus infection, starch metabolism and antioxidant defence system in potato (Solanum tuberosum L.). *Antioxidants, 12*, 1447. https://doi.org/10.3390/antiox12071447

Kumar, R., Lal, M. K., Tiwari, R. K., Kumar, A., Behera, B., Nayak, L., Dash, G. K., Sahoo, S. K., Jena, J., Lal, P., & Altaf, M. A. (2023b). Traditional and emerging climate-resilient agricultural practices for enhancing food production and nutritional quality. In T. Aftab (Ed.), *New frontiers in plant-environment interactions* (Environmental science and engineering). Springer. https://doi.org/10.1007/978-3-031-43729-8_21

Lahlali, R., Taoussi, M., Laasli, S.-E., Gachara, G., Ezzouggari, R., Belabess, Z., Aberkani, K., Assouguem, A., Meddich, A., El Jarroudi, M., & Barka, E. A. (2024). Effects of climate change on plant pathogens and host-pathogen interactions. *Crop and Environment, 3*(3), 159–170. https://doi.org/10.1016/j.crope.2024.05.003

Lal, M. K., Tiwari, R. K., Kumar, R., Naga, K. C., Kumar, A., Singh, B., Raigond, P., Dutt, S., Chourasia, K. N., Kumar, D., Parmar, V., & Changan, S. S. (2021). Effect of potato apical leaf curl disease on glycemic index and resistant starch of potato (Solanum tuberosum L.) tubers. *Food Chemistry, 359*, 129939. https://doi.org/10.1016/j.foodchem.2021.129939

Lee, C. (2018). Therapeutic modulation of virus-induced oxidative stress via the Nrf2-dependent antioxidative pathway. *Oxidative Medicine and Cellular Longevity, 2018*, 6208067. https://doi.org/10.1155/2018/6208067

Lei, J., Dai, P., Li, Y., Zhang, W., Zhou, G., Liu,C., & Liu, X. (2021). Heritable gene editing using FT mobile guide RNAs and DNA viruses. *Plant methods, 17*(1),20. https://doi.org/10.1186/s13007-021-00719-4

Lefeuvre, P., Martin, D. P., Elena, S. F., Shepherd, D. N., Roumagnac, P., & Varsani, A. (2019). Evolution and ecology of plant viruses. *Nature Reviews. Microbiology, 17*(10), 632–644. https://doi.org/10.1038/s41579-019-0232-3

Llorca, C. M., Potschin, M., & Zentgraf, U. (2014). bZIPs and WRKYs: Two large transcription factor families executing two different functional strategies. *Frontiers in Plant Science, 5*, 169. https://doi.org/10.3389/fpls.2014.00169

Martínez-Balerdi, M., Caballero, J., Aguirre, E., Caballero, P., & Beperet, I. (2025). Baculoviruses as MicrobialPesticides: Potential, Challenges, and Market Overview. *Viruses, 17*(7), 917. https://doi.org/10.3390/v17070917

Moriones, E., Praveen, S., & Chakraborty, S. (2017). Tomato leaf curl New Delhi virus: An emerging virus complex threatening vegetable and fiber crops. *Viruses, 9*(10), 264. https://doi.org/10.3390/v9100264

Moullec, F., Barrier, N., Drira, S., Guilhaumon, F., Marsaleix, P., Somot, S., Ulses, C., Velez, L., & Shin, Y.-J. (2019). An end-to-end model reveals losers and winners in a warming Mediterranean Sea. *Frontiers in Marine Science, 6*, 345.

Muhammad Aslam, M., Waseem, M., Jakada, B. H., Okal, E. J., Lei, Z., Saqib, H. S. A., Yuan, W., Xu, W., & Zhang, Q. (2022). Mechanisms of abscisic acid-mediated drought stress responses in plants. *International Journal of Molecular Sciences, 23*(3), 1084. https://doi.org/10.3390/ijms23031084

Naga, K. C., Subhash, S., Bhatnagar, A., Tiwari, R. K., Kumar, R., Verma, G., Sharma, S., & Chakrabarti, S. K. (2021a). Influence of host plants on virus acquisition and endosymbionts of whitefly, Bemisia tabaci (Gennadius). *Indian Journal of Entomology, 83*, 371–375. https://doi.org/10.5958/0974-8172.2020.00206.0

Naga, K. C., Sundaresha, S., Kumar, R., Tiwari, R. K., Subhash, S., Verma, G., Buckseth, T., Bairwa, A., Sharma, S., Katare, S., & Srivastava, R. M. (2021b). A new record of Asia II 5 genetic group of Bemisia tabaci (Gennadius) in the major potato growing areas of India and its relationship with tomato leaf curl New Delhi virus infecting potato. *3 Biotech, 11*, 421. https://doi.org/10.1007/s13205-021-02966-7

Nagendran, K., Venkataravanappa, V., Chauhan, N. S., Kodandaram, M. H., Rai, A. B., Singh, B., & Vennila, S. (2019). Viral diseases: A threat for tomato cultivation in Indo-Gangetic eastern plains of India. *Journal of Plant Pathology, 101*(1), 15–22. https://doi.org/10.1007/s42161-018-0124-9

Pagliarani, C., Moine, A., Chitarra, W., Nerva, L., Catoni, M., Tavazza, R., Matić, S., Vallino, M., Secchi, F., & Noris, E. (2022). The C4 protein of tomato yellow leaf curl Sardinia virus primes drought tolerance in tomato through morphological adjustments. *Horticulture Research, 9*, uhac164. https://doi.org/10.1093/hr/uhac164

Pantaleo, V., Vitali, M., Boccacci, P., Miozzi, L., Cuozzo, D., Chitarra, W., Mannini, F., Lovisolo, C., & Gambino, G. (2016). Novel functional microRNAs from virus-free and infected Vitis vinifera plants under water stress. *Scientific Reports, 6*, 20167. https://doi.org/10.1038/srep20167

Pianka, A.I. and Izhevsky, S.S. (1979). The introduction and application of beneficial organisms for pest control in theUSSR. In: Proceedings of the joint American-Soviet conference on use of beneficial organisms in the control ofcrop pests. *Wash.*, 31-35

Polston, J. E., & Anderson, P. K. (1997). The emergence of whitefly-transmitted geminiviruses in tomato in the Western Hemisphere. *Plant Disease, 81*(12), 1358–1369. https://doi.org/10.1094/pdis.1997.81.12.1358

Poudel, M., Mendes, R., Costa, L. A. S., Bueno, C. G., Meng, Y., Folimonova, S. Y., Garrett, K. A., & Martins, S. J. (2021). The role of plant-associated bacteria, fungi, and viruses in drought stress mitigation. *Frontiers in Microbiology, 12*, 743512. https://doi.org/10.3389/fmicb.2021.743512

Reyna, P., Suarez, F., Balzarini, M., & Rodriguez Pardina, P. (2023). Influence of climatic variables on incidence of whitefly-transmitted begomovirus in soybean and bean crops in North-Western Argentina. *Viruses, 15*(2), 462. https://www.mdpi.com/1999-4915/15/2/462

Rojas, M. R., Macedo, M. A., Maliano, M. R., Soto-Aguilar, M., Souza, J. O., Briddon, R. W., Kenyon, L., Rivera Bustamante, R. F., Zerbini, F. M., Adkins, S., Legg, J. P., Kvarnheden, A., Wintermantel, W. M., Sudarshana, M. R., Peterschmitt, M., Lapidot, M., Martin, D. P., Moriones, E., Inoue-Nagata, A. K., & Gilbertson, R. L. (2018). World management of geminiviruses. *Annual Review of Phytopathology, 56*, 637–677. https://doi.org/10.1146/annurev-phyto-080615-100327

Roossinck, M. J. (2011). The good viruses: Viral mutualistic symbioses. *Nature Reviews Microbiology, 9*(2), 99–108.

Roossinck, M. J. (2015a). A new look at plant viruses and their potential beneficial roles in crops. *Molecular Plant Pathology, 16*(4), 331–333. https://doi.org/10.1111/mpp.12241

Roossinck, M. J. (2015b). Plants, viruses and the environment: Ecology and mutualism. *Virology, 479–480*, 271–277. https://doi.org/10.1016/j.virol.2015.03.041

Roossinck, M. J., Martin, D. P., & Roumagnac, P. (2015). Plant virus metagenomics: Advances in virus discovery. *Phytopathology, 105*(6), 716–727. https://doi.org/10.1094/phyto-12-14-0356-rvw

Sachdev, S., Ansari, S. A., Ansari, M. I., Fujita, M., & Hasanuzzaman, M. (2021). Abiotic stress and reactive oxygen species: Generation, signaling, and defense mechanisms. *Antioxidants, 10*(2), 277. https://www.mdpi.com/2076-3921/10/2/277

Saharan, B. S., Brar, B., Duhan, J. S., Kumar, R., Marwaha, S., Rajput, V. D., & Minkina, T. (2022). Molecular and physiological mechanisms to mitigate abiotic stress conditions in plants. *Life, 12*(10), 1634. https://doi.org/10.3390/life12101634

Sarkar, M., Gupta, D., Singh, O. W., Paul, S., Kumar, R., Mandal, B., & Roy, A. (2024). Identification of pathogenicity determinants in ToLCNDV and their RNAi-based knock-down for disease management in Nicotiana benthamiana and tomato plants. *Frontiers in Microbiology, 15*, 1481523. https://doi.org/10.3389/fmicb.2024.1481523

Sharma, N., & Prasad, M. (2017). An insight into plant–tomato leaf curl New Delhi virus interaction. *The Nucleus, 60*(3), 335–348. https://doi.org/10.1007/s13237-017-0224-5

Sharma, N., Muthamilarasan, M., Dulani, P., & Prasad, M. (2021). Genomic dissection of ROS detoxifying enzyme encoding genes for their role in the antioxidative defense mechanism against tomato leaf curl New Delhi virus infection in tomato. *Genomics, 113*(3), 889–899. https://doi.org/10.1016/j.ygeno.2021.01.022

Sharma, V., Mohammed, S. A., Devi, N., Vats, G., Tuli, H. S., Saini, A. K., Dhir, Y. W., Dhir, S., & Singh, B. (2024). Unveiling the dynamic relationship of viruses and/or symbiotic bacteria with plant resilience in abiotic stress. *Stress Biology, 4*(1), 10. https://doi.org/10.1007/s44154-023-00126-w

Shteinberg, M., Mishra, R., Anfoka, G., Altaleb, M., Brotman, Y., Moshelion, M., Gorovits, R., & Czosnek, H. (2021). Tomato yellow leaf curl virus (TYLCV) promotes plant tolerance to drought. *Cells, 10*(11), 2875. https://doi.org/10.3390/cells10112875

Singh, R., Rai, N., Singh, M., Singh, R., & Kumar, P. (2015). Effect of climate change on tomato leaf curl virus (ToLCV) disease in tomato. *Indian Journal of Agricultural Sciences, 85*, 290–292. https://doi.org/10.56093/ijas.v85i2.46547

Stobbe, A. H., & Roossinck, M. J. (2014). Plant virus metagenomics: What we know and why we need to know more. *Frontiers in Plant Science, 5*, 150. https://doi.org/10.3389/fpls.2014.00150

Sujatha, S., Sindhura, K. A. V., Koti, P. S., Hiremath, S., Muttappagol, M., Vinay Kumar, H. D., Shankarappa, K. S., Venkataravanappa, V., Reddy, K. M. S., & Reddy, C. N. L. (2025). Influence of weather and seasonal factors on whitefly dynamics, associated endosymbiotic microbiomes, and Begomovirus transmission causing tomato leaf curl disease: Insights from a metagenomic perspective. *Frontiers in Microbiology, 16*, 1555058. https://doi.org/10.3389/fmicb.2025.1555058

Trebicki, P. (2020). Climate change and plant virus epidemiology. *Virus Research, 286*, 198059. https://doi.org/10.1016/j.virusres.2020.198059

Tsai, W.-A., Brosnan, C. A., Mitter, N., & Dietzgen, R. G. (2022). Perspectives on plant virus diseases in a climate change scenario of elevated temperatures. *Stress Biology, 2*(1), 37. https://doi.org/10.1007/s44154-022-00058-x

Vo, T. T. B., Tabassum, M., Nattanong, B., Qureshi, M. A., Im, H.-J., Parrella, G., Kil, E.-J., & Lee, S. (2025). The insidious threat: Assessing the dangers and spread of tomato leaf curl New Delhi virus. *The Plant Pathology Journal, 41*(1), 1–16. https://doi.org/10.5423/PPJ.RW.11.2024.0177

Westwood, J. H., McCann, L., Naish, M., Dixon, H., Murphy, A. M., Stancombe, M. A., Bennett, M. H., Powell, G., Webb, A. A., & Carr, J. P. (2013). A viral RNA silencing suppressor interferes with abscisic acid-mediated signalling and induces drought tolerance in Arabidopsis thaliana. *Molecular Plant Pathology, 14*(2), 158–170. https://doi.org/10.1111/j.1364-3703.2012.00840.x

Wu, X., & Ye, J. (2020). Manipulation of jasmonate signaling by plant viruses and their insect vectors. *Viruses, 12*(2), 148. https://doi.org/10.3390/v12020148

Wu, X., Chen, S., Zhang, Z., Zhou, W., Sun, T., Ning, K., Xu, M., Ke, X., & Xu, P. (2024). A viral small interfering RNA-host plant mRNA pathway modulates virus-induced drought tolerance by enhancing autophagy. *The Plant Cell, 36*(9), 3219–3236. https://doi.org/10.1093/plcell/koae158

Xu, P., Chen, F., Mannas, J. P., Feldman, T., Sumner, L. W., & Roossinck, M. J. (2008). Virus infection improves drought tolerance. *New Phytologist, 180*(4), 911–921. https://doi.org/10.1111/j.1469-8137.2008.02627.x

Xu, A., Zhao, Z., Chen, W., Zhang, H., Liao, Q., Chen, J., Carr, J. P., & Du, Z. (2013). Self-interaction of the cucumber mosaic virus 2b protein plays a vital role in the suppression of RNA silencing and the induction of viral symptoms. *Molecular Plant Pathology, 14*(8), 803–812. https://doi.org/10.1111/mpp.12051

Zerbini, F. M., Briddon, R. W., Idris, A., Martin, D. P., Moriones, E., Navas-Castillo, J., Rivera-Bustamante, R., Roumagnac, P., Varsani, A., & Consortium, I. R. (2017). ICTV virus taxonomy profile: Geminiviridae. *Journal of General Virology, 98*(2), 131–133. https://doi.org/10.1099/jgv.0.000738

Zhan, X., Zhang, F., Li, N., Xu, K., Wang, X., Gao, S., Yin, Y., Yuan, W., Chen, W., Ren, Z., Yao, M., & Wang, F. (2024). CRISPR/Cas: An emerging toolbox for engineering virus resistance in plants. *Plants, 13*(23), 3313. https://doi.org/10.3390/plants13233313

Zhang, X., & Yu, W. (2022). Heat shock proteins and viral infection. *Frontiers in Immunology, 13*, 947789. https://doi.org/10.3389/fimmu.2022.947789

Zhang, X., Yuan, Y. R., Pei, Y., Lin, S. S., Tuschl, T., Patel, D. J., & Chua, N. H. (2006). Cucumber mosaic virus-encoded 2b suppressor inhibits Arabidopsis Argonaute1 cleavage activity to counter plant defense. *Genes & Development, 20*(23), 3255–3268. https://doi.org/10.1101/gad.1495506

Zhang, H., Tan, X., He, Y., Xie, K., Li, L., Wang, R., Hong, G., Li, J., Li, J., Taliansky, M., MacFarlane, S., Yan, F., Chen, J., & Sun, Z. (2019). Rice black-streaked dwarf virus P10 acts as either a synergistic or antagonistic determinant during superinfection with related or unrelated virus. *Molecular Plant Pathology, 20*(5), 641–655. https://doi.org/10.1111/mpp.12782

RNA Interference (RNAi) and Other Molecular Approaches for Tomato Leaf Curl New Delhi Virus Control

13

Rohit Sharma, Tuyelee Das, Komal Devi, and Ashish Prasad

Abstract

Tomato leaf curl New Delhi virus (ToLCNDV) was initially identified in 1995, causing tomato infection. Within 29 years, it rapidly expanded its host range and distribution, causing infections in more than 19 countries and economically important crop families, including *Cucurbitaceae*, *Solanaceae*, *Malvaceae*, *Fabaceae*, and *Euphorbiaceae.* There are various traditional ways to control ToLCNDV infection, such as insecticides, insect-proof nets, crop rotation, and sowing resistant varieties. These methods are very effective but also have some flaws, such as being environmentally unfriendly, costly, requiring increased labor, and limited long-term effectiveness. Recently, RNA interference (RNAi) has emerged as a potent molecular approach to control this viral infection by degrading essential viral proteins' mRNA using cellular enzymes. This chapter highlights the mechanism of action, various applications, and challenges of the RNAi technique in plant protection against viral infection. This book chapter also accommodates the mechanism of action, advantages, and challenges of other molecular methods, such as CRISPR/Cas9 gene editing, virus-induced gene silencing (VIGS), host-mediated RNA silencing (HM-RNAi), and transgenic resistance plant development as potent tools for managing ToLCNDV infection. This book chapter highlights the innovative application of molecular techniques in addressing the challenges of ToLCNDV infection and mitigating food scarcity resulting from crop damage in affected areas.

R. Sharma · K. Devi · A. Prasad (✉)
Department of Botany, Kurukshetra University, Kurukshetra, Haryana, India
e-mail: ashishprasad0411@kuk.ac.in

T. Das
University Centre for Research and Development, Chandigarh University, Mohali, Punjab, India, 140413

R. Kumar et al. (eds.), *Tomato Leaf Curl New Delhi Virus (ToLCNDV)*,
https://doi.org/10.1007/978-981-95-2852-3_13

Keywords

ToLCNDV · RNAi · CRISPR · Tomato · Whitefly

1 Introduction

Viruses are infectious agents with nucleoprotein complexes and act as intracellular parasites that rely on the host cellular system. They lack key components for central dogma processes; therefore, they need host molecular machinery to complete essential biochemical processes and their infectious cycle (Nazarov et al., 2020). Some of them also require vectors to transmit to their host and spread infections. Viruses have small-sized, continuously mutating genomes that encode multifunctional proteins targeting the host defense system and hence prevent themselves (Gong et al., 2021). *Tobacco mosaic virus* (TMV) discovery unlocked new avenues for research in plant virology. The International Committee on Taxonomy of Viruses (ICTV) has listed over 2100 plant viruses so far (J. Wu et al., 2024). ToLCNDV is an economically significant plant virus belonging to the genus *Begomovirus*, the largest genus within the family *Geminiviridae* (Cai et al., 2023).

ToLCNDV made its initial appearance in 1995 in northern India, but it increased its host range and geographic distribution in the last two decades. Initially, it infected tomato plants in Asia, but it has now spread to Tunisia, Iran, Spain, southern Italy, etc. (Sharma & Prasad, 2017). Its infection is not only confined to members of *Solanaceae* but also infects a broad spectrum of prominent crop plant families, such as *Fabaceae*, *Malvaceae*, *Cucurbitaceae*, and *Euphorbiaceae* (Seal et al., 2006). This behavior of ToLCNDV is due to the combination of genetic adaptability, an efficient insect vector, and environmental factors. The primary vector of ToLCNDV is *Bemisia tabaci* (whitefly), which belongs to the order *Hemiptera* and family *Aleyrodidae*. ToLCNDV, a bipartite *Begomovirus*, is made of DNA-A and DNA-B circular single-stranded DNA. The virus's DNA components are roughly 2.7 kilobases long (Srivastava et al., 1995). The DNA-A segment has eight open reading frames (ORFs), three oriented in the virion-sense direction (AV1, AV2, and AV3) and five oriented in the complementary-sense direction (AC1, AC2, AC3, AC4, and AC5). Among these, AV1 and AV2 code for the coat protein (CP) and pre-coat protein, respectively. On the complementary strand, AC1, AC2, AC3, and AC4 encode the replication-associated protein (Rep), transcriptional activator protein (TrAP), replication enhancer protein (REn), and a viral effector, respectively (Kumar et al., 2021; Jeevalatha et al., 2021, 2023). Satellite DNAs may also be present along with the viral genome during natural infection, which can modulate the host range and its pathogenicity. Among these DNA satellites, beta satellites are primarily involved in ToLCNDV infection (Zaidi et al., 2017). ToLCNDV has caused severe production loss in numerous crops. Even in elite tomato cultivars, yellow mosaic disease causes 100% yield loss in India (Vo et al., 2023, b). ToLCNDV induces distinct symptoms. In tomatoes, these include chlorosis (yellowing), leaf curling, stunted growth, mosaic patterns, and poor fruit set (Ito et al., 2008). Figure 13.1 displays some of the

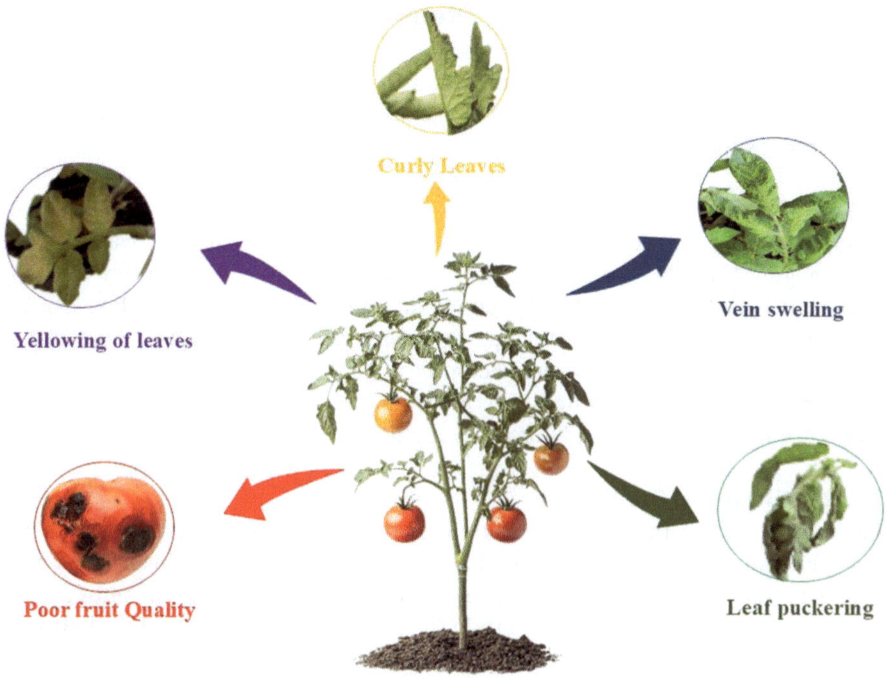

Fig. 13.1 Major symptoms of ToLCNDV in a host plant

important symptoms of ToLCNDV infection in tomato. Due to its symptomatic effects, expanding host range, and significant yield losses, ToLCNDV poses a major economic threat to agricultural production. Consequently, it has become a key subject of study for plant virologists aiming to understand its pathogenesis and develop novel, effective disease management and prevention strategies. This book chapter highlights the various conventional or primitive method, RNA interference, and other molecular techniques and their advantages and disadvantages in controlling ToLCNDV infection.

2 Conventional Method

Traditional methods to control ToLCNDV primarily focus on vector management, cultural practices, soil management, and using resistant varieties. The effectiveness of all these vector-controlling methods has been improved when these are used in combination with one another, and this combining strategy is known as integrated pest management (IPM). Conventional methods of controlling ToLCNDV infection have been discussed below:

2.1 Vector Management (Whitefly Control)

Vector management is the most important method for preventing virus spread. *B. tabaci* is one of the main vectors of the genus *Begomovirus*, including ToLCNDV, and serves as the principal host.

2.1.1 Physical Barriers

Complementing genetic resistance, several physical strategies have proven effective in managing ToLCNDV spread, especially whiteflies (*B. tabaci*), from infecting host plants. This is facilitated by employing row coverings, screenhouses, and insect-proof nets. This further helps to reduce vector exposure to host plants. Mulching can also help to prevent vector movement. Reflective mulches, for example, can be used to deter whiteflies by limiting their ability to detect or locate host plants (Dinsdale et al., 2010; Njoroge et al., 2017).

2.1.2 Chemical Control

Insecticides and pesticides have traditionally been the primary methods for managing whiteflies (Legg et al., 2014). However, the continuous application of these chemicals can adversely affect soil ecology and lead to the development of a resistant whitefly population (Chavan et al., 2015). So, there is a need for alternative solutions, such as the development of systemic insecticides and eco-friendly plant-based solutions. Recently, neonicotinoids and flupyradifurone types of systemic insecticides gained the attention of plant virologists. These exhibit the highest efficiency in reducing ToLCNDV infection by controlling *B. tabaci* (Anikina et al., 2023). A recent study in zucchini plants infected by ToLCNDV highlights that biostimulants such as Fulcrum, Crop Set, Redicrop 2000, and BioVeg 500 increase the biomass, number of leaves, and fruits (Donati et al., 2022). Another class of phytochemicals, such as essential oils (EOs) and hydrosols (HS), has recently gained the attention of the scientific community in plant-viral interactions. EOs and HS from different plants, such as *Origanum vulgare*, *Rosmarinum officinalis*, etc., play an important role in controlling ToLCNDV infection in *Cucurbita pepo* (Taglienti et al., 2022).

2.1.3 Biological Control

Techniques for controlling plant viruses have evolved, transitioning from traditional methods to more sophisticated biological approaches. Utilizing plant resistance systems and modifying inducible defenses are the main goals of contemporary methods. Nonconventional resistance techniques include introducing nucleic acid sequences, such as coat protein-mediated protection and antisense nucleic acids, into plants that disrupt the cycles of viral infection (Anikina et al., 2023). The overview of insect eggs and nymph-eating biological agents remains one of the best strategies to control the vector and virus simultaneously. For instance, the introduction of predatory mite, i.e., *Amblyseius swirskii*, known to feed on whitefly eggs and nymphs, significantly reduces the transmission of ToLCNDV in zucchini crops (Tellez et al., 2017). Recent developments in controlling plant viruses include ribosome-inactivating proteins, antiviral protein generation through transgenic, and pathogen-derived resistance strategies. One preventative strategy focuses on using the apical meristem technique in conjunction with chemotherapy and thermotherapy to check and recover planting material (Rubio et al., 2020). Current research optimizes the virus through the innate plant immunity route and has finally decoded the pathways related to plant resistance (Rubio et al., 2020).

2.1.4 Soil Management for Virus Suppression

Preventing soil-borne diseases and maintaining plant health is crucial, which can be achieved through organic modifications. TSWV-tomato pathosystem utilizes organic resources such as agricultural waste, alfalfa straw, compost, and biochar to reduce the occurrence of viral infections. These approaches improve disease suppression by boosting soil microbial diversity and activity. Conservation tillage, cover crops, and crop rotations all aid in disease management by improving microbial populations and soil structure (Mazzola, 2004). As shown with verticillium wilt and common potato scab, nitrogen-rich organic additions can release allelochemicals that limit soil-borne diseases. Organic additions and residue management take time to produce disease-resistant soils, but the benefits accumulate over time, enhancing plant productivity and soil health overall.

2.1.5 Development of a Resistant Variety

There are six key resistance loci, Ty-1 through Ty-6, that aid in the development of resistant plant types (Gill et al., 2019). Scientists exploit these loci to produce resistance cultivars to numerous plant viruses, including ToLCNDV. ToLCNDV-resistant variants consist of Ty-1, Ty-2, and Ty-3 loci. In tomatoes, a combination of multiple Ty-2 and Ty-3 genes can increase resistance to ToLCNDV and other begomoviruses. These resistant plant varieties are comparatively more effective than single-gene-resistant varieties (Nevame et al., 2018). Scientists identified additional genes, i.e., the R gene and quantitative trait loci (QTL), such as Cy-1, which help in forming resistant varieties (Sáez et al., 2017). Sharma et al. (2021) highlighted the development of a tomato resistance cultivar, H-88-78-1, against ToLCNDV infection, utilizing the R-genelSw5a (Sharma et al., 2021).

2.2 Advantages and Disadvantages of Conventional Methods

Conventional methods, including physical barriers and good cultural practices, are easily accessible and well-established, and they also help limit vector transmission and viral infection. The use of resistance varieties to reduce viral infection and yield loss is a very effective and long-standing strategy (Kamachi et al., 2007; Pandey et al., 2015). Along with advantages, conventional methods have various drawbacks, including chemical methods that ultimately increase environmental pollution and pesticides that develop resistance vectors (Vanderschuren et al., 2009). Crop rotation has been practiced for thousands of years and increases yield; however, it is not possible in monoculture systems. Successful implementation of crop rotation requires good land and cultural management changes (Krubphachaya et al., 2007). Another critical point is that resistant varieties often lack broad-spectrum resistance, and their resistance is limited to a particular pathogen. These varieties can quickly overcome the virus through rapid mutation (Sharma & Prasad, 2017). These conventional strategies don't address the root cause of viral transmission and spread and hence require continuous management practices (Basu et al., 2018). Consequently, while conventional methods have their importance, they often need more advanced and long-term strategies for solving ToLCNDV infection, such as RNA interference (RNAi) and other molecular techniques.

3 RNA Interference (RNAi) Based Methods

RNAi is a biological mechanism that causes both transcriptional gene silencing (TGS) and posttranscriptional gene silencing (PTGS). Double-stranded RNA (dsRNA) triggers gene silencing, which protects plants from viral infections by suppressing the expression of specific viral genes (Mello, 2007; Sundaresha et al., 2022). In plants, RNAi acts as a natural defense mechanism and was first discovered in the petunia flower in 1990. It was an accidental discovery where overexpression of chalcone synthase led to variegated flowers, leading to the concept of co-suppression (Fletcher et al., 2020; Hung & Slotkin, 2021). In plants, RNAi involves two primary types of small RNA (sRNA) molecules: microRNAs (miRNAs) and small interfering RNAs (siRNAs). These are important for gene silencing and preventing viral infection. Different types of sRNA are recently known in RNAi, some of which are discussed in Table 13.1. miRNA is 20–22 nucleotides in size and binds to the 3′ UTR of targeted mRNA for its regulation under different stress responses (Talmor-Neiman et al., 2006). The synthesis of miRNA requires RNA polymerase II, translating MIR genes into primary miRNA transcripts (pri-miRNA), which Drosha then processes to create precursor miRNA. Both of these processes occur in the nucleus. Pre-miRNA is synthesized and then processed by Dicer in the cytoplasm to form mature miRNA. These mature miRNAs subsequently incorporate into the RNA-induced silencing complex (RISC) for RNAi of other targeted mRNAs (J. Zhang et al., 2017). siRNAs are also similar in size to miRNAs, which is one of the main types of sRNA involved in RNAi. It mainly includes two types of siRNA, i.e., virus-derived siRNA vsiRNAs and virus-activated siRNAs (vasiRNAs). Dicer targets viral double-stranded RNAs (dsRNAs) to produce vsiRNAs, and future amplification is done by RNA-dependent RNA polymerase (RdRP) (Brodersen & Voinnet, 2006; Hamilton & Baulcombe, 1999). However, vasiRNAs are derived from host genes and play a crucial role in antiviral immunity (Cao et al., 2014). MiRNAs like miR319 and miR528, which target the host defense gene, can reduce plant immunity (Sundaresha et al., 2022). Dicer is an RNase III-type endonuclease and a key enzyme in RNAi. There are four important classes of Dicer: 1–4. Dicer 1 is involved in miRNA biogenesis, and Dicer 2, 3, and 4 are involved in siRNA biogenesis (Vaucheret & Voinnet, 2024). In response to plant defense mechanisms,

Table 13.1 Small RNAs that are involved in plant-virus interaction

Name	Function	Example	Ref
Virus-derived siRNAs (vsiRNAs)	Degrade viral RNA via RNA-induced silencing complex (RISC)	vsiRNAs from *Tobacco mosaic virus* TMV target the viral genome	Liu et al. (2021)
Natural antisense siRNAs (nat-siRNAs)	Regulate host genes involved in antiviral defense and stress responses	Antiviral nat-siRNAs from *Arabidopsis* infected with *Cucumber mosaic virus* (CMV)	Borsani et al. (2005); Vaucheret and Voinnet (2024)
Trans-acting siRNAs (ta-siRNAs)	Regulate resistance gene expression and plant immunity	TAS3-derived ta-siRNAs affect auxin signaling during viral infections	Luo et al. (2022)
Heterochromatic siRNAs (hc-siRNAs)	Involved in epigenetic silencing (DNA methylation, histone modifications)	hc-siRNAs influence *Potato virus* X (PVX) infection	Huang et al. (2019); B. Zhang et al. (2019)
Endogenous virus-activated siRNAs (vasiRNAs)	Host-derived siRNAs are activated by viral infection to regulate immune responses	vasiRNAs identified in *Arabidopsis* infected with *Turnip mosaic virus* (TuMV)	Deng et al. (2022)
miR168 (antiviral miRNA)	Regulates AGO1, a key RNA silencing component, affecting vsiRNA accumulation	In *Tomato yellow leaf curl virus* (TYLCV) infection. *Rice stripe virus* (RSV) suppresses host antiviral responses	Várallyay and Havelda (2013)
miR393 (miRNA regulating host immunity)	Targets TIR1, affecting auxin signaling and plant immune responses	*Turnip mosaic virus* (TuMV) infection in *Arabidopsis*	Jiang et al. (2022)
miR319 (virus-induced miRNA)	Modulates jasmonic acid (JA) signaling, which influences antiviral defense	*Tomato yellow leaf curl virus* (TYLCV) infection	C. Zhang et al. (2016)

viruses also produce viral suppressors of RNA silencing (VSRs) and increase their infection (Derrien et al., 2018).

3.1 Molecular Mechanism of RNAi-Mediated Gene Silencing

RNAi begins with the synthesis of mature sRNA from dsRNA precursors via dicer-like proteins. These mature sRNAs are integrated into RISC and then unwound, dependent on the strand's stability, with the strand remaining on the RISC complex as the guide strand (Hung & Slotkin, 2021; Sundaresha et al., 2022). Now, the RISC, having a guide strand, scans the target mRNA and, with the help of Argonaute (AGO), an essential component of the RISC complex, either damages the mRNA or suppresses its translation (Fletcher et al., 2020). RdRP plays a significant role in the amplification of sRNAs, which DCLs further process, and the continuous cycle of RNAi begins (Fig. 13.2).

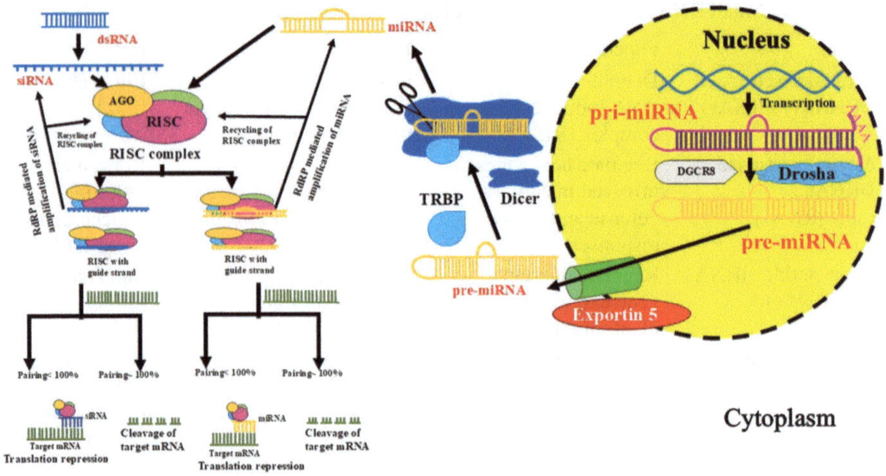

Fig. 13.2 Molecular mechanism of RNAi

4 Application of RNAi in Plants

4.1 Development of Resistant Plant Varieties

Plants can be infected by DNA and RNA viruses. RNAi is a powerful tool for generating virus-resistant plants by targeting viral genes, such as coat protein genes (Kamachi et al., 2007; Krubphachaya et al., 2007). RNAi targets different types of plant viruses. *Potato virus* Y (PVY) was the first virus that was suppressed through RNAi. RNAi technology is being used to generate resistant plant types such as *Potato spindle tuber viroid* (PSTVd) in tomatoes, *African cassava mosaic virus* (ACMV) in cassava, and many more (Singh et al., 2019; Vanderschuren et al., 2009). RNAi can develop resistant cultivars to other infections and pests. Suppression of the rice gene OsSSI2 via RNA interference (RNAi) enhanced resistance to the leaf blight bacterium *Xanthomonas oryzae* and the blast fungus *Magnaporthe grisea* (Dubrovina et al., 2019). RNAi is the most efficacious method for managing pests and insect vectors, since the expression of dsRNAs in maize plants leads to the development of resistance to the western corn rootworm (WCR) (Kamthan et al., 2015).

4.2 Improving the Yield and Productivity of Plants

RNAi can enhance plant production and nutritional quality by increasing agricultural and fruit plant yields through modifications to key agronomic features. Zhang et al. (2017) observed that RNAi-mediated silencing of the OsDWARF4 gene led to shorter plants with an erect leaf architecture, enhanced photosynthesis in the lower leaves, and increased yields under dense planting conditions in rice (J. Zhang et al., 2017). Whereas Qiao and Zhao (2011) reported that *OsGA20ox2* gene suppression resulted in semidwarf plants with increased panicle length, elevated seed yield per panicle, and increased grain weight, implying enhanced crop yield (Qiao & Zhao, 2011). However, in tomatoes, Davuluri et al. (2005) reported that when the DET1 gene was suppressed through RNAi, carotenoid and flavonoid levels were increased, thereby enhancing the nutritional value of the tomato (Davuluri et al., 2005). RNAi-targeting genes when modified light-controlled signaling pathways, resulting in higher accumulation of carotenoid and flavonoid content in tomato (Mamta & Rajam, 2017).

4.2.1 Advantages and Disadvantages of RNAi

RNAi offers several benefits to plants, including improved yield, quality, and nutrient value of economically important fruits and vegetables. Additionally, it provides defense mechanisms against various plant pathogens, such as viruses, bacteria, and fungi (Brodersen & Voinnet, 2006; Ding & Voinnet, 2007). These mechanisms involve the silencing of various genes of pathogens and hosts and help in the development of more resistant plant varieties (Brodersen & Voinnet, 2006; Tenllado et al., 2004). It also provides resistance to plants via a non-transgenic approach, addresses all the concerns of genetically modified organisms (GMOs), and serves as the main component for crop improvement. Every coin has two faces, and RNAi is no exception, having its drawbacks. It has different silencing efficiencies across plant species. GMOs produced by RNAi are not directly introduced in the field because of the varying regulations on GMOs across countries. Especially in India, GMOs are not accepted for cultivation in the field and for sale in the market (Christiaens et al., 2018). Additionally, VSRs produced by the viruses that suppress the RNAi machinery are a major concern for the scientific community (Tenllado et al., 2004).

4.2.2 Other Molecular Techniques of Controlling ToLCNDV

When it came to the development of resistance, genetic engineering technology proved to be more successful than traditional breeding techniques. These genes either encode the production of insect-toxic proteins (such as the B. *thuringiensis* bacteria's Cry and Vip proteins and/or proteinase inhibitors), potentially antiviral proteins (such as interferon and/or ribonucleases), or capsid proteins (CPs) of the virus itself. (Halder et al., 2022). In this age of agricultural revolution, technologies such as exogenous clustered regularly interspaced short palindromic repeats (CRISPR-Cas), host-induced gene silencing (HIGS), and RNA aptamer-based approach hold promise for lowering crop damage, reducing reliance on chemical pesticides, and enhancing agricultural sustainability in general.

4.3 CRISPR-Cas

CRISPR-Cas-mediated mechanism provides an effective way to develop plant resistance to viral infections by altering host characteristics. One of the key approaches involved modifying host components that are necessary for viral replication. Researchers can prevent the virus from finishing its life cycle inside the plant by altering or upsetting these susceptibility elements (Cao et al., 2020). CRISPR-edited *Nicotiana benthamiana* for improved defense against plant viruses is one such noteworthy case. Few studies have documented the modification of host UbEF1B and CCR4/NOT3, which is essential for geminivirus replication, by CRISPR, leading to resistance (Q. Lu et al., 2023; Masi et al., 2023; Yoon et al., 2020). The new field of CRISpen-Cas-based plant virus detection has enormous potential to transform plant viral infection diagnosis and treatment, provide early warnings of probable future epidemics, and assist with decision-making to limit further damage and spread. Researchers are actively working to overcome the current limits of sensitivity, specificity, and field deployment of CRISpen-Cas-based detection (Devi et al., 2024).

4.4 Artificial MicroRNAs (amiRNAs)

AmiRNAs are an important molecular technique that targets coat and pre-coat proteins of ToLCNDV, ultimately removing plant virus infections (Vu et al., 2013). ToLCNDV infection alters tomato host miRNA expression, resulting in the elevation of a large number of novel miRNAs. In *Arabidopsis thaliana*, amiRNA-mediated resistance to *Turnip mosaic virus* (TuMV) was preserved following coinfection with several viruses; however, this resistance was weakened by preinfection with some viruses (Carbonell & Daròs, 2017; Pradhan et al., 2015). Furthermore, amiRNAs and synthetic trans-acting small interfering RNAs (syn-tasiRNAs) are effective against viroids as well as viruses. Carbonell and Daròs (2017) found that amiRNAs and syn-tasiRNAs targeting *Potato spindle tuber viroid* (PSTVd) significantly reduced viroid accumulation in *Nicotiana benthamiana*, suggesting their potential application to regulate plant viroid infection (Carbonell & Daròs, 2017).

4.5 Host-Induced Gene Silencing (HIGS)

HIGS has emerged as an important alternative to chemical treatments. All HIGS techniques begin with the identification of a target gene whose RNAi-mediated silencing produces high lethality, long-term effectiveness, and minimal off-target impacts on non-target species and the environment (Zand Karimi & Innes, 2022). HIGS has been widely used to combat many diseases, such as downy mildew in lettuce by *Bremia lactucae* and Verticillium wilt in tomato and Arabidopsis by *Verticillium dahliae* (Govindarajulu et al., 2015; Song & Thomma, 2018). HIGS-based targeted gene selection, dsRNA generation, translocation, and the pathogen's capacity to absorb and digest RNA are some of the variables that affect how successful HIGS is (Zand Karimi & Innes, 2022). HIGS presents possibilities for long-term disease management; its effectiveness varies. HIGS is still a viable strategy for increasing crop resistance to a variety of pests and diseases, despite obstacles (Govindarajulu et al., 2015). Future research and innovation will be needed to manage viral diseases in agriculture and ensure food safety, environmental health, and food quality.

4.6 RNA Aptamer-Based Strategies for Plant Virus Control

Aptamers suppress viral replication and infection by binding to specific targets such as capsids, nucleoproteins, and motility proteins (Taliansky et al., 2021). Each of several peptides and synthetic zinc finger proteins with varied specificities and modes of action has been investigated to give virus resistance (Yadav et al., 2019). Using quenching GFP fluorescence in *N. benthamiana*, studies on RNA aptamers showed how endogenous protein expression is suppressed in plants. RNA-aptamers can suppress endogenous proteins, including pathogen effectors like HopU1 from *P. syringae*, potentially increasing plant resistance to phytopathogens (Abdeeva et al., 2019). A key drawback of implementing these technologies for widespread crop protection is addressing potential risks and improving efficacy.

4.7 Conclusion

ToLCNDV is an economically important virus that poses a threat to many commercially vital crops. ToLCNDV has a broad host range, transmission efficiency, and high adaptability to a changing climate (Prasad et al., 2022). Vector and soil management, as well as resistant varieties development, are conventional strategies with drawbacks, including environmental pollution, which leads to the formation of tolerant vectors and pests. Therefore, promising solutions, such as RNAi, CRISPR-Cas, etc., to control ToLCNDV infection are needed (Cao et al., 2020). RNAi has been used to silence important viral genes, and CRISPR-Cas technology allows precise genome editing, thereby increasing resistance and eliminating viral susceptible factors in plants and removing pathogenic genes from different pathogens, respectively (R. Lu et al., 2003; B. Zhang et al., 2019). Artificial microRNAs (amiRNAs) and RNA aptamers disrupt virus replication and growth, eventually preventing viral infection (Prasad et al., 2021). Key challenges include RNAi efficiency, regulatory restrictions, and the evolution of viral suppressors of RNA silencing (Singh et al., 2019). An integrated approach combining conventional and molecular techniques is needed for effective and sustainable management of ToLCNDV. Future research based on resistance mechanisms of plant and plant-virus interaction will be vital for future agricultural sustainability and food security (Basu et al., 2018).

References

Abdeeva, I. A., Maloshenok, L. G., Pogorelko, G. V., Mokrykova, M. V., & Bruskin, S. A. (2019). RNA-aptamers—As targeted inhibitors of protein functions in plants. *Journal of Plant Physiology, 232,* 127-129. https://doi.org/10.1016/j.jplph.2018.10.026

Anikina, I., Kamarova, A., Issayeva, K., Issakhanova, S., Mustafayeva, N., Insebayeva, M., Mukhamedzhanova, A., Khan, S. M., Ahmad, Z., Lho, L. H., Han, H., & Raposo, A. (2023). Plant protection from virus: A review of different approaches. *Frontiers in Plant Science, 14,* 1163270. https://doi.org/10.3389/fpls.2023.1163270

Basu, S., Kumar Kushwaha, N., Kumar Singh, A., Pankaj Sahu, P., Vinoth Kumar, R., & Chakraborty, S. (2018). Dynamics of a geminivirus-encoded pre-coat protein and host RNA-dependent RNA polymerase 1 in regulating symptom recovery in tobacco. *Journal of Experimental Botany, 69,* 2085-2102. https://doi.org/10.1093/jxb/ery043

Borsani, O., Zhu, J., Verslues, P. E., Sunkar, R., & Zhu, J. K. (2005). Endogenous siRNAs derived from a pair of natural cis-antisense transcripts regulate salt tolerance in Arabidopsis. *Cell, 123,* 1279-1291. https://doi.org/10.1016/j.cell.2005.11.035

Brodersen, P., & Voinnet, O. (2006). The diversity of RNA silencing pathways in plants. *Trends in Genetics, 22,* 268-280. https://doi.org/10.1016/j.tig.2006.03.003

Cai, L., Mei, Y., Ye, R., Deng, Y., Zhang, X., Hu, Z., Zhou, X., Zhang, M., & Yang, J. (2023). Tomato leaf curl New Delhi virus: An emerging plant begomovirus threatening cucurbit production. *aBIOTECH, 4,* 257-266. https://doi.org/10.1007/s42994-023-00118-4

Cao, Y., Zhou, H., Zhou, X., & Li, F. (2020). Control of plant viruses by CRISPR/Cas system-mediated adaptive immunity. *Frontiers in Microbiology, 11,* 593700. https://doi.org/10.3389/fmicb.2020.593700

Cao, M., Du, P., Wang, X., Yu, Y. Q., Qiu, Y. H., Li, W., Gal-On, A., Zhou, C., Li, Y., & Ding, S. W. (2014). Virus infection triggers widespread silencing of host genes by a distinct class of endogenous siRNAs in Arabidopsis. *Proceedings of the National Academy of Sciences of the United States of America, 111,* 14613–14618. https://doi.org/10.1073/pnas.1407131111

Carbonell, A., & Daròs, J. A. (2017). Artificial microRNAs and synthetic trans-acting small interfering RNAs interfere with viroid infection. *Molecular Plant Pathology, 18,* 746-753. https://doi.org/10.1111/mpp.12529

Chavan, R. D., Yeotikar, S. G., Gaikwad, B. B., & Dongarjal, R. P. (2015). Management of major pests of tomato with biopesticides. *Journal of Entomological Research, 39,* 213-217. https://doi.org/10.5958/0974-4576.2015.00022.5

Christiaens, O., Dzhambazova, T., Kostov, K., Arpaia, S., Joga, M. R., Urru, I., Sweet, J., & Smagghe, G. (2018). Literature review of baseline information on RNAi to support the environmental risk assessment of RNAi-based GM plants. *EFSA Supporting Publications, 15,* 1424E. https://doi.org/10.2903/sp.efsa.2018.en-1424

Davuluri, G. R., Van Tuinen, A., Fraser, P. D., Manfredonia, A., Newman, R., Burgess, D., Brummell, D. A., King, S. R., Palys, J., Uhlig, J., Bramley, P. M., Pennings, H. M. J., & Bowler, C. (2005). Fruit-specific RNAi-mediated suppression of DET1 enhances carotenoid and flavonoid content in tomatoes. *Nature Biotechnology, 23,* 890-895. https://doi.org/10.1038/nbt1108

Deng, Z., Ma, L., Zhang, P., & Zhu, H. (2022). Small RNAs participate in plant–virus interaction and their application in plant viral defense. *International Journal of Molecular Sciences, 23,* 696. https://doi.org/10.3390/ijms23020696

Derrien, B., Clavel, M., Baumberger, N., Iki, T., Sarazin, A., Hacquard, T., Ponce, M. R., Ziegler-Graff, V., Vaucheret, H., Micol, J. L., Voinnet, O., & Genschik, P. (2018). A suppressor screen for AGO1 degradation by the viral F-box P0 protein uncovers a role for AGO DUF1785 in sRNA duplex unwinding [OPEN]. *Plant Cell, 30,* 1353-1374. https://doi.org/10.1105/tpc.18.00111

Devi, B. M., Guruprasath, S., Balu, P., Chattopadhyay, A., Thilagar, S. S., Dhanabalan, K. V., Choudhary, M., Moparthi, S., & Jailani, A. A. K. (2024). Dissecting diagnostic and management strategies for plant viral diseases: What next? *Agriculture (Switzerland), 14,* 284. https://doi.org/10.3390/agriculture14020284

Ding, S. W., & Voinnet, O. (2007). Antiviral immunity directed by small RNAs. *Cell, 130,* 413-426. https://doi.org/10.1016/j.cell.2007.07.039

Dinsdale, A., Cook, L., Riginos, C., Buckley, Y. M., & De Barro, P. (2010). Refined global analysis of Bemisia tabaci (Hemiptera: Sternorrhyncha: Aleyrodoidea: Aleyrodidae) mitochondrial cytochrome oxidase 1 to identify species level genetic boundaries. *Annals of the Entomological Society of America, 103,* 196-208. https://doi.org/10.1603/AN09061

Donati, L., Bertin, S., Gentili, A., Luigi, M., Taglienti, A., Manglli, A., Tiberini, A., Brasili, E., Sciubba, F., Pasqua, G., & Ferretti, L. (2022). Effects of organic biostimulants added with zeolite on zucchini squash plants infected by tomato leaf curl New Delhi virus. *Viruses, 14,* 607. https://doi.org/10.3390/v14030607

Dubrovina, A. S., Aleynova, O. A., Kalachev, A. V., Suprun, A. R., Ogneva, Z. V., & Kiselev, K. V. (2019). Induction of transgene suppression in plants via external application of synthetic dsRNA. *International Journal of Molecular Sciences, 20,* 1585. https://doi.org/10.3390/ijms20071585

Fletcher, S. J., Reeves, P. T., Hoang, B. T., & Mitter, N. (2020). A perspective on RNAi-based biopesticides. *Frontiers in Plant Science, 11,* 51. https://doi.org/10.3389/fpls.2020.00051

Gill, U., Scott, J. W., Shekasteband, R., Ogundiwin, E., Schuit, C., Francis, D. M., Sim, S. C., Smith, H., & Hutton, S. F. (2019). Ty-6, a major begomovirus resistance gene on chromosome 10, is effective against tomato yellow leaf curl virus and tomato mottle virus. *Theoretical and Applied Genetics, 132,* 1543-1554. https://doi.org/10.1007/s00122-019-03298-0

Gong, P., Tan, H., Zhao, S., Li, H., Liu, H., Ma, Y., Zhang, X., Rong, J., Fu, X., Lozano-Durán, R., Li, F., & Zhou, X. (2021). Geminiviruses encode additional small proteins with specific subcellular localizations and virulence function. *Nature Communications, 12,* 4278. https://doi.org/10.1038/s41467-021-24617-4

Govindarajulu, M., Epstein, L., Wroblewski, T., & Michelmore, R. W. (2015). Host-induced gene silencing inhibits the biotrophic pathogen causing downy mildew of lettuce. *Plant Biotechnology Journal, 13,* 875-883. https://doi.org/10.1111/pbi.12307

Halder, K., Chaudhuri, A., Abdin, M. Z., Majee, M., & Datta, A. (2022). RNA interference for improving disease resistance in plants and its relevance in this clustered regularly interspaced short palindromic repeats-dominated era in terms of dsRNA-based biopesticides. *Frontiers in Plant Science, 13,* 885128. https://doi.org/10.3389/fpls.2022.885128

Hamilton, A. J., & Baulcombe, D. C. (1999). A species of small antisense RNA in posttranscriptional gene silencing in plants. *Science, 286,* 950-952. https://doi.org/10.1126/science.286.5441.950

Huang, C. Y., Wang, H., Hu, P., Hamby, R., & Jin, H. (2019). Small RNAs – Big players in plant-microbe interactions. *Cell Host and Microbe, 26,* 173-182. https://doi.org/10.1016/j.chom.2019.07.021

Hung, Y. H., & Slotkin, R. K. (2021). The initiation of RNA interference (RNAi) in plants. *Current Opinion in Plant Biology, 61,* 102014. https://doi.org/10.1016/j.pbi.2021.102014

Ito, T., Sharma, P., Kittipakorn, K., & Ikegami, M. (2008). Complete nucleotide sequence of a new isolate of tomato leaf curl New Delhi virus infecting cucumber, bottle gourd and muskmelon in Thailand. *Archives of Virology, 153,* 611-613. https://doi.org/10.1007/s00705-007-0029-y

Jeevalatha, A., Siddappa, S., Kumar, R., Tiwari, R. K., Lal, M. K., Sharma, S., Chakrabarty, S. K., & Singh, B. P. (2023). RNA-seq analysis reveals an early defense response to tomato leaf curl New Delhi virus in potato cultivar Kufri Bahar. *Functional & Integrative Genomics, 23,* 215. https://doi.org/10.1007/s10142-023-01138-5

Jeevalatha, A., Vanishree, G., Siddappa, S., Kumar, R., Kaundal, P., Kumar, A., & Chakrabarti, S. K. (2021). Molecular characterization and infectivity analysis of tomato leaf curl New Delhi virus isolates infecting potato. *3 Biotech, 11,* 203. https://doi.org/10.1007/s13205-021-02752-5

Jiang, J., Zhu, H., Li, N., Batley, J., & Wang, Y. (2022). The miR393-target module regulates plant development and responses to biotic and abiotic stresses. *International Journal of Molecular Sciences, 23,* 9477. https://doi.org/10.3390/ijms23169477

Kamachi, S., Mochizuki, A., Nishiguchi, M., & Tabei, Y. (2007). Transgenic Nicotiana benthamiana plants resistant to cucumber green mottle mosaic virus based on RNA silencing. *Plant Cell Reports, 26,* 1283-1288. https://doi.org/10.1007/s00299-007-0358-z

Kamthan, A., Chaudhuri, A., Kamthan, M., & Datta, A. (2015). Small RNAs in plants: Recent development and application for crop improvement. *Frontiers in Plant Science, 6,* 208. https://doi.org/10.3389/fpls.2015.00208

Krubphachaya, P., Juříček, M., & Kertbundit, S. (2007). Induction of RNA-mediated resistance to papaya ringspot virus type W. *Journal of Biochemistry and Molecular Biology, 40,* 404-411. https://doi.org/10.5483/bmbrep.2007.40.3.404

Kumar, R., Tiwari, R. K., Jeevalatha, A., Sundaresha, S., Shah, M. A., Sharma, S., Sagar, V., Kumar, M., & Chakrabarti, S. K. (2021). Potato apical leaf curl disease: Current status and perspectives on a disease caused by tomato leaf curl New Delhi virus. *Journal of Plant Diseases and Protection, 128*, 897–911. https://doi.org/10.1007/s41348-021-00463-w

Legg, J. P., Shirima, R., Tajebe, L. S., Guastella, D., Boniface, S., Jeremiah, S., Nsami, E., Chikoti, P., & Rapisarda, C. (2014). Biology and management of Bemisia whitefly vectors of cassava virus pandemics in Africa. *Pest Management Science, 70,* 1446-1453. https://doi.org/10.1002/ps.3793

Liu, P., Zhang, X., Zhang, F., Xu, M., Ye, Z., Wang, K., Liu, S., Han, X., Cheng, Y., Zhong, K., Zhang, T., Li, L., Ma, Y., Chen, M., Chen, J., & Yang, J. (2021). A virus-derived siRNA activates plant immunity by interfering with ROS scavenging. *Molecular Plant, 14,* 1088-1103. https://doi.org/10.1016/j.molp.2021.03.022

Lu, Q., Luo, X., Yang, X., Zhou, T., Zhang, Y., Lan, Y., Zhang, D., Zheng, L., Li, Y., Li, L., Zhang, S., & Liu, Y. (2023). CRISPR/Cas9-mediated gene editing of vacuolar ATPase subunit d mediates phytohormone biosynthesis and virus resistance in rice. *Frontiers in Plant Science, 14,* 1122978. https://doi.org/10.3389/fpls.2023.1122978

Lu, R., Martin-Hernandez, A. M., Peart, J. R., Malcuit, I., & Baulcombe, D. C. (2003). Virus-induced gene silencing in plants. *Methods, 30,* 296-303. https://doi.org/10.1016/S1046-2023(03)00037-9

Luo, L., Yang, X., Guo, M., Lan, T., Yu, Y., Mo, B., Chen, X., Gao, L., & Liu, L. (2022). TRANS-ACTING SIRNA3-derived short interfering RNAs confer cleavage of mRNAs in rice. *Plant Physiology, 188,* 347-362. https://doi.org/10.1093/plphys/kiab452

Mamta, B., & Rajam, M. V. (2017). RNAi technology: A new platform for crop pest control. *Physiology and Molecular Biology of Plants, 23,* 487-501. https://doi.org/10.1007/s12298-017-0443-x

Masi, A., Antonacci, A., Moccia, M., Frisulli, V., De Felice, M., De Falco, M., & Scognamiglio, V. (2023). CRISPR-Cas assisted diagnostics: A broad application biosensing approach. *TrAC Trends in Analytical Chemistry, 162* 117028. https://doi.org/10.1016/j.trac.2023.117028

Mazzola, M. (2004). Assessment and management of soil microbial community structure for disease suppression. *Annual Review of Phytopathology, 42,* 35-59. https://doi.org/10.1146/annurev.phyto.42.040803.140408

Mello, C. C. (2007). Return to the RNAi world: Rethinking gene expression and evolution. *Cell Death and Differentiation, 14,* 2013-2020. https://doi.org/10.1038/sj.cdd.4402252

Nazarov, P. A., Baleev, D. N., Ivanova, M. I., Sokolova, L. M., & Karakozova, M. V. (2020). Infectious plant diseases: Etiology, current status, problems and prospects in plant protection. *Acta Naturae, 12,* 46-59. https://doi.org/10.32607/actanaturae.11026

Nevame, A. Y. M., Xia, L., Nchongboh, C. G., Hasan, M. M., Alam, M. A., Yongbo, L., Wenting, Z., Yafei, H., Emon, R. M., Ismail, M. R., Efisue, A., Gang, S., Wenhu, L., & Longting, S. (2018). Development of a new molecular marker for the resistance to tomato yellow leaf curl virus. *BioMed Research International, 2018,* 8120281. https://doi.org/10.1155/2018/8120281

Njoroge, M. K., Mutisya, D. L., Miano, D. W., & Kilalo, D. C. (2017). Whitefly species efficiency in transmitting cassava mosaic and brown streak virus diseases. *Cogent Biology, 3,* 1311499. https://doi.org/10.1080/23312025.2017.1311499

Pandey, P., Sinha, R., Mysore, K. S., & Senthil-Kumar, M. (2015). Impact of concurrent drought stress and pathogen infection on plants. In: Mahalingam, R. (eds) *Combined Stresses in Plants.* Springer, Cham, 203-222. https://doi.org/10.1007/978-3-319-07899-1_10

Pradhan, B., Naqvi, A. R., Saraf, S., Mukherjee, S. K., & Dey, N. (2015). Prediction and characterization of tomato leaf curl New Delhi virus (ToLCNDV) responsive novel microRNAs in Solanum lycopersicum. *Virus Research, 195,* 183-195. https://doi.org/10.1016/j.virusres.2014.09.001

Prasad, A., Hari-Gowthem, G., Muthamilarasan, M., Hussain, Z., Yadav, P. K., Tripathi, S., & Prasad, M. (2021). Molecular characterization of SlATG18f in response to tomato leaf curl New Delhi virus infection in tomato and development of a CAPS marker for leaf curl disease tolerance. *Theoretical and Applied Genetics, 134,* 1463-1474. https://doi.org/10.1007/s00122-021-03783-5

Prasad, A., Sharma, N., Chirom, O., & Prasad, M. (2022). The sly-miR166-SlyHB module acts as a susceptibility factor during ToLCNDV infection. *Theoretical and Applied Genetics, 135,* 233-242. https://doi.org/10.1007/s00122-021-03962-4

Qiao, F., & Zhao, K. J. (2011). The influence of RNAi targeting of OsGA20ox2 gene on plant height in Rice. *Plant Molecular Biology Reporter, 29,* 952-960. https://doi.org/10.1007/s11105-011-0309-2

Rubio, L., Galipienso, L., & Ferriol, I. (2020). Detection of plant viruses and disease management: Relevance of genetic diversity and evolution. *Frontiers in Plant Science, 11,* 1092. https://doi.org/10.3389/fpls.2020.01092

Sáez, C., Esteras, C., Martínez, C., Ferriol, M., Dhillon, N. P. S., López, C., & Picó, B. (2017). Resistance to tomato leaf curl New Delhi virus in melon is controlled by a major QTL located in chromosome 11. *Plant Cell Reports, 36,* 1571-1584. https://doi.org/10.1007/s00299-017-2175-3

Seal, S. E., VandenBosch, F., & Jeger, M. J. (2006). Factors influencing begomovirus evolution and their increasing global significance: Implications for sustainable control. *Critical Reviews in Plant Sciences, 25,* 23-46. https://doi.org/10.1080/07352680500365257

Sharma, N., & Prasad, M. (2017). An insight into plant–tomato leaf curl New Delhi virus interaction. *The Nucleus, 60,* 335-348. https://doi.org/10.1007/s13237-017-0224-5

Sharma, N., Sahu, P. P., Prasad, A., Muthamilarasan, M., Waseem, M., Khan, Y., Thakur, J. K., Chakraborty, S., & Prasad, M. (2021). The Sw5a gene confers resistance to ToLCNDV and triggers an HR response after direct AC4 effector recognition. *Proceedings of the National Academy of Sciences of the United States of America, 118,* e2101833118. https://doi.org/10.1073/pnas.2101833118

Singh, K., Dardick, C., & Kundu, J. K. (2019). RNAi-mediated resistance against viruses in perennial fruit plants. *Plants, 8,* 359. https://doi.org/10.3390/plants8100359

Song, Y., & Thomma, B. P. H. J. (2018). Host-induced gene silencing compromises Verticillium wilt in tomato and Arabidopsis. *Molecular Plant Pathology, 19,* 77-89. https://doi.org/10.1111/mpp.12500

Srivastava, K. M., Hallan, V., Raizada, R. K., Chandra, G., Singh, B. P., & Sane, P. V. (1995). Molecular cloning of Indian tomato leaf curl vims genome following a simple method of concentrating the supercoiled replicative form of viral DNA. *Journal of Virological Methods, 51,* 297-304. https://doi.org/10.1016/0166-0934(94)00122-W

Sundaresha, S., Bairwa, A., Tomar, M., Kumar, R., Venkatasalam, E. P., Sagar, V., et al. (2022). In vitro method for synthesis of large-scale dsRNA molecule as a novel plant protection strategy. *Methods in molecular biology, 2408,* 211–226. https://doi.org/10.1007/978-1-0716-1875-2_14

Sundaresha, S., Jeevalatha, A., Kumar, R., Sood, S., Sharma, S., Bhardwaj, V., et al. (2022). RNA interference: A versatile tool to augment plant protection strategies in potato. In: Chakrabarti, S.K., Sharma, S., Shah, M.A. (eds) *Sustainable management of potato pests and diseases* (pp. 369–388). Springer, Singapore, 369-388. https://doi.org/10.1007/978-981-16-7695-6_15

Sundaresha, S., Sharma, S., Bairwa, A., Tomar, M., Kumar, R., Bhardwaj, V., et al. (2022). Spraying of dsRNA molecules derived from Phytophthora infestans, along with nanoclay carriers as a proof of concept for developing novel protection strategy for potato late blight. *Pest Management Science, 78,* 3183–3192. https://doi.org/10.1002/ps.6949

Taglienti, A., Donati, L., Ferretti, L., Tomassoli, L., Sapienza, F., Sabatino, M., Di Massimo, G., Fiorentino, S., Vecchiarelli, V., Nota, P., & Ragno, R. (2022). In vivo antiphytoviral activity of essential oils and hydrosols from Origanum vulgare, Thymus vulgaris, and Rosmarinus officinalis to control zucchini yellow mosaic virus and tomato leaf curl New Delhi virus in Cucurbita pepo L. *Frontiers in Microbiology, 13,* 840893. https://doi.org/10.3389/fmicb.2022.840893

Taliansky, M., Samarskaya, V., Zavriev, S. K., Fesenko, I., Kalinina, N. O., & Love, A. J. (2021). RNA-based technologies for engineering plant virus resistance. *Plants, 10,* 82. https://doi.org/10.3390/plants10010082

Talmor-Neiman, M., Stav, R., Klipcan, L., Buxdorf, K., Baulcombe, D. C., & Arazi, T. (2006). Identification of trans-acting siRNAs in moss and an RNA-dependent RNA polymerase required for their biogenesis. *Plant Journal, 48,* 511-521. https://doi.org/10.1111/j.1365-313X.2006.02895.x

Tellez, M. D. M., Simon, A., Rodriguez, E., & Janssen, D. (2017). Control of tomato leaf curl New Delhi virus in zucchini using the predatory mite Amblyseius swirskii. *Biological Control, 114,* 106-113. https://doi.org/10.1016/j.biocontrol.2017.08.008

Tenllado, F., Llave, C., & Díaz-Ruíz, J. R. (2004). RNA interference as a new biotechnological tool for the control of virus diseases in plants. *Virus Research, 102,* 85-96. https://doi.org/10.1016/j.virusres.2004.01.019

Vanderschuren, H., Alder, A., Zhang, P., & Gruissem, W. (2009). Dose-dependent RNAi-mediated geminivirus resistance in the tropical root crop cassava. *Plant Molecular Biology, 70,* 265-272. https://doi.org/10.1007/s11103-009-9472-3

Várallyay, É., & Havelda, Z. (2013). Unrelated viral suppressors of RNA silencing mediate the control of ARGONAUTE1 level. *Molecular Plant Pathology, 14,* 567-575. https://doi.org/10.1111/mpp.12029

Vaucheret, H., & Voinnet, O. (2024). The plant siRNA landscape. *Plant Cell, 36,* 246-275. https://doi.org/10.1093/plcell/koad253

Vo, T. T. B., Cho, W. K., Jo, Y., Lal, A., Nattanong, B., Qureshi, M. A., Tabssum, M., Troiano, E., Parrella, G., Kil, E. J., Lee, T. K., & Lee, S. (2023). Transcriptional analysis of the differences between ToLCNDV-India and ToLCNDV-ES leading to contrary symptom development in cucumber. *International Journal of Molecular Sciences, 24,* 2181. https://doi.org/10.3390/ijms24032181

Vo, T. T. B., Lal, A., Nattanong, B., Tabassum, M., Qureshi, M. A., Troiano, E., Parrella, G., Kil, E. J., & Lee, S. (2023). Coat protein is responsible for tomato leaf curl New Delhi virus pathogenicity in tomato. *Frontiers in Plant Science, 14,* 1206255. https://doi.org/10.3389/fpls.2023.1206255

Vu, T. V., Roy Choudhury, N., & Mukherjee, S. K. (2013). Transgenic tomato plants expressing artificial microRNAs for silencing the pre-coat and coat proteins of a begomovirus, tomato leaf curl New Delhi virus, show tolerance to virus infection. *Virus Research, 172,* 35-45. https://doi.org/10.1016/j.virusres.2012.12.008

Wu, J., Zhang, Y., Li, F., Zhang, X., Ye, J., Wei, T., Li, Z., Tao, X., Cui, F., Wang, X., Zhang, L., Yan, F., Li, S., Liu, Y., Li, D., Zhou, X., & Li, Y. (2024). Plant virology in the 21st century in China: Recent advances and future directions. *Journal of Integrative Plant Biology, 66,* 579-622. https://doi.org/10.1111/jipb.13580

Yadav, P. K., Kumar, S., Yadav, S., & Kumar, S. (2019). Role of aptamers in plant defense mechanism against viral diseases. In Yadav, G., Kumar, V., Aggarwal, N. (eds) *Aptamers.* Springer, Singapore, 169-174. https://doi.org/10.1007/978-981-13-8836-1_11

Yoon, Y. J., Venkatesh, J., Lee, J. H., Kim, J., Lee, H. E., Kim, D. S., & Kang, B. C. (2020). Genome editing of eIF4E1 in tomato confers resistance to pepper mottle virus. *Frontiers in Plant Science, 11,* 1098. https://doi.org/10.3389/fpls.2020.01098

Zaidi, S. S. E. A., Martin, D. P., Amin, I., Farooq, M., & Mansoor, S. (2017). Tomato leaf curl New Delhi virus: A widespread bipartite begomovirus in the territory of monopartite begomoviruses. *Molecular Plant Pathology, 18,* 901-911. https://doi.org/10.1111/mpp.12481

Zand Karimi, H., & Innes, R. W. (2022). Molecular mechanisms underlying host-induced gene silencing. *Plant Cell, 34,* 3183-3199. https://doi.org/10.1093/plcell/koac165

Zhang, B., Li, W., Zhang, J., Wang, L., & Wu, J. (2019). Roles of small RNAs in virus-plant interactions. *Viruses, 11,* 827. https://doi.org/10.3390/V11090827

Zhang, C., Ding, Z., Wu, K., Yang, L., Li, Y., Yang, Z., Shi, S., Liu, X., Zhao, S., Yang, Z., Wang, Y., Zheng, L., Wei, J., Du, Z., Zhang, A., Miao, H., Li, Y., Wu, Z., & Wu, J. (2016). Suppression of jasmonic acid-mediated defense by viral-inducible microRNA319 facilitates virus infection in rice. *Molecular Plant, 9,* 1302-1314. https://doi.org/10.1016/j.molp.2016.06.014

Zhang, J., Khan, S. A., Heckel, D. G., & Bock, R. (2017). Next-generation insect-resistant plants: RNAi-mediated crop protection. *Trends in Biotechnology, 35,* 871-882. https://doi.org/10.1016/j.tibtech.2017.04.009

Sustainable Strategies for the Management of Tomato Leaf Curl New Delhi Virus

14

G. Karthikeyan, K. Sivagnanapazham, K. Nagendran,
R. K. Mohana Pradeep, and N. Boopathi

Abstract

Tomato leaf curl New Delhi virus (ToLCNDV), classified under the genus *Begomovirus*, is a positive-sense single-stranded DNA (ssDNA) virus primarily transmitted by the whitefly *Bemisia tabaci*. In recent years, ToLCNDV has emerged as a significant virus due to its broad host range and severe impact on a wide spectrum of economically important crops. The virus is associated with considerable yield losses, thereby posing a critical threat to global food security. Effective field management of ToLCNDV remains challenging, primarily due to the ubiquity of its vector, the presence of multiple whitefly biotypes, the persistent mode of transmission, and the limited availability of resistant cultivars. This chapter provides a comprehensive overview of ToLCNDV, focusing on its vector-mediated transmission dynamics, early detection strategies, and surveillance approaches using advanced diagnostic platforms, such as ELISA, loop-mediated isothermal amplification (LAMP), RT-LAMP, and CRISPR/Cas12a-based assays. Furthermore, the chapter explores integrated management frameworks combining conventional practices with innovative tools, including host plant resistance, RNA interference (RNAi), microRNAs (miRNAs), grafting techniques, and genome editing technologies, like CRISPR/Cas systems.

G. Karthikeyan (✉) · K. Sivagnanapazham · R. K. Mohana Pradeep · N. Boopathi
Department of Plant Pathology, Tamil Nadu Agricultural University,
Coimbatore, Tamil Nadu, India
e-mail: gandhikarthikeyan@tnau.ac.in

K. Nagendran
Division of Crop Protection, ICAR-National Research Centre for Banana,
Tiruchirappalli, Tamil Nadu, India

R. Kumar et al. (eds.), *Tomato Leaf Curl New Delhi Virus (ToLCNDV)*,
https://doi.org/10.1007/978-981-95-2852-3_14

Keywords

Sustainable management · ToLCNDV · Resistant cultivar · Vector · Plant health

1 Introduction

In modern agriculture, plant viruses represent a significant constraint to crop productivity, causing extensive economic losses due to their detrimental effects on plant health (Jones, 2014). These viral infections not only compromise crop yields but also disrupt entire agricultural value chains, affecting food supply systems and endangering the livelihoods of farming communities—particularly in subsistence-based economies (Aworh, 2021; Kumar & Singh, 2023). The financial burden of managing plant viral diseases is substantial, often necessitating intensive inputs such as pesticides and herbicides (Castle et al., 2009). However, the excessive use of chemical agents raises critical environmental and public health concerns, including contamination of soil and water resources, harm to nontarget organisms, and health risks to agricultural workers and consumers (Ankit et al., 2020). These challenges highlight the urgency of addressing viral pathogens as key biological stressors threatening crop sustainability and food security.

Although there is no curative treatment for viral infections in plants, multiple preventive and control measures have proven effective in limiting virus spread (Kumar et al., 2022; Tatineni & Hein, 2023). Plant viruses typically enter host tissues through mechanical wounds or vector-mediated transmission, infecting individual cells where they hijack the cellular machinery for replication (Atabekov & Dorokhov, 1984). During replication, viruses may undergo mutation, recombination, or pseudo-recombination—mechanisms that enhance their genetic diversity and adaptability. Recombination involves the exchange of genetic material between distinct viral genomes, while pseudo-recombination refers to the reassortment of genetic components within the same virus. These evolutionary strategies enhance virulence, host range, and resistance-breaking potential, reinforcing the importance of studying viral adaptation mechanisms for adequate crop protection (Ghosh et al., 2022).

Within this context, the genus *Begomovirus*, belonging to the family *Geminiviridae*, is significant due to its global impact on agriculture. Among the 14 genera of *Geminiviridae* divided based on genomic organization, host range, and vector specificity, *Begomovirus* is the largest genus, known for its broad host spectrum and severe pathogenicity (Navas-Castillo et al., 2011; Brown et al., 2015). Begomoviruses possess either monopartite or bipartite genomes. The bipartite forms consist of two single-stranded circular DNA molecules: DNA-A, which encodes proteins for replication and encapsidation, and DNA-B, which is involved in viral movement. These viruses are transmitted in a circulative, persistent manner by the whitefly (*Bemisia tabaci*, *Gennadius*; order: *Hemiptera*, family: *Aleyrodidae*), infecting a broad range of dicotyledonous plant species (Naga et al., 2021a, b).

ToLCNDV, *Begomovirus solanum delhiense*, a bipartite begomovirus, was first identified in tomato (*Solanum lycopersicum*) crops in northern India in 1995 (Srivastava et al., 1995). It has since emerged as a highly virulent and rapidly spreading pathogen, affecting numerous solanaceous and cucurbitaceous crops. Each of its genomic components (DNA-A and DNA-B) measures approximately 2.7 kb and is essential for infection and systemic movement. Due to its broad host range and high transmissibility, ToLCNDV poses a serious threat to both open-field and protected cultivation systems. Efforts to manage ToLCNDV have increasingly focused on integrated disease management (IDM) strategies (Kumar et al., 2021). These approaches combine chemical, biological, and cultural practices to enhance overall control efficacy. IDM aims not only to suppress the viral pathogen and its vector but also to promote environmental sustainability and agricultural resilience. There is a growing consensus that the successful containment of ToLCNDV requires a holistic, interdisciplinary approach, integrating traditional and emerging technologies for durable disease resistance and long-term crop protection.

2 Vector Dynamics of ToLCNDV

A variety of transmission routes intensifies the threat of ToLCNDV. The virus can be transmitted through insect vectors, seeds, mechanical injury, and, more recently, by pollen. This broad transmission versatility enhances the spread of the virus and increases its adaptability across a wide range of ecological conditions. Such versatility complicates efforts to manage the disease effectively. Like other begomoviruses, ToLCNDV is primarily disseminated by the whiteflies (Padidam et al., 1995). Variations in the biology and behavior of different *B. tabaci* cryptic species influence virus emergence, host range, and transmission efficiency, thereby shaping the epidemiology of the disease. The specificity of virus-vector interactions is further demonstrated by the differential transmission capabilities of various begomoviruses within the *B. tabaci* species complex.

3 Diagnostic Tools and Surveillance

ToLCNDV has emerged as a significant threat to global vegetable production, underscoring the urgent need for rapid, sensitive, and accurate diagnostic tools to support effective disease management strategies. Over the years, a range of molecular and serological techniques has been developed and routinely employed to ensure the timely and reliable detection of ToLCNDV. Serological assays, particularly enzyme-linked immunosorbent assay (ELISA), are integral to ToLCNDV diagnostics due to their operational simplicity, reliability, and accuracy. The development of monoclonal and polyclonal antibodies, using purified virus particles or recombinant proteins expressed in *Escherichia coli*, has enabled efficient targeting of the viral coat protein, facilitating precise quantification of virus particles in test samples (Singh et al., 2020). Double antibody sandwich ELISA (DAS-ELISA) has demonstrated high efficiency in ToLCNDV detection (Figàs et al., 2017; Kumar et al., 2022). Furthermore, immuno-capture PCR (IC-PCR), which combines serological and molecular approaches, has improved diagnostic sensitivity (Naganur et al., 2023).

Among nucleic acid-based methods, polymerase chain reaction (PCR) and rolling circle amplification (RCA) are widely utilized for their high sensitivity and specificity in detecting and amplifying ToLCNDV genomes. RCA has revolutionized begomovirus diagnostics and genomic research by enabling efficient cloning and full-genome characterization of viral isolates (Inoue-Nagata et al., 2004; Haible et al., 2006). Jeevalatha et al. (2014) successfully applied RCA to detect ToLCNDV in potato mother and nucleus seed stocks. Phylogenetic and variability analyses based on complete genome sequences obtained via RCA revealed the uniqueness of ToLCNDV strains and provided evidence of recombination events (Naveen et al., 2025; Krishnan et al., 2025). RCA remains a powerful tool for tracking the emergence of novel viral strains, thereby reinforcing its role in comprehensive plant protection strategies. Moreover, the development of multiplex PCR assays allows simultaneous detection of ToLCNDV alongside related viruses such as *Tomato yellow leaf curl virus* (TYLCV) in plant samples (Hu et al., 2025).

Additional molecular techniques, such as nucleic acid spot hybridization (NASH), quantitative PCR (qPCR), and loop-mediated isothermal amplification (LAMP), have further enhanced the precision and speed of virus detection. NASH continues to serve as a reliable detection technique. The effectiveness of this method in identifying ToLCNDV associated with potato apical leaf curl disease was demonstrated by Venkatasalam et al. (2011). Alfaro-Fernández et al. (2016) developed two riboprobes targeting the replicase and coat protein regions of ToLCNDV, achieving high specificity with no cross-reactivity to closely related viruses. Quantitative PCR, including TaqMan-based assays, has facilitated accurate estimation of viral load in naturally and experimentally infected crops such as squash and tomato (Simón et al., 2018). qPCR has been adopted to quantify viral load (Vignesh et al., 2023) and to study transmission dynamics and tissue localization in cucurbit hosts (Sivagnanapazham et al., 2025). Janssen and Ruiz García (2025) further refined TaqMan qPCR by developing an assay explicitly tailored for the ToLCNDV-ES strain.

The LAMP assay has gained prominence as a rapid, field-deployable diagnostic method. It has been successfully employed for detecting ridge gourd yellow mosaic disease caused by ToLCNDV (Naganur et al., 2019; Jeevalatha et al., 2018), as well as for watermelon leaf curl and yellowing diseases (Venkataravanappa et al., 2020). Caruso et al. (2023) enhanced this approach by integrating it with a simplified sample preparation protocol for in-field diagnostics in zucchini, pepper, and tomato. Furthermore, coupling LAMP with the CRISPR-Cas12a system has enabled highly sensitive and specific detection, illustrating the promising potential of precision diagnostics in future plant virus management (Mahas et al., 2021). Beyond molecular and serological diagnostics, technological advancements in remote sensing (RS) and geographic information systems (GIS) have transformed plant disease surveillance. These tools offer large-scale spatial data crucial for early detection and monitoring of disease outbreaks. RS-based platforms, through the use of spectral indices, have shown promise in detecting plant health anomalies and mapping the spatial spread of ToLCNDV (Maji et al., 2025).

4 Management of ToLCNDV and Its Whitefly Vectors

Effective management of ToLCNDV presents considerable challenges due to several complex factors. These include its extensive host range (particularly within the *Solanaceae* and *Cucurbitaceae* families), the polyphagous and adaptable nature of its whitefly vector (*Bemisia tabaci*), the emergence of various whitefly biotypes, increasing resistance to insecticides, and the absence of durable resistance in many crop hosts (Nagendran et al., 2017). These challenges necessitate the development and implementation of innovative, integrated, and sustainable management strategies. Recent advances in molecular biology offer promising avenues for managing the whitefly-ToLCNDV pathosystem more effectively.

Integrated disease management (IDM) began in the early twentieth century as scientists searched for alternatives to chemical-based disease control. By the 1930s, they started using cultural practices like crop rotation and field sanitation to reduce disease by breaking the life cycle of plant pathogens. In the 1970s, focus shifted from heavy pesticide use to more eco-friendly and integrated approaches. Over time, combining different methods became an important strategy. IDM brings together multiple tools—such as resistant varieties, chemicals, biological agents, and good farming practices—to manage diseases effectively while protecting the environment (Jones, 2004; Razdan & Sabitha, 2009). It includes prevention, early detection, accurate diagnosis, treatment, and regular monitoring to reduce disease impact (Gurjar et al., 2018). The development of disease-resistant crop varieties through plant breeding became a key part of IDM. These varieties help manage diseases more sustainably and with long-term effectiveness. In today's world, where plant diseases are increasing, IDM plays a crucial role in maintaining healthy crops and stable food systems. For success, all components of IDM must be well understood and adapted to the specific crop and disease situation (Fig. 14.1). By combining different methods, IDM offers a reliable and sustainable solution to managing viral diseases like ToLCNDV.

4.1 Host-Plant Resistance

Breeding for disease resistance offers a sustainable and cost-effective strategy for managing plant pathogens, including ToLCNDV. Identifying and utilizing natural sources of resistance is key to this approach. Resistant cultivars, carrying genetic traits that confer tolerance or immunity, reduce reliance on chemical controls. Plant breeders employ both conventional and molecular methods to develop virus-resistant varieties. Extensive screening of wild relatives of tomato and cucurbits has

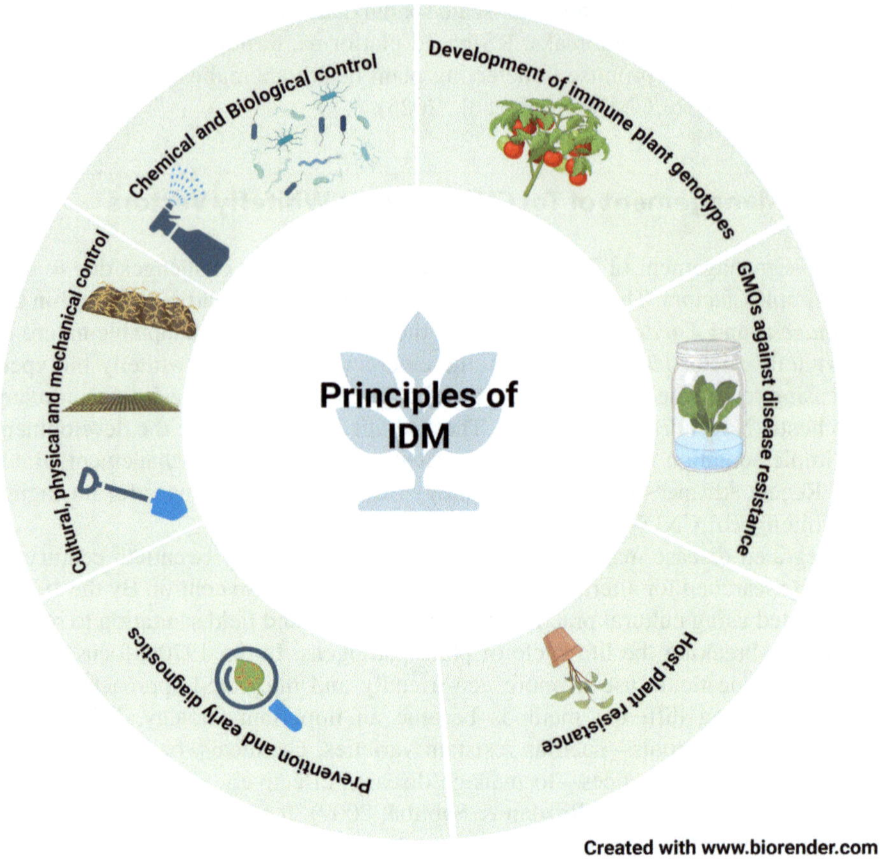

Fig. 14.1 Principles of integrated disease management against ToLCNDV

identified valuable resistance loci. In tomatoes, genes such as *Ty-1*, *Ty-2*, and *Ty-3*, which produce proteins that suppress viral replication, have been successfully introgressed into commercial cultivars (Lapidot et al., 2015; Prasanna et al., 2015; Sarkar et al., 2024; Yerasu et al., 2025). Similarly, the *Cy-1* locus in cucumber has shown strong resistance potential against ToLCNDV (Yamamoto et al., 2021).

Advanced tools like quantitative trait locus (QTL) mapping and genome-wide association studies (GWAS) have accelerated the discovery of resistance loci and the development of molecular markers for marker-assisted selection (MAS) (Hanson et al., 2016). Studies on wild *Solanum* species, including *S. habrochaites*, *S. chilense*, *S. peruvianum*, and *S. pimpinellifolium*, have highlighted natural resistance to begomoviruses, prompting further exploration. In Europe, a study evaluated ToLCNDV resistance in *Cucurbita pepo*, *C. moschata*, *C. maxima*, and related wild species (Sáez et al., 2016). Resistant *C. moschata* accessions were identified as promising candidates for breeding programs. ToLCNDV is also a major threat to luffa species, particularly *Luffa cylindrica* and *L. acutangula*, where yield losses can reach 100%. A recent field study screened 50 luffa genotypes and identified eight consistently resistant lines, also exhibiting desirable agronomic traits such as early maturity and high yield (Singh et al., 2024).

The integration of bioinformatics tools and high-throughput phenotyping platforms has further strengthened breeding efforts, enabling real-time tracking of resistance traits and enhancing understanding of host-pathogen interactions (Sinha et al., 2019). Combining these technologies with traditional breeding methods offers a robust framework for developing ToLCNDV-resistant cultivars. As research progresses, the development and deployment of resistant varieties are expected to play a pivotal role in managing ToLCNDV outbreaks (Table 14.1).

Table 14.1 ToLCNDV-resistant lines identified across various host species

Country	ToLCNDV resistance lines	References
Tomato		
Bangladesh	TLB111, TLB130, TLB133, and TLB182	Maruthi et al. (2005)
Pakistan	LA0317, LA0111, LO6221, LO6122, and LA2184	Akhtar et al. (2019)
India	H-88-78-1 (*SlSw5a* gene)	Sharma et al. (2021)
Potato		
India	Kufri Pukhraj	Jeevalatha et al. (2017)
Pumpkin		
Spain	PI 604506 and PI 381814	Sáez et al. (2020)
Taiwan	AVPU1426	Schafleitner et al. (2024)
Melon		
Spain	Indian *momordica* and wild *agrestis* (WM-7)	López et al. (2015)
France	PI 124112, PI 414723, WM9, WM7, AM 87, IC-274014, PI 282448, and PI 179901	Romay et al. (2019)
India	DSM 132, DSM 19, and DSM-11-7	Padmanabha et al. (2024)
Cucumber		
Spain	CGN23089, CGN23423, and CGN23633	Sáez et al. (2021)
Sponge gourd		
India	VRSG17-10 and VRSG17-33	Krishnan et al. (2023)
	DSG-6 and DSG-7	Islam et al. (2010)
Thailand	THLC 1, 3, 43, 43-9, 45-7-2, 45-7-5, 45-8, 46, 46-5, 56, 57, 60, 62, 70, 73, 83, 87, 96, 97, 99, 99-4, 104, 113, 115, 119, 120, 124, 125, 134, 134-6, 136, 138, 139, 141, 143, 147, 152, 153, 156, 168, 169, 170, 181, 190, 193, 195, 197, 198, 203, 204, 843, 213, 756, 387, 406, 414, 424, 455, 459, 463, 833, 781, 798, 799	Dhillon et al. (2020)

(continued)

Table 14.1 (continued)

Country	ToLCNDV resistance lines	References
India	IIHR-137, IIHR-138, and IIHR-Sel-1	Kaur et al. (2021)
	DSG-6, 7, 29, and DSGVRL-22, 23, 18, 3, 2	Singh et al. (2024)
Ridge gourd		
Thailand	THLA 2, 4, 20, 31, 36-41, 45, 48, 51, 54, 62-64, 67, 70, 70-10, 71, 75, 80, 88, 88-10, 89, 94, 96, 108, 109, 109-4, 111, 114, 114-9, 116, 119-121, 121-10, 123, 125-128, 130, 133-137, 139, 142-143, 145-147, 149, 151-152, 154-156, 158, 500, 267, 273, 275, 280, 303, 310, 311, 316, 317, 330, 335, 359, 399, 400, 416, 429, 449, and 462	Dhillon et al. (2020)
Squash		
USA	Cultivar Large Cheese from the USA (PI 604506), an Indian landrace (PI 381814) and a Japanese accession, BSUAL-252	Romero-Masegosa et al. (2021)

4.2 Cultural Control

The use of healthy planting material is a fundamental strategy in limiting the spread of any plant virus. Seed transmission of ToLCNDV has been confirmed in India and Italy (Sangeetha et al., 2018; Kil et al., 2020), underscoring the need for strict quarantine protocols when moving seed materials across regions. For vegetatively propagated crops like spine gourd, sourcing insect- and virus-free planting material is particularly important (Venkataravanappa et al., 2019). Cultural practices such as fallowing, adjusting planting dates, using physical barriers, and intercropping can effectively suppress whitefly populations, the primary vector of ToLCNDV. These methods aim to reduce the temporal overlap between the crop and virus-vector presence, thereby lowering infection risk. Crop selection also plays a vital role. For example, bitter gourd grown near tomato fields has shown susceptibility to ToLCNDV, potentially increasing disease spread (Kiran et al., 2021). Additionally, weed management is essential, as many weeds serve as alternative virus hosts and reservoirs.

4.2.1 Crop-Free Periods

The primary goal of implementing a crop-free period is to synchronize planting schedules across a region to disrupt the life cycle of both ToLCNDV and its vector, *Bemisia tabaci*. This production gap helps reduce vector populations and viral inoculum levels, thereby lowering the risk of widespread infection. Although *B. tabaci* can reproduce on over 500 plant species (Greathead, 1986) and complete elimination of all host plants is impractical, coordinated crop-free intervals can significantly limit large-scale vector movement between crops and interrupt virus transmission. The continuous, year-round cultivation of cucurbits has inadvertently created conditions conducive to virus epidemics. This has led to an increased incidence of several viruses in cucurbits, including TYLCV, *Zucchini yellow mosaic virus* (ZYMV), Cucumber mosaic virus (CMV), *Cucurbit aphid-borne yellows virus* (CABYV), and *Cucurbit yellow stunting disorder virus* (CYSDV) (Nagendran et al., 2020, 2023; Kumari et al., 2021). Several field studies also identified numerous weed species within the cropping ecosystem that host whiteflies and serve as alternate reservoirs for ToLCNDV. These suggest that cultivated crops themselves often act as the primary source of viral inoculum at the onset of epidemics. To mitigate these risks, it is advisable to separate *kharif* and *rabi* crops both temporally and spatially, reducing the carryover of whitefly populations and ToLCNDV between planting cycles.

4.2.2 Removal of Weeds

Weed management plays a vital role in reducing alternate hosts for *Bemisia tabaci* and minimizing potential sources of viral inoculum. Several weed species, including *Ecballium elaterium*, *Datura stramonium*, *Sonchus oleraceus*, and *Solanum nigrum*, have been identified as hosts for both *B. tabaci* and ToLCNDV (Juárez et al., 2019). The impact of weeds on vector population dynamics and viral outbreaks varies, depending on the cropping system and the specific plant-virus-vector interactions. Weeds can serve as reservoirs that sustain virus inoculum during crop-free periods, contributing to the persistence and spread of ToLCNDV. Recent studies have reported ToLCNDV infections in weed species from nine plant families, including *Amaranthaceae*, *Asparagaceae*, *Asteraceae*, *Convolvulaceae*, *Cucurbitaceae*, *Euphorbiaceae*, *Phyllanthaceae*, *Polygonaceae*, and *Solanaceae* (Vo et al., 2025). The wide host range of ToLCNDV highlights the remarkable adaptability of the virus and raises significant concerns about the potential for more frequent and severe outbreaks. This risk is particularly elevated due to the prevalence of weed species in both field and greenhouse conditions. Weeds often serve as alternative or intermediate hosts for crop-infecting viruses and create favorable habitats for vector populations, particularly whiteflies, owing to their perennial growth and resilience across seasons. These characteristics allow weeds to act as reservoirs for both the virus and its vectors, facilitating continuous transmission cycles. A thorough understanding of the complex interactions among ToLCNDV, weeds, and whitefly vectors is essential for developing effective field-level management strategies aimed at disrupting the virus transmission pathway and minimizing disease spread.

4.2.3 Avoidance in Space

Physical exclusion strategies aim to minimize whitefly contact with crops by either preventing insect entry or altering plant density to reduce vector movement. These methods can be highly effective in both field and greenhouse settings for reducing whitefly populations and limiting virus transmission. In some production systems, crops are cultivated entirely within insect-proof greenhouses or net structures. For example, in Israel, nearly all tomato crops are grown in enclosures made of solid plastic and/or fine-mesh screens to protect against high leaf curl disease pressure (Lapidot & Friedmann, 2002). Young plants are especially vulnerable to whitefly damage and virus infection, making early-stage protection critical. Seedlings and cuttings destined for field planting are often shielded using covered structures during their early growth phases. Row covers have also been widely used in cucurbits, bell peppers, and tomatoes to manage whitefly populations. These covers can reduce or delay the onset of whitefly-transmitted viruses, suppress symptoms such as squash silverleaf and tomato irregular ripening, and improve overall crop yields (Schuster et al., 2008).

4.2.4 Barrier's Crop

Most studies indicate that *Bemisia tabaci* adults typically fly within 2 m of the ground (Gerling & Horowitz, 1984). Consequently, tall physical barriers can obstruct or delay whitefly movement into crop fields, offering a practical management strategy. These barriers can be constructed from polyethene plastic or wood. In some cases, living barriers comprised of tall plant species are employed to reduce whitefly intrusion. Commonly used species include sorghum (*Sorghum bicolor*), Johnson grass (*Sorghum halepense*), corn (*Zea mays*), and elephant grass (*Pennisetum purpureum*). Gravena et al. (1984) reported reduced *B. tabaci* adult populations and increased predator abundance in tomato fields surrounded by sorghum barriers. Similarly, corn barriers in Mexican bell pepper fields effectively lowered viral infection rates. Whitefly-transmitted viruses are typically persistent or semi-persistent in nature (Nault, 1997), allowing infected whiteflies to remain infectious for hours, days, or even throughout their lifespan, irrespective of the number of non-host plants encountered before reaching a susceptible crop. Thus, while barrier plants may not reduce the proportion of infectious vectors, they can effectively reduce the total number of whiteflies entering the crop, thereby lowering overall infection pressure.

4.3 Behavioral Manipulation of Whitefly Vectors

These practices aim to disrupt whitefly host-searching behavior by interfering with their visual or olfactory cues.

4.3.1 Mulching

The primary objective of using mulches in whitefly management is to disrupt the ability of the insect to locate host plants. Inert ground covers, such as plastic, sawdust, straw, and rice husk, are effective in this regard, as they interfere with the visual cues' whiteflies use for host detection. Additionally, the reflective and sun-heated surfaces of these mulches can disorient or even increase mortality among the insects. Colored plastic mulches, particularly black-silver, transparent, white, and yellow, have been shown to significantly reduce whitefly populations and the incidence of whitefly-transmitted viruses (Smith et al., 2000; Priyanka et al., 2019; Nagendran et al., 2020). For example, in Jordan, the use of aluminum or silver reflective mulches led to a marked decrease in the incidence of TYLCV in tomato crops (Suwwan et al., 1988). These mulches are most effective during the early stages of crop development, before the canopy fully covers the soil surface, allowing maximum exposure of the mulch to light and pest interference.

4.3.2 Intercropping

Intercropping refers to the practice of growing two or more plant species nearby within the same field to create spatial crop arrangements. The goals of such crop associations include providing refuge for natural enemies and manipulating the host-seeking behavior of pests to protect the crop. Differential crop preference by pests can be exploited through the use of trap crops, which are more attractive to pests than the main crop. When placed near the target crop, trap crops draw pests away, thereby reducing pest pressure on the main crop. These trap crops may also be treated with insecticides to kill whiteflies during feeding or to prevent their movement to the primary crop. Al-Musa (1982) reported that intercropping cucumber between tomato rows significantly reduced TYLCV incidence. Cucumber, more attractive to whiteflies but not a host for TYLCV, served as an effective trap crop. Insecticide treatment of cucumber before senescence to prevent whitefly migration to tomatoes is widely followed in Sudan and other Middle Eastern countries. However, trap crops can shift from acting as sinks to becoming sources of infestation. Therefore, timely insecticide application is essential to prevent pest migration to the main crop. An effective trap crop should be highly attractive to whiteflies, maintain their population throughout its lifecycle, not act as a host for whitefly-transmitted viruses, and be a poor reproductive host for whiteflies.

4.4 Chemical Control

Bemisia tabaci is a highly efficient vector of ToLCNDV, and the application of insecticides has traditionally been the primary strategy for managing its populations (Legg et al., 2014). Systemic insecticides such as neonicotinoids (like imidacloprid, acetamiprid, and thiacloprid), cyantraniliprole, and other insecticides like dinotefuran were widely recommended for the control of whiteflies. However, excessive pesticide use led to the development of resistance and resurgence among whitefly populations. As a more sustainable alternative, natural plant biostimulants have shown promise in promoting plant health and resilience without adverse environmental effects. A combination of organic biostimulants (Fullcrhum Alert and BioVeg 500) and an inorganic corroborant (Clinogold, a zeolite-based product) enhanced plant fitness by improving growth parameters in both healthy and ToLCNDV-infected plants. Although no direct antiviral activity was observed, treated plants exhibited increased biomass and greater leaf, flower, and fruit production. The treatments also stimulated the accumulation of defense-related metabolites, including carbohydrates, phenylpropanoids, and free amino acids.

Plant-based insecticides or repellents are also effective tools for whitefly control. These compounds target whiteflies at various life stages, either through direct toxicity, disruption of feeding behavior, or interference with reproduction. Essential oils and hydrosols extracted from *Origanum vulgare* (oregano), *Thymus vulgaris* (thyme), and *Rosmarinus officinalis* (rosemary) have shown promising results when applied to *Cucurbita pepo* plants infected with ToLCNDV under greenhouse conditions (Taglienti et al., 2022). Their eco-friendly nature, potential antiviral properties, and ability to stimulate plant immune responses make them attractive alternatives to synthetic chemicals. In recent studies, the use of silica nanoparticles and botanical extracts, such as ginger (*Zingiber officinale*) *and horsemint* (*Mentha longifolia*), significantly reduced disease severity and enhanced systemic resistance against TYLCV. Treated plants displayed delayed symptom development with milder symptoms (El-Sawy et al., 2017).

4.5 Biological Control

Biological control of insect vectors offers an ecologically sound strategy to limit the spread of ToLCNDV by suppressing populations of *Bemisia tabaci*. One effective biocontrol agent is the predatory mite *Amblyseius swirskii*, which preys on *B. tabaci* eggs and first-instar nymphs (crawlers). Studies have demonstrated that the application of *A. swirskii* significantly reduces secondary transmission of ToLCNDV in zucchini crops (Del Mar Téllez et al., 2017). Jaber et al. (2018) evaluated the efficacy of endophytic fungal entomopathogens, including *Beauveria bassiana* and *Metarhizium brunneum*, as well as aqueous extracts of *Calotropis procera* and *Inula viscosa*, on the survival and development of *B. tabaci*. These treatments showed promising biocontrol potential. Furthermore, integrating chemical pesticides that are compatible with natural enemies, such as predatory mites, could help control *B. tabaci* adults, thereby reducing the primary transmission of ToLCNDV (Rodríguez et al., 2019). Endophytes, particularly beneficial microbial species, are emerging as sustainable tools in whitefly and virus management. For instance, *Bacillus velezensis* has been shown to induce systemic resistance in squash plants by enhancing the activity of defense-related enzymes, including β-1,3-glucanase, chitinase, polyphenol oxidase, and peroxidase. This microbial treatment also led to a reduction in whitefly population density, lowered egg-laying rates (2.28 eggs per female per day), and decreased egg hatchability by 5.7%, indicating its potential as a biocontrol agent (Soliman et al., 2022). Several biocontrol agents have been found to produce antiviral compounds, which are being explored for managing viral diseases in crop plants.

4.6 Novel Management Strategies for ToLCNDV

ToLCNDV has emerged as a significant threat (EPPO Alert List 2) to economically important crop plants, necessitating the development of innovative management strategies. Among the most promising approaches is RNA interference (RNAi), a vital, natural defense mechanism that targets explicitly and degrades viral genetic material. RNAi-based strategies have shown significant promise for disease management. This approach harnesses small RNA molecules—such as double-stranded RNAs (dsRNAs), microRNAs (miRNAs), and small interfering RNAs (siRNAs)—that are homologous to viral sequences, guiding the plant's silencing machinery to target and suppress viral replication specifically (Fig. 14.2). For instance, topical

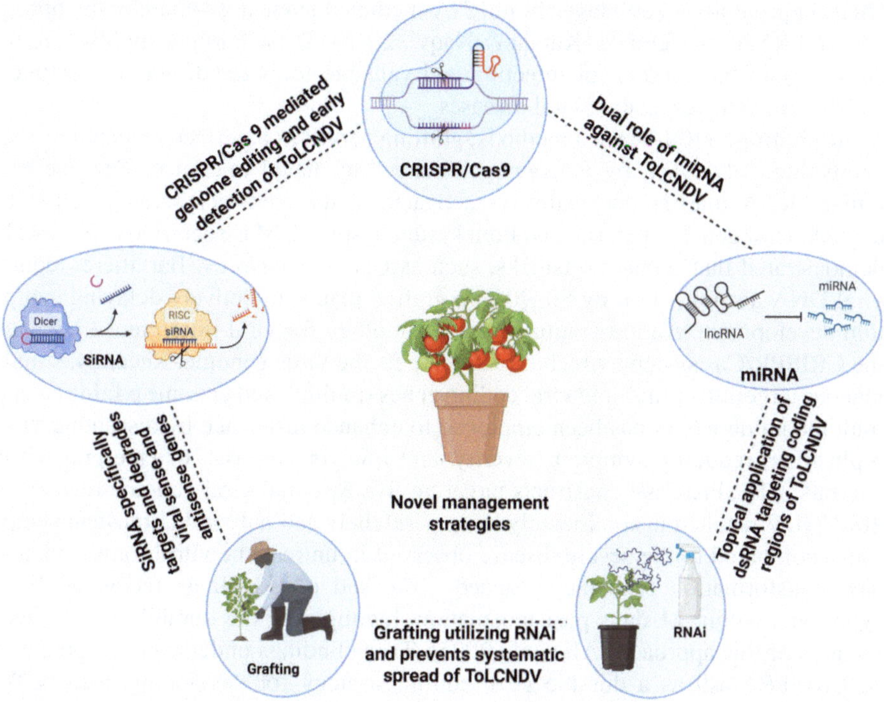

Fig. 14.2 Novel management tactics against ToLCNDV

application of chimeric dsRNA simultaneously targeting all viral coding regions of DNA-A limits the onset of severe symptoms and reduces the viral infection rate of ToLCNDV in zucchini (Frascati et al., 2024). Small interfering RNAs (siRNAs) play a central role in plant defense against invading viruses. In response to ToLCNDV infection, plants produce siRNA-derived molecules that specifically target viral RNA for degradation, thereby restricting viral replication and movement (Zaidi et al., 2017; Sahu et al., 2010). Studies on tolerant tomato cultivars, such as H-88-78-1, have demonstrated a higher accumulation of ToLCNDV-specific siRNAs in leaf tissues, which correlates with reduced viral titer and milder symptoms (Sahu et al., 2010).

In addition to siRNAs, microRNAs (miRNAs) also play a dual role in ToLCNDV interactions, acting both as components of the host defense mechanism and as contributors to viral pathogenesis. During infection, miR159/319 and miR172 are significantly upregulated in tomato leaves, correlating with leaf curling symptoms by targeting MYB transcription factors and TCP4, which regulate leaf morphology (Naqvi et al., 2010; Sharma & Prasad, 2020). Conversely, artificial miRNAs (amiRNAs) engineered to silence the ToLCNDV AC1 gene—critical for viral replication—reduce symptom severity by 40–80% in transgenic tomatoes, demonstrating their potential for engineered resistance (Al-Roshdi et al., 2023). However, the virus counteracts this defense via its AC2 protein, which binds to the host's Argonaute 1

(AGO1) protein—a key player in miRNA-mediated silencing—thereby disrupting antiviral RNA interference (Kumar & Naqvi, 2016). These findings highlight miRNAs as both biomarkers for infection and valuable tools for developing targeted resistance strategies against viral diseases.

In addition to RNA-based methods, grafting serves as an effective technique for combating ToLCNDV by leveraging rootstock-mediated tolerance. This method utilizes RNA interference pathways activated at the rootstock-scion junction to degrade viral genetic material and limit systemic spread. Mastrochirico et al. (2022) demonstrated that robust rootstocks, such as *Cucumis melo* cv. Barattiere, reduce viral DNA accumulation by 80–90% in grafted plants, effectively delaying symptom development. Another cutting-edge technology for viral interference involves the CRISPR/Cas system, which directly targets the virus genome. Recently, simultaneous targeting of multiple viral coding genes (multiplexed genome editing) using multiple guide RNAs has been employed to enhance resistance by disrupting viral replication, reducing symptom severity, and lowering the risk of emerging virus variants. CRISPR/Cas9 constructs targeting five essential viral genome regions of ToLCNDV resulted in significantly lower viral titers and only mild, transient symptoms, compared to the severe disease observed in untreated control tomato plants. The transformation efficiency reached 30%, and edited plants recovered from symptoms within 21 days post-inoculation, highlighting the durability and effectiveness of this approach (Ali et al., 2025). These findings underscore the potential of CRISPR/Cas9 as a durable gene editing strategy for developing ToLCNDV-resistant crop plants, offering a promising avenue for sustainable disease management in regions heavily impacted by this virus.

5 Conclusions and Future Perspectives

ToLCNDV is a major threat to many crops worldwide, largely due to its wide host range and efficient spread by whiteflies. Rising temperatures accelerate whitefly reproduction, increasing virus transmission risks. The ability of viruses to infect crops, weeds, and wild plants allows them to persist and spread across landscapes. This highlights the need to manage both crop and non-crop hosts in control strategies. Recent advances in breeding for resistance, especially in tomatoes and cucurbits, offer promising solutions. Molecular tools like RNA interference (RNAi), CRISPR interference (CRISPRi), marker-assisted selection, and genome editing provide sustainable alternatives to chemical controls. Integrating these molecular approaches with environmentally friendly methods, such as biological control of whiteflies and improved cultural practices, is essential. Success will depend on collaboration among researchers, breeders, and policymakers. A coordinated, multidisciplinary strategy is crucial to reduce the impact of ToLCNDV and protect global food security in horticulture.

References

Akhtar, K. P., Akram, A., Ullah, N., Saleem, M. Y., & Saeed, M. (2019). Evaluation of *Solanum* species for resistance to tomato leaf curl New Delhi virus using chip grafting assay. *Scientia Horticulturae, 256*, 108646. https://doi.org/10.1016/j.scienta.2019.108646

Alfaro-Fernández, A., Sánchez-Navarro, J. A., Landeira, M., Font, M. I., Hernández-Llópis, D., & Pallás, V. (2016).Evaluation of PCR and non-radioactive molecular hybridization techniques for the routine diagnosis of Tomato leafcurl New Delhi virus, Tomato yellow leaf curl virus and Tomato yellow leaf curl Sardinia virus. *Journal of PlantPathology*, 245–254

Ali, M. A., Khan, M. A. U., Naz, H., Malik, M. A. M., Rashid, U., Ashraf, N. M., Afroz, A., Shafiq, M., & Rao, A. Q. (2025). CAS9 mediated in-planta defence strategy against tomato leaf curl New Delhi virus (ToLCNDV) in tomato. *Journal of Phytopathology, 173*(1), e70026. https://doi.org/10.1111/jph.70026

Al-musa, A. (1982). Incidence, economic importance, and control of tomato yellow leaf curl in Jordan. *Plant Disease, 66*(7), 561. https://doi.org/10.1094/PD-66-561

Al-Roshdi, M. R., Ammara, U., Khan, J., Al-Sadi, A. M., & Shahid, M. S. (2023). Artificial microRNA-mediated resistance against Oman strain of tomato yellow leaf curl virus. *Frontiers in Plant Science, 14*, 1164921. https://doi.org/10.3389/fpls.2023.1164921

Ankit, Saha, L., Kishor, V., & Bauddh, K. (2020). Impacts of synthetic pesticides on soil health and non-targeted flora and fauna. In K. Bauddh et al. (Eds.), *Ecological and practical applications for sustainable agriculture* (pp. 65–88). Springer.

Atabekov, J. G., & Dorokhov, Y. L. (1984). Plant virus-specific transport function and resistance of plants to viruses. *Advances in Virus Research, 29*, 313–364. https://doi.org/10.1016/S0065-3527(08)60263-0

Aworh, O. C. (2021). Food safety issues in fresh produce supply chain with particular reference to sub-Saharan Africa. *Food Control, 123*, 107737. https://doi.org/10.1016/j.foodcont.2020.107737

Brown, J. K., Zerbini, F. M., Navas-Castillo, J., Moriones, E., Ramos-Sobrinho, R., Silva, J. C., et al. (2015). Revision of *Begomovirus* taxonomy based on pairwise sequence comparisons. *Archives of Virology, 160*(6), 1593–1619. https://doi.org/10.1007/s00705-015-2398-y

Caruso, A. G., Ragona, A., Bertacca, S., Montoya, M. A. M., Panno, S., & Davino, S. (2023). Development of an in-field real-time LAMP assay for rapid detection of tomato leaf curl New Delhi virus. *Plants, 12*(7), 1487. https://doi.org/10.3390/plants12071487

Castle, S., Palumbo, J., & Prabhaker, N. (2009). Newer insecticides for plant virus disease management. *Virus Research, 141*(2), 131–139. https://doi.org/10.1016/j.virusres.2008.07.030

Del Mar Tellez, M., Simon, A., Rodriguez, E., & Janssen, D. (2017). Control of *Tomato leaf curl New Delhi virus* in zucchini using the predatory mite *Amblyseius swirskii*. *Biological Control, 114*, 106–113. https://doi.org/10.1016/j.biocontrol.2017.07.007

Dhillon, N. P., Masud, M. A. T., Pruangwitayakun, S., Natheung, M., Lertlam, S., & Jarret, R. L. (2020). Evaluation of loofah lines for resistance to tomato leaf curl New Delhi virus and downy mildew, as well as key horticultural traits. *Agriculture, 10*(7), 298. https://doi.org/10.3390/agriculture10070298

El-Sawy, M. M., Elsharkawy, M. M., Abass, J. M., & Hagag, E. S. (2017). Inhibition of tomato yellow leaf curl virus by *Zingiberoffcinale* and *Menthalongifolia*. *International Journal of Antivirals & Antiretrovirology, 1*(1), 001–006.

Figàs, M. R., Alfaro-Fernández, A., Font, M. I., Borràs, D., Casanova, C., Hurtado, M., Plazas, M., Prohens, J., & Soler, S. (2017). Inoculation of cucumber, melon and zucchini varieties with tomato leaf curl New Delhi virus and evaluation of infection using different detection methods. *The Annals of Applied Biology, 170*(3), 405–414. https://doi.org/10.1111/aab.12353

Frascati, F., Rotunno, S., Accotto, G. P., Noris, E., Vaira, A. M., & Miozzi, L. (2024). Exogenous application of dsRNA for protection against tomato leaf curl New Delhi virus. *Viruses, 16*(3), 436. https://doi.org/10.3390/v16030436

Gerling, D. A. N., & Horowitz, A. R. (1984). Yellow traps for evaluating the population levels and dispersal patterns of *Bemisiatabaci* (Gennadius) (Homoptera: Aleyrodidae). *Annals of the Entomological Society of America, 77*(6), 753–759. https://doi.org/10.1093/aesa/77.6.753

Ghosh, D., Chakraborty, S., Kodamana, H., & Chakraborty, S. (2022). Application of machine learning in understanding plant virus pathogenesis: Trends and perspectives on emergence, diagnosis, host-virus interplay and management. *Virology Journal, 19*(1), 42. https://doi.org/10.1186/s12985-022-01751-7

Gravena, S., Churata-Masca, M. G. C., Arai, J., & Raga, A. (1984). Manejo integrado da mosca branca Bemisiatabaci (Gennadius, 1889) em cultivares de tomateiro de crescimento determinado visando redução de virose domosaico dourado. *Anais da Sociedade Entomológica do Brasil, 13*(1), 35–45

Greathead, A. H. (1986). Host plants. In *Bemisiatabaci: A literature survey on the cotton whitefly with an annotated bibliography* (pp. 17–25). CABI.

Gurjar, M. S., Saharan, M. S., & Aggarwal, R. (2018). Integrated disease management practices for sustainable agriculture under ICM approach. In A. K. Chaoudhary, R. S. Bana, & V. Pooniya (Eds.), *Integrated crop management practices for enhancing productivity, resource use efficiency, soil health, and livelihood practices* (pp. 113–120). Indian Council of Agricultural Research.

Haible, D., Kober, S., & Jeske, H. (2006). Rolling circle amplification revolutionizes diagnosis and genomics of geminiviruses. *Journal of Virological Methods, 135*(1), 9–16. https://doi.org/10.1016/j.jviromet.2006.01.023

Hanson, P., Lu, S. F., Wang, J. F., Chen, W., Kenyon, L., Tan, C. W., et al. (2016). Conventional and molecular marker-assisted selection and pyramiding of genes for multiple disease resistance in tomato. *Scientia Horticulturae, 201*, 346–354. https://doi.org/10.1016/j.scienta.2016.01.024

Hu, H., Zhang, J., Wu, X., Li, L., & Qian, Y. (2025). Development and application of a multiplex PCR assay for simultaneous detection of tomato yellow leaf curl virus and tomato leaf curl New Delhi virus. *Viruses, 17*(3), 322. https://doi.org/10.3390/v17030322

Inoue-Nagata, A. K., Albuquerque, L. C., Rocha, W. B., & Nagata, T. (2004). A simple method for cloning the complete begomovirus genome using the bacteriophage φ29 DNA polymerase. *Journal of Virological Methods, 116*(2), 209–211. https://doi.org/10.1016/j.jviromet.2003.11.010

Islam, S., Munshi, A. D., Mandal, B., Kumar, R., & Behera, T. K. (2010). Genetics of resistance in Luffa cylindrica Roem. against tomato leaf curl New Delhi virus. *Euphytica, 174*, 83–89. https://doi.org/10.1007/s10681-010-0194-z

Jaber, L. R., Araj, S. E., & Qasem, J. R. (2018). Compatibility of endophytic fungal entomopathogens with plant extracts for the management of sweetpotato whitefly *Bemisiatabaci* Gennadius (Homoptera: Aleyrodidae). *Biological Control, 117*, 164–171. https://doi.org/10.1016/j.biocontrol.2017.10.010

Janssen, D., & Ruiz García, L. (2025). Detection of tomato leaf curl New Delhi virus-ES by real-time quantitative PCR. In D. Wang (Ed.), *Geminiviruses: Methods and protocols* (Methods in molecular biology) (Vol. 2710, pp. 145–156). Springer. https://doi.org/10.1007/978-1-0716-3421-2_11

Jeevalatha, A., Singh, B. P., Kaundal, P., Kumar, R., & Raigond, B. (2014). RCA-PCR: A robust technique for the detection of *Tomato leaf curl New Delhi virus*-[potato] at ultra-low virus titre. *Potato Journal, 41*(1), 76–80.

Jeevalatha, A., Siddappa, S., Kumar, A., Kaundal, P., Guleria, A., Sharma, S., et al. (2017). An insight into differentially regulated genes in resistant and susceptible genotypes of potato in response to tomato leaf curl New Delhi virus-[potato] infection. *Virus Research, 232*, 22–33. https://doi.org/10.1016/j.virusres.2017.02.004

Jeevalatha, A., Kaundal, P., Kumar, R., Raigond, B., Kumar, R., Sharma, S., & Chakrabarti, S. K. (2018). Optimized loop-mediated isothermal amplification assay for tomato leaf curl New Delhi virus-[potato] detection in potato leaves and tubers. *European Journal of Plant Pathology, 150*, 565–573. https://doi.org/10.1007/s10658-017-1300-z

Jones, R. A. (2004). Using epidemiological information to develop effective integrated virus disease management strategies. *Virus Research, 100*(1), 5–30. https://doi.org/10.1016/j.virusres.2003.12.007

Jones, R. A. C. (2014). Plant virus ecology and epidemiology: Historical perspectives, recent progress and future prospects. *The Annals of Applied Biology, 164*(3), 320–347. https://doi.org/10.1111/aab.12105

Juárez, M., Rabadán, M. P., Martínez, L. D., Tayahi, M., Grande-Pérez, A., & Gómez, P. (2019). Natural hosts and genetic diversity of the emerging tomato leaf curl New Delhi virus in Spain. *Frontiers in Microbiology, 10*, 140. https://doi.org/10.3389/fmicb.2019.00140

Kaur, M., Varalakshmi, B., Pitchaimuthu, M., & Mahesha, B. (2021). Screening Luffa germplasm and advanced breeding lines for resistance to tomato leaf curl New Delhi virus. *Journal of General Plant Pathology, 87*, 287–294. https://doi.org/10.1007/s10327-021-01000-1

Kil, E. J., Vo, T. T. B., Fadhila, C., Ho, P. T., Lal, A., Troiano, E., & Lee, S. (2020). Seed transmission of tomato leaf curl New Delhi virus from zucchini squash in Italy. *Plants, 9*(5), 563. https://doi.org/10.3390/plants9050563

Kiran, G. V. M., Nagaraju, N., Girish, T. R., & Ashwini, B. N. (2021). Molecular investigations reveal bitter gourd crop is more susceptible to tomato leaf curl New Delhi virus infection in diverse crop cultivation practices. *3 Biotech, 11*, 1–9. https://doi.org/10.1007/s13205-021-02734-9

Krishnan, N., Kumari, S., Chaubey, T., Kumar, R. V., Chinnappa, M., Dubey, V., Pandey, K. K., Singh, J., & Singh, A. K. (2023). Study on mosaic disease of sponge gourd (*Luffa cylindrica* L.) caused by tomato leaf curl New Delhi virus. *Journal of Plant Pathology, 105*(2), 573–580. https://doi.org/10.1007/s42161-023-01325-0

Krishnan, N., Kumari, S., Kumar, R. V., Sakthivel, K., Dubey, V., Reddy, R., Singh, A. K., & Kumar, R. (2025). Prevalence of chilli leaf curl virus and tomato leaf curl New Delhi virus with chilli leaf curl disease in India. *Physiology and Molecular Biology of Plants*. https://doi.org/10.1007/s12298-025-01570-w

Kumar, V., & Naqvi, A. R. (2016). *Tomato leaf curl New Delhi virus* (ToLCNDV) encoded AC2 associates with host miRNA by directly interacting with AGO1. *Journal of RNAi and Gene Silencing, 12*, Article 534.

Kumar, R., & Singh, B. P. (2023). Diagnostics for production of disease-free planting material of potato. *International Journal of Innovative Horticulture, 12*, 31–49. https://doi.org/10.5958/2582-2527.2023.00003.9

Kumar, R., Tiwari, R. K., Jeevalatha, A., Sundaresha, S., Shah, M. A., Sharma, S., Sagar, V., Kumar, M., & Chakrabarti, S. K. (2021). Potato apical leaf curl disease: Current status and perspectives on a disease caused by tomato leaf curl New Delhi virus. *Journal of Plant Diseases and Protection, 128*, 897–911. https://doi.org/10.1007/s41348-021-00463-w

Kumar, R., Tiwari, R. K., Sundaresha, S., Kaundal, P., & Raigond, B. (2022). Potato viruses and their management. In S. K. Chakrabarti, S. Sharma, & M. A. Shah (Eds.), *Sustainable management of potato pests and diseases*. Springer. https://doi.org/10.1007/978-981-16-7695-6_12

Kumari, S., Nagendran, K., Dubey, V., Das, B., Pandey, K. K., & Singh, J. (2021). Investigations on annual spreading of viruses infecting cucurbit crops in Uttar Pradesh state India. *Scientific Reports, 11*(1), 1–17. https://doi.org/10.1038/s41598-021-97232-4

Lapidot, M., & Friedmann, M. (2002). Breeding for resistance to whitefly-transmitted geminiviruses. *Annals of Applied Biology, 140*, 109–127.

Lapidot, M., Karniel, U., Gelbart, D., Fogel, D., Evenor, D., Kutsher, Y., & Levin, I. (2015). A novel route controlling begomovirus resistance by the messenger RNA surveillance factor pelota. *PLoS Genetics, 11*(10), e1005538. https://doi.org/10.1371/journal.pgen.1005538

Legg, J. P., Shirima, R., Tajebe, L. S., Guastella, D., Boniface, S., Jeremiah, S., et al. (2014). Biology and management of *Bemisia* whitefly vectors of cassava virus pandemics in Africa. *Pest Management Science, 70*(10), 1446–1453. https://doi.org/10.1002/ps.3802

López, C., Ferriol, M., & Picó, M. B. (2015). Mechanical transmission of *Tomato leaf curl New Delhi virus* to cucurbit germplasm: Selection of tolerance sources in *Cucumismelo*. *Euphytica, 204*, 679–691. https://doi.org/10.1007/s10681-015-1537-5

Mahas, A., Hassan, N., Aman, R., Marsic, T., Wang, Q., Ali, Z., & Mahfouz, M. M. (2021). LAMP-coupled CRISPR–Cas12a module for rapid and sensitive detection of plant DNA viruses. *Viruses, 13*, 466. https://doi.org/10.3390/v13030466

Maji, A., Yadav, P., Aslam, S. R. B., & Choudhary, R. (2025). Tracing crop disease spread with remote sensing (RS) and geographic information systems (GIS). *Agriculture and Food: E-Newsletter, 7*, 592–595.

Maruthi, M. N., Alam, S. N., Kader, K. A., Rekha, A. R., Cork, A., & Colvin, J. (2005). Nucleotide sequencing, whitefly transmission, and screening tomato for resistance against two newly described begomoviruses in Bangladesh. *Phytopathology, 95*, 1472–1481. https://doi.org/10.1094/PHYTO-95-1472

Mastrochirico, M., Spanò, R., & Mascia, T. (2022). Grafting to manage infections of the emerging tomato leaf curl New Delhi virus in cucurbits. *Plants, 12*(1), 37. https://doi.org/10.3390/plants12010037

Naga, K. C., Subhash, S., Bhatnagar, A., Tiwari, R. K., Kumar, R., Verma, G., Sharma, S., & Chakrabarti, S. K. (2021a). Influence of host plants on virus acquisition and endosymbionts of whitefly, Bemisia tabaci (Gennadius). *Indian Journal Of Entomology, 83*, 371–375. https://doi.org/10.5958/0974-8172.2020.00206.0

Naga, K. C., Sundaresha, S., Kumar, R., Tiwari, R. K., Subhash, S., Verma, G., Buckseth, T., Bairwa, A., Sharma, S., Katare, S., & Srivastava, R. M. (2021b). A new record of Asia II 5 genetic group of Bemisia tabaci (Gennadius) in the major potato growing areas of India and its relationship with tomato leaf curl New Delhi virus infecting potato. *3 Biotech, 11*, 421. https://doi.org/10.1007/s13205-021-02966-7

Naganur, P., Premchand, U., Shankarappa, K. S., Mesta, R. K., Manjunatha, C., & Patil, C. V. (2019). Development of a loop-mediated isothermal amplification assay for detection of tomato leaf curl New Delhi virus in ridge gourd [*Luffa acutangula* (L.) Roxb.]. *International Journal of Current Microbiology and Applied Sciences, 8*, 2282–2295. https://doi.org/10.20546/ijcmas.2019.809.264

Naganur, P., Shankarappa, K. S., Mesta, R. K., Rao, C. D., Venkataravanappa, V., Maruthi, M. N., & Reddy, L. R. C. N. (2023). Detecting tomato leaf curl New Delhi virus causing ridge gourd yellow mosaic disease, and other begomoviruses by antibody-based methods. *Plants, 12*(3), 490. https://doi.org/10.3390/plants12030490

Nagendran, K., Mohankumar, S., Aravintharaj, R., Balaji, C. G., Manoranjitham, S. K., Singh, A. K., Rai, A. B., Singh, B., & Karthikeyan, G. (2017). The occurrence and distribution of major viruses infecting cucurbits in Tamil Nadu state, India. *Crop Protection, 99*, 10–16. https://doi.org/10.1016/j.cropro.2017.05.006

Nagendran, K., Kumari, S., Dubey, V., Kumar, R., & Pandey, K. K. (2020). Development and evaluation of integrated disease management module for leaf curl disease in chilli. *Vegetable Science, 47*(1), 135–137.

Nagendran, K., Kumari, S., Pandey, S., Karmakar, P., Chaubey, T., Kumar, R., Vignesh, S., Karthikeyan, G., & Behera, T. K. (2023). Emergence of yellowing disease in cucurbitaceous vegetables caused by Crinivirus and Polerovirus in India. *Virology, 587*, 109876.

Naqvi, A. R., Haq, Q. M., & Mukherjee, S. K. (2010). MicroRNA profiling of tomato leaf curl New Delhi virus (ToLCNDV) infected tomato leaves indicates that deregulation of miR159/319 and miR172 might be linked with leaf curl disease. *Virology Journal, 7*, 1–16. https://doi.org/10.1186/1743-422X-7-1

Nault, L. R. (1997). Arthropod transmission of plant viruses: A new synthesis. *Annals of the Entomological Society of America, 90*, 521–541. https://doi.org/10.1093/aesa/90.5.521

Navas-Castillo, J., Fiallo-Olivé, E., & Sánchez-Campos, S. (2011). Emerging virus diseases transmitted by whiteflies. *Annual Review of Phytopathology, 49*, 219–248. https://doi.org/10.1146/annurev-phyto-072910-095235

Naveen, M., Shanmugam, P. S., Srinivasan, R., Shanthi, M., Murugan, M., Sivagnanapazham, K., Angappan, K., Oliva, R., Karthikeyan, G., & Boopathi, N. M. (2025). Recombination, phylogenetic insights and co-existence dynamics of begomoviruses and satellite DNAs: Implications for tomato viral disease management. *Molecular Biology Reports, 52*(1), 390. https://doi.org/10.1007/s11033-025-10499-y

Padidam, M., Beachy, R. N., & Fauquet, C. M. (1995). Tomato leaf curl geminivirus from India has a bipartite genome and coat protein is not essential for infectivity. *The Journal of General Virology, 76*, 25–35. https://doi.org/10.1099/0022-1317-76-1-25

Padmanabha, K., Choudhary, H., Mishra, G. P., Mandal, B., Solanke, A. U., Mishra, D. C., & Yadav, R. K. (2024). Identifying new sources of resistance to tomato leaf curl New Delhi virus from Indian melon germplasm by designing an improved method of field screening. *Genetic Resources and Crop Evolution, 71*, 1911–1933. https://doi.org/10.1007/s10722-023-01580-6

Prasanna, H. C., Sinha, D. P., Rai, G. K., Krishna, R., Kashyap, S. P., Singh, N. K., Singh, M., & Malathi, V. G. (2015). Pyramiding Ty-2 and Ty-3 genes for resistance to monopartite and bipartite tomato leaf curl viruses of India. *Plant Pathology, 64*(2), 256–264.

Priyanka, R., Nagendran, K., Aravintharaj, R., Balaji, C. G., Mohankumar, S., Renukadevi, P., & Karthikeyan, G. (2019). Characterization and management of watermelon bud necrosis virus infecting watermelon in India. *European Journal of Plant Pathology, 153*(3), 759–770. https://doi.org/10.1007/s10658-018-1589-2

Razdan, V. K., & Sabitha, M. (2009). Integrated disease management: Concepts and practices. In R. Peshin & A. K. Dhawan (Eds.), *Integrated pest management: Innovation-development process* (Vol. 1, pp. 369–389). Springer. https://doi.org/10.1007/978-1-4020-8992-3_12

Rodríguez, E., Téllez, M. M., & Janssen, D. (2019). Whitefly control strategies against tomato leaf curl New Delhi virus in greenhouse zucchini. *International Journal of Environmental Research and Public Health, 16*(15), 2673. https://doi.org/10.3390/ijerph16152673

Romay, G., Pitrat, M., Lecoq, H., Wipf-Scheibel, C., Millot, P., Girardot, G., & Desbiez, C. (2019). Resistance against melon chlorotic mosaic virus and tomato leaf curl New Delhi virus in melon. *Plant Disease, 103*(11), 2913–2919. https://doi.org/10.1094/PDIS-11-18-2015-RE

Romero-Masegosa, J., Martínez, C., Aguado, E., García, A., Cebrián, G., Iglesias-Moya, J., & Jamilena, M. (2021). Response of *Cucurbita* spp. to tomato leaf curl New Delhi virus inoculation and identification of a dominant source of resistance in *Cucurbita moschata*. *Plant Pathology, 70*(1), 206–218. https://doi.org/10.1111/ppa.13278

Sáez, C., Martínez, C., Ferriol, M., Manzano, S., Velasco, L., Jamilena, M., et al. (2016). Resistance to tomato leaf curl New Delhi virus in *Cucurbita* spp. *The Annals of Applied Biology, 169*(1), 91–105. https://doi.org/10.1111/aab.12256

Sáez, C., Martínez, C., Montero-Pau, J., Esteras, C., Sifres, A., Blanca, J., et al. (2020). A major QTL located in chromosome 8 of *Cucurbita moschata* is responsible for resistance to tomato leaf curl New Delhi virus. *Frontiers in Plant Science, 11*, 207. https://doi.org/10.3389/fpls.2020.00207

Sáez, C., Ambrosio, L. G., Miguel, S. M., Valcárcel, J. V., Díez, M. J., Picó, B., & López, C. (2021). Resistant sources and genetic control of resistance to ToLCNDV in cucumber. *Microorganisms, 9*(5), 913. https://doi.org/10.3390/microorganisms9050913

Sahu, P. P., Rai, N. K., Chakraborty, S., Singh, M., Chandrappa, P. H., Ramesh, B., Chattopadhyay, D., & Prasad, M. (2010). Tomato cultivar tolerant to tomato leaf curl New Delhi virus infection induces virus-specific short interfering RNA accumulation and defence-associated host gene expression. *Molecular Plant Pathology, 11*(4), 531–544. https://doi.org/10.1111/j.1364-3703.2010.00630.x

Sangeetha, B., Malathi, V. G., Alice, D., Suganthy, M., & Renukadevi, P. (2018). A distinct seed-transmissible strain of tomato leaf curl New Delhi virus infecting Chayote in India. *Virus Research, 258*, 81–91. https://doi.org/10.1016/j.virusres.2018.09.009

Sarkar, M., Gupta, D., Singh, O. W., Paul, S., Kumar, R., Mandal, B., & Roy, A. (2024). Identification of pathogenicity determinants in ToLCNDV and their RNAi-based knockdown for disease management in Nicotiana benthamiana and tomato plants. *Frontiers in Microbiology, 15*, 1481523. https://doi.org/10.3389/fmicb.2024.1481523

Schafleitner, R., Chen-Yu, L., Laenoi, S., Shu-Mei, H., Srimat, S., Gi-An, L., et al. (2024). Molecular markers associated with resistance to squash leaf curl China virus and tomato leaf curl New Delhi virus in tropical pumpkin (*Cucurbita moschata* Duchesne ex Poir.) breeding line AVPU1426. *Scientific Reports, 14*(1), 6793. https://doi.org/10.1038/s41598-024-42209-z

Schuster, D. J., Thompson, S., & Kring, J. B. (2008). Biological and cultural management practices for tomato yellow leaf curl virus. *Plant Health Progress, 9*(1), 31.

Sharma, N., & Prasad, M. (2020). Silencing AC1 of tomato leaf curl virus using artificial microRNA confers resistance to leaf curl disease in transgenic tomato. *Plant Cell Reports, 39*, 1565–1579. https://doi.org/10.1007/s00299-020-02584-2

Sharma, N., Sahu, P. P., Prasad, A., Muthamilarasan, M., Waseem, M., Khan, Y., et al. (2021). The Sw5a gene confers resistance to ToLCNDV and triggers an HR response after direct AC4 effector recognition. *Proceedings of the National Academy of Sciences of the USA, 118*(33), e2101833118. https://doi.org/10.1073/pnas.2101833118

Simón, A., Ruiz, L., Velasco, L., & Janssen, D. (2018). Absolute quantification of tomato leaf curl New Delhi virus Spain strain, ToLCNDV-ES: Virus accumulation in a host-specific manner. *Plant Disease, 102*(1), 165–171. https://doi.org/10.1094/PDIS-06-17-0840-RE

Singh, S., Nirmalkar, V. K., & Awasthi, L. P. (2020). Recent advances in begomovirus research in India. In *Applied plant virology* (pp. 493–513). Academic Press. https://doi.org/10.1016/B978-0-12-818654-1.00035-9

Singh, J., Munshi, A. D., Singh, D., Meena, B. R., Singh, A. K., Nagar, A., & Mahajani, K. (2024). Identification of new stable resistant sources and assessing agro-morphological performance of sponge gourd germplasm against tomato leaf curl New Delhi virus incidence. *Frontiers in Plant Science, 15*, 1373352. https://doi.org/10.3389/fpls.2024.1373352

Sinha, D., et al. (2019). Advances in phenotyping and bioinformatics for crop resistance. *Plant Physiology and Biochemistry, 143*, 47–59. https://doi.org/10.1016/j.plaphy.2019.07.027

Sivagnanapazham, K., Karthikeyan, G., Pavithran, S., Harish, S., Murugan, M., Latha, T. K. S., & Devi, N. (2025). Transmission dynamics and localization of tomato leaf curl New Delhi virus in cucurbits via sweet potato whitefly, *Bemisiatabaci* (Gennadius). *Molecular Biology Reports, 52*(1), 1–14. https://doi.org/10.1007/s11033-025-10467-6

Smith, H. A., Koenig, R. L., McAuslane, H. J., & McSorley, R. (2000). Effect of silver reflective mulch and a summer squash trap crop on densities of immature *Bemisiaargentifolii* (Homoptera: Aleyrodidae) on organic bean. *Journal of Economic Entomology, 93*(3), 726–731. https://doi.org/10.1603/0022-0493-93.3.726

Soliman, A., Matar, S., & Abo-Zaid, G. (2022). Production of *Bacillus velezensis* strain GB1 as a biocontrol agent and its impact on *Bemisiatabaci* by inducing systemic resistance in a squash plant. *Horticulturae, 8*(6), 511. https://doi.org/10.3390/horticulturae8060511

Srivastava, K. M., Hallan, V., Raizada, R. K., Chandra, G., Singh, B. P., & Sane, P. V. (1995). Molecular cloning of Indian tomato leaf curl virus genome following a simple method of concentrating the supercoiled replicative form of viral DNA. *Journal of Virological Methods, 51*(2–3), 297–304. https://doi.org/10.1016/0166-0934(94)00124-Y

Suwwan, M. A., Akkawi, M., Al-Musa, A. M., & Mansour, A. (1988). Tomato performance and incidence of tomato yellow leaf curl (TYLC) virus as affected by type of mulch. *Scientia Horticulturae, 37*(1–2), 39–45. https://doi.org/10.1016/0304-4238(88)90185-0

Taglienti, A., Donati, L., Ferretti, L., Tomassoli, L., Sapienza, F., Sabatino, M., et al. (2022). In vivo antiphytoviral activity of essential oils and hydrosols from *Origanumvulgare*, *Thymus vulgaris*, and *Rosmarinusofficinalis* to control zucchini yellow mosaic virus and tomato leaf curl New Delhi virus in *Cucurbita pepo* L. *Frontiers in Microbiology, 13*, 840893. https://doi.org/10.3389/fmicb.2022.840893

Tatineni, S., & Hein, G. L. (2023). Plant viruses of agricultural importance: Current and future perspectives of virus disease management strategies. *Phytopathology, 113*(2), 117–141. https://doi.org/10.1094/PHYTO-08-22-0303-RVW

Venkataravanappa, V., Reddy, C. L., Shankarappa, K. S., & Reddy, M. K. (2019). Association of tomato leaf curl New Delhi virus, betasatellite, and alphasatellite with mosaic disease of spine gourd (*Momordicadioica* Roxb. Willd) in India. *Iranian Journal of Biotechnology, 17*(1), e2134. https://doi.org/10.30498/IJB.2019.2134

Venkataravanappa, V., Ashwathappa, K. V., Reddy, C. L., Shankarappa, K. S., & Reddy, M. K. (2020). Characterization of tomato leaf curl New Delhi virus associated with leaf curl and yellowing disease of watermelon and development of LAMP assay for its detection. *3 Biotech, 10*, 1–12. https://doi.org/10.1007/s13205-020-02245-x

Venkatasalam, E. P., Singh, S., Sivalingam, P. N., Malathi, V. G., & Garg, I. D. (2011). Polymerase chain reaction and nucleic acid spot hybridisation detection of Begomovirus(es) associated with apical leaf curl disease of potato. *Archives of Phytopathology and Plant Protection, 44*(10), 987–992. https://doi.org/10.1080/03235401003633741

Vignesh, S., Renukadevi, P., Nagendran, K., Senthil, N., Kumar, R. V., Swarna Priya, R., Behera, T. K., & Karthikeyan, G. (2023). A distinct strain of tomato leaf curl New Delhi virus that causes mosaic disease in ash gourd and other cucurbitaceous crops. *Frontiers in Microbiology, 14*, 1268333.

Vo, T. T. B., Tabassum, M., Nattanong, B., Qureshi, M. A., Im, H. J., Parrella, G., Kil, E. J., & Lee, S. (2025). The insidious threat: Assessing the dangers and spread of tomato leaf curl New Delhi virus. *The Plant Pathology Journal, 41*(1), 1–16.

Yamamoto, H., Wakita, Y., Kitaoka, T., Fujishiro, K., Kesumawati, E., & Koeda, S. (2021). Southeast Asian isolate of the tomato leaf curl New Delhi virus shows higher pathogenicity against tomato and cucurbit crops compared to that of the Mediterranean isolate. *Horticultural Journal, 90*(3), 314–325. https://doi.org/10.2503/hrj.90.314

Yerasu, S. R., Prasanna, H. C., Krishnan, N., Maurya, S., Panwar, H. S., Reddy, S. K., Tiwari, J. K., Rai, N., & Behera, T. K. (2025). Marker assisted pyramiding of major resistance genes of tomato leaf curl and late blight diseases for stabilising tomato production. *Physiology and Molecular Biology of Plants, 31*, 105–118. https://doi.org/10.1007/s12298-025-01548-8

Zaidi, S., Martin, D. P., Amin, I., Farooq, M., & Mansoor, S. (2017). Tomato leaf curl New Delhi virus: A widespread bipartite begomovirus in the territory of monopartite begomoviruses. *Molecular Plant Pathology, 18*(7), 901–911. https://doi.org/10.1111/mpp.12481

Genomic Diversity and Host Adaptation of Tomato Leaf Curl New Delhi Virus

15

P. Renukadevi, M. Gayathri, S. Kousalya, R. Sharanya,
A. Kokila, N. Saranya, and V. G. Malathi

Abstract

The chapter summarises the genetic propensity of ToLCNDV, justifying its emergence as a phytopathogen of global importance. Its expanding geographic distribution, adaptability to new hosts, and new environments are surely pointers to the global conquest of this pathogen. Frequencies of mutation and recombination, along with the ability to capture satellite DNAs and trans replication components of other begomoviruses, make it an evolutionarily upward-moving virus. To manage this virus, a good support system with complete information and knowledge on various factors contributing to its success is required.

Keywords

Genetic diversity · ToLCNDV · Environment · DNA A · DNA B · Begomovirus

P. Renukadevi (✉) · M. Gayathri · S. Kousalya · R. Sharanya · A. Kokila
Department of Plant Pathology, Centre for Plant Protection Studies, Tamil Nadu Agricultural University, Coimbatore, Tamil Nadu, India
e-mail: renukadevi.p@tnau.ac.in

N. Saranya
Department of Plant Molecular Biology and Bioinformatics, Centre for Plant Molecular Biology and Biotechnology, Tamil Nadu Agricultural University, Coimbatore, Tamil Nadu, India
e-mail: saranya.n@tnau.ac.in

V. G. Malathi
Retired Scientist, Division of Plant Pathology, ICAR-Indian Agricultural Research Institute, New Delhi, Delhi, India

R. Kumar et al. (eds.), *Tomato Leaf Curl New Delhi Virus (ToLCNDV)*,
https://doi.org/10.1007/978-981-95-2852-3_15

1 Introduction

Tomato leaf curl New Delhi virus (ToLCNDV) has emerged as the most devastating virus species in the genus *Begomovirus*, within the family *Geminiviridae*. The virus has caused maximum impact on the agricultural productivity of various crops, particularly in the families *Solanaceae* and *Cucurbitaceae* (Varma & Malathi, 2003). The virus was first identified in the early 1950s and has rapidly disseminated across several countries in Asia, Europe, and North Africa, causing considerable economic losses to farmers and affecting food and nutritional security. ToLCNDV infects a wide range of host plants with symptoms manifesting severe leaf curl, leaf lamina deformation, enation, stunted growth, and reduced fruit yield, thereby raising serious challenges in crop management and production (Kumar et al., 2023; Sarkar et al., 2024).

ToLCNDV is primarily transmitted by the whitefly *Bemisia tabaci* (Genn.), an omnivorous pest that facilitates the virus spread across diverse ecological zones (Naga et al., 2021a, b). The genome of ToLCNDV is bipartite, comprising two distinct single-stranded circular DNA components (DNA A, DNA B) encapsidated independently within a paired quasi-icosahedral particle of size 28–34 nm. Besides DNA A and DNA B, it is also associated with satellite DNAs of 1.3 kb size (Kumar et al., 2021; Jeevalatha et al., 2021, 2023). Given its widespread incidence and the ability to cause devastating losses, ToLCNDV has become a focal point of study in plant virology and crop health management. As climate change and global trade continue to shape the dynamics of plant health, innovative approaches supported by the background data and information are essential to address the challenges posed by ToLCNDV and ensure the sustainability of the affected cropping system. The chapter aims to outline the challenges posed by ToLCNDV's genetic propensities to adapt to new hosts and environments. The chapter focuses mainly on the genetic diversity of the virus and its impact on host adaptations. The readers may see articles, such as Malathi et al. (2017); Zaidi et al. (2017); Moriones et al. (2017), for information on viral pathogenicity and, molecular interaction between the virus and host.

2 Discovery

There is a grave misconception that ToLCNDV was discovered in India only in the 1990s. Contrastingly, the tomato leaf curl disease was recognized as a threat to the cultivation of the crop as early as 1948 in northern India (Vasudeva & Sam Raj, 1948). Vasudeva and Samraj demonstrated the vector to be *Bemisia tabaci* and transmitted the virus causing the leaf curl disease to several solanaceous hosts. The disease spread to central India in 1950, and subsequently, from 1958 to 1990, the leaf curl disease spread all over India. It will be appropriate to state that the molecular characterization of the virus causing leaf curl disease in New Delhi (Padidam et al., 1995) and Lucknow (Srivastava et al., 1995) was accomplished in the 1990s. The name ToLCNDV was proposed to refer to the virus causing severe leaf curl disease in northern India. The genus *Begomovirus of* the family *Geminiviridae* is the largest one, encompassing nearly 480 species, infecting only dicotyledonous plants. The members are transmitted through the whitefly vector, *Bemisia tabaci*. The begomoviruses affect a large number of food, feed, fiber, and ornamental crops, causing economic losses. The ToLCNDV is one of the begomoviruses, emerging as a pathogen of global importance. The latest taxonomical classification and binomial nomenclature of the ToLCNDV are given below. For smooth presentation in this chapter, common English names and abbreviations are used to refer to the viruses.

The nomenclature and classification (https://ictv.global/taxonomy/2024):

Realm:	*Monodnaviria*
Kingdom:	*Shotokuvirae*
Phylum:	*Cressdnaviricota*
Class:	*Repensiviricetes*
Order:	*Geplafuvirales*
Family:	*Geminiviridae*
Genus:	*Begomovirus*
Species:	*Begomovirus solanumdelhiense*

3 Distribution

Initially, natural infection of ToLCNDV was reported only in solanaceous crops in India. In 1980, the virus was recorded in cucurbitaceous hosts (Mandal, 2010). The jump to cucurbits coincided with its spread to other countries. From 2000 onwards, the distribution of ToLCNDV in wax gourd, cantaloupe, and melon was reported by Samretwanich et al. (2000), in Thailand, in bitter gourd in Pakistan (Tahir & Haider, 2005); ToLCNDV infection in cucumber, bottle gourd, and muskmelon in Thailand was noted by Ito et al. (2008). Slowly, further spread of ToLCNDV happened westward, including Algeria, Spain, Morocco, Tunisia, Italy, Greece, and Iran (Cai et al., 2023). Further, ToLCNDV has been described to affect the productivity of cucurbits crops in Malaysia, Cambodia, Laos, Indonesia, France, and China (Chen et al., 2021). A graphic representation of the distribution of ToLCNDV globally is provided in Fig. 15.1.

4 Symptomatology

Typical symptoms caused in tomatoes include leaf curling, puckering of leaves, veinal yellowing, stunting, excessive branching, and pale yellowing to deep yellowing of leaves (Vasudeva & Sam Raj, 1948). Extreme distortion of leaves is one of the severe symptoms including leaf curling, yellow mosaic, stunted growth and fruit deformation were observed in hosts viz., (a) bitter gourd, (b) brinjal, (c)pumpkin, (d) ash gourd, (e) chayote, (f) carrot, (g) physalis and (h) silver leaf night shade (Fig. 15.2). Flowers may drop off, and fruit yield may be less and smaller in size.

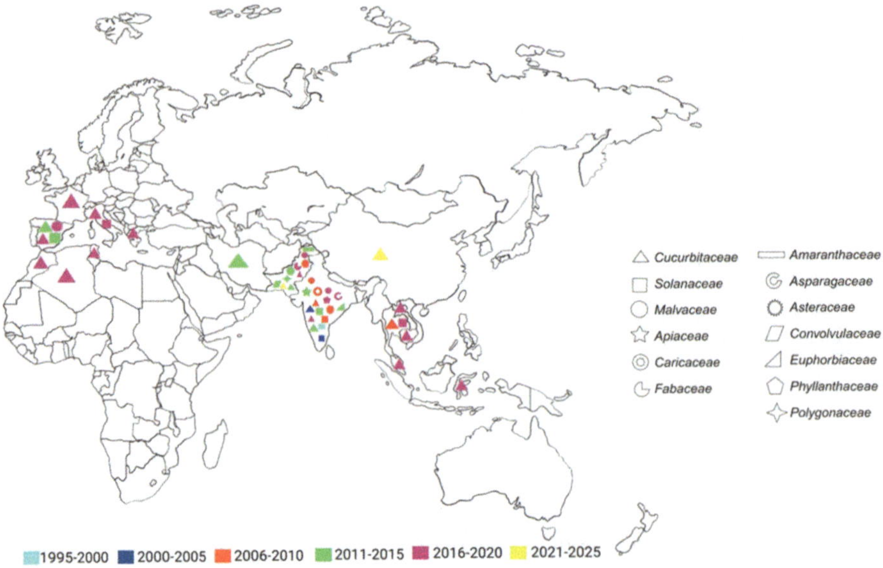

Fig. 15.1 Worldwide distribution of ToLCNDV in different crops and countries

Fig. 15.2 Symptoms of ToLCNDV infection in different host species

Sometimes, upward rolling of margins and islands of golden-colored areas scattered amidst the normal green tissue are also observed. Thickening and distortion of leaves with vein clearing appear prominent in some cucurbitaceous hosts. The type of symptom produced is dependent on the host, genotype cultivated, and the developmental stage at which the infection occurs. Two generic terms, tomato leaf curl virus and tomato yellow leaf curl virus, are used to refer to the viruses associated with the diseases in tomato in Asia. This gives the impression that the yellow leaf curl symptom is exclusively caused by tomato yellow leaf curl virus described from

Israel, and not caused by other tomato leaf curl viruses, which is a misconception. All the mono or bipartite begomoviruses cause yellow leaf curl symptoms, and the intensity of yellowing symptoms produced depends on the genotype cultivated and the climatic conditions (Varma & Malathi, 2003). In chili pepper, symptoms are distorted and upward curling of leaves, yellow mosaic, and bushy appearance are common (Chakraborty et al., 2008; Reddy et al., 2024). In cucurbits, extreme distortion of leaves, corrugated appearance of enations, stunting, and malformed fruits are observed.

5 Yield Loss

The disease incidence of nearly 100% led to total failure of the crop tomato and sponge gourd (Kumar et al., 2015; Sohrab et al., 2003). Nagendran et al. (2017) reported a 60% loss with maximum disease incidence in chayote. Sangeetha et al. (2018), investigated the loss due to ToLCNDV in hilly areas and found that it was around 69.9% when the disease incidence was 50.3%. Yield losses in cucurbitaceous hosts are much more severe in the Mediterranean region. Losses of up to 20% in the open field of melon and zucchini, affecting production in Central Spain, were reported by Cai et al. (2023), who described tremendous economic loss due to ToLCNDV in cucumber, zucchini, watermelon, and green beans. Nearly 80% of the pumpkin crop was lost in Italy (Crespo et al., 2020; Panno et al., 2019; Saez et al., 2017). In Southeastern areas of China, ToLCNDV infection covered nearly 650 hectares, resulting in an economic loss of almost $15 million.

6 Transmission Mode

The whitefly Bemesia tabaci transmits ToLCNDV, a typical member of the begomoviruses group. It is necessary to emphasise here that some of the ToLCNDV isolates are highly mechanically transmissible. In recent years, there have been increasing reports on seed transmission and pollen transmission of ToLCNDV. At least theoretically, ToLCNDV has been shown to have four different modes of transmission: vector, mechanical, seed and pollen. This could be a significant factor contributing to its wide host range.

7 Genome Organization

The genome of ToLCNDV consists of two (bipartite) circular single stranded DNA (ssDNA) molecules (2700 bp) which are encapsidated within a para icosahedral twinned particle of size (22 × 38 nm) (Böttcher et al., 2004), the geminate particles of ToLCNDV had been captured in EM of leaf dip preparations of leaves of infected tomato and cucurbitaceous hosts. The two genomic components of the bipartite begomoviruses are designated as DNA A and DNA B (Fig. 15.3). The DNA A has

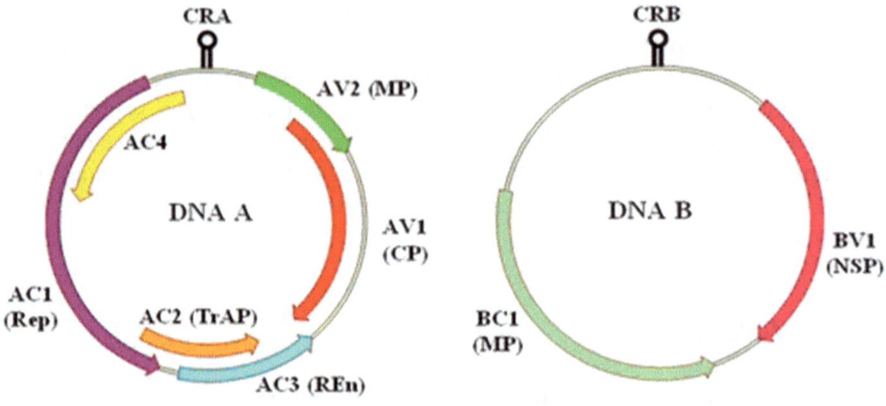

Fig. 15.3 Genome organization of ToLCNDV. Genome organization of DNA A and DNA B components of ToLCNDV. Open reading frames (ORFs) are denoted as either being encoded in the virion-sense (V) or complementary-sense (C) strand, preceded by component designation (A or B). CRA: common region A; CRB: common region B; CP: coat protein; MP: movement protein; Rep: replication initiation protein; TrAP: transcription activator protein; REn: replication enhancer protein; BV1: nuclear shuttle protein; BC1: cell to cell movement protein

two open reading frames (ORFs) in the virion sense, ORF AV1 encodes for the coat protein essential for encapsidation and vector transmission and ORF AV2 encodes the pre-coat protein facilitating the movement in monopartite genome, functions as an efficient post transcriptional gene silencing (PTGS) suppressor in the bipartite genome. There are four ORFs in the complementary sense: ORF AC1 stands for replication initiator protein, ORF AC2 encodes the transcription activator protein or TrAP and a PTGS suppressor; ORF AC3 encodes a replication enhancer protein; ORF AC4 encodes a small protein considered to be a symptom determinant protein. In ToLCNDV, like any other Old World begomovirus, an additional ORF AV3 had been detected in ToLCNDV first by Padidam et al. (1996), and subsequently by other workers too. Likewise, a complementary sense ORF AC5 has also been identified in several isolates of ToLCNDV. The function of AV3 was demonstrated to be an RNA silencing suppressor localised to the Golgi body (Gong et al., 2021). Li et al. (2015) described the role of the AC5 protein as a silencing suppressor in MYMV. These ORFs are present in the majority of ToLCNDV isolates; however, little information has been generated on the functions of these proteins in ToLCNDV.

There are two ORFs, one in each viral and complementary sense in DNA B. ORF BV1 in the viral sense encodes for the nuclear shuttle protein, and ORF BC1 in the complementary sense encodes for the movement protein; both proteins facilitate the intra- and intercellular movement of the viral genome. The ORFs are diagonally arranged, leaving a non-coding intergenic region between the 5′ of ORF AC1 and the 5′ of AV2, in DNA A and 5′ of ORF BC1 and ORF BV1 in DNA B. The segment of the IR that is common and almost identical between DNA A and DNA B is called the common region (CR). The length of the common region (CR) between the A and B components in ToLCNDV is approximately 150–165 nucleotides. The CR contains an origin of replication including a stem loop structure containing nonanucleotide TAATATTAC sequences, which is the site for circular DNA cleavage and initiation of rolling circle replication and joining of viral DNA during replication.

The whitefly vector places the virion particles in the deep-seated phloem parenchyma or companion cells. The particles then become uncoated, and the single-stranded viral genome is released from the virions, entering the nucleus. The ssDNA genome is converted into dsDNA, a process in which the virus is dependent on the host for primers, DNA polymerase and other DNA machinery of the host. The dsDNA is the replicative intermediate and also the transcription template, from which transcripts of the replication initiator protein and other viral protein-encoding RNA are transcribed. The Rep protein reprograms the cell cycles and recruits host DNA replication machinery. Replication occurs through the rolling circle model, where RCA generates new ssDNA copies of the viral genome. The replication is also viewed as recombination-dependent, mediating the interchange of fragments between viral genomes in the case of mixed infection. The viral genome moves intra- and intercellularly, invades new cells, alters the environment of each cell to facilitate its replication, and finally assembles into virions, which the vector will acquire. During this process, they reprogramme host transcription, alter gene expression, interfere with hormonal signalling, and manipulate plant development and physiology, resulting in the expression of severe symptoms and yield loss. Though ToLCNDV is a bipartite begomovirus, it is often associated with satellite molecules related to single-stranded DNA satellites of 1.3–1.4 kb length, which are referred to as alpha or betasatellite (Sivalingam & Varma, 2012). The betasatellites contribute to the severity of disease expression, and the role of the alphasatellite in viral pathogenicity is not yet clear.

8 Genome Comparisons

Complete genome sequences are available at the GenBank/NCBI database for more than 680 ToLCNDV isolates. The DNA A and B components are 2700/2600 nucleotides in length, respectively. The details of the coding region of the isolates of ToLCNDV representing the three groups, IND (isolates from the Indian subcontinent), OAS (isolates from other Asian countries), and MED (isolates from Mediterranean and North African countries), categorised in this study, are given in Table 15.1. The CR region spanned around 170 nucleotides in these representative isolates. The identity between DNA A and DNA B in the CR region is 99%. CR comprised conserved nonanucleotide sequence TAATATTAC and the Rep binding iteron sequences, which were identified to be GGTGC, which occurs twice upstream of the TATAA box.

Table 15.1 Coding features of DNA A and DNA B components of ToLCNDV isolates from three groups

DNA A								
Group	*AV3*	*AV2*	*AV1*	*AC5*	*AC4*	*AC3*	*AC2*	*AC1*
MED MH577731 Melon Spain	Data not given	119–457	279–1049	309–671	2250–2426	1046–1456	1191–1595	1498–2583
OAS OQ190941 Melon China	Data not given	120–458	280–1050	310–795	2251–2427	1047–1457	1177–1596	1499–2584
IND U15015 Tomato India	48–434	127–465	287–1057	317–802	2258–2434	1054–1464	1184–1603	1506–2591
DNA B								
Group	*BV1*	*BC1*						
MED KF891467 Zucchini	430–1236	1290–2135						
OAS OR157980 Melon China	442–1248	1305–2150						
IND U15017 Tomato India	405–1250	1276–2154						

9 Comparison of the DNA A Component

To look into diversity of ToLCNDV genome based on DNA A complete nucleotide sequence, in the present communication, we compared the DNA A sequence of nearly 118 isolates, taking 20 sequences from the Mediterranean and north African region (MED group), nearly 72 of Indian Subcontinent region including countries, India, Nepal, Pakistan, Sri Lanka, Bangladesh and 26 isolates from other Asian regions (OAS) group. The identity was computed based on multiple alignment of nucleotides by the MUSCLE programme. The results are summarised below. Only 40 isolates share >91% identity in the complete nucleotide sequences of DNA A. About 70 isolates are on the borderline of the threshold, ranging from 89.5% to 90.9%. They align with any one or few of the ToLCNDV isolates, which share 91% identity with them, and are therefore justified in being included in the species ToLCNDV. However, there are a few (15) sequences, which share identity much below the species demarcation threshold of 91%, set for the differentiation of begomovirus species (Zerbini et al., 2017).

This divergence in the genome is well understood when the heatmap is generated based on SDT version 1.3 (Fig. 15.4). It is evident from map A that ToLCNDV isolates, as listed in Table 15.2, share far less than the threshold value and need to be removed from the species ToLCNDV. The deviation exhibited by the sequence are clear from the hue of blue colour. From the colour code it is evident that ToLCNDV isolates are more grouped in lighter shades of orange than dark red, which denotes <98.5 identity. The relationship between the isolates is discussed groupwise in the following sections.

9.1 ToLCNDV Isolates from the MED Group (Mediterranean and North Africa)

It is surprising to find that 20 ToLCNDV isolates from the MED group taken for comparison share 99.2–99.5% identity among themselves. Unlike the heterogeneity observed among the isolates in the OSA and IND groups, this group is highly homogeneous. These isolates are considered to belong to the new strain of ToLCNDV, and the name ES was proposed (Fortes et al., 2016). These isolates share 89% identity in DNA A with the majority of OAS isolates. But the identity with ToLCNDV from Cambodia and Laos is only 86%. Its relationship with isolates from the IND group is interesting, with the type isolate ToLCNDV from New Delhi (U15015) showing an identity of only 89–90%, which is almost bordering on the species demarcation threshold value. Identity with other ToLCNDV isolates from the IND group is between 89% and 90%. The identity with five ridge gourd isolates from IND KT426907, KT426906, KT426905, KT426904, KT426903 and chayote isolate KP19047, and six ash gourd isolates KX363562, KX363563, KX363564, KX363565. KX363566 and KX363567, the MED isolates, exhibited identity ranging from 92% to 93%, justifying their inclusion in the species ToLCNDV (Fig. 15.5).

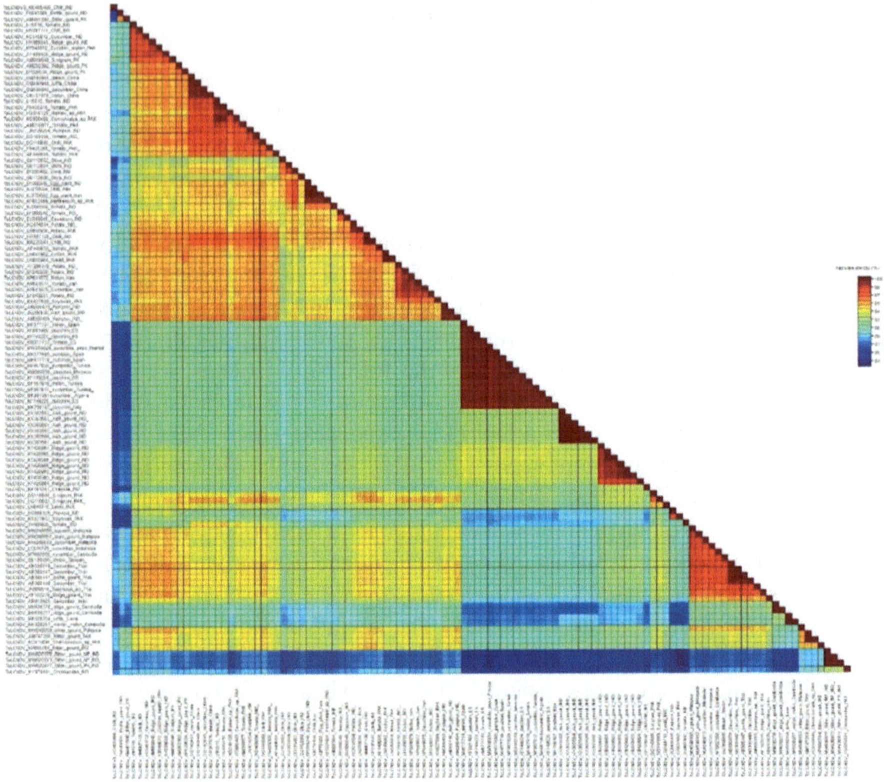

Fig. 15.4 Heat map of pairwise genome scores and nucleotide identity plot of full-length genomes (Complete nucleotide sequence of 118 DNA A components) of ToLCNDV: Two-dimensional percentage pairwise nucleotide identity plot (Calculated with pairwise deletion of gaps) of full-length genomes of selected begomoviruses of DNA A. The plot was generated by using the SDTv1.0 program (http://web.cbio.uct.ac.za/SDT)

Table 15.2 ToLCNDV isolates are listed as Outlier based on identity in the DNA A component with the type isolate

Hosts	Country	Accession number	Percentage identity with type isolate
Bottle gourd	Pakistan	FN645905	83
Tomato	India	KP195265	85
Chilli	India	KC465466	81
Bitter gourd	Pakistan	AM491590	82
Bitter gourd	India	MW620973	81
Bitter gourd	India	MW620975	80
Bitter gourd	India	MW620977	81
Crossandra	India	KY979591	78

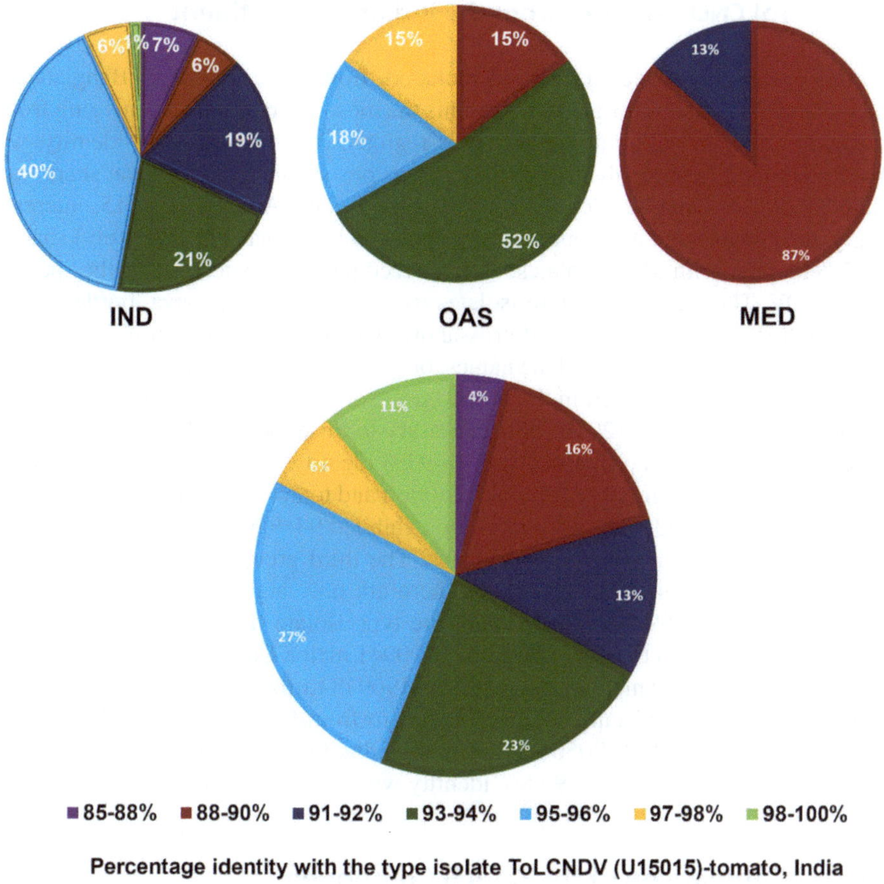

Fig. 15.5 Graphic representation of percentage identity in DNA A of ToLCNDV isolate with the type isolate

9.2 ToLCNDV Isolates from the OAS Group (From Other Asian Countries)

In this group, the isolates from Iran, China, Thailand, Indonesia, Laos and Cambodia are included. Among the 26 isolates taken for comparison, the isolates within the OAS group show divergence. The isolates from Thailand share 99% identity among them, but share only 89% identity with MED. Their identity with isolates from Malaysia and Indonesia is from 90 to 92, but with isolates from Laos and Cambodia, it is 89%. Surprisingly, all four isolates from China share 98% identity with the type isolate from India. With the majority of Indian isolates, identity is from 89% to 94%. Malaysia and Indonesia share 98–99% identity between them, 89% with MED and 89–92% with the IND group (Fig. 15.5).

9.3 ToLCNDV Isolates from the Indian Subcontinent

In this group, 70 isolates have been included, from India, Pakistan, Bangladesh, Nepal and Sri Lanka. This group represents the most diverse isolates emerging from different crops and different agroclimatic regions. Based on pairwise identity, we have attempted to group them. The first group consists of 15 isolates that share less than 91 (90.1) sequence identity with the type isolate. Among these 15, interestingly, the chayote and ash gourd isolates (one and six isolates, respectively) shared 91.2% identity with MED isolates. So, it is recognized as belonging to the species ToLCNDV. There are eight more isolates from bitter gourd, papaya, bottle gourd, chilli, bitter gourd (Pakistan) and crossandra, which show less than 87% identity. They are listed as Outliers, and the names, origin of the country, and accession numbers of the isolates included in the outlier group are given in Table 15.2. As recommended by Fortes et al. (2016), it is necessary to raise them to the distinct species level. The second group that showed 91–93% identity with the type isolate comprised 21 isolates originating from okra, cotton and tomato from India and Pakistan (Chauhan et al., 2023). The second group exhibited 94–95% identity with the type isolate, U15015, and included 29 isolates. The third group included five isolates sharing 94–96% identity. Surprisingly, there are only eight isolates in the fourth group showing 96–99.8% identity with the type isolate (Fig. 15.5). Surprisingly, four of them are from the OAS group: OQ190941 melon China, OQ190945 cucumber China, OR157979 melon China, and OQ190948 Luffa China. The other four are chilli (DQ11680), Pakistan, AM258977 tomato from Pakistan, FN435309 tomato from Pakistan, and DQ169050 tomato from India. Only one isolate, FN435310, a tomato from Pakistan, shares 99.8% identity with the type isolate.

10 Comparison of the DNA B Component

The comparison of the DNA B component of ToLCNDV revealed a picture aligning with the results of the DNA A component. The identity in the DNA B component was computed based on multiple alignment of the complete nucleotide sequence in the MUSCLE programme. The identity is discussed in a groupwise manner.

10.1 ToLCNDV Isolates from the MED Group (Mediterranean and North Africa)

About 21 isolates are included in the analysis. It was interesting to see that all the isolates shared 98–99% identity among themselves, as in the case of the DNA A component. However, its identity with isolates from the OAS group was very low, about 76%. The relationship with the IND group is significantly less, with the isolates sharing only 78–79% identity with the type isolate U15017. The majority of the IND group isolates show identity with the isolates ranging from 76% to 78% (Fig. 15.6).

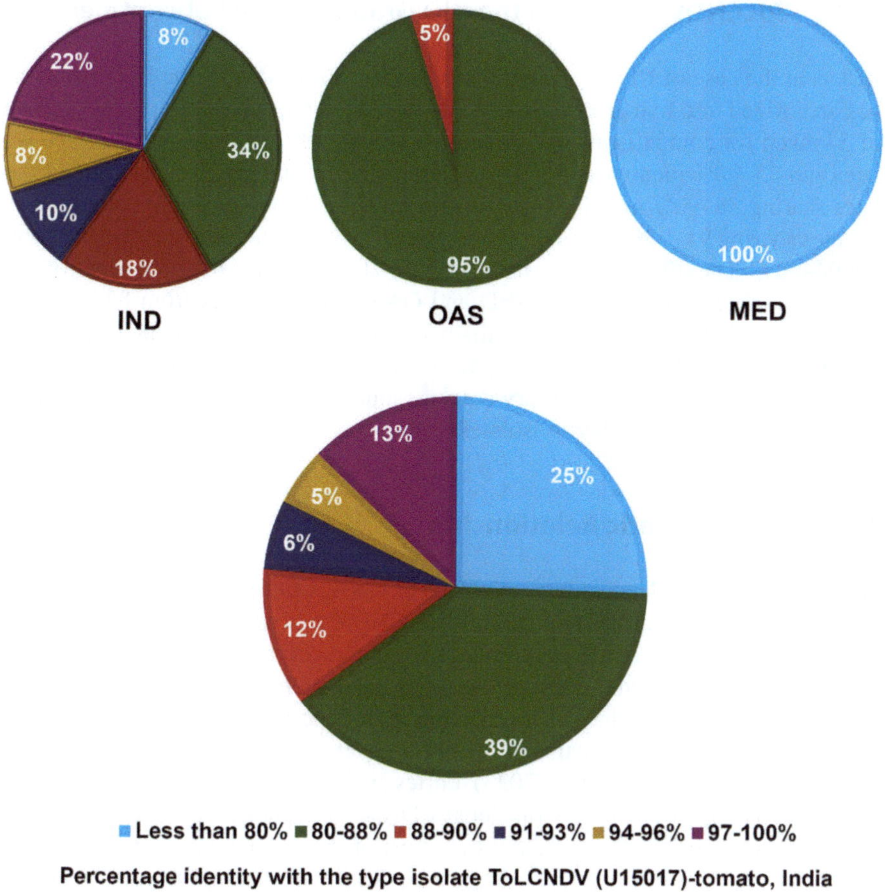

Fig. 15.6 Graphic representation of percentage identity in DNA B of ToLCNDV isolate with the type isolate

10.2 ToLCNDV Isolates of the OAS Group (Other Asian Countries)

We have taken bout 23 isolates of this group. Among them, they share 93–96% identity. The isolates from these countries share 83% identity with the type isolate U15017. Surprisingly, the two melon isolates from China share 96% identity with the type isolate. Just as in the case of DNA A, isolates from China in DNA B are also closer to the type isolate. With the majority of other isolates from the IND group, an identity of 79–83% is observed. With MED isolates, the incidence was significantly lower, at only 76% (Fig. 15.6).

10.3 ToLCNDV Isolates of the IND Group (Indian Subcontinent)

Just as in the case of DNA A, leaving the Outlier, ToLCNDV of the IND group can be categorized into five groups. Group I comprises a majority of the isolates (20 out of 54 taken for comparison), which share less than 88% identity with the type isolates and 85–90% identity with other isolates. The second group consists of 11 isolates sharing 88–90% identity with the type isolates. The third group comprises six isolates sharing 91–93% identity, and the fourth group includes five isolates exhibiting 94–96%. The fifth group consists of 13 isolates that share 97–98% identity with the type isolates. Identity with MED and OAS isolates ranges from 83% to 92% (Fig. 15.6). Based on genome comparison of the three groups it can be inferred that the entry of ToLCNDV isolate into MED countries is a single event and only one isolate is involved; On the contrary entry to south east Asian country may be multiple involving more number of isolates.

11 Phylogenetic Relationship

At present, the complete genome sequence of DNA A of 680 isolates of ToLCNDV is available. From the place of origin, ToLCNDV has spread both westward and eastward. From the original host tomato, it has jumped to unimaginable hosts like soybean and mung bean. Several researchers have attempted to link the context of the host species and geographical location with genome information, a technique known as phylogeographical analysis (Fortes et al., 2016; Zaidi et al., 2017; Moriones et al., 2017; Cai et al., 2023). Fortes et al. (2016) compared six ToLCNDV isolates from Spain with 129 full-length DNA A sequences of ToLCNDV from other Asian countries. They showed that most of the Asian isolates were nested in different clusters but formed a single group, despite originating from various geographical regions, host species, and collection at different time points. The isolates from Spain cluster together (sharing 99% sequence identity among them) and form a single monophyletic group distinct from other Asian isolates and exhibit less than 94% identity with ToLCNDV isolates from other countries. Moriones et al. (2017) and Zaidi et al. (2017) shared the same opinion, suggesting that ToLCNDV-ES isolates belonged to a different topology group within the same clade, which also included Asian isolates. Cai et al. (2023) analysed the relationship among the ToLCNDV isolates infecting cucurbitaceous hosts and suggested their grouping into two main clades: an Asian and European clade. They also indicated that the grouping is not dependent on host species but is more related to the region from which the isolates originate. The ToLCNDV isolates from China, Pakistan, Malaysia, Laos, and Cambodia were distributed within the Asian clade along with ToLCNDV isolates from the Indian subcontinent.

In a study, we re-examined the phylogenetic relationship of ToLCNDV isolates, taking representatives from MED, OAS and IND groups. One hundred ten isolates were studied. From Fig. 15.7, it is clear that no separate European or Asian clades were distinguished. On the contrary, two Clades, Clade 1 consisting of only one

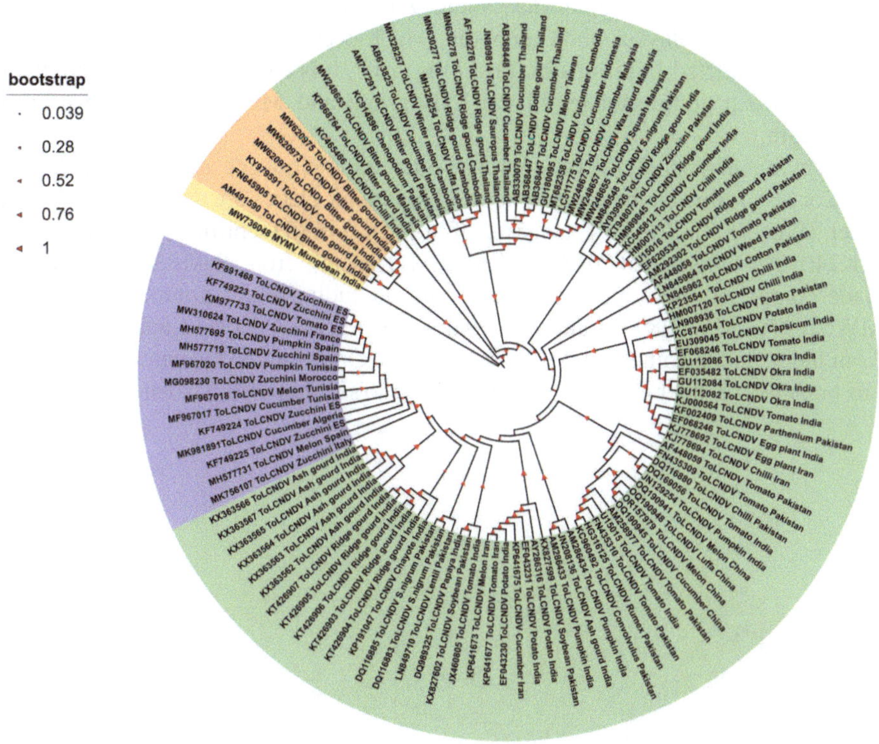

Fig. 15.7 Phylogenetic dendrogram based on complete nucleotide sequences of DNA A component of ToLCNDV isolates: Phylogenetic tree generated from aligned full-length DNA A sequences of ToLCNDV isolates retrieved from NCBI database. The tree was generated using MLmethod, which aligns sequences in MEGA 11 with the MUSCLE programme. The codes at the nodes indicate the strength of bootstrap values (1000 replicates). MYMV DNA A is used as out group

isolate, KC465466 chilli and Clade 2 consisting of six isolates, AM491599 bitter gourd Pakistan, FN645905 bottle gourd India, KY979591 crossandra India, MW620977, MW620975 bitter gourd isolates from India. These two clades deviate much separately from all other ToLCNDV isolates. The third major clade comprises 113 isolates having members from all three groups, MED, OAS, and IND. It is interesting to see that MED isolates and ash gourd isolates of India occupy branches of the same subclade, indicating that MED isolates are closely related to ash gourd and chayote isolates of India and perhaps share a common origin. It may be recalled here that MED isolates share a high percent identity with the ashgourd isolates. The presence of multiple clades within tomato, chilli and ridge gourd isolates suggests ongoing genetic evolution and potential host-driven selection. The clustering of isolates from different regions indicates the possibility of transboundary movement of virus transmission. The grouping of isolates from other hosts in one cluster also suggests selective pressure on the virus isolate due to the whitefly genotype involved in virus transmission in this region. It is noteworthy to record here that OAS isolates were distributed in various clades, along with Indian subcontinent isolates from China, Taiwan, Malaysia, and Cambodia, which occupied the topology group with their closely related IND isolates. From Fig. 15.7, it is clear that the chayote, ridge gourd and ash gourd isolates of the IND group and MED isolates had a common origin.

A phylogenetic study on the complete nucleotide sequence alignment of DNA B more or less followed the same pattern. Three major clades are deviating from each other (Fig. 15.8). Clade 1 consists of the DQ116882 chilli, LN54628 cotton, MK240307 tomato, AJ620188 *Solanum nigrum*, all from Pakistan. The second clade comprises two topology groups: one, including bitter gourd (MW628978, MW628976, MW628974) and crossandra isolates (MK 395549 and KY979592) from India, and the other, the MED isolates. The OAS and IND isolates occupy various branches of the third clade. It is interesting to see that melon isolates from China

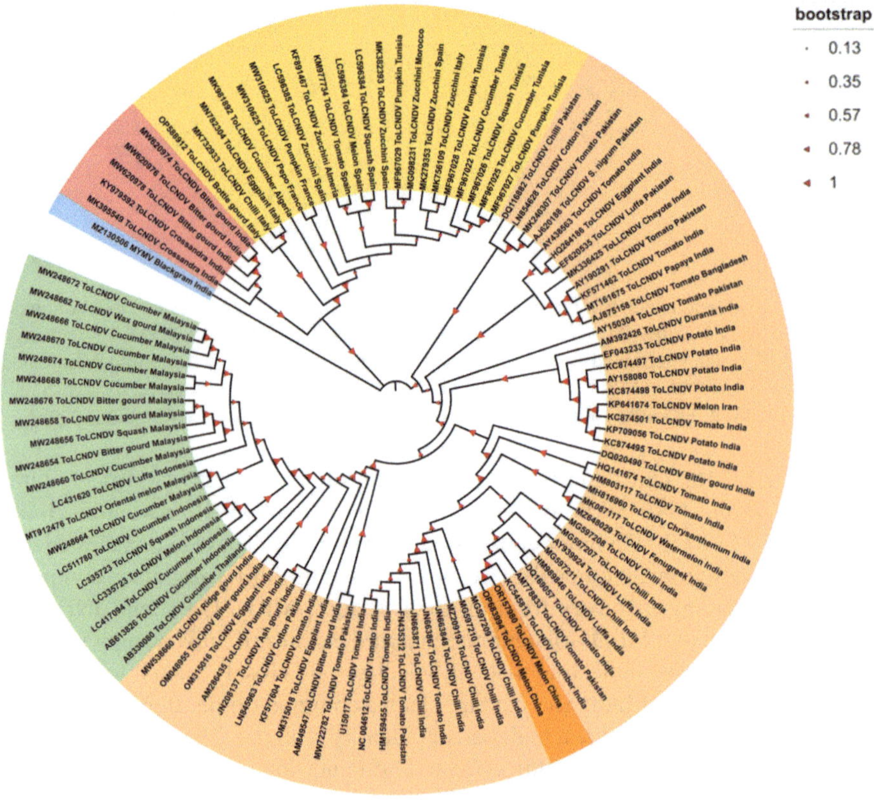

Fig. 15.8 Phylogenetic dendrogram based on complete nucleotide sequences of DNA B component of ToLCNDV isolates: Phylogenetic tree generated from aligned full length DNA B sequences of ToLCNDV isolates retrived from NCBI database. Tree was generated by MLmethod by aligning the sequences in MEGA 11 using MUSCLE programme. The codes at the nodes indicate the strength of boot straps values (1000 replicates). MYMV DNA B is used as out group

are grouped along with tomato and Luffa isolates of India. Whereas, isolates from Thailand and Malaysia occupy a topology group of different clades. The phylogenetic patterns highlight the genetic diversification within ToLCNDV. The presence of multiple clades within tomato, chilli, and ridge gourd isolates suggests ongoing genetic evolution and potential host-driven selection (Mazumder et al., 2024). Interestingly, some isolates from different regions appear closely related, indicating possible cross-border virus transmission. These findings reinforce the need for continuous genomic surveillance to monitor the emergence of new variants. Overall, the high bootstrap support in key branches confirms the robustness of the evolutionary inference, while the clustering patterns provide valuable insights into the virus's adaptation and spread.

12 Mutation Frequency Analysis

Variants were detected by aligning genomes to reference sequences of type isolate (U15015 for DNA A and U15017 for DNA B) using tools such as BWA or Bowtie2 for indexing (Li & Durbin, 2009; Langmead & Salzberg, 2012), followed by conversion of SAM files to sorted BAM files, indexing, and variant calling with bcftools (Danecek et al., 2021). High-quality variants were filtered and annotated using SnpEff to identify biologically significant mutations in the amino acid, as well as mutation frequency (Cingolani et al., 2012). Mutation frequency was calculated using the formula: Mutation frequency = (Number of Variants) ÷ (Genome Length). Mutation frequency plots were generated using the ggplot2 (Wickham & Sievert, 2009) package. The mutation frequency analysis of DNA A and DNA B of (ToLCNDV) provides valuable insights into the virus's evolutionary processes, pathogenicity, and adaptability. The analysis of mutation frequencies in DNA A (Fig. 15.9) and DNA B (Fig. 15.10) revealed distinct patterns. For DNA A, the most frequent mutations are C–T (2476 counts), T–C (2612 counts), A–G (1897 counts), and G–A (1577 counts). In contrast, DNA B exhibited high occurrences of C–T (1339 counts), T–C (1337 counts), A–T (1042 counts), and A–G (860 counts). Overall, transitions (purine-purine or pyrimidine-pyrimidine) were more common than transversions (purine-pyrimidine), which is consistent with observations in other DNA viruses. The dominance of C–T and T–C transitions in both DNA A and DNA B is suggestive of cytosine deamination, where cytosine is converted to thymine. This process can occur spontaneously or be facilitated by enzymatic mechanisms such as the APOBEC family of cytidine deaminases, which has been documented to introduce hypermutations in various DNA viruses (García-Andrés et al., 2006; Salazar-Martínez et al., 2021). Furthermore, the higher transition-to-transversion ratio in DNA A (due to elevated C–T and T–C mutations) compared to DNA B may indicate different selective pressures or functional constraints between the two components. Since DNA A contains essential genes for replication and encapsidation, it may be subject to greater selective constraints. At the same time, DNA B may be more prone to genetic variation aimed at optimising host-virus interactions (Padidam et al., 1996) (Tables 15.3 and 15.4). However, the greater

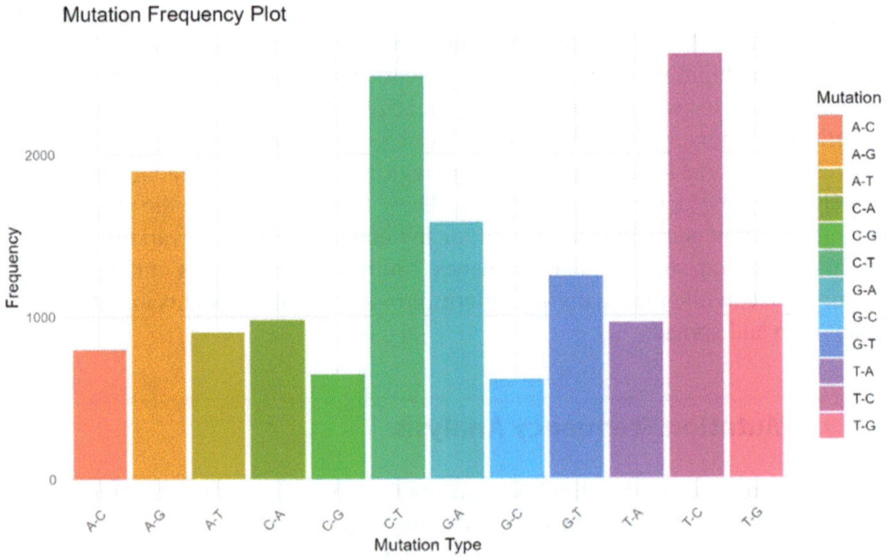

Fig. 15.9 Mutation frequency plot of nucleotide changes in DNA A of ToLCNDV

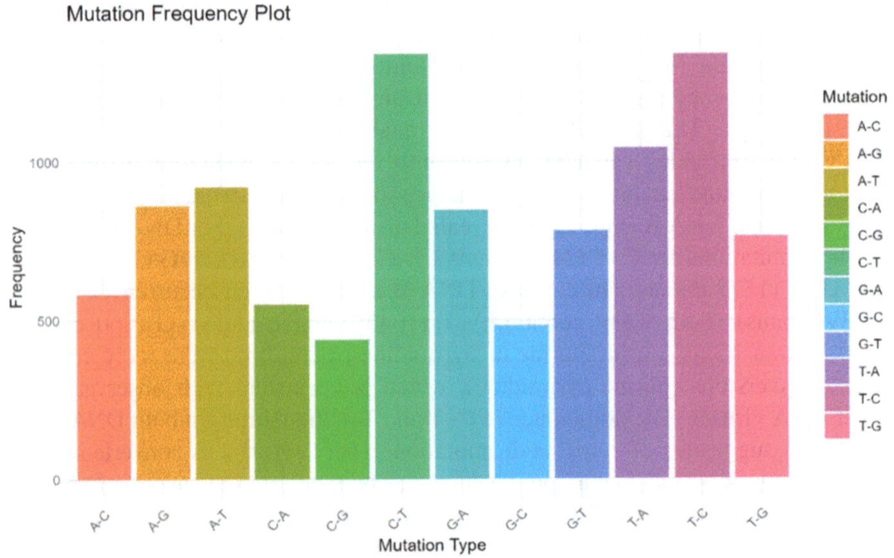

Fig. 15.10 Mutation frequency plot of nucleotide changes in DNA B of ToLCNDV

overall mutation counts in DNA A suggest higher replication fidelity in DNA B, potentially due to differential selective pressures or functional requirements between the two genomic components. The most common mutations in DNA A include V–I and P–S, both occurring at a frequency of 14. Additionally, R–K and I–V mutations

Table 15.3 Comparative table representing the nucleotide mutation frequency change in DNA A and DNA B of ToLCNDV

Mutation type	DNA A count	DNA B count
A–T	901	920
A–G	1897	860
A–C	798	583
T–A	958	1042
T–G	1064	763
T–C	2612	1337
G–A	1577	846
G–T	1249	782
G–C	609	483
C–A	978	551
C–T	2476	1339
C–G	643	440

Table 15.4 Amino acid mutation frequency in DNA A and DNA B encoded proteins

DNA A		DNA B	
Mutation	Frequency	Mutation	Frequency
V–I	14	R–G	20
R–K	10	T–A	14
P–S	14	V–I	14
I–V	9	K–R	10

are observed with frequencies of 10 and 9, respectively. The V–I mutation is a conservative change between hydrophobic residues (valine to isoleucine), which is less likely to cause significant structural changes but may still influence the protein's stability or interaction with other molecules. The P–S mutation involves a shift from a rigid, non-polar proline to a more flexible, polar serine, potentially affecting protein folding or function. Studies have highlighted that similar mutations in other begomoviruses have been associated with host adaptation and virus survival (Zaidi et al., 2016, 2017). The relatively high frequency of these mutations suggests that they may play a significant role in viral evolution and host adaptation.

In DNA B, the most frequent mutation is R–G, with a frequency of 20. This change involves a transition from a positively charged amino acid (Arginine) to a non-polar amino acid (Glycine), which could result in significant structural alterations. The second most common mutation is T–A (frequency of 14), where a polar threonine is replaced by a non-polar alanine, potentially affecting protein folding or stability. The V–I mutation (14), which is also found in DNA A, indicates a possible evolutionary conservation mechanism where this substitution maintains hydrophobicity and structural integrity. Lastly, K–R (10) involves a conservative change between positively charged residues, likely resulting in minimal structural changes but potentially affecting binding interactions. These mutations have been associated with the rolling-circle replication mechanism of begomoviruses, where errors during replication contribute to mutation accumulation (Seal et al., 2006; Zaidi et al., 2017). In terms of mutation frequency, DNA A exhibited a frequency of 0.004. At the same time, DNA B showed a higher mutation frequency of 0.010, indicating a

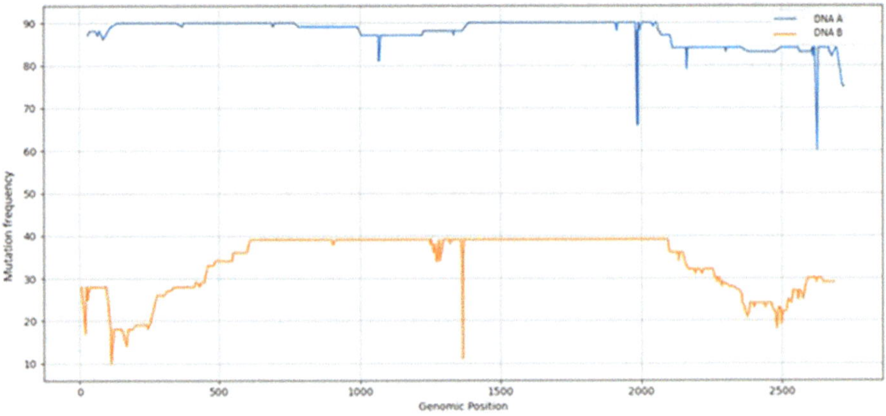

Fig. 15.11 Whole sequence variant analysis of DNA A and DNA B sequences of ToLCNDV: Mutation frequencies were calculated along the entire length of the sequences and plotted to visualize the distribution and density of variants. Variants are represented by their frequency at each position, allowing comparison of mutation hotspots between DNA A and DNA B

greater degree of variability and possible structural variations in DNA B. This suggests that DNA B (Fig. 15.11) may be more prone to structural variations and rapid evolutionary changes. Such differences between the two components highlight the complex nature of begomovirus evolution and adaptation across various hosts and geographical regions.

13 Comparison of Viral Proteins

The ability of the virus to infect the host in a new environment depends on the dynamic interaction between the sets of host gene products and the virus-encoded proteins. We investigated how much of the variability in the virus isolates is due to mutation and resultant changes in the amino acid composition of the viral-encoded protein. The predicted amino acid sequences of ToLCNDV belonging to MED, OAS, and IND groups were analysed based on multiple alignment and identity between them. The striking differences are noticed in ORFS AC1, AC2, and AC4. In the case of REP, Renukadevi et al. (2024) clearly showed the differences in RCR motifs exhibited by three ToLCNDV isolates. There was also the C-terminal extension in the ORF AC2 and AC4 in selected ToLCNDV isolates, as discussed by them.

14 Recombination

Recombination is a crucial evolutionary mechanism that enables begomoviruses to generate genetic diversity (Lefeuvre & Moriones, 2015), which is essential for their adaptability and survival in the face of host resistance and environmental changes. The occurrence of recombination among begomoviruses can lead to the emergence of novel viral strains with altered pathogenic profiles or expanded host ranges, complicating efforts in disease management (Bhardwaj et al., 2019). Based on two-dimensional electrophoresis and electron microscopy studies, Jeske et al. (2001) suggested that the high frequency of recombination in begomoviruses is due to their recombination-dependent replication. They provided a model for recombinant-dependent replication, which must be happening whenever host machinery is not conducive to RCR. According to them, incomplete ssDNA interacts with closed circular (ccc) DNA at homologous sites, leading to homologous recombination. This is followed by loop migration, ssDNA elongation, and complementary strand synthesis, resulting in dsDNA. The inherent feature of recombination-dependent replication is the shuffling of fragments between virus isolates infecting a cell together. Several factors facilitate recombination events in begomoviruses, such as mixed infection, which offers opportunities for recombination and stabilisation of recombinants. ToLCNDV is often present in mixed infections, co-exists with other begomoviruses such as chilli leaf curl virus, tomato leaf curl Karnataka virus and tomato leaf curl Palampur virus in a host (Jyothsna et al. 2013a, b; Kanakala et al., 2013). The co-existence of the virus isolates in a cell satisfies the conditions required for recombination, where a sequence of DNA is interchanged between two viruses. Examining the history of recombination in ToLCNDV reveals a different picture. Prasanna and Rai (2007) described that up to 2007, the recombination frequency in ToLCNDV isolates was much lower. They suggested that of the six isolates of ToLCNDV analysed, three isolates showed recombination, and there is a non-random distribution of events, with the highest frequency mapped in the N-terminal region of Rep.

In the last decade, the RDP programme, a powerful tool with five algorithms, has been utilised by geminivirus researchers to predict recombination events. A recombination analysis involving 115 ToLCNDV isolates representing the MED, OAS and IND groups was performed in RDP version 4.101 for the present communication. Among the isolates, the frequencies of recombination vary, as given in Table 15.5 and Fig. 15.12. In DNA A, a total of 42 events are detected, of which only 28 events are significant in more than four algorithms in the RDP programme. IN DNA B 41 events are detected, of which only 17 are substantial. It is surprising to see that all the MED isolates show traces of recombination in DNA A. Out of 27 OAS isolates analysed, recombination is not detected in five. They are AB330079, AB68447, AB368448, from cucumber, Thailand, one melon isolate (GU1180095) from Taiwan and one isolate from the weed Sauropus, JN809814 from Thailand. Among the IND group of 71 isolates analysed, recombination events are not detected only in two isolates, KC914896, a chenopodium isolate and a bitter gourd isolate, AM74791 from Pakistan (Table 15.6, Fig. 15.13).

Table 15.5 Significant recombination events detected in DNA A component of TOLCNDV isolates

Event no.	Recombinant	Breakpoint starting nt co-ordinate	Breakpoint ending nt co-ordinate	Major parent	Minor parent	Significance					
						RDP	GENECONV	BOOTSCAN	MAXCHI	SISCAN	3SEQ
1	JX460805 Tomato India	744	1337	HG316125 Rumex sp. Pakistan	MW620975 Bitter gourd India	+	+	+	+	+	+
2	KC465466 Chilli India	2344	854	AY939926 Ridge gourd India	DQ989325 Papaya India	+	+	+	+	+	+
3	KY979591 Crossandra India	2141	111	MW248655 Squash Malaysia	DQ989325	+	+	+	+	+	+
19	WN849710 Lentil Pakistan	2619	29	HG316125 Rumex Pakistan	AB368448 Papaya India	–	–	+	+	+	+
29 (22 isolates)	AY939921 Ridge gourd India	1950	2564	DQ116885 S. nigrum Pakistan	MT682358 Cucumber Cambodia	+	+	+	+	+	+
35	MH57771 Melon Spain	1505	2207	KT4269	AY9399	+	+	+	+	+	+
36 (74 isolates)	LN908936 Potato Pakistan	43–111	276	KF002409 Parthenium Pakistan	MN630278 Ridge gourd Cambodia	–	+	–	+	–	–
41 (29 isolate was retyped by altering it)	KT426907 Ridge gourd India	1218	1517	GU112082 Okra India	KY9795 Crossandra India	–	–	–	–	+	–

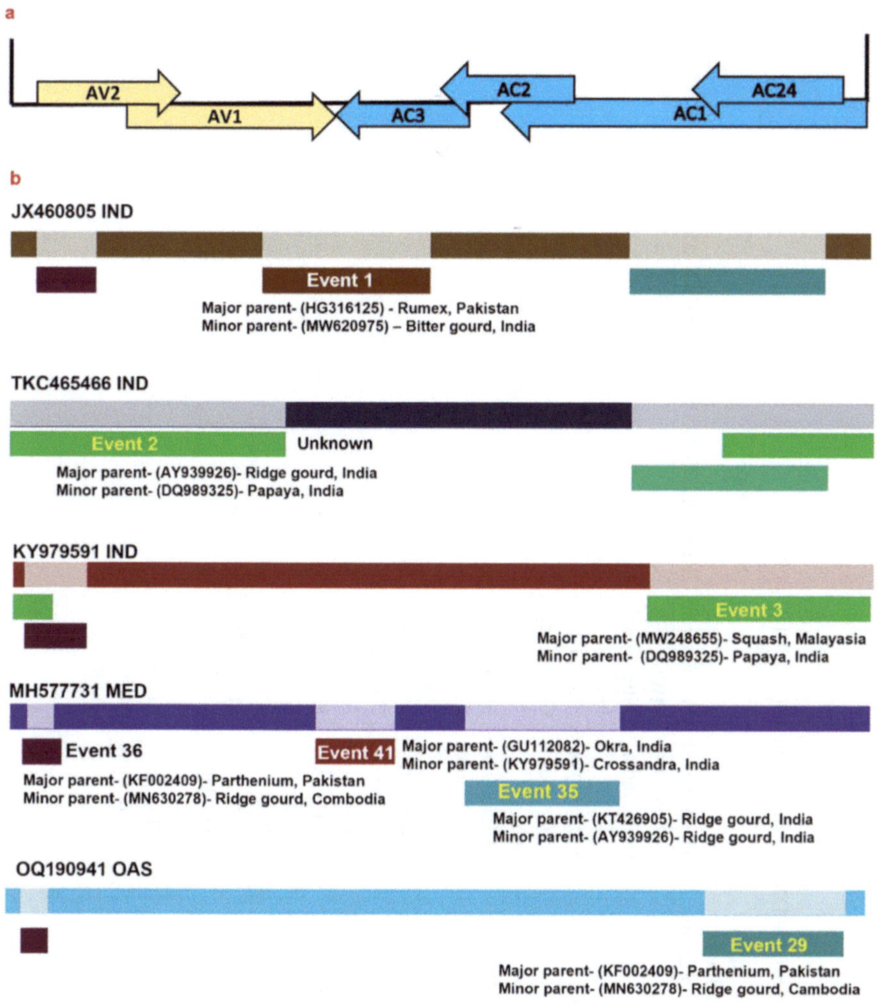

Fig. 15.12 Graphic representation of recombination analysis of TOLCNDV DNA A component. (**a**) Genome organization, (**b**) recombination events

Salient points that emerge from the critical appraisal of the recombination events are:

1. The events involve either upstream of ORF AC1 or BC1 and the intergenic region. As predicted earlier, the non-coding regions that have an origin of replication are hot spots of recombination. Our analyses too confirm it.
2. Some of the events are highly repetitive, observed in many isolates, underlying the significance of these regions in the emergence of variants. This region may contribute to sustenance and successful establishments in new hosts.

Table 15.6 Significant recombination events detected in DNA B component of TOLCNDV isolates

Event no.	Recombinant	Breakpoint starting nt co-ordinate	Breakpoint ending nt co-ordinate	Major parent	Minor parent	Significance					
						RDP	GENECONV	BOOTSCAN	MAXCHI	SISCAN	3SEQ
1	MK087117 Watermelon, India	1026	2013	HM989846 Luffa, India	MW620978 Bitter gourd, India	+	+	+	+	+	+
2	MZ648029 Fenugreek, India	2054	660	MW620978 Bitter gourd, India	MZ209193 Chilli, India	+	+	+	+	+	+
3	KY979592 Crossandra, India	698	2482	MW620978 Bitter gourd, India	MK395549 Crossandra, India	+	+	+	+	+	+
7 (67 isolates)	KF571462 Tomato, India	2558	14	MW620978 Bitter gourd, India	MK981892 Cucumber, Algeria	+	+	+	+	+	
23 (16 isolates)	LC335723 Melon, Indonesia	585	890	HM159455 Tomato, India	K874495 Potato, India	–	+	–	+	+	+
25 (20 isolates)	KP641674 Melon, Iran	466	1015	KC545813 Cucumber, India	MG597211 Chilli, India	+	+	–	–	+	+
34 (23 isolates)	OP683994 Melon, China	96	289	AY158080 Potato, India	LC3357293 Melon, Indonesia	+	–	–	–	–	+

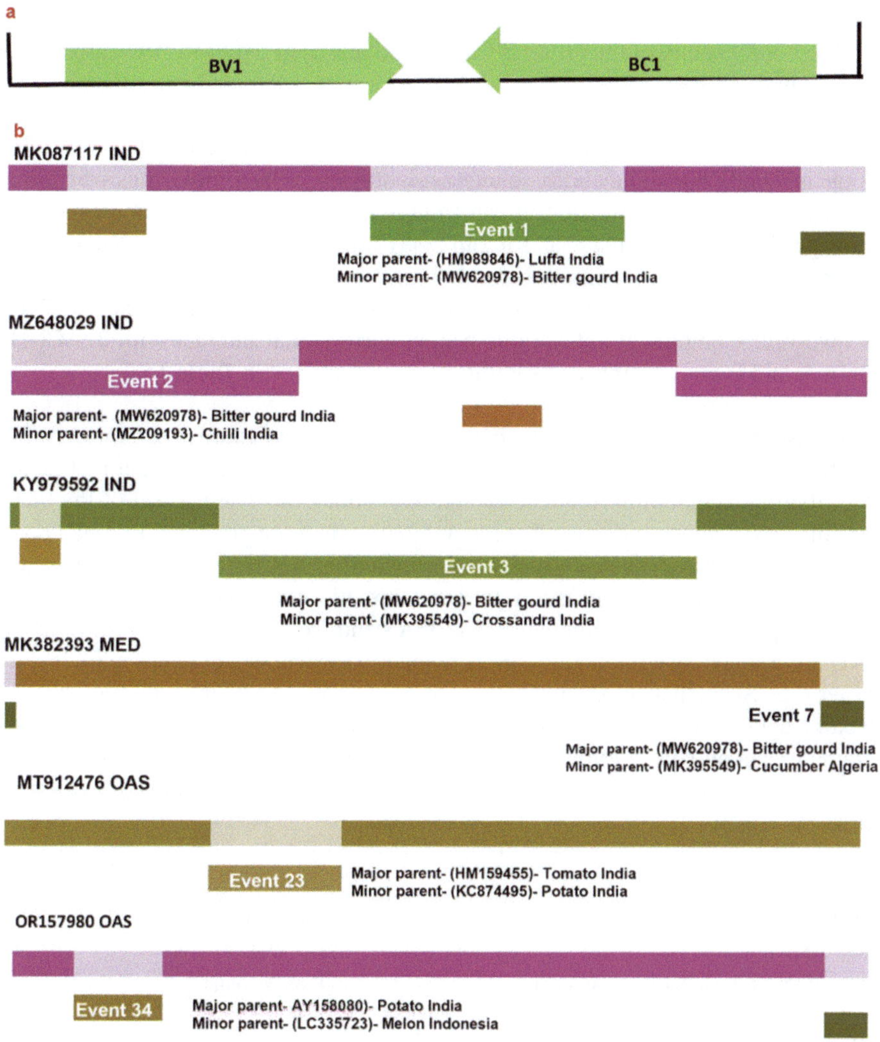

Fig. 15.13 Graphic representation of recombination analysis of TOLCNDV DNA B. (**a**) Genome organization, (**b**) recombination events

3. In at least four events, we have shown the involvement of weeds in the exchange of genetic material. Theoretically, the role of weeds in facilitating mixed infection and recombination has been speculated for a long time; in the present communication, we have provided evidence for it.
4. The prediction of events involving isolates from a distant country, lacking physical continuity of land mass, is intriguing. We infer that, based on this analysis, we can predict the presence of similar isoates in the vicinity of the recombinant. These parent sequences could well be present undetected in all the hosts around. A metagenomic study would reveal the sequences.

5. The study also highlighted the potential changes occurring in the MED group. The homogeneity of the group is fragile, and just as in IND and OAS, group variants may emerge.

The emergence of ToLCNDV as a super virulent pathogen with a wide host range is mainly due to its ability to interact with various begomoviruses while present together in mixed infection. It can capture the betasatellites when present with monopartite begomoviruses, and it can *trans* replicate the DNA B component of other begomoviruses. Its pseudorecombination with ToLCPalV has also been demonstrated. Table 15.6 provides details on the various interactions in which ToLCNDV is involved. Components of various begomoviruses that interact with ToLCNDV DNA A and B components and satellites are summarised in Table 15.7.

Conclusion and Future Perspectives It is necessary to address certain outstanding fundamental questions. First, it is essential to resolve the combination of factors contributing to mutation frequencies and recombination. The perusal of the literature discussed above indicates that recombination is more prevalent in Asian countries. Several factors play an active role, including the introduction of new host genotypes, changes in cultivation patterns, climatic variation, and disturbances in vector populations. If we compare MED, OAS and IND, the isolates from OAS and IND are not stabilised; a heterologous population exists which keeps evolving at a faster rate. Surveillance on the emergence of variants and their correlation with all factors using advanced data processing algorithms and machine learning programmes can only lay a foundation for understanding the frequency of mutation and recombination in an ecosystem, which alone can lead to a sustainable management strategy.

Table 15.7 ToLCNDV interaction with A or B or betasatellites components of various begomovirus

ToLCNDV component	Interacting virus	Interacting betasatellite	Field host	Experimental host	Reference
DNA A	ChLCV-A and ToLCGuV-B	TbLCB	*Capsicum annum*, *Capsicum chinense*	*Nicotiana benthamiana*, *Capsicum annum*	Singh et al. (2016)
DNA A	ZYMV		*Momordica charantia*		Sharma et al. (2015)
DNA A	ToLCPalV-B		*Cucumis sativus*	*Solanum lycopersicum*	Kanakala et al. (2013)
DNA A	SLCCNV and ToLCPalV		*Cucurbita moschata*		Jaiswal et al. (2012)
DNA A		RaLCB		*Nicotiana benthamiana*, *Solanum lycopersicum*	Singh et al. (2012)

(continued)

Table 15.7 (continued)

ToLCNDV component	Interacting virus	Interacting betasatellite	Field host	Experimental host	Reference
DNA A	CYVMV	CYVMB		*Nicotiana benthamiana*, *Solanum lycopersicum*	Singh et al. (2012)
DNA A		ToLCRnB		*Nicotiana benthamiana*	Kumari et al. (2011)
DNA A	ToLCGuV-B			*Nicotiana benthamiana*, *Nicotiana tabacum*, *Solanum lycopersicum*	Chakraborty et al. (2008)
DNA B	BYVMV		*Abelmoschus esculentus*		Venkataravanappa et al. (2015)
DNA B	ToLCGuV-A		*Gossypium hirsutum*		Zaidi et al. (2016)
DNA B	ToLCGuV-A			*Nicotiana benthamiana*, *Solanum lycopersicum*	Jyothsna et al. (2013b)
DNA B	ToLCGuV-A	TYLCTHB		*Nicotiana benthamiana*, *Solanum lycopersicum*	Jyothsna et al. (2013b)
DNA B	ToLCPalV-A			*Solanum lycopersicum*, *Cucumis sativus*	Kanakala et al. (2013)
DNA B	RLCuV			*Nicotiana benthamiana*, *Solanum lycopersicum*	Singh et al. (2012)
DNA B	CYVMV			*Nicotiana benthamiana*, *Solanum lycopersicum*	Singh et al. (2012)
DNA B	ToLCRnV			*Nicotiana benthamiana*, *Solanum lycopersicum*	Kumari et al. (2011)
DNA B	ToLCPalV-A		*Cucurbita melo*	*Nicotiana benthamiana*	Malik et al. (2011)
DNA B	PepLCLV	ChLCB	*Capsicum annum*	*Nicotiana benthamiana*, *Nicotiana tabacum*, *Capsicum annum*	Shafiq et al. (2010)

(continued)

Table 15.7 (continued)

ToLCNDV component	Interacting virus	Interacting betasatellite	Field host	Experimental host	Reference
DNA-B	PepLCLV			*Nicotiana benthamiana*, *Nicotiana tabacum*	Shafiq et al. (2010)
DNA-B	ToLCGuV-A			*Nicotiana benthamiana*, *Nicotiana tabacum*, *Solanum lycopersicum*	Chakraborty et al. (2008)
DNA A and DNA B	CLCuKnV-Bur	CLCuMuB	*Gossypium hirsutum*		Zaidi et al. (2016)
DNA A and DNA B		ChLCB		*Nicotiana benthamiana*	Akhter et al. (2014)
DNA A and DNA B		CLCuMub	*Solanum lycopersicum*	*Nicotiana benthamiana*, *Solanum lycopersicum*	Jyothsna et al. (2013b)
DNA A and DNA B		TYLCTHB	*Solanum lycopersicum*		Jyothsna et al. (2013b)
DNA A and DNA B		LuLDB	*Luffa cylindrica*	*Nicotiana benthamiana*, *Solanum lycopersicum*	Jyothsna et al. (2013a)
DNA A and DNA B	ToLCpalV-A and ToLCPalV-B		*Solanum lycopersicum*	*Solanum lycopersicum*	Kanakala et al. (2013)
DNA A and DNA B	CYVMV			*Nicotiana benthamiana*, *Solanum lycopersicum*	Singh et al. (2012)
DNA A and DNA B	CYVMV	CYVMB		*Nicotiana benthamiana*, *Solanum lycopersicum*	Singh et al. (2012)

References

Akhter, A., Akhtar, S., Saeed, M. and Mansoor, S. (2014). Chili leaf curl betasatellite enhances symptoms inducedby Tomato leaf curl New Delhi virus, a bipartite begomovirus. *International Journal of Agriculture and Biology, 16*(6), 1225–1228

Bhardwaj, V., Sood, S., Kumar, A., Vanishree, G., Sharma, S., Sundaresha, S., Raigond, B., Kumar, R., Bairwa, A., Lal, M., & Chakrabarti, S. K. (2019). Efficiency and reliability of marker-assisted selection for resistance to major biotic stresses in potato. *Potato Journal, 46*, 56–66.

Böttcher, B., Unseld, S., Ceulemans, H., et al. (2004). Geminate structures of African cassava mosaic virus. *Journal of Virology, 78*(13), 6758–6765.

Cai, L., Mei, Y., Ye, R., Deng, Y., et al. (2023). Tomato leaf curl New Delhi virus: An emerging plant begomovirus threatening cucurbit production. *aBIOTECH, 4*(3), 257–266.

Chakraborty, S., Vanitharani, R., Chattopadhyay, B., et al. (2008). Supervirulent pseudorecombination and asymmetric synergism between genomic components of two distinct species of begomovirus associated with severe tomato leaf curl disease in India. *Journal of General Virology, 89*(3), 818–828.

Chauhan, P. K., Mehta, N., Chauhan, R. S., Kumar, A., Singh, H., Lal, M. K., Tiwari, R. K., & Kumar, R. (2023). Utilization of primary and secondary biochemical compounds in cotton as diagnostic markers for measuring resistance to cotton leaf curl virus (CLCuV). *Frontiers in Plant Science, 14*, 1185337. https://doi.org/10.3389/fpls.2023.1185337

Chen, Y.-J., Lai, H.-C., Lin, C.-C., et al. (2021). Genetic diversity, pathogenicity and pseudorecombination of cucurbit-infecting begomoviruses in Malaysia. *Plants, 10*(11), 2396.

Cingolani, P., Platts, A., Wang, L. L., et al. (2012). A program for annotating and predicting the effects of single nucleotide polymorphisms, SnpEff: SNPs in the genome of Drosophila melanogaster strain w^{1118}; iso-2; iso-3. *Fly, 6*(2), 80–92.

Crespo, O., Robles, C., Ruiz, L., et al. (2020). Antagonism of Cucumber green mottle mosaic virus against tomato leaf curl New Delhi virus in zucchini and cucumber. *Annals of Applied Biology, 176*(2), 147–157.

Danecek, P., Bonfield, J. K., Liddle, J., et al. (2021). Twelve years of SAMtools and BCFtools. *GigaScience, 10*(2), giab008.

Fortes, I. M., Sánchez-Campos, S., Fiallo-Olivé, E., et al. (2016). A novel strain of tomato leaf curl New Delhi virus has spread to the Mediterranean basin. *Viruses, 8*(11), 307.

García-Andrés, S., Monci, F., Navas-Castillo, J., et al. (2006). Begomovirus genetic diversity in the native plant reservoir Solanum nigrum: Evidence for the presence of a new virus species of recombinant nature. *Virology, 350*(2), 433–442.

Gong, P., Tan, H., Zhao, S., et al. (2021). Geminiviruses encode additional small proteins with specific subcellular localizations and virulence function. *Nature Communications, 12*(1), 4278.

Ito, T., Sharma, P., Kittipakorn, K., et al. (2008). Complete nucleotide sequence of a new isolate of tomato leaf curl New Delhi virus infecting cucumber, bottle gourd and muskmelon in Thailand. *Archives of Virology, 153*, 611–613.

Jaiswal, N., Saritha, R.K., Datta, D., Singh, M., Dubey, R.S., Rai, A.B. and Rai, M. (2012). Mixed infections ofbegomoviruses in pumpkins with yellow vein mosaic disease in north India. *Archives of Phytopathology and Plantprotection, 45 (8)*, 938–941

Jeevalatha, A., Siddappa, S., Kumar, R., Tiwari, R. K., Lal, M. K., Sharma, S., Chakrabarty, S. K., & Singh, B. P. (2023). RNA-seq analysis reveals an early defense response to tomato leaf curl New Delhi virus in potato cultivar Kufri Bahar. *Functional & Integrative Genomics, 23*, 215. https://doi.org/10.1007/s10142-023-01138-5

Jeevalatha, A., Vanishree, G., Siddappa, S., Kumar, R., Kaundal, P., Kumar, A., & Chakrabarti, S. K. (2021). Molecular characterization and infectivity analysis of tomato leaf curl New Delhi virus isolates infecting potato. *3 Biotech, 11*, 203. https://doi.org/10.1007/s13205-021-02752-5

Jeske, H., Lutgemeier, M., & Preiß, W. (2001). DNA forms indicate rolling circle and recombination-dependent replication of Abutilon mosaic virus. *The EMBO Journal, 2021*, 6158–6167. https://doi.org/10.1093/emboj/20.21.6158

Jyothsna, P., Haq, Q., Singh, P., et al. (2013a). Infection of tomato leaf curl New Delhi virus (ToLCNDV), a bipartite begomovirus with betasatellites, results in enhanced level of helper virus components and antagonistic interaction between DNA B and betasatellites. *Applied Microbiology and Biotechnology, 97*, 5457–5471.

Jyothsna, P., Rawat, R., & Malathi, V. (2013b). Predominance of tomato leaf curl Gujarat virus as a monopartite begomovirus: Association with tomato yellow leaf curl Thailand betasatellite. *Archives of Virology, 158*, 217–224.

Kanakala, S., Jyothsna, P., Shukla, R., et al. (2013). Asymmetric synergism and heteroencapsidation between two bipartite begomoviruses, tomato leaf curl New Delhi virus and tomato leaf curl Palampur virus. *Virus Research, 174*(1–2), 126–136.

Kumar, R. V., Singh, A. K., Singh, A. K., et al. (2015). Complexity of begomovirus and betasatellite populations associated with chilli leaf curl disease in India. *Journal of General Virology, 96*(10), 3143–3158.

Kumar, R., Lal, M. K., Tiwari, R. K., Kumar, A., Behera, B., Nayak, L., Dash, G. K., Sahoo, S. K., Jena, J., Lal, P., & Altaf, M. A. (2023). Traditional and emerging climate-resilient agricultural practices for enhancing food production and nutritional quality. In T. Aftab (Ed.), *New frontiers in plant-environment interactions* (Environmental science and engineering). Springer. https://doi.org/10.1007/978-3-031-43729-8_21

Kumar, R., Tiwari, R. K., Jeevalatha, A., Sundaresha, S., Shah, M. A., Sharma, S., Sagar, V., Kumar, M., & Chakrabarti, S. K. (2021). Potato apical leaf curl disease: Current status and perspectives on a disease caused by tomato leaf curl New Delhi virus. *Journal of Plant Diseases and Protection, 128*, 897–911. https://doi.org/10.1007/s41348-021-00463-w

Kumari, P., Singh, A.K., Sharma, V.K., Chattopadhyay, B. and Chakraborty, S. (2011). A novel recombinanttomato-infecting begomovirus capable of transcomplementing heterologous DNA-B components. *Archives ofvirology, 156* (5), 769–783.

Kumari, P., Singh, A.K.,Venkataravanappa, V., Lakshminarayana Reddy, C.N., Jalali, S. and Krishna Reddy, M. (2015). Association ofTomato leaf curl New Delhi virus DNA-B with bhendi yellow vein mosaic virus in okra showing yellow vein mosaicdisease symptoms. *Acta Virol, 59* (2), 125–139

Langmead, B., & Salzberg, S. L. (2012). Fast gapped-read alignment with Bowtie 2. *Nature Methods, 9*(4), 357–359.

Lefeuvre, P., & Moriones, E. (2015). Recombination as a motor of host switches and virus emergence: Geminiviruses as case studies. *Current Opinion in Virology, 10*, 14–19.

Li, H., & Durbin, R. (2009). Fast and accurate short read alignment with Burrows–Wheeler transform. *Bioinformatics, 25*(14), 1754–1760.

Li, F., Xiongbiao, X., Huang, C., Zhouhang, G., Cao, L., Tao, H., Ding, M., Li, Z., & Zhou, X. (2015). The AC 5 protein encoded by Mungbean yellow mosaic India virus is a pathogenicity determinant that suppresses RNA silencing-based antiviral defenses. *New Phytologist, 208*(2), 555–569.

Malathi, V., Renukadevi, P., Chakraborty, S., et al. (2017). Begomoviruses and their satellites occurring in India: Distribution, diversity and pathogenesis. In *A century of plant virology in India* (pp. 75–177). Springer.

Malik, A.H., Briddon, R.W. and Mansoor, S. (2011). Infectious clones of Tomato leaf curl Palampur virus with adefective DNA B and their pseudo-recombination with Tomato leaf curl New Delhi virus. *Virology journal, 8* (1), 173

Mandal, B. (2010). Emerging geminiviral diseases and their management. In *Emergence of begomovirus diseases in cucurbits in India* (pp. 167–181). Nova Science Publishers Inc.

Mazumder, P., Mondal, F., Sarkar, M., Majumdar, A., Biswas, K. K., Sharma, S. K., Lal, M. K., Tiwari, R. K., Kumar, R., & Roy, A. (2024). Biological and molecular characterization of a new isolate of tomato mottle mosaic virus causing severe shoestring and fruit deformities in tomato plants in India. *Plants, 13*, 2811. https://doi.org/10.3390/plants13192811

Moriones, E., Praveen, S., & Chakraborty, S. (2017). Tomato leaf curl New Delhi virus: An emerging virus complex threatening vegetable and fiber crops. *Viruses, 9*(10), 264.

Naga, K. C., Subhash, S., Bhatnagar, A., Tiwari, R. K., Kumar, R., Verma, G., Sharma, S., & Chakrabarti, S. K. (2021a). Influence of host plants on virus acquisition and endosymbionts of whitefly, Bemisia tabaci (Gennadius). *Indian Journal of Entomology, 83*, 371–375. https://doi.org/10.5958/0974-8172.2020.00206.0

Naga, K. C., Sundaresha, S., Kumar, R., Tiwari, R. K., Subhash, S., Verma, G., Buckseth, T., Bairwa, A., Sharma, S., Katare, S., & Srivastava, R. M. (2021b). A new record of Asia II 5 genetic group of Bemisia tabaci (Gennadius) in the major potato growing areas of India and its relationship with tomato leaf curl New Delhi virus infecting potato. *3 Biotech, 11*, 421. https://doi.org/10.1007/s13205-021-02966-7

Nagendran, K., Mohankumar, S., Aravintharaj, R., et al. (2017). The occurrence and distribution of major viruses infecting cucurbits in Tamil Nadu state, India. *Crop Protection, 99*, 10–16.

Padidam, M., Beachy, R. N., & Fauquet, C. M. (1995). Tomato leaf curl geminivirus from India has a bipartite genome and coat protein is not essential for infectivity. *Journal of General Virology, 76*(1), 25–35.

Padidam, M., Beachy, R. N., & Fauquet, C. M. (1996). The role of AV2 ("precoat") and coat protein in viral replication and movement in tomato leaf curl geminivirus. *Virology, 224*(2), 390–404.

Panno, S., Caruso, A., Troiano, E., et al. (2019). Emergence of tomato leaf curl New Delhi virus in Italy: Estimation of incidence and genetic diversity. *Plant Pathology, 68*(3), 601–608.

Prasanna, H., & Rai, M. (2007). Detection and frequency of recombination in tomato-infecting begomoviruses of South and Southeast Asia. *Virology Journal, 4*, 1–10.

Reddy, M. K., Nagendran, K., Kumari, S., et al. (2024). Global scenario of begomovirus diseases in vegetable crops. *Vegetable Science, 51*, 43–53.

Renukadevi, P., Devi, R. G., Jothika, C., et al. (2024). Genomic distinctiveness and recombination in tomato leaf curl New Delhi virus (ToLCNDV-BG) isolates infecting bitter gourd. *3 Biotech, 14*(8), 184.

Saez, C., Esteras, C., Martínez, C., et al. (2017). Resistance to tomato leaf curl New Delhi virus in melon is controlled by a major QTL located in chromosome 11. *Plant Cell Reports, 36*, 1571–1584.

Salazar-Martínez, J., Martínez, J., & Lelasco-Velasco, J. (2021). Host-induced editing mechanisms in geminiviruses. *Viruses, 133*, 459.

Samretwanich, K., Chiemsombat, P., Kittipakorn, K., et al. (2000). Yellow leaf disease of cantaloupe and wax gourd from Thailand caused by tomato leaf curl virus. *Plant Disease, 84*(2), 200–200.

Sangeetha, B., Malathi, V., Alice, D., et al. (2018). A distinct seed-transmissible strain of tomato leaf curl New Delhi virus infecting Chayote in India. *Virus Research, 258*, 81–91.

Sarkar, M., Gupta, D., Singh, O. W., Paul, S., Kumar, R., Mandal, B., & Roy, A. (2024). Identification of pathogenicity determinants in ToLCNDV and their RNAi-based knockdown for disease management in Nicotiana benthamiana and tomato plants. *Frontiers in Microbiology, 15*, 1481523. https://doi.org/10.3389/fmicb.2024.1481523

Seal, S., VandenBosch, F., & Jeger, M. (2006). Factors influencing begomovirus evolution and their increasing global significance: Implications for sustainable control. *Critical Reviews in Plant Sciences, 25*(1), 23–46.

Shafiq, M., Asad, S., Zafar, Y., Briddon, R.W. and Mansoor, S. (2010). Pepper leaf curl Lahore virus requires theDNA B component of Tomato leaf curl New Delhi virus to cause leaf curl symptoms. *Virology journal, 7* (1), 367

Sharma, S., Kang, S. and Sharma, A. (2015). First report of mixed infection of Zucchini yellow mosaic virus andTomato leaf curl New Delhi virus in bittergourd in India. *Journal of Plant Pathology, 97* (2), 397

Singh, A.K., Kushwaha, N. and Chakraborty, S. (2016). Synergistic interaction among begomoviruses leads to thesuppression of host defense-related gene expression and breakdown of resistance in chilli. *Applied microbiologyand biotechnology, 100* (9), 4035–4049

Singh, A., Chattopadhyay, B. and Chakraborty, S. (2012). Biology and interactions of two distinct monopartitebegomoviruses and betasatellites associated with radish leaf curl disease in India. *Virology journal, 9* (1), 43

Sivalingam, P., & Varma, A. (2012). Role of betasatellite in the pathogenesis of a bipartite begomovirus affecting tomato in India. *Archives of Virology, 157*, 1081–1092.

Sohrab, S., Mandal, B., Pant, R., et al. (2003). First report of association of tomato leaf curl virus-New Delhi with yellow mosaic disease of Luffa cylindrica in India. *Plant Disease, 87*(9), 1148–1148.

Srivastava, K., Hallan, V., Raizada, R., et al. (1995). Molecular cloning of Indian tomato leaf curl vims genome following a simple method of concentrating the supercoiled replicative form of viral DNA. *Journal of Virological Methods, 51*(2–3), 297–304.
Tahir, M., & Haider, M. (2005). First report of tomato leaf curl New Delhi virus infecting bitter gourd in Pakistan. *Plant Pathology, 54*(6), 807.
Varma, A., & Malathi, V. (2003). Emerging geminivirus problems: A serious threat to crop production. *Annals of Applied Biology, 142*(2), 145–164.
Vasudeva, R., & Sam Raj, J. (1948). A leaf-curl disease of tomato. *Phytopathology, 38*, 364–369.
Venkataravanappa, V., Lakshminarayana Reddy, C.N., Jalali, S. and Krishna Reddy, M. (2015). Association ofTomato leaf curl New Delhi virus DNA-B with bhendi yellow vein mosaic virus in okra showing yellow vein mosaicdisease symptoms. *Acta Virol, 59* (2), 125–139
Wickham, H., & Sievert, C. (2009). *ggplot2: Elegant graphics for data analysis* (Vol. 10). Springer.
Zaidi, S. S. E. A., Martin, D. P., Amin, I., et al. (2017). Tomato leaf curl New Delhi virus: A widespread bipartite begomovirus in the territory of monopartite begomoviruses. *Molecular Plant Pathology, 18*(7), 901–911.
Zaidi, S. S.-E.-A., Shafiq, M., Amin, I., et al. (2016). Frequent occurrence of tomato leaf curl New Delhi virus in cotton leaf curl disease affected cotton in Pakistan. *PLoS One, 11*(5), e0155520.
Zerbini, F. M., Briddon, R. W., Idris, A., et al. (2017). ICTV virus taxonomy profile: Geminiviridae. *Journal of General Virology, 98*(2), 131–133.

Molecular Arsenal of Tomato Leaf Curl New Delhi Virus: Viral Suppressors of RNA Silencing and Their Roles in Pathogenicity

16

Mehulee Sarkar, Firoz Mondal, Anik Majumdar, Nimmala Sree Valli, Bishal Roy, Namrata Verma, Yeluru Mohanbabu, Ravinder Kumar, and Anirban Roy

Abstract

Tomato leaf curl New Delhi virus (ToLCNDV), a bipartite begomovirus, is a major plant pathogen causing severe yield losses in tomato and other economically important crops. Like many begomoviruses, ToLCNDV overcomes the host's antiviral RNA silencing machinery through the action of viral suppressors of RNA silencing (VSRs). This chapter provides a detailed account of three key ToLCNDV-encoded VSRs—AV2 (pre-coat protein), AC2 (transcriptional activator protein), and AC4—highlighting their structural features, mechanisms of RNA silencing suppression at both transcriptional and posttranscriptional levels, and their roles as pathogenicity determinants. AV2 primarily disrupts host RNA-dependent RNA polymerase 1 (RDR1)-mediated antiviral defense, AC2 interferes with the microRNA pathway and AGO1-associated RISC function, while AC4 targets AGO4 and modulates host DNA methylation machinery. Beyond silencing suppression, these proteins modulate host transcription, hormone signaling, and defense responses, thereby enhancing viral replication, movement, and symptom expression. The chapter also outlines experimental approaches for VSR characterization, including transient expression, reversal of silencing, and stable expression assays. Understanding the molecular arsenal of ToLCNDV provides critical insights for devising RNAi-based and genetic resistance strategies against begomovirus infections.

M. Sarkar · F. Mondal · A. Majumdar · N. S. Valli · B. Roy · N. Verma · Y. Mohanbabu · R. Kumar · A. Roy (✉)
Advanced Centre for Plant Virology, Division of Plant Pathology, ICAR-Indian Agricultural Research Institute, New Delhi, India

R. Kumar et al. (eds.), *Tomato Leaf Curl New Delhi Virus (ToLCNDV)*,
https://doi.org/10.1007/978-981-95-2852-3_16

Keywords

Viral suppressors of RNA silencing (VSRs) · Begomoviruses · Pathogenicity determinants · RNA interference (RNAi) · AV2–AC2–AC4 proteins

1 Introduction

RNA silencing, commonly referred to as RNA interference (RNAi), is a natural cellular mechanism that regulates gene expression at both the transcriptional and post-transcriptional levels. Moreover, it serves as a dynamic biological tool that provides a robust defense against plant pathogens, particularly viruses. However, viruses, being the masters of molecular manipulation, have developed strategies to counteract this robust plant defense. Many viruses produce proteins that disrupt various stages of the RNA silencing pathway, thereby suppressing its antiviral effects and facilitating efficient replication and spread of the virus. These proteins are commonly known as viral suppressors of RNA silencing (VSRs). Begomoviruses, belonging to the family *Geminiviridae* are whitefly-transmitted ssDNA viruses (2.5–3 kb), and are among the most destructive plant pathogens, causing significant yield losses in crops worldwide (Kumar et al., 2021; Jeevalatha et al., 2021, 2023). These viruses encode multifunctional proteins that not only facilitate viral replication and movement but also act as VSRs to suppress the host RNA silencing pathway. Host RNA silencing can be categorized into at least three overlapping pathways: (i) cytoplasmic gene silencing mediated by siRNA (posttranscriptional gene silencing), (ii) microRNA-mediated silencing that controls messenger RNA expression, and (iii) gene silencing dependent on DNA methylation (transcriptional gene silencing) (Baulcombe, 2004).

Begomoviruses encode different VSR proteins with distinct structures and functions that interfere with host gene-silencing activity (Rojas et al., 2001; Zrachya et al., 2007). One of the most extensively studied VSRs in begomoviruses is the C2 protein (also referred to as AC2 or TrAP, transcriptional activator protein). The C2 protein of *Tomato leaf curl virus* (TLCV) has been shown to suppress RNA silencing by blocking methylation machinery (Ramesh et al., 2017), while the AC2 protein of the *Tomato golden mosaic virus* triggers the expression of rgsCaM, a calmodulin-like protein that regulates RNA silencing (Yong Chung et al., 2014). The C2 protein of *Beet severe curly top virus* (BSCTV) hinders DNA methylation-mediated gene silencing by blocking the 26S proteasome-driven degradation of SAMDC1 (Zhang et al., 2011). Another well-characterized VSR is the V2 protein (also referred to as the pre-coat protein). The V2 protein of *Tomato yellow leaf curl virus* (TYLCV) suppresses the host's suppressor of gene silencing 3 (AtSGS3), a crucial component of the host's antiviral silencing system (Glick et al., 2008). In contrast, the V2 protein of *Tomato yellow leaf China virus* (TYLCCNV) inhibits RNA silencing by binding to siRNA molecules and blocking methylation-driven gene silencing, a key component of transcriptional gene silencing pathways (Wang et al., 2014, 2019). VSRs like AC4 and C4 act as suppressors of both transcriptional and posttranscriptional gene silencing by interacting with single-stranded si-/miRNAs (Chellappan et al., 2005) or AGO4 (Vinutha et al., 2018). For example, AC4 acts both upstream and downstream of siRNA strand unwinding, inhibiting the incorporation of siRNA into the RNA-induced silencing complex (Hanley-Bowdoin et al., 2013; Ramesh et al., 2017). The transient expression of βC1 protein of monopartite begomoviruses also inhibits PTGS by binding to both dsDNAs and ssDNAs (Cui et al., 2005; Kon et al., 2007; Shukla et al., 2013). βC1 protein of TYLCCNV interacts with proteins such as MYC2, a transcription factor involved in jasmonic acid signaling, to manipulate host defense responses (Li et al., 2014). Thus, begomoviruses employ diverse strategies to counteract host RNA silencing machinery.

Begomovirus solanumdelhiense (*Tomato leaf curl New Delhi virus* (ToLCNDV)) is one of the most important species of the *Begomovirus* genus, known for causing economically important diseases in tomatoes and other vegetables. It was first described in 1995, infecting tomatoes (*Solanum lycopersicum*) in northern India (Padidam et al., 1995). It is a bipartite begomovirus consisting of two circular single-stranded DNA genome components (DNA-A and DNA-B). In recent years, it has rapidly expanded its host range by infecting economically important crops from various families, including *Solanaceae*, *Cucurbitaceae*, *Malvaceae*, *Euphorbiaceae*, and *Fabaceae* (Cai et al., 2023). Like other begomoviruses, ToLCNDV encodes a set of viral suppressor proteins (VSRs) that play a crucial role in overcoming the host's RNA silencing defense mechanisms. These VSRs, including proteins like AV2, AC2, and AC4, interfere with the host's antiviral RNA silencing machinery, allowing the virus to replicate and spread more efficiently. The suppressor proteins collectively help ToLCNDV evade the host's defense system, contributing to the virus's persistence and virulence in infected plants.

2 General Mechanism of Antiviral RNAi

A general antiviral RNAi pathway involves—(i) viral infection: when the plant is infected with a virus, double-stranded RNA (dsRNA) intermediates are produced during viral replication or transcription. (ii) dsRNA recognition and processing: the plant recognizes the viral dsRNAs as foreign macromolecules. It employs the dicer-like proteins (DCLs) to cleave the dsRNAs into several small interfering RNAs (siRNAs), typically 21–24 nucleotides in length. (iii) RISC formation: siRNAs are loaded into the Argonaute (AGO) proteins to form the RNA-induced silencing complex (RISC). (iv) target recognition: the RISC complex uses the siRNA as a guide to recognize and bind to complementary viral DNA and/or RNA sequences. (v) viral DNA methylation and/or RNA degradation: the guide siRNA either directs viral DNA methylation through RNA-dependent DNA methylation (RdDM) pathway (transcriptional gene silencing or TGS) or AGO protein in the RISC cleaves the target viral RNA, leading to its degradation (posttranscriptional gene silencing or PTGS) (Sundaresha et al., 2022a, b, c).

One of the fundamental aspects of the begomovirus-plant interaction is the plant's innate defense via RNA silencing and its inhibition by virus-encoded RNA silencing suppressor proteins. Among the eight different proteins encoded by the bipartite ToLCNDV genome, the pre-coat protein (AV2), the TrAP (AC2), and the AC4 proteins are reported to possess the VSR activities along with several other functions.

3 AV2: Pre-coat Protein

3.1 Structural Features

The ORF-V2/AV2 of begomoviruses codes for the pre-coat protein. A 112 amino acid long (~13 kDa) AV2 is commonly reported in most of the ToLCNDV isolates, with a few isolates having smaller sizes of 109 and 94 amino acids. The amino acid composition is dominated by hydrophobic amino acids (~35%). The protein has a characteristic zinc-finger-like WCCH domain (Pfam PF03716) toward the C-terminal region (79–101 aa) of the protein, which is expected to have an important role in VSR activity of the protein. Another important feature of the AV2 includes a protein kinase C (PKC) motif towards the N-terminal region (58–60 aa) (Sarkar et al., 2021). This motif is proven to be essential for silencing suppression, and subcellular localization, as well as for the pathogenicity of AV2 of another bipartite begomovirus, East African cassava mosaic Cameroon virus (EACMCV) (Chowda-Reddy et al., 2008). In contrast to AV2 of EACMCV, many other tomato-infecting begomoviruses, like *Tomato leaf curl Palampur virus* (ToLCPalV), and *tomato yellow leaf curl virus* (TYLCV), where AV2 localizes either to the cytoplasmic periphery with low nuclear presence or the perinuclear region and cytoplasm (Chowda-Reddy et al., 2008; Hak et al., 2015; Roshan et al., 2017); the subcellular localization of ToLCNDV AV2 is reported to be at both the nucleus and the cytoplasmic membrane (Sarkar et al., 2021).

3.2 AV2 as a Viral Suppressor of RNA Silencing (VSR)

AV2 of begomoviruses is a multifunctional protein. In monopartite begomoviruses, although V2 has an important role as VSR, it acts mainly as a movement protein facilitating cell-to-cell and long-distance movement of the virus. However, in bipartite begomoviruses like ToLCNDV, the movement function is predominantly contributed by the protein coded by ORF-BV1 of DNA-B. The AV2 protein of such bipartite begomoviruses acts primarily as a strong RNA silencing suppressor against plant innate immunity. The very first experimental evidence for the silencing suppressor activity of begomovirus-encoded V2/AV2 was provided by Zrachya et al. (2007), when the V2 protein of *Tomato yellow leaf curl virus*-Israel (TYLCV-Isr) was shown to possess the VSR function). Two amino acid residues (Cys-84 and Cys-86) have been established to be the key determinants of the VSR functionality of TYLCV-Isr V2. Interestingly, these residues were found conserved in the ToLCNDV AV2 protein when Vu et al. (2013) provided the first evidence of the silencing suppressor activity of ToLCNDV-AV2. The over-expression of the ToLCNDV-AV2 protein in tobacco (*Nicotiana tabacum* L. cv. Xanthi and *N. benthamiana* cv. 16c reverses the silencing effects of GFP, establishing this protein as a strong viral suppressor of RNA silencing (VSR) (Vu et al., 2013; Basu et al., 2021, p. 20). Furthermore, VSR activity of ToLCNDV-AV2 appears early during infection with a greater reduction of siRNAs at the early time point (3 dpi) compared to AC2, another VSR of ToLCNDV (Sundaresan et al., 2020). AV2 of several other tomato-infecting begomoviruses, like *Tomato leaf curl Palampur virus* (ToLCPalV), *Tomato leaf curl Gujarat virus* (ToLCGuV), *Croton yellow vein mosaic virus* (CYVMV), *Tomato leaf curl Java virus* (ToLCJV), etc., are also evidenced to disrupt the host RNA silencing-based antiviral defense (Sharma & Ikegami, 2010; Roshan et al., 2018; Basu et al., 2021; Zhai et al., 2022).

3.3 Mechanisms of Silencing Suppression by AV2

Although studies showed that the pre-coat protein or AV2 of ToLCNDV acts mainly as a suppressor of posttranscriptional gene silencing (PTGS), the exact mechanism of the same is still an evolving subject. A study reported that AV2 of ToLCNDV can selectively downregulate the *RNA-DEPENDENT RNA POLYMERASE 1* (RDR1), leading to inhibition of antiviral resistance in *N. tabacum.* AV2 mutated ToLCNDV also failed to infect transgenic *N. benthamiana* expressing RDR1 (Basu et al., 2018). In other begomoviruses, V2/AV2 is known for suppressing antiviral RNA silencing during the downstream events of transcriptional gene silencing (TGS), such as the amplification phase, and interfering with host methylation activities (Luna et al., 2017, 2020; Wang et al., 2018). The V2 protein of the TYLCV reduces viral small interfering RNA (vsiRNA) amplification by inhibiting the cofactor of RDR6, i.e., suppressor of gene silencing 3 (SGS3) (Glick et al., 2008). It may also sequester RDR6/SGS3 dsRNA intermediates, further disrupting this amplification (Luna et al., 2017). Similarly, the V2 protein from the beet curly top virus (BCTV) suppresses PTGS by impairing the RDR6/SGS3 pathway (Luna et al., 2020). However, in beet curly top Iran virus (BCTIV), the VSR activity of V2 was independent of (RDR6) (Ebrahimi et al., 2023). Moreover, V2 interacts with histone deacetylase 6 (NbHDA6), disrupting the recruitment of methyltransferase 1 (MET1) and decreasing the methylation of the viral genome, which increases TYLCV pathogenicity (Wang et al., 2018). The *Cotton leaf curl Multan virus* (CLCuMV) V2 interacts with long dsRNA to inhibit dicer-like (DCL) processing (Amin et al., 2011).

3.4 AV2 as a Pathogenicity Determinant

A pathogenicity determinant in begomoviruses refers to a specific viral factor (usually a protein) that directly contributes to the virus's ability to cause disease in its host plants. They can influence several aspects of the virus infection, including symptom determination, viral replication, host manipulation, and other related processes. Begomovirus VSRs often demonstrate the ability to determine pathogenicity, as their capacity to suppress the host's antiviral defenses frequently results in successful infection and spread of the virus. In the case of ToLCNDV, pre-coat protein or AV2 has been proven as a crucial pathogenicity determinant of the virus, playing a vital role in disease causation. The transient overexpression of AV2 enables the successful development of systemic symptoms in *N. benthamiana* plants, even in the absence of the complete genome of the parent virus. This ability of AV2 can be attributed to its transcripts' capacity to move systemically in the host. Also, delayed symptom development and reduction in AV2 transcript level in hairpin RNAi (hpRNAi) mediated AV2 knocked down plants confirm its role in disease establishment by ToLCNDV (Sarkar et al., 2024). Moreover, AV2 is found to be indispensable for the complete pathogenesis of ToLCNDV as the AV2 mutant virus inoculated *N. benthamiana* plants showed symptom remission at the later stage of virus inoculation (2 wpi) along with the increased accumulation of vsiRNAs. A similar trend was observed in another begomovirus, ToLCGuV. Also, AV2 of ToLCNDV can selectively block the symptom recovery in chimeric ToLCGuV-infected *N. tabacum* plants (Basu et al., 2018). In the case of TYLCV, site-directed mutation at amino acid position 71 of V2 (V2S71A) led to a delay of virus infection associated with mild symptoms in *N. benthamiana*, indicating its involvement in pathogenesis (Zhao et al., 2018). Mutations targeting the V2 coding region result in altered disease severity in other begomoviruses like tomato curly stunt virus (ToCSV), soybean geminivirus A (SGVA), etc. (Li et al., 2022; Zwolinski et al., 2023). Furthermore, transgenic *N. tabacum* plants expressing antisense AV2 transcript prevent symptom development even after ToLCNDV infection, which confirms the necessity of AV2 in pathogenesis (Mubin et al., 2007).

4 AC2: Transcriptional Activator Protein

4.1 Structural Features

The AC2 protein, also known as the transcriptional activator protein (TrAP) of ToLCNDV, is 15 kDa in size and comprises 139 amino acid residues. This protein is essential for the expression of the AV1 protein (coat protein) (Haley et al., 1992). This protein contains three distinct conserved regions: an N-terminal basic domain with a nuclear localization signal (NLS), a centrally located DNA-binding domain with a zinc-finger motif, and a C-terminal acidic activator domain (Trinks et al., 2005). ScanProsite analysis showed an N-glycosylation site (amino acids 9–12), a cAMP-and cGMP-dependent protein kinase phosphorylation site (amino acids 23–26), a nuclear localization signal (amino acids 6–54), a protein kinase C phosphorylation site (amino acids 26–28; 52–54 and 61–63), and a casein kinase II phosphorylation site (amino acids 90–92; 61–64, and 124–127) (Sarkar et al., 2021). In comparison, detailed experiments on the AC2 protein of the *Tomato golden mosaic virus* (129 amino acid residues) showed a nuclear localization signal (amino acids 17–31), a zinc finger-like domain (amino acids 36–53), and a transcriptional activation domain (amino acids 115–129) (Hartitz et al., 1999). Subcellular localization experiments using confocal microscopy have shown that AC2 of ToLCNDV is localized to the nucleus (Sarkar et al., 2021).

4.2 AC2 as a Viral Suppressor of RNA Silencing (VSR)

Geminiviruses encode viral suppressor of RNA silencing (VSRs) proteins that can interfere with both transcriptional gene silencing (TGS) and posttranscriptional gene silencing (PTGS). VSRs can bind to miRNAs and siRNAs; however, the AC2 protein encoded by ToLCNDV does not bind to any form of RNA (Lakatos et al., 2006). Studies have shown the possible involvement of AC2 in the deregulation of the host miRNA pathway, where it interacts with miR172 and miR319. Furthermore, ex vivo and in vitro assays confirmed that this interaction is mediated by the direct association of AGO1 with AC2 (Kumar & Naqvi, 2016). Thus, AC2 of ToLCNDV associates with miRNA-loaded AGO1 protein in RISC and leads to suppression of the host RNAi machinery, that is, interfering with PTGS. Another study revealed that AC2 interacts with tomato ATG8f and mediates its degradation through the autophagy pathway, and the presence of SlATG8f inhibits the silencing activity of AC2 (Prasad & Prasad, 2022). Moreover, reversal of GFP silencing assays showed that AC2 suppressed gene silencing, where it was reported that siRNA accumulation was reduced to 9% in ToLCNDV+AC2 infiltrated plants (Basu et al., 2018). Interestingly, the C2 protein of tomato golden mosaic virus, another tomato-infecting begomovirus, targets adenosine kinase (ADK), which is involved in the S-adenosylmethionine-dependent methyltransferase activity. This interaction causes methylation and suppresses transcriptional gene silencing. Suppression of ADK also prevents the phosphorylation of cytokinin nucleosides and inhibits their conversion into less active nucleotides. This results in the accumulation of bioactive cytokinins, which are crucial for cell cycle progression, a process that begomoviruses rely on for their genomic replication. Research has also demonstrated that TGMV-infected *Arabidopsis* plants exhibit increased ARR5 activity, a primary cytokinin-responsive promoter (Baliji et al., 2010).

4.3 AC2 as a Pathogenicity Determinant

The AC2 protein of ToLCNDV has been reported to be involved in pathogenesis through mutational experiments, where a nonsense mutation (ΔAC2) caused a delay in the development of mild symptoms in *N. benthamiana* and *N. tabacum* plants, compared to plants inoculated with the wild-type virus (Basu et al., 2018). Furthermore, its involvement in symptom development was confirmed using transient overexpression and knockdown assays. Transient overexpression of AC2 resulted in downward curling symptoms in approximately 73% of inoculated plants

30 days postinoculation. Their subsequent knockdown through hpRNAi constructs caused a delay in symptom appearance and mild curling (Sarkar et al., 2024). The AC2 protein of ToLCNDV has been reported to counteract the HR response induced by the nuclear shuttle protein (NSP). The central region of AC2, which consists of the NLS and zinc-finger domains, is linked to this function (Hussain et al., 2005).

5 AC4: Multifunctional Protein

5.1 Structural Features

AC4 in bipartite viruses and C4 in monopartite viruses is the smallest ORF embedded within the coding region of the AC1 ORF and is the least conserved segment of all geminiviruses. The AC4 protein of ToLCNDV is 6 kDa in size and comprises 58 amino acid residues. ScanProsite analysis showed N-terminal myristoylation site (amino acids 2–7, 14–19, and 31–36), a N-glycosylation site (amino acids 15–18), protein kinase C phosphorylation site (amino acids 17–19, 35–37, 38–40, and 52–54) and casein kinase II phosphorylation site (amino acids 38–41 and 53–56). Cellular experiments showed that AC4 localizes in the host membrane under a confocal microscope (Sarkar et al., 2021). In *Tomato leaf curl Yunnan virus* (TLCYnV), C4 protein has been reported as a nucleocytoplasmic shuttle protein. Phosphorylation of C4 by NbSKη, a member of plant SHAGGY-like kinases in the nucleus, facilitates myristoylation of viral protein. The addition of a hydrophobic myristoyl group to glycine makes the protein hydrophobic, enabling it to bind to cellular membranes (Mei et al., 2021). After these changes, the complex shuttles back to the cytoplasm via its interaction with exportin-a, which serves as a nuclear export signal (NES) (Fondong, 2019).

5.2 AC4 as a Viral Suppressor of RNA Silencing (VSR)

AC4 protein encoded by ToLCNDV employs multiple counter-defense strategies at both the PTGS and TGS levels of RNA silencing. At the TGS level, AC4 hinders the host methylation pathway to prevent viral DNA methylation by sequestering host AGO4 protein. AGO4 plays a role in the methylation of invading viral nucleic acids by recruiting 24-nt siRNAs (Mallory and Vaucheret, 2010). The ability of the AC4 protein to reverse GFP silencing in *N. benthamiana* indicates that it also functions at the PTGS level. Furthermore, the study reported that the Asn50 in the SKNT motif at the C-terminal region is essential for its RNA silencing suppressor function (Vinutha et al., 2018). Similarly, in another study, ectopic expression of AC4 misregulated DNA demethylases and methyltransferases, reconfirming that AC4 suppresses TGS (Chowdhury et al., 2022).

5.3 AC4 as a Pathogenicity Determinant

Apart from being a potent silencing suppressor of ToLCNDV, the AC4 protein also acts as a pathogenicity and/or symptom determinant of the virus. Transgenic tomato plants constitutively expressing the ToLCNDV-AC4 protein exhibit severe morphological abnormalities like stunted and aberrant growth, reduction in the size of leaf lamina, typical leaf curl symptoms, and abnormal root morphology. These abnormalities are reported to be caused by AC4's ability to interact with host factors such as tryptophan amino transferase (TAR1) like protein, senescence-associated gene (SAG) protein, TOi54, and cytochrome P450 monooxygenase, thus attenuating host auxin biosynthesis and metabolism (Thimmegowda et al., 2020). Its ability to induce systemic symptoms even after the absence of other genomic components of the virus can be linked to the systemic movement of AC4 transcripts upon transient expression in *N. benthamiana.* Delayed symptom development associated with lower AV4 transcript levels in hpRNAi-mediated AC4 knocked-down plants ascertains ToLCNDV AC4 as a pathogenicity and symptom determinant of the virus (Sarkar et al., 2024). Leaf vein patterning-related TORNADO1 (*SlTRN1*) gene transcription is repressed in developed mature leaves; misregulation due to virus infection potentially alters the leaf venation pattern. Upon transient overexpression of ToLCNDV-AC4, levels of DNA methyltransferases are downregulated with concomitant upregulation of DNA demethylases (*slDML1*, *slDML2*), leading to hypomethylated *SlTRN1* promoter. This, along with higher WRKY16 expression and recognition of W-box elements, causes augmentation of *SlTRN1* gene expression, which ultimately leads to several venation abnormalities and contributes to the typical leaf curl symptom manifestation (Chowdhury et al., 2022).

6 Assay for Viral Suppressor of RNA Silencing (VSRs)

There are three main approaches used to identify VSR activity:

(i) Transient expression assays

The assay involves the co-infiltration of two agrobacterial strains (one that induces silencing of the reporter gene (e.g., green fluorescent protein (GFP) and the other is a candidate suppressor gene) into a plant leaf (*N. benthamiana*) and examined over time to silence the reporter gene. Interference of the candidate suppressor with RNA silencing causes the expression of GFP, and the patch remains bright green. However, if the candidate gene is not a suppressor, the path remains red under ultraviolet (UV) light (Voinnet et al., 2000).

(ii) Reversal of silencing assay

The general approach involves introducing the candidate virus into a plant that has been silenced and observing whether the silenced characteristics are restored. This technique usually uses a GFP-expressing transgenic line (*N. benthamiana*). At the four-leaf stage, it was infiltrated with GFP-expressing *Agrobacterium*, which led to local and systemic silencing. The plants were then inoculated with the candidate virus. Reversal of GFP expression in plants indicates suppressor activity of the virus (Voinnet et al., 1999).

(iii) Stable expression assays

In this method, a transgenic line expressing a candidate suppressor gene is crossed with a series of transgenic lines silenced for the reporter gene. This approach offers an examination of the effect of a suppressor on different types of transgene-induced silencing and provides insights into the mechanism of suppression (Anandalakshmi et al., 1998).

References

Amin, I., Hussain, K., Akbergenov, R., et al. (2011). Suppressors of RNA silencing encoded by the components of the cotton leaf curl begomovirus-betasatellite complex. *Molecular Plant-Microbe Interactions, 24*, 973–983. https://doi.org/10.1094/MPMI-01-11-0001

Anandalakshmi, R., Pruss, G. J., Ge, X., et al. (1998). A viral suppressor of gene silencing in plants. *Proceedings of the National Academy of Sciences, 95*, 13079–13084. https://doi.org/10.1073/pnas.95.22.13079

Baliji, S., Lacatus, G., & Sunter, G. (2010). The interaction between geminivirus pathogenicity proteins and adenosine kinase leads to increased expression of primary cytokinin-responsive genes. *Virology, 402*, 238–247. https://doi.org/10.1016/j.virol.2010.03.023

Basu, S., Kumar Kushwaha, N., Kumar Singh, A., et al. (2018). Dynamics of a geminivirus-encoded pre-coat protein and host RNA-dependent RNA polymerase 1 in regulating symptom recovery in tobacco. *Journal of Experimental Botany, 69*, 2085–2102. https://doi.org/10.1093/jxb/ery043

Basu, S., Singh, A. K., Singh, D., et al. (2021). Role of viral suppressors governing asymmetric synergism between tomato-infecting begomoviruses. *Applied Microbiology and Biotechnology, 105*, 1107–1121. https://doi.org/10.1007/s00253-020-11070-4

Baulcombe, D. (2004). RNA silencing in plants. *Nature, 431*, 356–363. https://doi.org/10.1038/nature02874

Cai, L., Mei, Y., Ye, R., et al. (2023). Tomato leaf curl New Delhi virus: An emerging plant begomovirus threatening cucurbit production. *aBIOTECH, 4*, 257–266. https://doi.org/10.1007/s42994-023-00118-4

Chellappan, P., Vanitharani, R., & Fauquet, C. M. (2005). MicroRNA-binding viral protein interferes with Arabidopsis development. *Proceedings of the National Academy of Sciences, 102*, 10381–10386. https://doi.org/10.1073/pnas.0504439102

Chowda-Reddy, R. V., Achenjang, F., Felton, C., et al. (2008). Role of a geminivirus AV2 protein putative protein kinase C motif on subcellular localization and pathogenicity. *Virus Research, 135*, 115–124. https://doi.org/10.1016/j.virusres.2008.02.014

Chowdhury, S., Mukherjee, A., Basak, S., et al. (2022). Disruption of tomato TGS machinery by ToLCNDV causes reprogramming of vascular tissue-specific TORNADO1 gene expression. *Planta, 256*, 78. https://doi.org/10.1007/s00425-022-03985-1

Cui, X., Li, G., Wang, D., et al. (2005). A Begomovirus DNAβ-encoded protein binds DNA, functions as a suppressor of RNA silencing, and targets the cell nucleus. *Journal of Virology, 79*, 10764–10775. https://doi.org/10.1128/jvi.79.16.10764-10775.2005

Ebrahimi, S., Eini, O., Baßler, A., et al. (2023). Beet curly top Iran virus rep and V2 suppress post-transcriptional gene silencing via distinct modes of action. *Viruses, 15*, 1996. https://doi.org/10.3390/v15101996

Fondong, V. N. (2019). The ever-expanding role of C4/AC4 in geminivirus infection: Punching above its weight? *Molecular Plant, 12*, 145–147. https://doi.org/10.1016/j.molp.2018.12.006

Glick, E., Zrachya, A., Levy, Y., et al. (2008). Interaction with host SGS3 is required for suppression of RNA silencing by tomato yellow leaf curl virus V2 protein. *Proceedings of the National Academy of Sciences, 105*, 157–161. https://doi.org/10.1073/pnas.0709036105

Hak, H., Levy, Y., Chandran, S. A., et al. (2015). TYLCV-is movement *in planta* does not require V2 protein. *Virology, 477*, 56–60. https://doi.org/10.1016/j.virol.2015.01.007

Haley, A., Zhan, X., Richardson, K., et al. (1992). Regulation of the activities of African cassava mosaic virus promoters by the AC1, AC2, and AC3 gene products. *Virology, 188*, 905–909. https://doi.org/10.1016/0042-6822(92)90551-Y

Hanley-Bowdoin, L., Bejarano, E. R., Robertson, D., & Mansoor, S. (2013). Geminiviruses: Masters at redirecting and reprogramming plant processes. *Nature Reviews. Microbiology, 11*, 777–788. https://doi.org/10.1038/nrmicro3117

Hartitz, M. D., Sunter, G., & Bisaro, D. M. (1999). The tomato golden mosaic virus transactivator (TrAP) is a single-stranded DNA and zinc-binding phosphoprotein with an acidic activation domain. *Virology, 263*, 1–14. https://doi.org/10.1006/viro.1999.9925

Hussain, M., Mansoor, S., Iram, S., et al. (2005). The nuclear shuttle protein of tomato leaf curl New Delhi virus is a pathogenicity determinant. *Journal of Virology, 79*, 4434–4439. https://doi.org/10.1128/JVI.79.7.4434-4439.2005

Jeevalatha, A., Vanishree, G., Siddappa, S., Kumar, R., Kaundal, P., Kumar, A., & Chakrabarti, S. K. (2021). Molecular characterization and infectivity analysis of tomato leaf curl New Delhi virus isolates infecting potato. *3 Biotech, 11*, 203. https://doi.org/10.1007/s13205-021-02752-5

Jeevalatha, A., Siddappa, S., Kumar, R., Tiwari, R. K., Lal, M. K., Sharma, S., Chakrabarty, S. K., & Singh, B. P. (2023). RNA-seq analysis reveals an early defense response to tomato leaf curl New Delhi virus in potato cultivar Kufri Bahar. *Functional & Integrative Genomics, 23*, 215. https://doi.org/10.1007/s10142-023-01138-5

Kon, T., Sharma, P., & Ikegami, M. (2007). Suppressor of RNA silencing encoded by the monopartite tomato leaf curl Java begomovirus. *Archives of Virology, 152*, 1273–1282. https://doi.org/10.1007/s00705-007-0957-6

Kumar, V., & Naqvi, A. R. (2016). Tomato leaf curl New Delhi virus (ToLCNDV) encoded AC2 associates with host mirnas by directly interacting with AGO1. *Journal of RNA and Genomics, 12*, 515–520.

Kumar, R., Tiwari, R. K., Jeevalatha, A., Sundaresha, S., Shah, M. A., Sharma, S., Sagar, V., Kumar, M., & Chakrabarti, S. K. (2021). Potato apical leaf curl disease: Current status and perspectives on a disease caused by tomato leaf curl New Delhi virus. *Journal of Plant Diseases and Protection, 128*, 897–911. https://doi.org/10.1007/s41348-021-00463-w

Lakatos, L., Csorba, T., Pantaleo, V., et al. (2006). Small RNA binding is a common strategy to suppress RNA silencing by several viral suppressors. *The EMBO Journal, 25*, 2768–2780. https://doi.org/10.1038/sj.emboj.7601164

Li, R., Weldegergis, B. T., Li, J., et al. (2014). Virulence factors of geminivirus interact with MYC2 to subvert plant resistance and promote vector performance. *The Plant Cell, 26*, 4991–5008. https://doi.org/10.1105/tpc.114.133181

Li, Q., Zhang, Y., Lu, W., et al. (2022). Identification and characterization of a new geminivirus from soybean plants and determination of V2 as a pathogenicity factor and silencing suppressor. *BMC Plant Biology, 22*, 362. https://doi.org/10.1186/s12870-022-03745-z

Luna, A. P., Rodríguez-Negrete, E. A., Morilla, G., et al. (2017). V2 from a curtovirus is a suppressor of post-transcriptional gene silencing. *Journal of General Virology, 98*, 2607–2614. https://doi.org/10.1099/jgv.0.000933

Luna, A. P., Romero-Rodríguez, B., Rosas-Díaz, T., et al. (2020). Characterization of Curtovirus V2 protein, a functional homolog of begomovirus V2. *Frontiers in Plant Science, 11*, 835. https://doi.org/10.3389/fpls.2020.00835

Mallory, A., & Vaucheret, H. (2010). Form, function, and regulation of ARGONAUTE proteins. *The Plant Cell, 22*, 3879–3889. https://doi.org/10.1105/tpc.110.080671

Mei, Y., Wang, Y., Hu, T., et al. (2021). The C4 protein encoded by tomato leaf curl Yunnan virus interferes with mitogen-activated protein kinase cascade-related defense responses through inhibiting the dissociation of the ERECTA/BKI1 complex. *EBSCOhost*. https://openurl.ebsco.com/contentitem/doi:10.1111%2Fnph.17387?sid=ebsco:plink:crawler&id=ebsco:doi:10.1111%2Fnph.17387. Accessed November 5, 2024.

Mubin, M., Mansoor, S., Hussain, M., & Zafar, Y. (2007). Silencing of the AV2 gene by antisense RNA protects transgenic plants against a bipartite begomovirus. *Virology Journal, 4*, 10. https://doi.org/10.1186/1743-422X-4-10

Padidam, M., Beachy, R. N., & Fauquet, C. M. (1995). Classification and identification of geminiviruses using sequence comparisons. *The Journal of General Virology, 76*(Pt 2), 249–263. https://doi.org/10.1099/0022-1317-76-2-249

Prasad, A., & Prasad, M. (2022). Interaction of ToLCNDV TrAP with SlATG8f marks it susceptible to degradation by autophagy. *Cellular and Molecular Life Sciences, 79*, 241. https://doi.org/10.1007/s00018-022-04281-7

Ramesh, S. V., Sahu, P. P., Prasad, M., et al. (2017). Geminiviruses and plant hosts: A closer examination of the molecular arms race. *Viruses, 9*, 256. https://doi.org/10.3390/v9090256

Rojas, M. R., Jiang, H., Salati, R., et al. (2001). Functional analysis of proteins involved in movement of the monopartite begomovirus, tomato yellow leaf curl virus. *Virology, 291*, 110–125. https://doi.org/10.1006/viro.2001.1194

Roshan, P., Kulshreshtha, A., & Hallan, V. (2017). Identification of host cellular targets of AC4 and AV2 proteins of tomato leaf curl palampur virus and their sub-cellular localization studies. *Virus, 28*, 390–400. https://doi.org/10.1007/s13337-017-0405-5

Roshan, P., Kulshreshtha, A., Kumar, S., et al. (2018). AV2 protein of tomato leaf curl Palampur virus promotes systemic necrosis in *Nicotiana benthamiana* and interacts with host Catalase2. *Scientific Reports, 8*, 1273. https://doi.org/10.1038/s41598-018-19292-3

Sarkar, M., Aggarwal, S., Mukherjee, S. K., et al. (2021). Sub-cellular localization of suppressor proteins of tomato leaf curl New Delhi virus. *Virus, 32*, 298–304. https://doi.org/10.1007/s13337-021-00651-0

Sarkar, M., Gupta, D., Singh, O. W., et al. (2024). Identification of pathogenicity determinants in ToLCNDV and their RNAi-based knockdown for disease management in *Nicotiana benthamiana* and tomato plants. *Frontiers in Microbiology, 15*, 1481523. https://doi.org/10.3389/fmicb.2024.1481523

Sharma, P., & Ikegami, M. (2010). Tomato leaf curl Java virus V2 protein is a determinant of virulence, hypersensitive response and suppression of posttranscriptional gene silencing. *Virology, 396*, 85–93. https://doi.org/10.1016/j.virol.2009.10.012

Shukla, R., Dalal, S., & Malathi, V. (2013). Suppressors of RNA silencing encoded by tomato leaf curl betasatellites. *Journal of Biosciences, 38*, 45–51. https://doi.org/10.1007/s12038-012-9291-6

Sundaresan, G., Das, S.S., Tripathi, A., Mukherjee, S.K. and Sanan-Mishra, N. (2020). Evaluating the strength of RNA silencing suppressor proteins encoded by two geminiviruses using assay based on reversal of GFP silencing. *Australasian Plant Pathology, 49*(2), pp. 95–106.

Sundaresha, S., Jeevalatha, A., Kumar, R., Sood, S., Sharma, S., Bhardwaj, V., et al. (2022a). RNA interference: A versatile tool to augment plant protection strategies in potato. In *Sustainable management of potato pests and diseases* (pp. 369–388). Singapore.

Sundaresha, S., Sharma, S., Bairwa, A., Tomar, M., Kumar, R., Bhardwaj, V., et al. (2022b). Spraying of dsRNA molecules derived from Phytophthora infestans, along with nanoclay carriers as a proof of concept for developing novel protection strategy for potato late blight. *Pest Management Science, 78*(7), 3183–3192.

Sundaresha, S., Bairwa, A., Tomar, M., Kumar, R., Venkatasalam, E. P., Sagar, V., et al. (2022c). In vitro method for synthesis of large-scale dsRNA molecule as a novel plant protection strategy. In *Plant gene silencing: Methods and protocols* (pp. 211–226). Springer US.

Thimmegowda, V., Vanchinathan, S., Bansal, N., et al. (2020). Tomato auxin biosynthesis/signaling is reprogrammed by the geminivirus to enhance its pathogenicity. *Planta, 252*, 51. https://doi.org/10.1007/s00425-020-03452-9

Trinks, D., Rajeswaran, R., Shivaprasad, P. V., et al. (2005). Suppression of RNA silencing by a geminivirus nuclear protein, AC2, correlates with transactivation of host genes. *Journal of Virology, 79*, 2517–2527. https://doi.org/10.1128/jvi.79.4.2517-2527.2005

Vinutha, T., Kumar, G., Garg, V., et al. (2018). Tomato geminivirus encoded RNAi suppressor protein, AC4 interacts with host AGO4 and precludes viral DNA methylation. *Gene, 678*, 184–195. https://doi.org/10.1016/j.gene.2018.08.009

Voinnet, O., Pinto, Y. M., & Baulcombe, D. C. (1999). Suppression of gene silencing: A general strategy used by diverse DNA and RNA viruses of plants. *Proceedings of the National Academy of Sciences, 96*, 14147–14152. https://doi.org/10.1073/pnas.96.24.14147

Voinnet, O., Lederer, C., & Baulcombe, D. C. (2000). A viral movement protein prevents spread of the gene silencing signal in *Nicotiana benthamiana. Cell, 103*, 157–167. https://doi.org/10.1016/s0092-8674(00)00095-7

Van Vu, T., Choudhury, N.R. and Mukherjee, S.K., (2013). Transgenic tomato plants expressing artificial microRNAs for silencing the pre-coat and coat proteins of a begomovirus, Tomato leaf curl New Delhi virus, show tolerance to virus infection. *Virus Research*, 172(1-2), pp. 35–45.

Wang, B., Li, F., Huang, C., et al. (2014). V2 of tomato yellow leaf curl virus can suppress methylation-mediated transcriptional gene silencing in plants. *The Journal of General Virology, 95*, 225–230. https://doi.org/10.1099/vir.0.055798-0

Wang, B., Yang, X., Wang, Y., et al. (2018). Tomato yellow leaf curl virus V2 interacts with host histone deacetylase 6 to suppress methylation-mediated transcriptional gene silencing in plants. *Journal of Virology, 92*. https://doi.org/10.1128/jvi.00036-18

Wang, Y., Wu, Y., Gong, Q., et al. (2019). Geminiviral V2 protein suppresses transcriptional gene silencing through interaction with AGO4. *Journal of Virology, 93*. https://doi.org/10.1128/jvi.01675-18

Yong Chung, H., Lacatus, G., & Sunter, G. (2014). Geminivirus AL2 protein induces expression of, and interacts with, a calmodulin-like gene, an endogenous regulator of gene silencing. *Virology, 460–461*, 108–118. https://doi.org/10.1016/j.virol.2014.04.034

Zhai, Y., Roy, A., Peng, H., et al. (2022). Identification and functional analysis of four RNA silencing suppressors in begomovirus Croton yellow vein mosaic virus. *Frontiers in Plant Science, 12*, 768800.

Zhang, Z., Chen, H., Huang, X., et al. (2011). BSCTV C2 attenuates the degradation of SAMDC1 to suppress DNA methylation-mediated gene silencing in Arabidopsis. *The Plant Cell, 23*, 273–288. https://doi.org/10.1105/tpc.110.081695

Zhao, W., Ji, Y., Wu, S., et al. (2018). Single amino acid in V2 encoded by TYLCV is responsible for its self-interaction, aggregates and pathogenicity. *Scientific Reports, 8*, 3561. https://doi.org/10.1038/s41598-018-21446-2

Zrachya, A., Glick, E., Levy, Y., et al. (2007). Suppressor of RNA silencing encoded by tomato yellow leaf curl virus-Israel. *Virology, 358*, 159–165. https://doi.org/10.1016/j.virol.2006.08.016

Zwolinski, A. M., Brigden, A., & Rey, M. E. C. (2023). Differences in the 3′ intergenic region and the V2 protein of two sequence variants of tomato curly stunt virus play an important role in disease pathology in *Nicotiana benthamiana. PLoS One, 18*, e0286149. https://doi.org/10.1371/journal.pone.0286149

RNA Vaccination in Plants: A Non-transgenic Approach to Combat Plant Viruses with Special Reference to Tomato Leaf Curl New Delhi Virus

17

Prantik Mazumder, Anik Majumdar, Bishal Roy, Yeluru Mohanbabu, Nimmala Sree Valli, Mehulee Sarkar, Firoz Mondal, Namrata Verma, Ravinder Kumar, and Anirban Roy

Abstract

Tomato leaf curl New Delhi virus (ToLCNDV) (*Begomovirus*, family: *Geminiviridae*), which is transmitted by whiteflies, poses a serious challenge to global production of tomato and leads to severe yield losses. Conventional management strategies, such as chemical pesticides and the use of resistant varieties, have limitations due to environmental concerns, potential resistance breakdown, and variable effectiveness. Recently double-stranded RNA (dsRNA) technique has emerged as a novel, environmentally sustainable approach for managing viral diseases. This chapter explores the potential of RNA interference (RNAi) mechanisms through the exogenous spraying of dsRNA molecules that target key genes of viruses essential for the replication and infection of ToLCNDV. Additionally, the chapter assesses various methods for delivering dsRNAs into the plant system, addresses safety concerns related to dsRNA application, and discusses prospects for dsRNA-based approaches to achieve durable and effective management of ToLCNDV.

Keywords

dsRNA · RNAi · ToLCNDV · Sustainable

P. Mazumder · A. Majumdar · B. Roy · Y. Mohanbabu · N. S. Valli · M. Sarkar · F. Mondal · N. Verma · R. Kumar · A. Roy (✉)
Advanced Centre for Plant Virology, Division of Plant Pathology, ICAR-Indian Agricultural Research Institute, New Delhi, India

R. Kumar et al. (eds.), *Tomato Leaf Curl New Delhi Virus (ToLCNDV)*,
https://doi.org/10.1007/978-981-95-2852-3_17

1 Introduction

Tomato yellow leaf curl disease (TYLCD) and tomato leaf curl disease (ToLCD) are destructive viral diseases of tomatoes worldwide. The causative agents of TYLCD are over 13 species of begomoviruses, whereas ToLCD is more diverse, with over 80 species associated with it. The most widely spread of these viruses is *Begomovirus coheni* (commonly known as tomato yellow leaf curl virus, TYLCV), a monopartite begomovirus (2.7 kb) available throughout the globe. However, TYLCV has not been reported in the Indian subcontinent. In India, ToLCD was first reported from the northern parts of the country by Vasudeva and Samraj (1948). According to Kumar and Kumar (2018), there are 14 species of begomoviruses associated with ToLCD in India. In India, the most widespread begomovirus is *Begomovirus solanumdelhiense* (Malathi et al., 2017). The losses in yield that may occur as a result of infection range from 17.6% to 99.7% for different stages of infection and varying severity of disease (Butter & Rataul, 1981; Kalloo, 1996). Winter-planted tomatoes have been observed to incur a yield loss of 52.5–100% as documented by Saikia and Muniyappa (1989), along with 6.4–52.2% yield loss in summer-planted crops. More recently, Zaidi et al. (2017) estimate that ToLCNDV infection could lead to complete yield loss in tomatoes.

Managing TYLCD and ToLCD is a daunting task, as these diseases continue their yearly incursions into new regions of tomato cultivation (Cohen & Lapidot, 2007; Moriones & Navas-Castillo, 2000). Integrated control of these diseases in the fields hinges largely on insecticide application aimed at curbing vector populations, alongside breeding cultivars with resistance to the viruses. This combination tends to fall short, underscoring the necessity for new and more effective strategies. Plant viruses have recently been addressed using RNA interference (RNAi) techniques (Jin et al., 2021). As a plant antiviral strategy, RNA interference works by degrading the viral messenger RNAs using sequence-specific double-stranded RNAs (dsRNA) aimed at viral mRNAs. RNAi has proven to be a potent natural antiviral system and has found application in the production of transgenic plants made resistant to a wide range of viruses (Sundaresha et al., 2022a, b). Unfortunately, crops modified using genetic engineering draw negative attention due to public concerns, forcing researchers to look for non-transgenic methods. One such revolutionary method is the direct coating of RNA molecules, referred to as 'RNA vaccination'. This method triggers RNAi in plants, enabling them to defend against virus infections. This makes topical dsRNA application a promising and sustainable approach, especially for protecting economically significant crops like tomatoes. This method can successfully neutralise the virus by specifically targeting its genome, halting its movement within the plant, and delaying the onset of disease symptoms. Through this chapter, the recent works that highlight the different levels of protection achieved against ToLCNDV through the exogenous application of dsRNA have been discussed.

2 RNAi Mechanism in Plants Against Viruses

Given the persistent prevalence of viral infections, new sustainable approaches to disease management are required (Tatineni & Hein, 2023). RNAi vaccination is a new method of disease management that involves the application of dsRNA molecules that target the virus. With the help of complementary RNAs, this strategy exploits the RNA interference phenomenon, a naturally occurring evolutionary mechanism that works through sequence-dependent mRNA degradation (Svoboda, 2020; Mazumder et al., 2024). Plants have been known to use RNA interference as an innate immune defense mechanism to recover from viral infections since it was first described by Fire et al. (1998). This defense mechanism not only equips plants to combat ongoing infections but also provides resistance against infections by closely related viruses (Ratcliff et al., 1997). In retaliation, viruses use counterattacking strategies using silencing suppressors to evade this immunity (Baulcombe, 2002). The RNAi pathway generates small interfering RNAs (siRNAs), which can be grouped into categories such as their source, size, structure, and function (Kim, 2005). Various small RNA (sRNA) types exist, such as siRNAs, microRNAs (miRNAs), trans-acting siRNAs (ta-siRNAs), and phased siRNAs (pha-siRNAs). siRNAs primarily originate from double-stranded RNAs (dsRNAs), which may arise from viral replication, harnessed by plant RNA-dependent RNA polymerases (RDRs), or form from single-stranded RNAs (ssRNAs) adopting hairpin-like structures (Sundaresha et al., 2022c). These dsRNAs are recognized and cleaved into siRNAs by Dicer-like (DCL) proteins. There are four types of DCL proteins (DCL1–4) in plants, each responsible for generating siRNAs of specific lengths. DCL4 primarily produces 21-nucleotide siRNAs, which play a critical function in RNA virus defense (Bouché et al., 2006; Deleris et al., 2006) (Fig. 17.1), while also contributing to tasiRNA biogenesis for gene regulation (Gasciolli et al., 2005). Meanwhile, DCL2 and DCL3 generate 22- and 24-nucleotide siRNAs each. The 22-nucleotide siRNAs are manifested in host defense against viruses (Garcia-Ruiz et al., 2010; Wang et al., 2011), whereas 24-nucleotide siRNAs control the regulation of transposable elements, repetitive DNA sequences (Xie et al., 2004), and play a role in DNA virus defense through transcriptional gene silencing (TGS) (Fig. 17.1). The function of DCL3 in defense against RNA viruses is still not well understood (Diaz-Pendon et al., 2007). DCL proteins process siRNA duplexes for incorporation into the RNA-induced silencing complex (RISC). In RISC, one of the siRNA strands acts as a guide to locate and degrade complementary RNA sequences; this action involves an argonaute (AGO) protein, which is a pivotal component of RISC (Rosa et al., 2018). There are many types of AGO proteins in plants (Zhang et al., 2015), and AGO2 is particularly significant in siRNA-directed antiviral defense (Jaubert et al., 2011). Furthermore, siRNAs are not only involved in RNA silencing, but also in RNA-directed DNA methylation (RdDM), which further bolsters antiviral defense (Raja et al., 2008). While 21-nucleotide siRNAs are the main actors in RNA interference, the full-length 24-nucleotide siRNAs, together with AGO4, move to

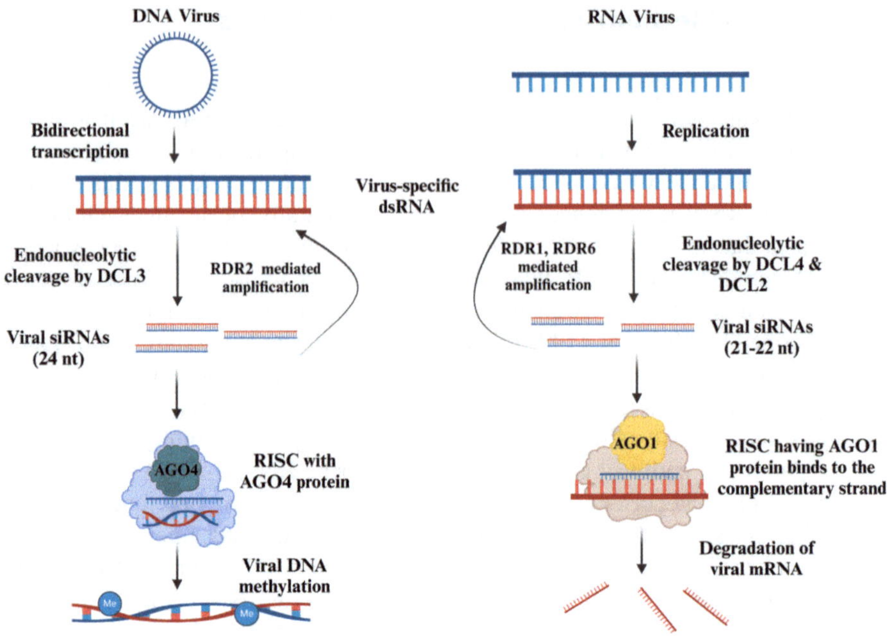

Fig. 17.1 Small interfering RNA-mediated antiviral defense in plants. Double-stranded RNA precursors, formed either as replicative intermediates of viral RNAs or through bidirectional transcription of circular viral DNA, are converted into vsiRNAs of 21, 22, or 24 nucleotides by plant Dicer-like proteins DCL4, DCL2, and DCL3, respectively. In many cases, effective RNAi against viruses also relies on the production of secondary vsiRNAs, which are derived from virus-specific long dsRNA precursors synthesized by host RDR1, RDR6, and RDR2. Various plant Argonaute (AGO) proteins mediate siRNA-directed antiviral defense by either facilitating RNA cleavage within the RISC or initiating the RdDM pathway, resulting in the transcriptional gene silencing of viral DNA genomes. (This figure was created using BioRender)

the nucleus to initiate the classic RdDM pathway. This pathway primarily targets transcripts produced by the host's DNA-dependent RNA polymerase, and with the help of cofactors like DNA methyltransferase 2, it sparks de novo DNA methylation (Zhong et al., 2014). To strengthen the silencing pathway even further, DCLs subsequently identify these methylation sites and convert them into siRNAs.

3 Success Stories of dsRNA-Based Management Strategy Against Plant Viruses

Several studies show that topical application of dsRNA molecules that target certain viruses can successfully confer resistance against a variety of RNA viruses from different genera (Tenllado & Díaz-Ruíz, 2001; Tenllado et al., 2004), including *Tobamovirus* (Tenllado & Díaz-Ruíz, 2001; Yin et al., 2009; Sun et al., 2010; Mitter et al., 2017; Niehl et al., 2018); *Potyvirus* (Gan et al., 2010; Sun et al., 2010; Kaldis et al., 2018; Worrall et al., 2019; Vadlamudi et al., 2020); *Cucumovirus* (Mitter et al., 2017; Namgial et al., 2019; Holeva et al., 2021; Routhu et al., 2022); *Potexvirus* (Lau et al., 2014); *Sobemovirus* (Konakalla et al., 2019) and *Closterovirus* (Borah et al., 2023). Using dsRNA to target the 2b gene of CMV had induced a resistant response in *Nicotiana* plants (Namgial et al., 2019). Similarly, approximately 60% resistance was found in "Bhut Jolokia" pepper (Borah et al., 2018). Routhu et al. (2022) reported that dsRNA targeting the coat protein of CMV protected ghost pepper from the virus successfully. Moreover, a foliar spray of dsRNA derived from the *NSs* gene of GBNV enhanced plant resistance (Gupta et al., 2021). Although not specified for DNA viruses, previous work by Konakalla et al. (2016) demonstrated that dsRNA targeting the *p126* and *CP* genes in TMV provided adequate protection against the virus. This dsRNA strategy has been characterized as a promising non-transgenic approach for managing plant diseases caused by RNA viruses. Still, it needs to be validated for DNA viruses as a possible management tool, especially within the family of *Geminiviridae*. For example, tomato plants achieved 45–60% protection against ToLCV through the exogenous application of dsRNA (Namgial et al., 2019). In another study, it was shown that applying a dsRNA cocktail on *Nicotiana benthamiana* reduced disease incidence by 66.7% against the chilli leaf curl virus (ChiLCV) (Singh et al., 2022). Furthermore, on infected black gram plants, an application of dsRNA against the *CP* gene and the *AC1* gene (the replication initiator protein) of the mungbean yellow mosaic virus (MYMV) substantially affected both symptom expression and virus titer (Kamesh Krishnamoorthy et al., 2023). Results like these highlight the reported observations of the exogenous dsRNA application as a reasonable alternative to control both DNA and RNA plant viruses and thus provide a means for better and safer agricultural practices.

3.1 In Vitro and In Vivo Produced dsRNA in the Management of Plant Viruses

There are two fundamentals methods for generating dsRNA: the in vivo method using engineered bacteria (Yin et al., 2009; Gan et al., 2010; Sun et al., 2010; Mitter et al., 2017; Gupta et al., 2021; Singh et al., 2022; Routhu et al., 2022) and the in vitro method using RNA polymerase (Konakalla et al., 2016; Konakalla et al., 2021; Niehl et al., 2018; Namgial et al., 2019; Vadlamudi et al., 2020; Borah et al., 2024). The dsRNA that is produced in vitro has been shown to control several viral diseases effectively. For example, dsRNA targeting the 500 bp *CP* gene of pea seed-borne mosaic virus (PSbMV) reduced virus load by 50–65% (Lau et al., 2014). In another study, a 666 bp segment of the *p126* gene and a 480 bp segment of the *CP* gene of TMV were used as targets, resulting in the *N. benthamiana* exhibiting a 65% reduction in virus load (Konakalla et al., 2016). dsRNA synthesized in vitro and produced using the RibomaxTM Express large-scale RNA production system effectively controlled the monopartite TYLCV. dsRNAs designed against the *C4* and *V2* genes of TYLCV were produced in vitro and applied topically onto tomato plants infected with TYLCV. Plant treatment with dsC4 and dsV2 significantly reduced incidence to 23% and 46%, respectively, as compared to the untreated control (Melita et al., 2021). When using genetically engineered bacterial systems for in vivo production, the protocol starts with amplifying a viral genomic fragment or target sequence by PCR with site-specific primers. This target fragment is then directionally cloned into *Xba* I and *Hind* III restriction sites (the restriction sites are chosen with consideration to the absence of these sites within the target sequence) of a specialized vector, L4440. This vector has advantageous features because it contains bacteriophage T7 promoter sites on either end of an MCS, allowing for optimal dsRNA production (Voloudakis et al., 2015). The resultant plasmid construct is then transformed into *E. coli* HT115 (DE3) cells using the standard heat shock transformation method. These cells are mutant strains lacking RNase III, an enzyme known for its ability to cut dsRNA, ensuring that viral targets remain intact (Tenllado et al., 2003). Pure cultures of HT115 cells containing the desired insert in the L4440 vector are grown overnight in Luria Bertani Broth media supplemented with appropriate antibiotics (Amp-100 μg/ml and Tet-12.5 μg/ml) at 37 °C with shaking at approximately 220 rpm. The overnight culture is incubated at 37 °C with shaking until the optical density at 600 nm reaches 0.4. IPTG is added to the mix, triggering the T7 promoter, which kicks off the transcription process and starts the production of double-stranded RNA right inside the bacterial cells. After incubation for 4 hours to allow ample synthesis of dsRNA, the cells are harvested by centrifugation at 6000 g for 5 min, followed by purifying the dsRNA using the 70% v/v ethanol method, all according to established protocols (Posiri et al., 2013; Singh et al., 2022; Gupta et al., 2021) (Fig. 17.2).

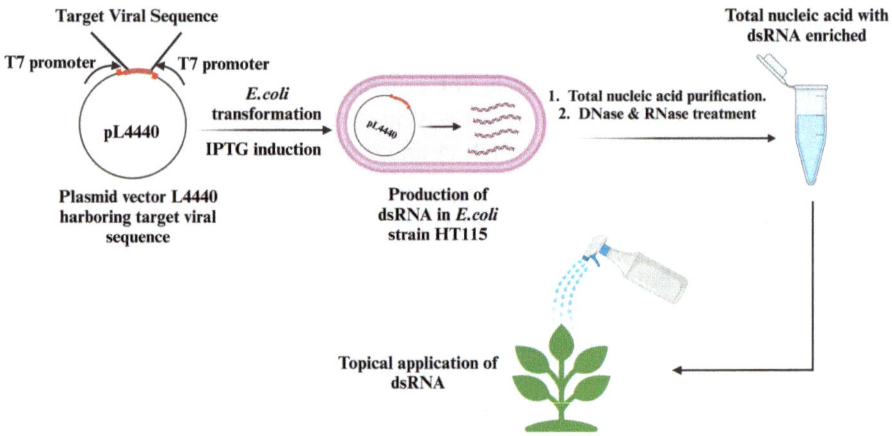

Fig. 17.2 In vivo production of dsRNAs in the bacterial system followed by topical application in plants. (This figure was created using BioRender)

4 Methods of Delivering dsRNA into the Plant System Against Begomoviruses

Delivering dsRNA into plant cells to harness RNA interference (RNAi) poses quite a challenge, primarily due to the physical barriers within the plants (Bennett et al., 2020). Different methods have been invented to overcome these challenges, such as mechanical rubbing, pressure spraying, infiltration, injection, and nano-carrier conjugation (Tenllado & Díaz-Ruíz, 2001; Numata et al., 2014; Mitter et al., 2017; Dalakouras et al., 2018; Dubrovina & Kiselev, 2019). For instance, Namgial et al. (2019) mechanically executed (rubbing with carborandum powder, followed by dsRNA rubbing) the in-vitro-produced dsRNAs on the leaf of 35-day-old tomato plants. Likewise, Melita et al. (2021) explored the protective potential of dsRNAs against TYLCV. The dsRNA was applied to leaves dusted with carborundum before TYLCV agroinfiltration. It's interesting to note that Singh et al. (2022) used a different strategy that involved spraying *Nicotiana benthamiana* plants with a dsRNA-celite suspension. The outcomes, however, have been less encouraging regarding the ToLCNDV. According to Delgado-Martín et al. (2022), rubbing or directly spraying zucchini plants with dsRNAs against ToLCNDV did not produce any protective benefits. Frascati et al.'s work from 2024, however, showed that dsRNAs applied in addition to the viral inoculum could provide some protection for zucchini plants. Therefore, the effectiveness of topically sprayed dsRNA depends on its durability and plant mobility.

4.1 Exogenous Application of dsRNA to Manage ToLCNDV Infection in Plants

Recent research has indicated that dsRNA-mediated non-transgenic approaches help control ToLCNDV disease. Sarkar et al. (2024) investigated the exogenous applications of dsRNAs targeting pathogenicity determinants from ToLCNDV: ORFs *AV2*, *AC2*, and *AC4*. The results indicated that the individual applications of dsRNA (dsAV2, dsAC2, or dsAC4) as well as in the combination (dsCk) resulted in a significant reduction of viral load and severity of disease in *Nicotiana benthamiana* and tomato plants, with the lowest severity of disease seen in the plants treated with dsCk, followed by the plants treated with dsAC4. Thus, targeting all three of the pathogenicity determinants using dsRNA showed the best control efforts. Sangwan et al. (2023) examined the size-dependent internalization of positively charged amine-functionalized mesoporous silica nanoparticles (amino-MSNPs) to facilitate the delivery of dsRNA derived from ToLCNDV *AC2* ORF (dsAC2). The results demonstrated that the dsRNA-loaded nanoparticles had a definite effect on viral load reduction. Additionally, the amino-MSNPs protected dsRNA molecules from nuclease degradation and facilitated efficient uptake by the leaves and roots of *Nicotiana benthamiana*. Real-time gene expression analysis showed that plants treated with nanoparticles of different sizes (~10 nm, ~32 nm, and ~66 nm) exhibited 5-, 11-, and 3-fold reductions in ToLCNDV levels, respectively, as compared to plants treated with naked dsRNA. The size-dependent internalization of amino-MSNPs and their effectiveness in delivering dsRNA were highlighted in this study, offering a viable method for easy topical treatments to protect plants from ToLCNDV infections. However, a few contrasting results have also been observed regarding the dsRNA-based management of ToLCNDV. Using either agroinfiltration or surface spray, no antiviral effect against ToLCNDV has been detected (Delgado-Martín et al., 2022). In another study, a chimeric dsRNA designed to target all of the coding regions of ToLCNDV DNA-A was evaluated as an exogenous application in zucchini plants. The viral genome of DNA-A encodes viral proteins necessary for the infection process, including replication, encapsidation, silencing suppressor, and movement; therefore, it was selected to target with its corresponding dsRNA for exogenous application. The results showed a lower number of infected plants and a delayed onset of symptom initiation. Additionally, treated plants indicated a tendency to decrease viral titer with the chimeric dsRNA (Frascati et al., 2024). This research suggests that targeting different ORFs with chimeric dsRNAs, which code for proteins involved in multiple phases of the viral life cycle, could be an effective approach. This could expand the possible applications of this technology against economically important viruses in the *Geminiviridae* family. All of this information suggests that exogenous dsRNA treatment could provide protection against ToLCNDV infections in plants. Nevertheless, further research is required to determine how plants process exogenous dsRNAs. With this knowledge, applied dsRNAs may be more efficient in suppressing plant viruses, which would make RNAi-based vaccination a sustainable crop protection option.

5 Safety Concerns of dsRNA Application

Despite its promising qualities, there are still some "gray areas" that need to be addressed in the management of plant viruses via RNA interference. Although dsRNA has a high specificity, it may have negative consequences if it has partial homology with other genes (Zhang et al., 2020; Scacheri et al., 2004). One of the main issues is the off-target effects, which might result in unforeseen abnormalities in other nontarget organisms by silencing genes other than the targeted ones. Innate immune system receptors' recognition of dsRNA in animals can set off immunological reactions that result in symptoms like inflammation and allergies. The competition between siRNA and miRNA pathways might result in off-target effects, where the siRNA hybridizes with other off-target mRNAs and causes unrelated mRNA degradation (Bridge et al., 2003; Khan et al., 2009; Chen et al., 2021; Jackson et al., 2003; Birmingham et al., 2006). Furthermore, non-specific dsRNA exposure can trigger sequence-independent responses in nontarget organisms (NTOs), which is a matter of great concern from an environmental point of view (Lundgren & Duan, 2013). A high concentration of dsRNA has the potential to activate endogenous RNA interference (RNAi) pathways. This unintended gene silencing may lead to detrimental consequences for the nontarget organisms, causing potential developmental abnormalities. Another significant concern surrounding dsRNA application relates to the formulations that enhance their stability and delivery within the target organism. NTOs exposed to these formulated dsRNAs may be highly affected due to their limited capacity to degrade them using their intrinsic nucleases. In the case of different dsRNA-based formulations, many exhibits delayed effects, which require a more extended period for observation of the evaluation parameters (Dietz-Pfeilstetter et al., 2021; Sundaresha et al., 2022b, c). The composition of the formulation has a profound impact on the availability, stability, efficacy, and toxicity (De Schutter et al., 2022). Thus, a detailed analysis is needed that can provide us with the implications of these dsRNA-based products on the environment, domestic animals, other related species, weeds, and humans. Most countries have yet to implement proper guidelines and safety standards regarding dsRNA-based products. To enhance safety, it is essential to develop dsRNA-based formulations that are sustainable for plants, humans, and the environment.

6 Conclusions

The application of exogenous dsRNA presents a promising strategy for managing plant viruses; however, it may not be effective in all cases, particularly with DNA viruses such as begomoviruses. While this approach has some promise in conferring protection to plants against the infection of ToLCNDV, the level of protection is relatively low and only truly effective at the very start of the infection process, which is widely linked to the life cycle of DNA viruses, particularly geminiviruses (Jeevalatha et al., 2021, 2023; Frascati et al., 2024). The main underlying reason for the inconsistencies in managing DNA viruses with exogenous dsRNA is that there are only low levels of 24-nucleotide-class virus-derived small interfering RNAs (vsiRNAs) production (Rego-Machado et al., 2020). Although there is potential for better management of DNA viruses using dsRNA application. For instance, it has been demonstrated that transgenic plants that express particular microRNAs specific to the ToLCNDV showed improved resistance against the virus in tomatoes (Van Vu et al., 2013). Better applications of exogenous dsRNA for these viruses may result from increasing the amount of dsRNA or altering the formulations (Delgado-Martín et al., 2022). To conclude, it can be suggested that ToLCNDV and other DNA viruses can be effectively managed by either continuous application or by maintaining the uptake of dsRNA.

References

Baulcombe, D. (2002). Viral suppression of systemic silencing. *Trends in Microbiology, 10*, 306–308.

Bennett, M., Deikman, J., Hendrix, B., & Iandolino, A. (2020). Barriers to efficient foliar uptake of dsRNA and molecular barriers to dsRNA activity in plant cells. *Frontiers in Plant Science, 11*, 816.

Birmingham, A., Anderson, E. M., Reynolds, A., Ilsley-Tyree, D., Leake, D., Fedorov, Y., & Khvorova, A. (2006). 3′ UTR seed matches, but not overall identity, are associated with RNAi off-targets. *Nature Methods, 3*(3), 199–204.

Borah, M., Berbati, M., Reppa, C., Holeva, M., Nath, P. D., & Voloudakis, A. (2018). RNA-based vaccination of Bhut Jolokia pepper (*Capsicum chinense* Jacq.) against cucumber mosaic virus. *Virus, 29*, 207–211.

Borah, M., Nath, P. D., Chaudhury, S. P., Biswas, K. K., Patil, B. L., & Voloudakis, A. (2024). Topical application of dsRNAs targeting citrus tristeza virus (CTV) reduces its titer in the CTV infected sweet orange (*Citrus sinensis*). *European Journal of Plant Pathology, 168*(2), 273–278.

Bouché, N., Lauressergues, D., Gasciolli, V., & Vaucheret, H. (2006). An antagonistic function for Arabidopsis DCL2 in development and a new function for DCL4 in generating viral siRNAs. *The EMBO Journal, 25*(14), 3347–3356.

Bridge, A. J., Pebernard, S., Ducraux, A., Nicoulaz, A. L., & Iggo, R. (2003). Induction of an interferon response by RNAi vectors in mammalian cells. *Nature Genetics, 34*(3), 263–264.

Borah, A., Bhausaheb, M.A., Rana, M. and Srivastava, S., 2023. Mechanism of Arthropod-mediated Transmission of Plant Viruses-A Review. *Journal of Advanced Zoology, 44*.

Butter, N. S., & Rataul, H. S. (1981). Nature and extent of losses in tomatoes due to tomato leaf curl virus (TLCV) transmitted by whitefly, *Bemisia tabaci* Gen. (Hemiptera, Aleyrodidae). *Indian Journal of Ecology, 8*, 299–300.

Chen, J., Peng, Y., Zhang, H., Wang, K., Zhao, C., Zhu, G., & Han, Z. (2021). Off-target effects of RNAi correlate with the mismatch rate between dsRNA and non-target mRNA. *RNA Biology, 18*(11), 1747–1759.

Cohen, S., & Lapidot, M. (2007). Appearance and expansion of TYLCV: A historical point of view. In *Tomato yellow leaf curl virus disease: Management, molecular biology, breeding for resistance* (pp. 3–12). Springer.

Dalakouras, A., Jarausch, W., Buchholz, G., Bassler, A., Braun, M., & Manthey, W. M. (2018). Delivery of hairpin RNAs and small RNAs into woody and herbaceous plants by trunk injection and petiole absorption. *Frontiers in Plant Science, 9*, 1253.

De Schutter, K., Taning, C. N. T., Van Daele, L., Van Damme, E. J., Dubruel, P., & Smagghe, G. (2022). RNAi-based biocontrol products: Market status, regulatory aspects, and risk assessment. *Frontiers in Insect Science, 1*, 818037.

Deleris, A., Gallego-Bartolome, J., Bao, J., Kasschau, K. D., Carrington, J. C., & Voinnet, O. (2006). Hierarchical action and inhibition of plant Dicer-like proteins in antiviral defense. *Science, 313*(5783), 68–71.

Delgado-Martín, J., Ruiz, L., Janssen, D., & Velasco, L. (2022). Exogenous application of dsRNA for the control of viruses in cucurbits. *Frontiers in Plant Science, 13*, 895953.

Diaz-Pendon, J. A., Li, F., Li, W.-X., & Ding, S.-W. (2007). Suppression of antiviral silencing by cucumber mosaic virus 2b protein in Arabidopsis is associated with drastically reduced accumulation of three classes of viral small interfering RNAs. *Plant Cell, 19*(6), 2053–2063.

Dietz-Pfeilstetter, A., Mendelsohn, M., Gathmann, A., & Klinkenbuß, D. (2021). Considerations and regulatory approaches in the USA and in the EU for dsRNA-based externally applied pesticides for plant protection. *Frontiers in Plant Science, 12*, 682387.

Dubrovina, A. S., & Kiselev, K. V. (2019). Exogenous RNAs for gene regulation and plant resistance. *International Journal of Molecular Sciences, 20*(9), 2282.

Fire, A., Xu, S., Montgomery, M. K., Kostas, S. A., Driver, S. E., & Mello, C. C. (1998). Potent and specific genetic interference by double-stranded RNA in *Caenorhabditis elegans*. *Nature, 391*(6669), 806–811.

Frascati, F., Rotunno, S., Accotto, G. P., Noris, E., Vaira, A. M., & Miozzi, L. (2024). Exogenous application of dsRNA for protection against tomato leaf curl New Delhi virus. *Viruses, 16*(3), 436.

Gan, D., Zhang, J., Jiang, H., Jiang, T., Zhu, S., & Cheng, B. (2010). Bacterially expressed dsRNA protects maize against SCMV infection. *Plant Cell Reports, 29*, 1261–1268.

Garcia-Ruiz, H., Takeda, A., Chapman, E. J., Sullivan, C. M., Fahlgren, N., Brempelis, K. J., & Carrington, J. C. (2010). Arabidopsis RNA-dependent RNA polymerases and dicer-like proteins in antiviral defense and small interfering RNA biogenesis during Turnip mosaic virus infection. *Plant Cell, 22*(2), 481–496.

Gasciolli, V., Mallory, A. C., Bartel, D. P., & Vaucheret, H. (2005). Partially redundant functions of Arabidopsis DICER-like enzymes and a role for DCL4 in producing trans-acting siRNAs. *Current Biology, 15*(16), 1494–1500.

Gupta, D., Singh, O. W., Basavaraj, Y. B., Roy, A., Mukherjee, S. K., & Mandal, B. (2021). Direct foliar application of dsRNA derived from the full-length gene of NSs of groundnut bud necrosis virus limits virus accumulation and symptom expression. *Frontiers in Plant Science, 12*, 734618.

Holeva, M. C., Sklavounos, A., Rajeswaran, R., Pooggin, M. M., & Voloudakis, A. E. (2021). Topical application of double-stranded RNA targeting 2b and CP genes of cucumber mosaic virus protects plants against local and systemic viral infection. *Plants, 10*(5), 963.

Jackson, A. L., Bartz, S. R., Schelter, J., Kobayashi, S. V., Burchard, J., Mao, M., & Linsley, P. S. (2003). Expression profiling reveals off-target gene regulation by RNAi. *Nature Biotechnology, 21*(6), 635–637.

Jaubert, M., Bhattacharjee, S., Mello, A. F., Perry, K. L., & Moffett, P. (2011). ARGONAUTE2 mediates RNA-silencing antiviral defenses against potato virus X in Arabidopsis. *Plant Physiology, 156*(3), 1556–1564.

Jeevalatha, A., Vanishree, G., Siddappa, S., Kumar, R., Kaundal, P., Kumar, A., & Chakrabarti, S. K. (2021). Molecular characterization and infectivity analysis of tomato leaf curl New Delhi virus isolates infecting potato. *3 Biotech, 11*, 203. https://doi.org/10.1007/s13205-021-02752-5

Jeevalatha, A., Siddappa, S., Kumar, R., Tiwari, R. K., Lal, M. K., Sharma, S., Chakrabarty, S. K., & Singh, B. P. (2023). RNA-seq analysis reveals an early defense response to tomato leaf curl New Delhi virus in potato cultivar Kufri Bahar. *Functional & Integrative Genomics, 23*, 215. https://doi.org/10.1007/s10142-023-01138-5

Jin, Y., Zhao, J. H., & Guo, H. S. (2021). Recent advances in understanding plant antiviral RNAi and viral suppressors of RNAi. *Current Opinion in Virology, 46*, 65–72.

Kaldis, A., Berbati, M., Melita, O., Reppa, C., Holeva, M., Otten, P., & Voloudakis, A. (2018). Exogenously applied dsRNA molecules deriving from the Zucchini yellow mosaic virus (ZYMV) genome move systemically and protect cucurbits against ZYMV. *Molecular Plant Pathology, 19*(4), 883–895.

Kalloo, G. (1996). Leaf curl virus of tomato and chili in India. In *Proceeding of the Phase I Final Workshop of the South Asian Vegetable Research Networks*, January 23–28, pp. 242–259.

Kamesh Krishnamoorthy, K., Malathi, V. G., Renukadevi, P., Kumar, S. M., Raveendran, M., Sudha, M., & Karthikeyan, G. (2023). Exogenous delivery of dsRNA for management of mungbean yellow mosaic virus on blackgram. *Planta, 258*(5), 94.

Khan, A. A., Betel, D., Miller, M. L., Sander, C., Leslie, C. S., & Marks, D. S. (2009). Transfection of small RNAs globally perturbs gene regulation by endogenous microRNAs. *Nature Biotechnology, 27*(6), 549–555.

Kim, V. N. (2005). Small RNAs: Classification, biogenesis, and function. *Molecules and Cells, 19*(1), 1–15.

Konakalla, N. C., Kaldis, A., Berbati, M., Masarapu, H., & Voloudakis, A. E. (2016). Exogenous application of double-stranded RNA molecules from TMV p126 and CP genes confers resistance against TMV in tobacco. *Planta, 244*(4), 961–969.

Konakalla, N. C., Kaldis, A., Masarapu, H., & Voloudakis, A. E. (2019). Topical application of double-stranded RNA molecules deriving from Sesbania mosaic virus (SeMV) CP and MP genes protects *Sesbania* plants against SeMV. *European Journal of Plant Pathology, 155*, 1345–1352.

Konakalla, N. C., Bag, S., Deraniyagala, A. S., Culbreath, A. K., & Pappu, H. R. (2021). Induction of plant resistance in tobacco (*Nicotiana tabacum*) against tomato spotted wilt orthotospovirus through foliar application of dsRNA. *Viruses, 13*(4), 662.

Kumar, P., & Kumar, M. (2018). Leaf curl disease: A significant constraint in the production of tomato in India. In *Advances in plant pathology* (p. 93). IntechOpen.

Lau, S. E., Mazumdar, P., Hee, T. W., Song, A. L. A., Othman, R. Y., & Harikrishna, J. A. (2014). Crude extracts of bacterially-expressed dsRNA protect orchid plants against Cymbidium mosaic virus during transplantation from in vitro culture. *The Journal of Horticultural Science and Biotechnology, 89*(5), 569–576.

Lundgren, J. G., & Duan, J. J. (2013). RNAi-based insecticidal crops: Potential effects on nontarget species. *Bioscience, 63*(8), 657–665.

Malathi, V. G., Renukadevi, P., Chakraborty, S., Biswas, K. K., Roy, A., Sivalingam, P. N., et al. (2017). Begomoviruses and their satellites occurring in India: Distribution, diversity and pathogenesis. In B. Mandal, G. P. Rao, V. K. Baranwal, & R. K. Jain (Eds.), *A century of plant virology in India* (pp. 75–177). Springer.

Mazumder, P., Mondal, F., Sarkar, M., Majumdar, A., Biswas, K. K., Sharma, S. K., Lal, M. K., Tiwari, R. K., Kumar, R., & Roy, A. (2024). Biological and molecular characterization of a new isolate of tomato mottle mosaic virus causing severe shoestring and fruit deformities in tomato plants in India. *Plants, 13*, 2811. https://doi.org/10.3390/plants13192811

Melita, O., Kaldis, A., Berbati, M., Reppa, C., Holeva, M., Lapidot, M., & Voloudakis, A. (2021). Topical application of double-stranded RNA molecules deriving from tomato yellow leaf curl virus reduces cognate virus infection in tomato. *Biologia Plantarum, 65*(1), 100.

Mitter, N., Worrall, E. A., Robinson, K. E., Xu, Z. P., & Carroll, B. J. (2017). Induction of virus resistance by exogenous application of double-stranded RNA. *Current Opinion in Virology, 26*, 49–55.

Moriones, E., & Navas-Castillo, J. (2000). Tomato yellow leaf curl virus, an emerging virus complex causing epidemics worldwide. *Virus Research, 71*(1–2), 123–134.

Namgial, T., Kaldis, A., Chakraborty, S., & Voloudakis, A. (2019). Topical application of double-stranded RNA molecules containing sequences of tomato leaf curl virus and cucumber mosaic virus confers protection against the cognate viruses. *Physiological and Molecular Plant Pathology, 108*, 101432.

Niehl, A., Soininen, M., Poranen, M. M., & Heinlein, M. (2018). Synthetic biology approach for plant protection using dsRNA. *Plant Biotechnology Journal, 16*(9), 1679–1687.

Numata, K., Ohtani, M., Yoshizumi, T., Demura, T., & Kodama, Y. (2014). Local gene silencing in plants via synthetic dsRNA and carrier peptide. *Plant Biotechnology Journal, 12*(8), 1027–1034.

Posiri, P., Ongvarrasopone, C., & Panyim, S. (2013). A simple one-step method for producing dsRNA from *E. Coli* to inhibit shrimp virus replication. *Journal of Virological Methods, 188*(1–2), 64–69.

Raja, P., Sanville, B. C., Buchmann, R. C., & Bisaro, D. M. (2008). Viral genome methylation as an epigenetic defense against geminiviruses. *Journal of Virology, 82*(18), 8997–9007.

Ratcliff, F., Harrison, B. D., & Baulcombe, D. C. (1997). A similarity between viral defense and gene silencing in plants. *Science, 276*(5318), 1558–1560.

Rosa, C., Kuo, Y. W., Wuriyanghan, H., & Falk, B. W. (2018). RNA interference mechanisms and applications in plant pathology. *Annual Review of Phytopathology, 56*(1), 581–610.

Routhu, G. K., Borah, M., Siddappa, S., & Nath, P. D. (2022). Exogenous application of coat protein-specific dsRNA inhibits cognate cucumber mosaic virus (CMV) of ghost pepper. *Journal of Plant Diseases and Protection, 129*(2), 293–300.

Saikia, A. K., & Muniyappa, V. (1989). Epidemiology and control of tomato leaf curl virus in southern India. *Tropical Agriculture (Trinidad), 66*, 350–354.

Sangwan, A., Gupta, D., Singh, O. W., Roy, A., Mukherjee, S. K., Mandal, B., & Singh, N. (2023). Size variations of mesoporous silica nanoparticle control uptake efficiency and delivery of AC2-derived dsRNA for protection against tomato leaf curl New Delhi virus. *Plant Cell Reports, 42*(10), 1571–1587.

Sarkar, M., Gupta, D., Singh, O. W., Paul, S., Kumar, R., Mandal, B., & Roy, A. (2024). Identification of pathogenicity determinants in ToLCNDV and their RNAi-based knockdown for disease management in *Nicotiana benthamiana* and tomato plants. *Frontiers in Microbiology, 15*, 1481523. https://doi.org/10.3389/fmicb.2024.1481523

Scacheri, P. C., Rozenblatt-Rosen, O., Caplen, N. J., Wolfsberg, T. G., Umayam, L., Lee, J. C., et al. (2004). Short interfering RNAs can induce unexpected and divergent changes in the levels of untargeted proteins in mammalian cells. *Proceedings of the National Academy of Sciences USA, 101*(7), 1892–1897.

Singh, O. W., Gupta, D., Joshi, B., Roy, A., Mukherjee, S. K., & Mandal, B. (2022). Spray application of a cocktail of dsRNAs reduces infection of chilli leaf curl virus in *Nicotiana benthamiana. Journal of Plant Diseases and Protection, 129*, 433–438.

Sun, Z. N., Yin, G. H., Song, Y. Z., An, H. L., Zhu, C. X., & Wen, F. J. (2010). Bacterially expressed double-stranded RNAs against hot-spot sequences of tobacco mosaic virus or potato virus Y genome have different ability to protect tobacco from viral infection. *Applied Biochemistry and Biotechnology, 162*, 1901–1914.

Sundaresha, S., Bairwa, A., Tomar, M., Kumar, R., Venkatasalam, E. P., Sagar, V., et al. (2022a). In vitro method for synthesis of large-scale dsRNA molecule as a novel plant protection strategy. In *Plant gene silencing: Methods and protocols* (pp. 211–226). Springer US.

Sundaresha, S., Jeevalatha, A., Kumar, R., Sood, S., Sharma, S., Bhardwaj, V., et al. (2022b). RNA interference: A versatile tool to augment plant protection strategies in potato. In *Sustainable management of potato pests and diseases* (pp. 369–388). Springer Singapore.

Sundaresha, S., Sharma, S., Bairwa, A., Tomar, M., Kumar, R., Bhardwaj, V., et al. (2022c). Spraying of dsRNA molecules derived from Phytophthora infestans, along with nanoclay carriers as a proof of concept for developing novel protection strategy for potato late blight. *Pest Management Science, 78*(7), 3183–3192.

Svoboda, P. (2020). Key mechanistic principles and considerations concerning RNA interference. *Frontiers in Plant Science, 11*, 1237.

Tatineni, S., & Hein, G. L. (2023). Plant viruses of agricultural importance: Current and future perspectives of virus disease management strategies. *Phytopathology, 113*(2), 117–141.

Tenllado, F., & Díaz-Ruíz, J. R. (2001). Double-stranded RNA-mediated interference with plant virus infection. *Journal of Virology, 75*(24), 12288–12297.

Tenllado, F., Martínez-García, B., Vargas, M., & Díaz-Ruíz, J. R. (2003). Crude extracts of bacterially expressed dsRNA can be used to protect plants against virus infections. *BMC Biotechnology, 3*, 1–11.

Tenllado, F., Llave, C., & Díaz-Ruíz, J. R. (2004). RNA interference as a new biotechnological tool for the control of virus diseases in plants. *Virus Research, 102*(1), 85–96.

Vadlamudi, T., Patil, B. L., Kaldis, A., Gopal, D. V. R. S., Mishra, R., Berbati, M., & Voloudakis, A. (2020). DsRNA-mediated protection against two isolates of papaya ringspot virus through topical application of dsRNA in papaya. *Journal of Virological Methods, 275*, 113750.

Van Vu, T., Choudhury, N. R., & Mukherjee, S. K. (2013). Transgenic tomato plants expressing artificial microRNAs for silencing the pre-coat and coat proteins of a begomovirus, tomato leaf curl New Delhi virus, show tolerance to virus infection. *Virus Research, 172*(1–2), 35–45.

Vasudeva, R. S., & Samraj, J. (1948). A leaf curl disease of tomato. *Phytopathology, 38*, 364–369.

Voloudakis, A. E., Holeva, M. C., Sarin, L. P., Bamford, D. H., Vargas, M., Poranen, M. M., & Tenllado, F. (2015). Efficient double-stranded RNA production methods for utilization in plant virus control. In A. Wang & X. Zhou (Eds.), *Plant virology protocols: New approaches to detect viruses and host responses* (pp. 255–274). Springer.

Wang, X. B., Jovel, J., Udomporn, P., Wang, Y., Wu, Q., Li, W. X., et al. (2011). The 21-nucleotide, but not 22-nucleotide, viral secondary small interfering RNAs direct potent antiviral defense by two cooperative argonautes in *Arabidopsis thaliana. Plant Cell, 23*(4), 1625–1638.

Worrall, E. A., Bravo-Cazar, A., Nilon, A. T., Fletcher, S. J., Robinson, K. E., Carr, J. P., & Mitter, N. (2019). Exogenous application of RNAi-inducing double-stranded RNA inhibits aphid-mediated transmission of a plant virus. *Frontiers in Plant Science, 10*, 265.

Xie, Z., Johansen, L. K., Gustafson, A. M., Kasschau, K. D., Lellis, A. D., Zilberman, D., et al. (2004). Genetic and functional diversification of small RNA pathways in plants. *PLoS Biology, 2*(5), e104.

Yin, G., Sun, Z., Liu, N., Zhang, L., Song, Y., Zhu, C., & Wen, F. (2009). Production of double-stranded RNA for interference with TMV infection utilizing a bacterial prokaryotic expression system. *Applied Microbiology and Biotechnology, 84*, 323–333.

Zaidi, S. S. E. A., Martin, D. P., Amin, I., Farooq, M., & Mansoor, S. (2017). Tomato leaf curl New Delhi virus: A widespread bipartite begomovirus in the territory of monopartite begomoviruses. *Molecular Plant Pathology, 18*, 901–911.

Zhang, S., Liu, Y., & Yu, B. (2015). New insights into pri-miRNA processing and accumulation in plants. *Wiley Interdisciplinary Reviews: RNA, 6*(5), 533–545.

Zhang, H., Zhang, H., Demirer, G. S., González-Grandío, E., Fan, C., & Landry, M. P. (2020). Engineering DNA nanostructures for siRNA delivery in plants. *Nature Protocols, 15*(9), 3064–3087.

Zhong, X., Du, J., Hale, C. J., Gallego-Bartolome, J., Feng, S., Vashisht, A. A., et al. (2014). Molecular mechanism of action of plant DRM de novo DNA methyltransferases. *Cell, 157*(5), 1050–1060.

Evolutionary Analysis of Tomato Leaf Curl New Delhi Virus

18

Abhilasha Singh, Swati Bhuria, Niranjana Prem Minipreman, Chinnakaruppan Marimuthu, Nikeshun Vivekananthan, Yogita Maheshwari, Susheel Kumar, and Vijayanandraj Selvaraj

Abstract

Tomato leaf curl New Delhi virus (ToLCNDV), a bipartite begomovirus (family Geminiviridae), has emerged as a significant threat to global agriculture due to its broad host range, rapid geographical expansion, and high genetic variability. This study integrates phylogenetic, pairwise divergence, and selection pressure analyses to elucidate the evolutionary dynamics of ToLCNDV. Full-length DNA-A and DNA-B sequences, along with the CRESS domain of the replicase gene, were evaluated using the sequences obtained from NCBI across 24 countries. Phylogenetic clustering revealed 19 DNA-A and 24 CRESS-based clades, with strong geographical and host-specific patterns such as Pakistan-dominated tomato isolates, cucurbit-adapted strains in Europe/Asia. Pairwise divergence highlighted extreme genetic diversity, with CRESS domain similarities ranging from 32.2% to 100%, compared to 56.4–100% for DNA-A. DNA-B exhibited higher nucleotide diversity (Pi = 0.114) than DNA-A (Pi = 0.063), reflecting divergent evolutionary constraints. Negative Tajima's D and Fu and Li's values across all genes indicated pervasive purifying selection and population expansion. These findings underscore that ToLCNDV evolutionary plasticity might be driven by host adaptation, recombination, and global trade-mediated spread, necessitating vigilant surveillance and tailored management strategies.

A. Singh · S. Bhuria · N. P. Minipreman · Y. Maheshwari · S. Kumar · V. Selvaraj (✉)
Plant Molecular Virology Laboratory, CSIR – National Botanical Research Institute, Lucknow, Uttar Pradesh, India

Academy of Scientific and Innovative Research (AcSIR), Ghaziabad, Uttar Pradesh, India
e-mail: vijayanandraj.s@nbri.res.in

C. Marimuthu · N. Vivekananthan
Plant Molecular Virology Laboratory, CSIR – National Botanical Research Institute, Lucknow, Uttar Pradesh, India

R. Kumar et al. (eds.), *Tomato Leaf Curl New Delhi Virus (ToLCNDV)*,
https://doi.org/10.1007/978-981-95-2852-3_18

Keywords

ToLCNDV · Begomovirus · Evolution · Phylogenetic divergence · CRESS domain

1 Introduction

Tomato leaf curl New Delhi virus (ToLCNDV) is a bipartite begomovirus belonging to the genus *Begomovirus* in the family *Geminiviridae*. The whitefly *Bemisia tabaci* primarily transmits it through a circulative and persistent mechanism (Varma and Malathi, 2003; Pratap et al., 2011; Kanakala et al., 2013; Kumar et al., 2021; Naga et al., 2021), and has also been shown to spread via mechanical transmission (López et al., 2015). Recognized as a serious phytopathogen, ToLCNDV inflicts considerable damage on tomato and several other high-value crops (Padidam et al., 1995; Jeevalatha et al., 2021, 2023). In recent years, the virus has broadened both its host range and geographic distribution, emerging as a growing concern for agricultural productivity worldwide (Navas-Castillo et al., 2011; Moriones et al., 2017). Although it was first identified in solanaceous crops, subsequent reports have confirmed its presence in a wide array of hosts, including members of the *Cucurbitaceae*, *Fabaceae*, *Malvaceae*, and *Euphorbiaceae* families. Its impact has been particularly pronounced in the Indian subcontinent, where it has been responsible for major yield reductions. In addition to the Indian subcontinent, the virus has also been reported in several other Asian countries, such as Pakistan, Bangladesh, Thailand, Indonesia, and Taiwan. In recent years, the virus has continued to spread into Iran and further into the western Mediterranean region, with major outbreaks reported in Spain, Tunisia, and Italy, particularly affecting cucurbit crops (Mnari-Hattab et al., 2015; Panno et al., 2016; Moriones et al., 2017; Juárez et al., 2019). These findings underscore the virus's growing adaptability and its continued threat to cucurbit production in new regions.

The genetic diversity of ToLCNDV is significant, with numerous full-length DNA-A sequences available for analysis. Various molecular techniques, including polymerase chain reaction (PCR), rolling circle amplification (RCA), and restriction fragment length polymorphism (RFLP) analysis, have played a key role in shedding light on the virus's evolutionary patterns (Brown et al., 2015). Phylogenetic analyses indicate that ToLCNDV isolates from Asia form a closely related group. At the same time, those in the Mediterranean region constitute a distinct strain, likely due to a founder effect and population bottleneck during transmission (Moriones et al., 2017). Research on host adaptation has revealed that the virus behaves differently in various plant species. For instance, *Nicotiana benthamiana* is highly susceptible, while plants like pepper show resistance. These differences may be linked to how effectively the virus hijacks host cellular processes and evades plant immune responses such as RNA silencing (Kushwaha et al., 2015). The virus's capacity to jump to new hosts appears to be driven by its genetic flexibility, frequent recombination events, and interactions with plant defense mechanisms, factors that collectively make ToLCNDV a highly adaptable plant pathogen (Maule et al., 2002;

Gillette et al., 1998; Petty et al., 2000). Understanding the evolutionary mechanisms behind the diversification and adaptation of ToLCNDV is essential for developing effective management strategies, especially as the virus continues to expand its host range and spread to new regions. This chapter focuses on the genetic diversity, host adaptation, and evolutionary patterns of ToLCNDV, offering insights into the factors driving its continued emergence.

2 Geographical Distribution and Spread of ToLCNDV

The global distribution of ToLCNDV was analyzed using sequences available in the NCBI database (Fig. 18.1). A total of 2112 sequences with country information were obtained, with India (760 sequences) and Pakistan (681 sequences) having the highest number of entries. This reflects significant research activity and the agricultural importance of ToLCNDV in these regions. Spain (257 sequences) and Italy (94 sequences) also contributed a notable number of sequences, indicating substantial viral incidence in parts of Europe. Moderate contributions were observed from China (69 sequences), Indonesia (44 sequences), Malaysia (42 sequences), Tunisia (32 sequences), Bangladesh (31 sequences), Morocco (22 sequences), and Cambodia (19 sequences). In comparison, Iran (14 sequences), Thailand (11 sequences), and Nepal (10 sequences) showed relatively fewer entries. A minor fraction of sequences was reported from Jordan, France, Taiwan, Sri Lanka, Laos, Greece, Estonia, Algeria, Turkey, Nigeria, and the Democratic Republic of Congo, with each contributing fewer than ten sequences. Conversely, no ToLCNDV sequences were reported from North or South America, despite the significant tomato cultivation in these regions. This absence may be attributed to stringent biosecurity measures or the dominance of other begomoviruses, potentially limiting the occurrence of ToLCNDV. The data highlight considerable variation in ToLCNDV disease incidence, research intensity, and sequence reporting across regions, with a strong

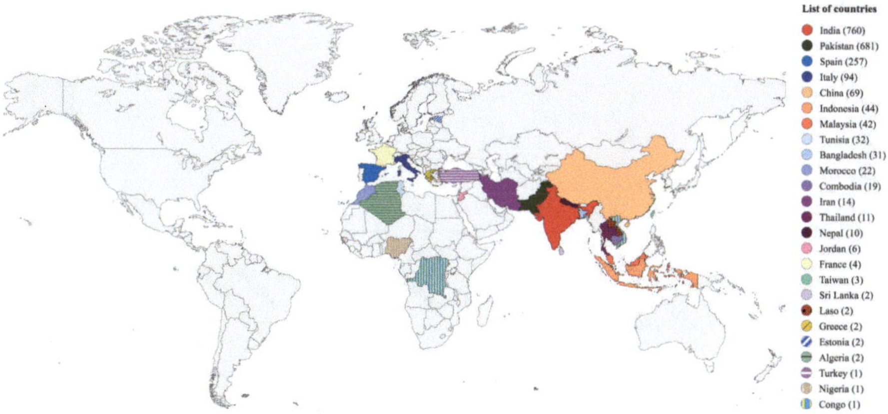

Fig. 18.1 Global distribution of *Tomato leaf curl New Delhi virus*

presence in South Asia, particularly India and Pakistan, as well as Southern Europe, such as Spain and Italy. Factors like climate, agricultural practices, and vector prevalence likely influence these patterns. At the same time, low reporting in regions like Africa and the Middle East may indicate gaps in surveillance or research. The movement of planting material and trade in agricultural commodities also play a key role in the virus's spread. This variation highlights the need for enhanced surveillance, especially in underrepresented areas, to better understand the virus's genetic diversity and epidemiology.

3 Host Range and Adaptation of ToLCNDV

The distribution of ToLCNDV across various plant families was analyzed using 2125 NCBI sequence data entries with host information. (Fig. 18.2). The highest number of sequences was reported from the *Solanaceae* family (1036 sequences), followed closely by *Cucurbitaceae* (951 sequences), highlighting these families as the primary hosts of ToLCNDV. *Malvaceae* (53 sequences) and Fabaceae (19 sequences) also exhibited notable representation, suggesting the virus's adaptability across diverse plant taxa. Moderate sequence counts were observed in *Caricaceae* (15 sequences), *Asteraceae* (13 sequences), *Verbenaceae* (8 sequences), and *Apocynaceae* (5 sequences), while several other families, including Piperaceae (4 sequences), *Euphorbiaceae* (3 sequences), and *Convolvulaceae* (2 sequences), contributed minimally, indicating limited occurrences or sporadic infections. Interestingly, only a single sequence was retrieved from plant families such as *Moraceae*, *Oleaceae*, *Papaveraceae*, *Pedaliaceae*, and *Phyllanthaceae*, suggesting rare or incidental infections. The dominance of *Solanaceae* and *Cucurbitaceae* in ToLCNDV sequences underscores their susceptibility and the significant economic

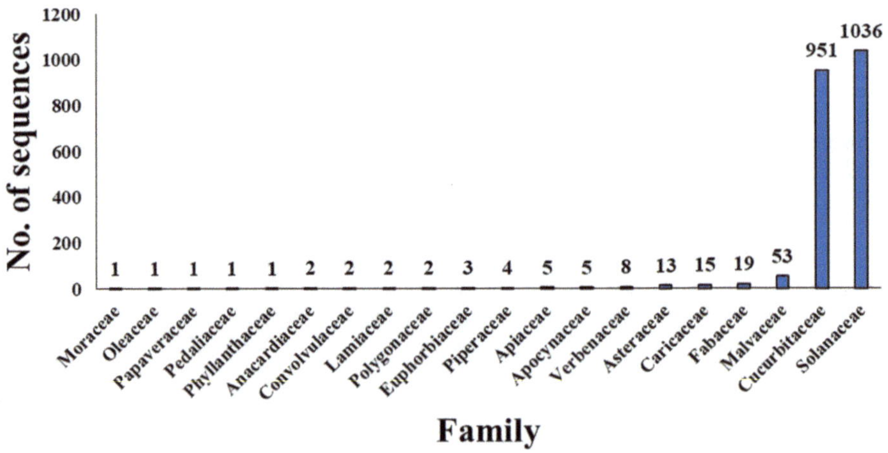

Fig. 18.2 Occurrence and distribution of *Tomato leaf curl New Delhi virus* across different plant families

impact of the virus on crops such as tomato, potato, cucumbers, and melons. The broader host range, spanning 18 plant families, reflects the adaptability of ToLCNDV to diverse plant species, raising concerns about its potential for further spread (Kumar et al., 2022; Chauhan et al., 2023; Sarkar et al., 2024).

4 Phylogenetic Analysis of Full-Length DNA-A Sequences

The maximum likelihood (ML) phylogenetic analysis of ToLCNDV, DNA-A, based on 764 full-length DNA-A sequences, grouped the virus into 19 distinct clades (Gp_1–Gp_19) (Fig. 18.3). Pakistan contributed the highest sequences, particularly in Gp_1 (100%, n = 228), with additional presence in Gp_2, Gp_3, Gp_5, and Gp_6, primarily infecting *Solanum lycopersicum* (tomato). India exhibited the greatest diversity, spanning 12 clades (Gp_4, Gp_6–Gp_10, Gp_12–Gp_13,

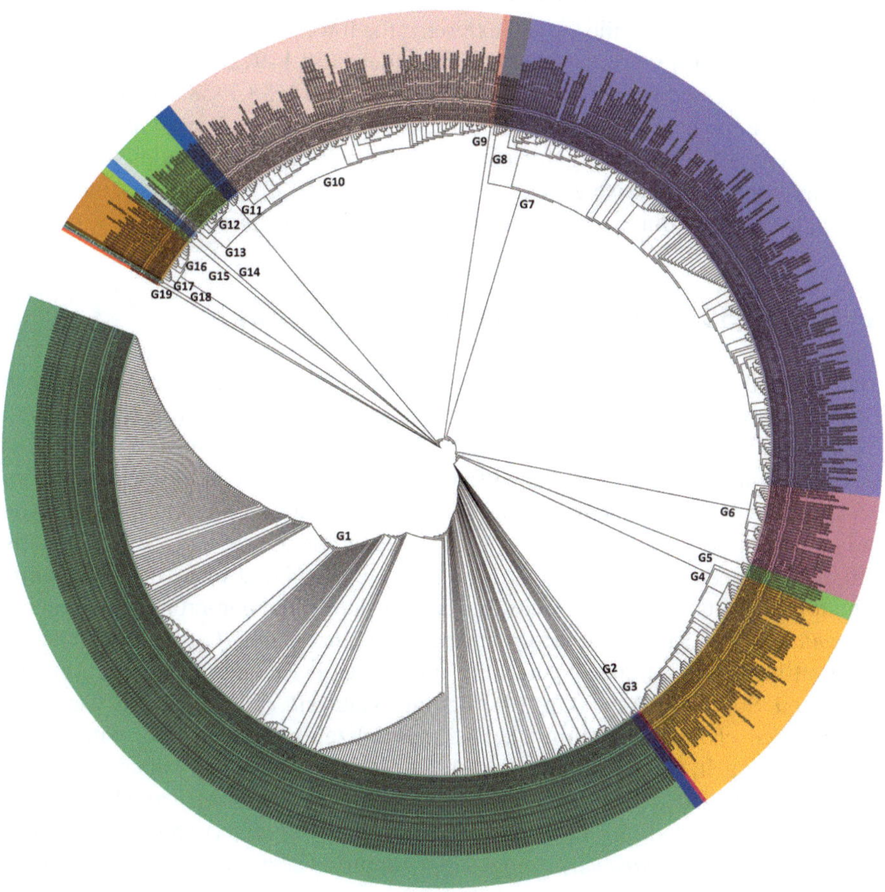

Fig. 18.3 Phylogenetic analysis of full-length ToLCNDV DNA-A sequences

Gp_15–Gp_19), infecting various hosts including cucurbits, tomato, and papaya. Other key regions included China (Gp_4), Spain (Gp_7), Thailand (Gp_10), and Italy (Gp_7), with cucurbits being a dominant host. Tomato was the most frequent host, while cucurbits (*Cucumis sativus*, Luffa, *Momordica charantia*) were widespread, particularly in Southeast Asian clades like Gp_10. Capsicum isolates appeared in multiple clades, primarily from India, while unusual hosts like cotton (*Gossypium hirsutum*) and papaya (*Carica papaya*) were identified in Gp_5, Gp_10, and Gp_12. The highest host-country association was seen in *S. lycopersicum* from Pakistan (Gp_1) and cucurbits in Spain (GP_7). The phylogeny highlights the regional dominance of ToLCNDV in India and Pakistan, with its spread into Southeast Asia and Europe likely facilitated by vector transmission and agricultural trade.

The pairwise genetic distance analysis of 19 groups (Gp_1–Gp_19) of ToLCNDV, DNA-A sequences, conducted using MEGA software, revealed an overall mean distance of 93%, indicating high genetic divergence (Table 18.1). Within-group distances ranged from 93.0% to 100%, with Gp_1, Gp_13, and Gp_15 showing complete identity (100%), while Gp_10 exhibited the lowest within-group homogeneity (93.0%). Between-group distances varied widely (58.1–97.3%), with the highest divergence observed between Gp_12 and Gp_17/Gp_18 (58.1–58.4%) and the closest relationships between Gp_1 and Gp_2 (97.3%) and Gp_1 and Gp_3 (96.4%). Notably, Gp_12 displayed significant divergence from multiple groups. These findings highlight the extensive genetic diversity among ToLCNDV isolates, underscoring the need for continuous surveillance to track viral evolution.

5 CRESS Domain-Based Phylogeny of DNA-A

The molecular phylogeny based on the CRESS domain of the replicase gene in DNA-A, used for classifying ssDNA viruses, resolved the analyzed sequences into 24 distinct clades (Gp_1–Gp_24) through Maximum Likelihood analysis (Fig. 18.4). Geographically, Pakistan dominated the dataset, particularly in Gp_1, which comprised exclusively Pakistani isolates (n = 228) infecting *Solanum lycopersicum*, indicating a highly localized, host-specific lineage. India exhibited the broadest distribution, appearing in 18 clades (Gp_2–Gp_24), with diverse hosts including tomato, cucurbits (*Cucumis sativus*, Luffa), capsicum, and non-crop species such as *Duranta erecta* (Gp_9). Secondary hotspots included China (Gp_3), dominated by cucurbit-infecting variants, Spain/Italy (Gp_5) with Mediterranean cucurbit isolates, and Southeast Asia (Gp_14), linking Malaysia, Thailand, and Cambodia (*Momordica charantia*, cucurbits). Host specialization was evident, with tomato predominating in Pakistan/India (Gp_1, Gp_2, Gp_13), while cucurbits defined transboundary clades (Gp_3 in China, Gp_5 in Europe/North Africa). Unusual host associations included Gp_12 (Pakistan: *Gossypium hirsutum*) and Gp_17 (India: *Ocimum kilimandscharicum*), suggesting adaptive host jumps. The phylogeny revealed strong geographical structuring, exemplified by Pakistan's tomato-restricted Gp_1, alongside host-driven divergence, particularly in cucurbit-infecting

Table 18.1 Pairwise genetic distances among ToLCNDV DNA-A

	Within	Between groups																		
Groups	group	Gp_1	Gp_2	Gp_4	Gp_5	Gp_6	Gp_7	Gp_8	Gp_9	Gp_10	Gp_11	Gp_12	Gp_13	Gp_14	Gp_15	Gp_16	Gp_3	Gp_17	Gp_18	Gp_19
Gp 1	99.9	100.0																		
Gp 2	99.2	97.3	100.0																	
Gp 3	na	96.4	96.3	100.0																
Gp 4	97.1	96.1	95.6	95.0	100.0															
Gp 5	97.9	95.0	94.5	94.1	94.0	100.0														
Gp 6	94.8	90.9	90.4	90.5	90.6	90.5	100.0													
Gp 7	94.0	96.6	96.3	96.1	96.0	94.6	91.6	100.0												
Gp 8	99.1	95.7	95.5	95.5	94.9	93.9	91.1	96.4	100.0											
Gp 9	98.7	93.3	92.8	92.8	92.6	92.1	89.2	93.2	93.2	100.0										
Gp 10	93.0	96.8	96.3	96.0	95.5	94.7	91.2	96.5	95.8	94.2	100.0									
Gp 11	99.7	85.3	84.9	84.8	84.6	83.7	80.9	85.0	85.1	84.2	85.4	100.0								
Gp 12	93.0	92.2	91.6	91.6	91.3	90.4	87.1	91.8	91.7	90.5	92.2	84.5	100.0							
Gp 13	100.0	90.9	90.6	90.6	90.1	89.4	86.6	91.1	90.6	88.4	91.2	81.4	87.4	100.0						
Gp 14	95.1	84.0	83.7	84.2	82.7	81.6	78.9	83.8	83.7	80.7	84.1	73.7	78.6	80.3	100.0					
Gp 15	100.0	88.4	87.5	87.7	87.8	87.2	85.7	88.2	88.4	87.6	88.6	83.8	90.2	85.5	79.3	100.0				
Gp 16	94.2	97.2	96.5	95.5	95.4	94.2	90.2	96.0	94.9	92.7	96.1	84.8	91.1	90.5	83.9	87.4	100.0			
Gp 17	na	63.4	63.3	62.9	63.2	62.8	63.0	62.7	64.2	63.5	63.0	58.4	62.8	61.5	56.4	63.2	62.1	100.0		
Gp 18	na	66.2	66.8	66.3	65.5	65.3	63.6	65.6	66.6	65.6	65.8	58.1	64.3	64.8	60.3	64.1	65.8	64.5	100.0	
Gp 19	na	86.8	86.5	86.6	86.6	85.7	83.7	86.9	86.9	85.3	86.8	80.2	84.3	85.7	78.3	85.5	86.1	64.4	64.7	100.0

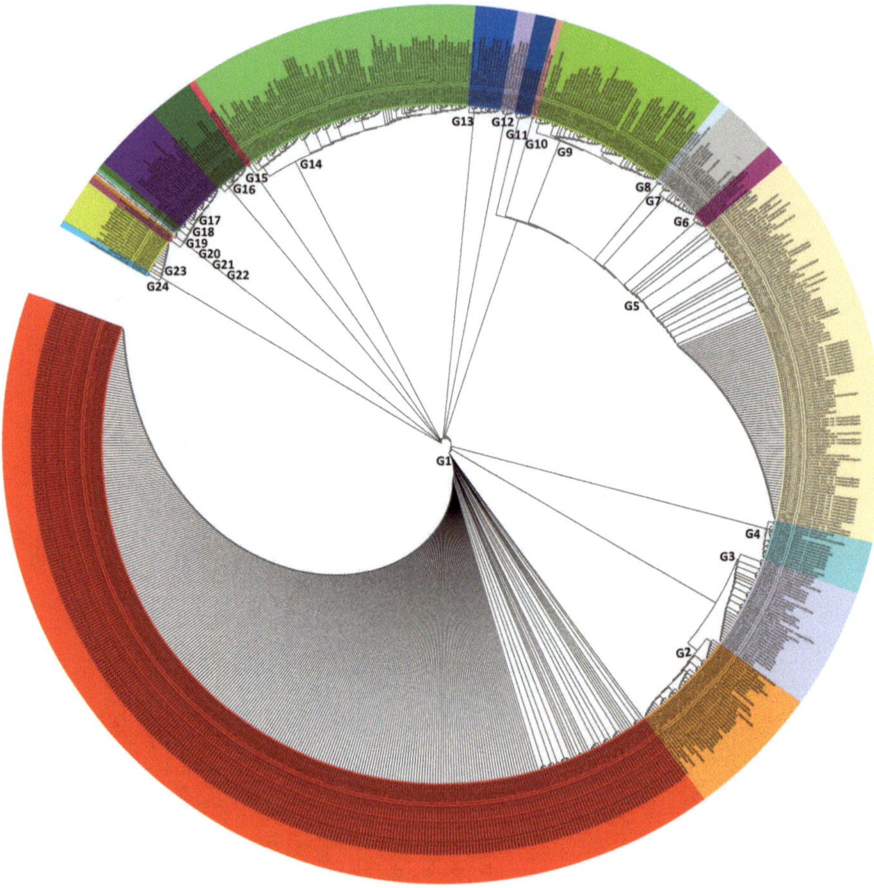

Fig. 18.4 Phylogenetic analysis of ToLCNDV DNA-A based on the CRESS domain of the replicase gene

lineages (Gp_3, Gp_5), which displayed wider dispersal, potentially linked to agricultural trade. This highlights how regional host preferences and human-mediated spread shape ToLCNDV evolution.

The pairwise sequence comparison of the CRESS domain within the replicase gene revealed high within-group similarity among the studied groups, ranging from 93.1% (Gp_16) to 100% (Gp_11, Gp_15, Gp_18, Gp_24), indicating strong genetic conservation within each group (Table 18.2). Notably, Gp_1 exhibited 99.9% similarity, while Gp_5 and Gp_10 also showed near-complete homogeneity of 99.7%. Between-group comparisons demonstrated varying levels of divergence, with the highest similarities observed among Gp_1, Gp_2, and Gp_3 (95.9–96.3%), suggesting a close evolutionary relationship. In contrast, Gp_15 displayed the lowest intergroup similarity (57.3–82.8%), particularly with Gp_20 (57.3%), indicating substantial genetic divergence. The Gp_23 and Gp_24 showed exceptionally low similarity (32.2–70.9%) with most groups, suggesting possible recombination or

Table 18.2 Pairwise genetic distances among ToLCNDV CRESS domain

Groups	Within group	Between groups Gp 1	Gp 2	Gp 3	Gp 4	Gp 5	Gp 6	Gp 7	Gp 8	Gp 9	Gp 10	Gp 11	Gp 12	Gp 13	Gp 14	Gp 15	Gp 16	Gp 17	Gp 18	Gp 19	Gp 20	Gp 21	Gp 22	Gp 23	Gp 24
Gp 1	99.9	100.0																							
Gp 2	96.9	95.9	100.0																						
Gp 3	98.9	96.1	96.3	100.0																					
Gp 4	98.2	94.6	93.5	94.1	100.0																				
Gp 5	99.7	92.9	92.3	92.9	92.3	100.0																			
Gp 6	94.6	89.5	89.4	89.6	89.1	93.1	100.0																		
Gp 7	98.0	92.3	92.6	92.9	92.7	95.8	92.8	100.0																	
Gp 8	98.2	93.2	92.8	93.0	93.0	95.6	92.2	96.0	100.0																
Gp 9	95.5	91.7	91.4	91.4	92.0	94.0	90.9	94.0	94.8	100.0															
Gp 10	99.7	93.2	92.8	93.4	92.7	93.5	91.0	93.4	93.7	93.0	100.0														
Gp 11	100.0	94.7	94.1	94.3	95.2	95.3	92.6	95.2	96.1	95.1	94.9	100.0													
Gp 12	97.4	94.7	93.2	93.6	93.6	92.7	90.6	92.7	93.5	92.6	92.0	96.1	100.0												
Gp 13	96.8	94.2	93.7	94.5	93.9	92.8	89.8	92.6	92.6	91.5	93.9	93.7	93.3	100.0											
Gp 14	95.4	93.2	93.3	93.1	92.9	92.5	90.4	93.2	93.3	92.0	92.7	94.5	93.5	93.8	100.0										
Gp 15	100.0	81.1	82.8	81.5	79.6	78.3	76.2	79.4	78.4	78.3	77.6	80.7	79.7	79.8	80.7	100.0									
Gp 16	93.1	86.2	86.5	86.3	85.7	85.4	84.0	86.0	84.8	84.7	85.1	86.8	86.4	87.4	87.4	74.1	100.0								
Gp 17	95.6	78.5	77.8	77.5	78.1	77.4	75.3	79.1	78.5	78.5	78.4	78.4	79.3	79.9	80.1	69.1	78.3	100.0							
Gp 18	100.0	78.0	78.0	77.8	77.5	78.5	74.6	79.4	79.7	79.3	79.0	79.3	79.7	79.9	79.7	63.4	75.6	85.9	100.0						
Gp 19	100.0	78.6	78.8	78.3	77.0	75.6	75.2	77.0	74.2	75.3	76.7	76.0	75.7	78.5	77.8	66.6	80.0	75.5	68.1	100.0					
Gp 20	100.0	71.3	69.7	70.3	71.7	73.0	71.2	72.4	71.6	71.0	70.5	71.3	71.5	72.1	72.6	57.3	71.0	72.4	70.2	72.9	100.0				
Gp 21	100.0	77.2	76.5	75.2	77.2	76.5	76.4	76.0	77.0	76.8	75.7	77.1	78.6	77.2	76.7	70.7	75.3	76.3	69.8	72.0	71.0	100.0			
Gp 22	96.0	92.5	92.3	91.4	91.0	91.0	90.4	91.3	92.9	91.5	91.2	92.9	92.3	92.6	92.9	78.7	85.5	78.6	78.5	77.8	73.7	79.5	100.0		
Gp 23	96.6	43.5	45.2	44.7	46.0	42.0	45.5	44.7	46.5	43.4	45.2	45.5	46.5	44.2	47.3	32.2	41.1	39.9	33.4	40.5	43.3	41.3	48.5	100.0	
Gp 24	100.0	58.6	58.6	58.5	60.1	58.1	56.9	58.1	59.9	57.5	58.7	60.5	60.9	59.4	60.4	41.2	51.5	46.5	47.0	37.2	45.6	45.6	57.7	70.9	100.0

distant evolutionary origins. Overall, these findings highlight both conserved and divergent evolutionary patterns.

6 Comparative Phylogeny of DNA-A and CRESS Domains

The phylogenetic analysis of full-length DNA-A (19 clades) and the CRESS domain of the replicase gene (24 clades) revealed distinct evolutionary patterns, with DNA-A providing broader, more stable groupings reflective of long-term evolutionary relationships and strong geographical clustering (Pakistan-specific tomato isolates in Gp_1, intercontinental spread in Gp_7), while the CRESS domain offered higher resolution, splitting some DNA-A clades into finer subdivisions and detecting new host-associated lineages (Gp_20–Gp_24). The DNA-A tree emphasized geographical structuring and moderate host specificity, whereas the CRESS phylogeny highlighted recent diversification, stronger host adaptation (Gp_8 for *Momordica charantia*, Gp_11 for Capsicum), and regional spread (Gp_5 in Europe, Gp_14 in Southeast Asia). Together, these approaches complement each other: DNA-A delineates major strain classifications, while the CRESS domain captures emerging variants and host shifts, providing a more dynamic view of viral evolution and dispersal.

The pairwise distance matrices reveal distinct evolutionary patterns between the full-length DNA-A component (19 groups) and the CRESS domain of the Rep gene (24 groups) in ToLCNDV. While both genomic regions show substantial diversity, the CRESS domain exhibits more extreme divergence, with between-group distances ranging from 32.2–100% compared to 56.4–100% in DNA-A. Notably, the CRESS domain contains hyper-divergent groups (Gp 15, 20, 23, 24) showing <60% similarity to others, a pattern not observed in DNA-A where the most divergent groups (Gp 17–19) maintained >56.4% similarity. Both regions display high within-group conservation, with 4/24 CRESS groups and 6/19 DNA-A groups showing more than 99% identity. The overall mean distance of 91.6% for CRESS contrasts with DNA-A's 93%, indicating slightly higher conservation in the full-length component. Both matrices identify conserved clusters with >95% inter-group similarity, but the CRESS domain shows more pronounced evolutionary splits, particularly in groups 15–24, which form a distinct cluster with <80% similarity to other groups. These differences likely reflect varying selection pressures, with the CRESS domain's critical replication function potentially driving both extreme conservation in some lineages and adaptive diversification in others, while the full DNA-A component maintains more balanced evolutionary constraints across its multiple functional domains.

7 Phylogenetic Analysis of Full-Length DNA-B Sequences

The maximum likelihood phylogenetic analysis of TolCNDV DNA-B full-length sequences revealed five distinct groups with significant geographical and host-specific patterns (Fig. 18.5). Group 1, the largest cluster, predominantly

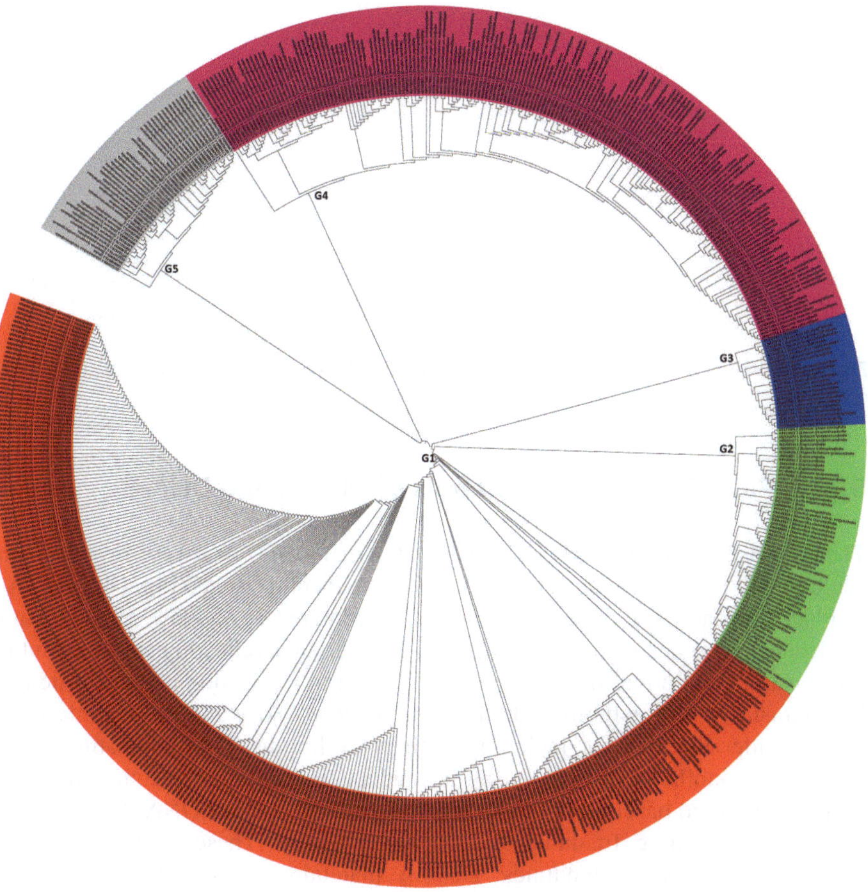

Fig. 18.5 Phylogenetic analysis of full-length ToLCNDV DNA-B sequences

comprised isolates from India and Pakistan, with spillover into Southeast Asia (Thailand, Malaysia, Indonesia), infecting diverse hosts, including solanaceous (tomato, eggplant) and cucurbit (bitter gourd, cucumber) crops, highlighting its broad adaptability. Group 2 featured Indian and emerging Chinese strains, suggesting recent spread, while Group 3 was primarily associated with potato in India, indicating niche adaptation. Group 4, centered in Pakistan, included European isolates (Spain, Italy), likely introduced via trade, and exhibited a wide host range. Group 5, concentrated in Bangladesh, Pakistan and India, included unique hosts like papaya and weeds.

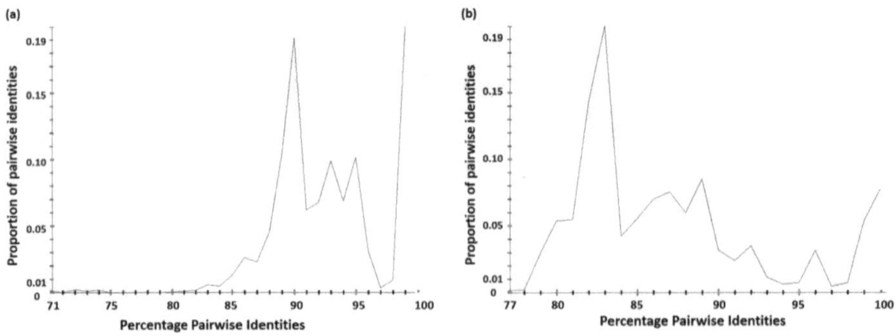

Fig. 18.6 Pairwise distance matrix plots of ToLCNDV. (**a**) DNA-A component. (**b**) DNA-B component

8 Pairwise Distance Analysis of DNA-A and DNA-B

The pairwise distance matrix plots generated using SDT software reveal distinct evolutionary patterns between DNA-A and DNA-B components (Fig. 18.6). DNA-A displays a broad peak of sequence similarity centered at 85–95%, demonstrating moderate divergence with a tail of more distant variants, reflecting stable evolutionary constraints imposed by its essential role in viral replication and encapsidation. In contrast, DNA-B exhibits a sharper peak between 80% and 85% similarity but with a wider distribution, including outliers as low as 77%, indicating greater sequence variability likely driven by host adaptation pressures. Both components share a conserved baseline (minimal identities below 71% for DNA-A and 77% for DNA-B), confirming fundamental genomic stability. These results underscore divergent evolutionary dynamics: DNA-A maintains conserved core functions, while DNA-B's heightened variability facilitates host-specific diversification, highlighting their complementary roles in viral evolution.

9 Nucleotide Diversity (Pi) in DNA-A and DNA-B

The nucleotide diversity (Pi) plots from DnaSP reveal contrasting patterns of genetic variation between DNA-A and DNA-B: DNA-A exhibits significant fluctuations in Pi values, indicating a mosaic of conserved and hypervariable regions likely reflecting varying selective pressures or higher mutation rates, while DNA-B shows a more consistent but still variable pattern with a general trend of higher diversity towards the 5′ end and a gradual decrease towards the 3′ end, suggesting stronger selective constraints and a more uniform distribution of nucleotide diversity, potentially due to its essential roles in viral function (Fig. 18.7).

The nucleotide diversity (Pi) analysis reveals significant variation across the TolCNDV genome segments. Notably, DNA-B exhibits the highest Pi value (0.11441), indicating substantial genetic diversity within this component, which

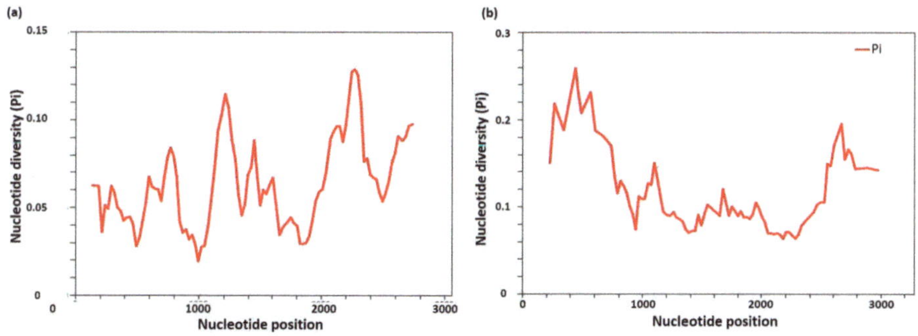

Fig. 18.7 Nucleotide diversity (Pi) of ToLCNDV. (**a**) DNA-A component. (**b**) DNA-B component

contrasts with the lower diversity observed in DNA-A (0.06315). Within DNA-A, the individual open reading frames (ORFs) show varying levels of Pi, ranging from 0.04437 in AV2 to 0.07534 in AC3, suggesting differential evolutionary rates across these genes. This variability in Pi across genome segments and ORFs highlights the dynamic nature of TolCNDV evolution, likely influenced by factors such as mutation rates, recombination, and selective pressures (Table 18.3).

10 Selection Pressure and Evolutionary Constraints

The negative values for Tajima's D, Fu and Li's D, and Fu and Li's F statistics across all genome segments and ORFs strongly suggest an excess of rare alleles, indicative of population expansion or purifying selection. Specifically, the consistently negative Tajima's D values, ranging from −2.19657 in AC2 to −1.31709 in DNA-B, point toward a recent selective sweep or demographic expansion (Table 18.3). Similarly, the negative Fu and Li's D and F values reinforce this pattern, further supporting the presence of purifying selection or rapid population growth. These results imply that TolCNDV populations have likely undergone recent evolutionary events that have shaped their current genetic structure, with purifying selection playing a significant role in maintaining functional constraints.

11 Discussion

The evolutionary analysis of ToLCNDV presented in this chapter underscores its dynamic nature, shaped by geographical expansion, host adaptation, and genetic recombination. Our findings align with previous studies that have highlighted the significant genetic diversity and adaptability of ToLCNDV, which have contributed to its emergence as a major threat to global agriculture (Moriones et al., 2017; Juárez et al., 2019). The extensive genetic diversity observed in the CRESS domain (32.2–100% similarity), compared to the more conserved full-length DNA-A (56.4–100%), reflects the virus's capacity for rapid evolution, likely driven by

Table 18.3 Nucleotide diversity and selection pressure analysis of ToLCNDV genome segments

	Total no. of mutations	Nucleotide diversity (Pi)	No. of haplotypes (H)	Haplotype diversity (Hd)	No. of nucleotide differences (K)	Tajima's D	Fu and Li's D	Fu and Li's F
DNA-A	3305	0.06315	529	0.9462	158.822	−1.94992	−4.80359	−3.51191
AC1	1416	0.06836	417	0.9042	71.849	−1.8877	−2.59441	−2.48641
AC2	654	0.05837	301	0.8496	23.464	−2.19657	−9.60086	−6.04197
AC3	497	0.07534	308	0.85	29.459	−1.69172	−5.69833	−3.88943
AC4	240	0.06109	208	0.8334	10.814	−1.9709	−1.98295	−2.31616
AV1	893	0.04543	360	0.8871	34.8	−2.13547	−7.64655	−5.02597
AV2	410	0.04437	261	0.8303	14.642	−2.18879	−5.90911	−4.35968
DNA-B	2940	0.11441	419	0.9797	239.108	−1.31709	−3.24879	−2.43393
MP (BC1)	887	0.08522	319	0.9514	70.988	−1.33853	−4.31794	−3.00358
NSP (BV1)	1060	0.09956	324	0.9568	78.551	−1.46396	−4.05614	−2.94996

recombination events and host-specific selection pressures. This hypervariability in the CRESS domain may influence host-specific replication efficiency, while the relative stability of DNA-A suggests functional constraints on essential viral processes such as replication and encapsidation. Geographical and host-specific patterns emerged prominently, with Pakistan and India serving as hotspots for tomato-adapted strains (Gp_1 in DNA-A), while Mediterranean and Southeast Asian variants (Gp_5, Gp_14 in CRESS domain) predominantly infect cucurbits. This supports the hypothesis that ToLCNDV has a remarkable capacity for host shifts (Moriones et al., 2017), likely facilitated by agricultural trade, particularly the movement of cucurbit planting material (Panno et al., 2016; Mnari-Hattab et al., 2015). The absence of ToLCNDV sequences from the Americas, despite extensive tomato cultivation in these regions, may indicate successful biosecurity measures or competition from native begomoviruses. This geographical structuring underscores the role of human activity in the spread of ToLCNDV and highlights the need for stringent quarantine protocols to prevent further dissemination.

The higher nucleotide diversity observed in DNA-B (Pi = 0.114) compared to DNA-A (Pi = 0.063) reinforces its critical role in host adaptation and viral movement (Jyothsna et al., 2013). This disparity in evolutionary rates suggests that DNA-B evolves under relaxed selective constraints, enabling rapid adaptation to new hosts, while DNA-A remains more conserved due to its essential functions in viral replication and pathogenesis. The negative Tajima's D and Fu and Li's values across all genome segments indicate pervasive purifying selection and recent population expansion, likely driven by intensified agricultural practices and increasing whitefly vector populations. The detection of ToLCNDV in nontraditional hosts such as cotton, papaya, and Ocimum raises significant concerns about its expanding threat to diverse crops. The highly divergent Gp_23 and Gp_24 CRESS clades (32.2% similarity) could represent recombinant lineages or ancient divergences, necessitating further experimental validation to understand their biological significance. Such extreme divergence highlights the potential for ToLCNDV to continue evolving and adapting to new hosts, posing ongoing challenges for disease management. The genetic plasticity of ToLCNDV, driven by recombination, host shifts, and population expansion, calls for dynamic and integrated management strategies. Resistance breeding programs should target conserved genomic regions to exploit evolutionary constraints that limit viral escape. Surveillance efforts must prioritize high-diversity hotspots, particularly the CRESS domain and DNA-B, to detect emerging strains early and mitigate their impact. The virus's geographical structuring, such as the distinct Mediterranean cucurbit strains, suggests the need for region-specific interventions tailored to local epidemiological conditions. Additionally, the virus's host-shift potential underscores the importance of crop diversification to reduce adaptive selection pressures and limit the spread of virulent variants.

12 Conclusion

The evolutionary analysis of ToLCNDV reveals a highly adaptable pathogen shaped by genetic recombination, host jumps, and global spread, with distinct geographical hotspots (South Asia and Europe) and host-specific lineages (tomato- and cucurbit-dominated clades). Extreme genetic divergence in the CRESS domain (32.2–100% similarity) contrasts with more conserved DNA-A (56.4–100%), while DNA-B's higher nucleotide diversity (Pi = 0.114) underscores its role in host adaptation. Purifying selection signals (negative Tajima's D) and population expansion highlight rapid diversification, with recombination driving virulence and host range expansion to atypical crops like cotton, parthenium, Ocimum, and papaya. These findings emphasize the urgent need for genome-based surveillance, resistant crop development, and biosecurity measures to mitigate the escalating threat of ToLCNDV to global food security.

References

Brown, J. K., Zerbini, F. M., Navas-Castillo, J., Moriones, E., Ramos-Sobrinho, R., Silva, J. C., et al. (2015). Revision of begomovirus taxonomy based on pairwise sequence comparisons. *Archives of Virology, 160*, 1593–1619. https://doi.org/10.1007/s00705-015-2398-y

Chauhan, P. K., Mehta, N., Chauhan, R. S., Kumar, A., Singh, H., Lal, M. K., Tiwari, R. K., & Kumar, R. (2023). Utilization of primary and secondary biochemical compounds in cotton as diagnostic markers for measuring resistance to cotton leaf curl virus (CLCuV). *Frontiers in Plant Science, 14*, 1185337. https://doi.org/10.3389/fpls.2023.1185337

Gillette, W. K., Meade, T. J., Jeffrey, J. L., & Petty, I. T. (1998). Genetic determinants of host-specificity in bipartite geminivirus DNA A components. *Virology, 251*(2), 361–369. https://doi.org/10.1006/viro.1998.9424

Jeevalatha, A., Siddappa, S., Kumar, R., Tiwari, R. K., Lal, M. K., Sharma, S., Chakrabarty, S. K., & Singh, B. P. (2023). RNA-seq analysis reveals an early defense response to tomato leaf curl New Delhi virus in potato cultivar Kufri Bahar. *Functional & Integrative Genomics, 23*, 215. https://doi.org/10.1007/s10142-023-01138-5

Jeevalatha, A., Vanishree, G., Siddappa, S., Kumar, R., Kaundal, P., Kumar, A., & Chakrabarti, S. K. (2021). Molecular characterization and infectivity analysis of tomato leaf curl New Delhi virus isolates infecting potato. *3 Biotech, 11*, 203. https://doi.org/10.1007/s13205-021-02752-5

Juárez, M., Rabadán, M. P., Martínez, L. D., Tayahi, M., Grande-Pérez, A., & Gómez, P. (2019). Natural hosts and genetic diversity of the emerging tomato leaf curl New Delhi virus in Spain. *Frontiers in Microbiology, 10*, 140. https://doi.org/10.3389/fmicb.2019.00140

Jyothsna, P., Haq, Q. M. I., Singh, P., Sumiya, K. V., Praveen, S., Rawat, R., Briddon, R. W., & Malathi, V. G. (2013). Infection of tomato leaf curl New Delhi virus (ToLCNDV), a bipartite begomovirus with betasatellites, results in enhanced level of helper virus components and antagonistic interaction between DNA B and betasatellites. *Applied Microbiology and Biotechnology, 97*, 5457–5471. https://doi.org/10.1007/s00253-012-4685-9

Kanakala, S., Jyothsna, P., Shukla, R., Tiwari, N., Veer, B. S., Swarnalatha, P., Krishnareddy, M., & Malathi, V. G. (2013). Asymmetric synergism and heteroencapsidation between two bipartite begomoviruses, tomato leaf curl New Delhi virus and tomato leaf curl Palampur virus. *Virus Research, 174*(1–2), 126–136. https://doi.org/10.1016/j.virusres.2013.03.011

Kumar, R., Tiwari, R. K., Jeevalatha, A., Sundaresha, S., Shah, M. A., Sharma, S., Sagar, V., Kumar, M., & Chakrabarti, S. K. (2021). Potato apical leaf curl disease: Current status and perspectives on a disease caused by tomato leaf curl New Delhi virus. *Journal of Plant Diseases and Protection, 128*, 897–911. https://doi.org/10.1007/s41348-021-00463-w

Kumar, R., Tiwari, R. K., Sundaresha, S., Kaundal, P., & Raigond, B. (2022). Potato viruses and their management. In S. K. Chakrabarti, S. Sharma, & M. A. Shah (Eds.), *Sustainable management of potato pests and diseases*. Springer. https://doi.org/10.1007/978-981-16-7695-6_12

Kushwaha, N., Singh, A. K., Basu, S., & Chakraborty, S. (2015). Differential response of diverse solanaceous hosts to tomato leaf curl New Delhi virus infection indicates coordinated action of NBS-LRR and RNAi-mediated host defense. *Archives of Virology, 160*, 1499–1509. https://doi.org/10.1007/s00705-015-2399-x

López, C., Ferriol, M., & Picó, M. B. (2015). Mechanical transmission of tomato leaf curl New Delhi virus to cucurbit germplasm: Selection of tolerance sources in Cucumis melo. *Euphytica, 204*, 679–691. https://doi.org/10.1007/s10681-015-1371-x

Maule, A., Leh, V., & Lederer, C. (2002). The dialogue between viruses and hosts in compatible interactions. *Current Opinion in Plant Biology, 5*(4), 279–284. https://doi.org/10.1016/S1369-5266(02)00272-8

Mnari-Hattab, M., Zammouri, S., Belkadhi, M. S., Doña, D. B., Ben Nahia, E., & Hajlaoui, M. (2015). First report of tomato leaf curl New Delhi virus infecting cucurbits in Tunisia. *New Disease Reports, 31*(21), 21. https://doi.org/10.5197/j.2044-0588.2015.031.021

Moriones, E., Praveen, S., & Chakraborty, S. (2017). Tomato leaf curl New Delhi virus: An emerging virus complex threatening vegetable and fiber crops. *Viruses, 9*(10), 264. https://doi.org/10.3390/v9100264

Naga, K. C., Sundaresha, S., Kumar, R., Tiwari, R. K., Subhash, S., Verma, G., Buckseth, T., Bairwa, A., Sharma, S., Katare, S., & Srivastava, R. M. (2021). A new record of Asia II 5 genetic group of Bemisia tabaci (Gennadius) in the major potato growing areas of India and its relationship with tomato leaf curl New Delhi virus infecting potato. *3 Biotech, 11*, 421. https://doi.org/10.1007/s13205-021-02966-7

Navas-Castillo, J., Fiallo-Olivé, E., & Sánchez-Campos, S. (2011). Emerging virus diseases transmitted by whiteflies. *Annual Review of Phytopathology, 49*(1), 219–248. https://doi.org/10.1146/annurev-phyto-072910-095235

Padidam, M., Beachy, R. N., & Fauquet, C. M. (1995). Tomato leaf curl geminivirus from India has a bipartite genome and coat protein is not essential for infectivity. *Journal of General Virology, 76*(1), 25–35. https://doi.org/10.1099/0022-1317-76-1-25

Panno, S., Iacono, G., Davino, M., Marchione, S., Zappardo, V., Bella, P., Tomassoli, L., Accotto, G. P., & Davino, S. (2016). First report of tomato leaf curl New Delhi virus affecting zucchini squash in an important horticultural area of southern Italy. *New Disease Reports, 33*(1), 6. https://doi.org/10.5197/j.2044-0588.2016.033.006

Petty, I. T., Carter, S. C., Morra, M. R., Jeffrey, J. L., & Olivey, H. E. (2000). Bipartite geminivirus host adaptation determined cooperatively by coding and noncoding sequences of the genome. *Virology, 277*(2), 429–438. https://doi.org/10.1006/viro.2000.0620

Pratap, D., Kashikar, A. R., & Mukherjee, S. K. (2011). Molecular characterization and infectivity of a tomato leaf curl New Delhi virus variant associated with newly emerging yellow mosaic disease of eggplant in India. *Virology Journal, 8*, 1–13. https://doi.org/10.1186/1743-422X-8-305

Sarkar, M., Gupta, D., Singh, O. W., Paul, S., Kumar, R., Mandal, B., & Roy, A. (2024). Identification of pathogenicity determinants in ToLCNDV and their RNAi-based knockdown for disease management in Nicotiana benthamiana and tomato plants. *Frontiers in Microbiology, 15*, 1481523. https://doi.org/10.3389/fmicb.2024.1481523

Varma, A., & Malathi, V. G. (2003). Emerging geminivirus problems: A serious threat to crop production. *Annals of Applied Biology, 142*(2), 145–164. https://doi.org/10.1111/j.1744-7348.2003.tb00240.x

MIX
Papier aus verantwortungsvollen Quellen
Paper from responsible sources
FSC® C105338

If you have any concerns about our products,
you can contact us on
ProductSafety@springernature.com

In case Publisher is established outside the EU,
the EU authorized representative is:
Springer Nature Customer Service Center GmbH
Europaplatz 3, 69115 Heidelberg, Germany

Printed by Libri Plureos GmbH
in Hamburg, Germany